SCIENTIFIC FARM ANIMAL PRODUCTION

THIRD EDITION

SCIENTIFIC FARM ANIMAL PRODUCTION

AN INTRODUCTION TO ANIMAL SCIENCE

ROBERT E. TAYLOR

Department of Animal Sciences
Colorado State University
Fort Collins, Colorado

RALPH BOGART

Professor Emeritus
Department of Animal Science
Oregon State University
Corvallis, Oregon

MACMILLAN PUBLISHING COMPANY
New York
Collier Macmillan Publishers
London

Macmillan Publishing Company
866 Third Avenue, New York, New York 10022

Collier Macmillan Canada, Inc.

Library of Congress Cataloging in Publication Data

Taylor, Robert E. (Robert Ellis), 1934–
  Scientific farm animal production.

  Bogart's name appears first on the earlier edition.
  Includes bibliographies and index.
  1. Livestock.   I. Bogart, Ralph, 1908–   . II. Title.
SF61.T38 1988   636   88-2618
ISBN 0-02-311750-8

Printing: 4 5 6 7 8   Year: 9 0 1 2 3 4 5 6 7

The authors dedicate this book to the hundreds of students they have taught. The positive and rewarding interactions with these students have provided the stimulus to write this book. We hope the reading and studying of *Scientific Farm Animal Production* will be a motivational influence to many students as they seek to enhance their education.

PREFACE

Scientific Farm Animal Production is distinguished by an appropriate combination of both breadth and depth of livestock and poultry production and their respective industries. The book gives an overview of the biological principles applicable to the Animal Sciences with chapters on reproduction, genetics, nutrition, lactation, end products, and others. The book also covers the breeding, feeding and management of beef cattle, dairy cattle, horses, sheep, swine, poultry, and goats. Although books have been written on each of these separate chapters, the authors have highlighted the significant biological principles, scientific relationships, and management practices in a condensed but informative manner.

TARGET AUDIENCE

This book is designed as a text for the introductory Animal Science course typically taught at universities and junior or community colleges. It is also a valuable reference book for livestock producers, vocational agriculture instructors, and others desiring an overview of livestock production principles and management. The book is basic and sufficiently simple for the urban student with limited livestock experience, yet challenging for the student who has a livestock production background.

The book is designed to accommodate several instructional approaches to teaching the introductory course: (1) the life-cycle biological principles approach, including such areas as end products, reproduction, breeding, nutrition, and animal health; (2) the species approach (teaching the course primarily by the various species); or (3) a combination of the previous two. The latter appears to be the most popular teaching approach by covering principles in lecture and combining principles and species into laboratory exercises.

KEY FEATURES

Chapters 1–8 cover animal products, Chapters 9–20 discuss the biological principles while livestock and poultry management practices are presented in Chapters 21–33.

The glossary of the terms, used throughout the book, has been expanded so students can readily become familiar with animal science terminology. The bold lettered words in the text are included in the glossary.

Many illustrations in the form of photographs and line drawings are used throughout the book to communicate key points and major relationships. If "a picture is worth a thousand words", the 345 individual photographs and drawings expand the usefulness of the book beyond its 579 pages.

Selected references are provided for each chapter to direct students into greater depth and breadth as they become intrigued with certain topics. Instructors can also use the references to expand their knowledge in current background material. Also included in the selected references section are references to visuals that relate to the specific chapter.

Instructors are encouraged to review these visuals and use those which will enrich their course.

CHANGES IN THIS EDITION

This book has been thoroughly revised to bring it up-to-date with new technical and applied information. Four new chapters have been added: Chapter 3 (By-Products of Meat Animals), Chapter 5 (Poultry and Egg Products), Chapter 20 (Animal Health) and Chapter 35 (Animal Welfare). Other chapters have been expanded to better cover the dynamic animal industries. Also added to the third edition is a greater emphasis on economics, marketing, and the international dimensions of livestock, poultry, and food production. Dietary information about meat, milk, and eggs addresses current consumer concerns and offers scientific facts to combat popular and often misleading myths.

The number of colored photographs has been doubled from the previous edition. Most noteworthy are the colored pictures of livestock breeds, which will improve the students' ability to distinguish the various breeds of livestock.

Detailed attention has been given to comments from more than 100 reviewers of the 2nd and 3rd editions. Many of these reviewers currently teach the introductory Animal Science course.

ABOUT THE AUTHORS

The talents and backgrounds of the two authors have combined to give the book a strong scientific foundation tempered with practical production experience. Dr. Ralph Bogart was raised on a general livestock farm in Missouri. He received his B.S. degree in agriculture with a major in animal husbandry from the University of Missouri, his M.S. degree with a major in genetics from Kansas State, and his Ph.D. degree with a major in genetics and physiology from Cornell. He taught introductory animal husbandry classes as well as genetics and physiology at the University of Missouri for eight years, and continued his teaching at Oregon State University.

As teacher and researcher, Dr. Bogart has worked with students in the classroom as well as livestock producers throughout the world. Dr. Bogart has received both the Animal Breeding and Genetics Award and the Animal Industry Service Award from The American Society of Animal Science for his contribution to the livestock industry.

Dr. Taylor was raised on an Idaho livestock operation where several livestock species were produced. He received a B.S. degree in animal husbandry and a Master's degree in animal production from Utah State University. This background, combined with his Ph.D. work in animal breeding and physiology from Oklahoma State University, has provided much depth to his knowledge of livestock production. He has had practical production experience with beef cattle, dairy cattle, horses, poultry, sheep and swine.

Dr. Taylor received teaching awards at Iowa State University (where he also managed a swine herd) and at Colorado State University. He also received the Distinguished Teaching Award from The American Society of Animal Science in recognition of his ability to organize and present materials to students. Many of his concepts for effective teaching are used in this book.

ACKNOWLEDGMENTS

Appreciation is expressed to those individuals and organizations who have reviewed all or part of the second and third editions and offered suggestions for the revision. The following individuals and organizations made a major contribution to the third edition:

R. L. Belyea (University of Missouri)
J. G. Butler (National Wool Growers Association)
J. W. Edwards (Texas A&M University)
Egg Council
G. S. Gieger (University of Missouri)
G. F. W. Haenlein (University of Delaware)
J. C. Heird (Colorado State University)
G. W. Jesse (University of Missouri)
D. C. Jordan (Colorado State University)
M. D. Kenealy (Iowa State University)
D. H. Keisler (University of Missouri)
D. Lincicome (USDA)
W. E. Loch (University of Missouri)
H. W. Miller (Mississippi State University)
R. E. Morrow (University of Missouri)
R. G. Mortimer (Colorado State University)
National Dairy Council (E. W. Speckmann and Staff)
S. W. Neel (National Live Stock and Meat Board)
K. G. Odde (Colorado State University)
V. A. Rich (Colorado State University)
S. P. Schmidt (Auburn University)
G. E. Stoddard (Utah State University)
J. D. Tatum (Colorado State University)
Turkey Foundation
F. W. Williams, Jr. (USDA)
Dennis Giddings (major contributor to the illustrations)
Shirley Mitchell (typing the revised third edition)

Appreciation is also expressed to the many instructors of the Introductory Animal Science Course who made helpful suggestions when they responded to a questionnaire.

The authors are grateful for the contributions of Gary Ostedt, Beth Anderson, John Molyneux, Dan Tabata, and Laura Ierardi at Macmillan Publishing Company. They have made the transition from the second edition to the third edition a relatively easy one for the authors.

CONTENTS

CHAPTER 1

Animal Contributions to Human Needs

Our basic human needs are food, shelter, clothing, fuel, and emotional well-being. Animals and animal products supply many of these basic needs and contribute to a high standard of living associated with a high consumption of animal products.

Since the domestication of dogs, horses, cattle and sheep, between 6,000 and 10,000 years ago, wide differences have developed between people in various regions of the world in using agricultural technology to improve their standard of living; but in all societies, domestic animals are a source of food, commercial products, and companionship for people.

Of particular importance among the multitude of benefits that domestic animals provide for humans are food; clothing; slaughter by-products used for various chemical purposes and animal feeds; power; manure for fuel (Fig. 1.1), buildings, and fertilizer; information on human disease through studies of experimental animals; and pleasure for those who keep animals. Table 1.1 shows the major domesticated animal species, their approximate numbers, and how they are used by people throughout the world.

CONTRIBUTIONS TO FOOD NEEDS

When opportunity exists, most humans consume both plant and animal products (Fig. 1.2). Meat is nearly always consumed in quantity when it is available. Its availability in most countries is closely related to the economic status of the people and their agricultural technology. Vegetarianism in countries such as India may be the long-term result of intense population pressures and scarcity of feed for animals because of competition between humans and animals for food. Rising population pressures, particularly in developing regions (Southeast Asia, Africa, and Latin America), force people to consume foods primarily of plant origin. Some major groups in human society practice vegetarianism for ethical reasons. In the Buddhist philosophy and some religions of India, for example, all animal life is considered sacred.

Vegetarianism has never been practiced to a large extent in the United States. The

FIGURE 1.1.
A load of cow-dung cakes enroute to a market in India.
Courtesy of R. E. McDowell, Cornell University.

vegetarian movement in this country developed in the 19th century and was based on the view that eating only plant products was healthier than including meat in the diet. Actually, medical surveys of vegetarians have often shown evidence of anemia and poor health. The development of vegetarianism in England and in the United States was also closely linked to temperance movements and to conservative attitudes toward sexuality. Vegetarianism was thought to cool "animal passions."

The contribution of animal products to the per-capita calorie and protein supply in food is shown in Table 1.2. Animal products comprise approximately 16% of the calories and 34% of the protein in the total world food supply. Large differences exist between

FIGURE 1.2.
Animal food products, such as meat, milk, and eggs, are highly preferred foods in countries with high standards of living. Courtesy of American Egg Board.

TABLE 1.1.   Selected Animal Species—Their Numbers and Uses in the World

Animal Species	World Numbers (millions)	Leading Countries or Areas With Numbers[a] (millions)	Primary Uses
Ruminants			
Cattle	1,269	India (182), Brazil (134), USSR (121), U.S. (110)	Meat, milk, hides
Sheep	1,121	USSR (143), China (95), Turkey (40), India (41), U.S. (10),	Wool, meat, milk, hides
Goats	460	India (82), China (63), Pakistan (30), U.S. (1.6)	Milk, meat, hair, hides
Buffalo	129	India (64), China (20), Pakistan (13)	Draft, milk, meat, hides, bones
Camel	17	Somolia (5.7), Sudan (2.8)	Packing, riding, draft, meat, milk, hides
Yak	13	USSR, Tibet	Packing, riding, draft, meat, milk, hides
Llama	13	South America	Packing, riding, draft, meat, milk, hides
Nonruminants			
Chickens	8,287	China (1,361), USSR (1,090), U.S. (1,050), Brazil (450)	Meat, eggs, feathers
Swine	792	China (313), USSR (78), U.S. (54)	Meat
Turkeys	216	USSR (66), U.S. (53), Italy (21)	Meat, eggs, feathers
Ducks	169	Bangladesh (20), Vietnam (20), U.S. (6)	Meat, eggs, feathers
Horses	65	China (11), U.S. (11), Mexico (6), USSR (6)	Draft, packing, riding, meat, companionship
Asses	40	China (10), Ethiopia (3.9), U.S. (0.002)	Draft, packing, riding, meat, companionship
Mules	15	China (4.8) Mexico (3.1), U.S. (0.002)	Draft, packing, riding, meat, companionship

[a]U.S. numbers given for comparison—may not always be among the leading countries.
Source: Adapted from several sources, including 1985 FAO Production Yearbook.

the developed countries and developing countries in both total daily supply of calories (3,394 vs. 2,388) and protein (56.5 vs. 11.5 g). The comparison of developed and developing countries for the contribution of animal products in calories is 30% versus 8%, respectively, and the comparison for protein is 57% versus 20%.

Table 1.2 also shows calorie and protein data for selected countries in both developed and developing market economies. It is interesting to compare countries that have larger and smaller contributions of calories and protein from animal products to total calories (Iceland and Denmark vs. Nigeria and Bangladesh) and to total protein (Iceland and Denmark vs. India and China). The United States ranks very high, compared to other countries, in the contribution of animal products to the available calories and protein.

TABLE 1.2.   **Animal Product Contribution to Per-Capita Calorie and Protein Supply (selected countries, 1981–83)**

Country	Total	Per-Capita Calorie Supply (calories per day) From Animal Products Calories	From Animal Products Percent	Total	Per-Capita Protein Supply From Animal Products Grams	From Animal Products Percent
Developed Market Economies						
United States	3,647	1,280	35	105.8	70.9	67
United Kingdom	3,162	1,160	37	86.6	52.1	60
West Germany	3,431	1,292	38	91.2	58.5	64
Canada	3,421	1,280	37	97.8	62.2	64
Australia	3,382	1,154	34	97.0	63.1	65
Denmark	3,564	1,582	44	100.0	67.5	68
Israel	3,062	669	22	103.1	54.1	52
Japan	2,858	604	21	92.3	50.6	55
Iceland	3,142	1,429	45	123.4	96.9	78
USSR	3,426	883	26	100.9	51.8	51
Poland	3,301	1,030	31	103.4	54.5	53
Developing Market Economies						
Mexico	2,966	407	14	76.2	24.4	32
Brazil	2,584	389	15	60.6	22.4	37
Egypt	3,186	227	7	81.5	14.4	18
Turkey	3,150	304	10	84.3	19.8	23
Nigeria	2,203	99	4	49.3	9.8	20
China	2,602	178	7	61.2	7.9	13
India	2,088	113	5	50.8	5.8	11
Kenya	2,026	219	11	55.5	14.2	26
Bangladesh	1,878	63	3	40.3	4.9	12
All Developed Countries	3,398	1,036	30	99.4	57.1	57
All Developing Countries	2,409	194	8	58.5	11.5	20
World Total	2,665	412	15	69.1	23.3	34

Source: 1985 FAO Production Yearbook.

Selected recommended daily intakes (recommended daily allowance) of calories and protein are given in Table 1.3. Although the data in Table 1.3 do not represent an average of the U.S. population, comparison of these data with those in Table 1.2 is interesting. Some countries have a larger supply of calories and protein than needed, whereas other countries have an inadequate calorie and protein supply. This assumes an equal distribution of the available supply, which in reality does not occur.

The large differences among countries in the importance of animal products in their food supply can be partially explained by available resources and development of those resources. Most countries that have a small percentage of their population involved in agriculture have higher standards of living and have a higher per-capita consumption of animal products. Comparing Table 1.4 with Table 1.2, note that the countries in Table 1.4

TABLE 1.3. Recommended Daily Caloric and Protein Intake for Selected Males and Females in United States

Sex	Age (years)	Weight (lb)	Height	Average Daily Calories	Protein (g/day)
Female	23–50	120	5 ft 4 in	2,000	44
Male	23–50	155	5 ft 10 in	2,700	56

TABLE 1.4. Population Involved in Agriculture in Selected Countries (countries ranked by percent of population in agriculture)

Country	1985 Population (millions)	1985 Population in Agriculture[a] (millions)	Percent of Population Economically Active in Agriculture (1985)[b]	Number of Tractors (1983) (thousands)
Developed Market Economies				
United States	239	7	3	4,550
United Kingdom	57	1	3	529
Canada	25	1	4	658
West Germany	61	3	5	1,472
Israel	4	0.2	5	28
Australia	16	1	6	332
Denmark	5	0.3	6	183
Japan	121	10	8	1,584
Iceland	0.2	0.02	8	14
Ireland	4	0.6	15	148
USSR	278	49	18	2,697
Poland	37	9	24	757
Developing Market Economies				
Mexico	79	26	33	170
Egypt	47	20	42	41
Turkey	49	26	52	512
Nigeria	95	63	67	9
India	759	521	69	503
Bangladesh	101	72	71	5
China	1,060	763	72	842
Kenya	21	16	79	6
Nepal	17	15	93	0.5
All Developed Countries	1,210	128	10	19,170
All Developing Countries	3,626	2,201	61	3,746
World Total	4,837	2,328	48	22,916

[a]"Agricultural population" is defined as all persons depending for their livelihood on agriculture. This comprises all persons actively engaged in agriculture and their nonworking dependents.
[b]Includes all economically active persons engaged principally in agriculture, forestry, hunting, or fishing.
Source: 1985 FAO Production Yearbook.

are listed by percentage of their population involved in agriculture. The countries in Table 1.2 are listed in the same order as in Table 1.4.

Developed countries have approximately 10% of their population economically active in agriculture, whereas more than 50% of the population in developing countries are involved in agriculture (Table 1.4). Agriculture mechanization has been largely responsible for increased food production and allowing many people to work in other industries. This facilitates the provision of many goods and services and thus raises the standard of living in that country.

The tremendous increase in the productivity of U.S. agriculture and the relative cost of food is vividly demonstrated in Table 1.5. The most dramatic change occurred after World War II, when productivity increased more than fivefold in 30 years. During that time, the abundant production of feed grains provided a marked stimulus in increasing livestock production, thus providing large amounts of animal products for the human population.

Releasing people from producing their own food in the United States has given them the opportunity to improve their per-capita incomes. The increased per-capita income associated with an abundance of animal products has resulted in the decrease in relative costs of some animal products with time (Table 1.6).

U.S. consumers allocate a smaller share (15%) of their disposable income for food than people in other countries. In contrast, people of India and China spend 55–65% of their incomes for food.

Table 1.7 shows that cereal grains are the most important source of energy in world diets. The energy derived from cereal grains, however, is twice as important in developing countries (as a group, because there are exceptions) as in developed countries. Table 1.7 also shows that meat and milk are the major animal products contributing to the world supply of calories and protein.

Most of the world meat supply comes from cattle, buffalo, swine, sheep, goats, and horses. There are, however, 20 or more additional species, unfamiliar to most Americans,

TABLE 1.5.   Persons Supplied by One Farm Worker in the United States and the Percent of Income Spent on Food

Year	No. Persons Supplied Per Farm Worker	Percent Personal Disposable Income Spent on Food
1820	4	—
1850	4	—
1880	6	—
1910	7	—
1940	11	—
1950	16	30.0
1960	26	20.2
1970	48	17.3
1980	76	16.6
1986	80	14.7

Source: USDA.

TABLE 1.6. Amount of Food Purchased by an Hours Pay (average U.S. worker)

Food Item	1950	1984
Frying chicken	2.5 lb	10.3 lb
Milk	8.0 qt	14.8 qt
Eggs	2.4 doz	8.3 doz
Pork	2.7 lb	5.1 lb
White bread	10.1 lb	15.4 lb

Source: USDA.

that collectively contribute about 3 million metric tons of edible protein per year or approximately 10% of the estimated total protein from all meats. These include the alpaca, llama, yak, deer, elk, antelope, kangaroo, rabbit, guinea pig, capybara, fowl other than chicken (duck, turkey, goose, guinea fowl, pigeon), and wild game exclusive of birds. For example, the USSR cans more than 50,000 metric tons of reindeer meat per year, and in West Germany the annual sales of local venison exceed $1 million. Peru derives more than 5% of its meat from the guinea pig.

Meat is important as a food for two scientifically based reasons. The first is that the assortment of amino acids in animal protein more closely matches the needs of the human body than does the assortment of amino acids in plant protein. The second is that vitamin B_{12}, which is required in human nutrition, may be obtained in adequate quantities from consumption of meat or other animal products but not from consumption of plants.

Milk is one of the largest single sources of food from animals. In the United States, 99% of the milk comes from cattle, but on a worldwide basis, milk from other species is important, too; the domestic buffalo, sheep, goat, alpaca, camel, reindeer, and yak supply significant amounts of milk in certain countries. Milk and products made from milk contribute protein, energy, vitamins, and minerals for humans.

TABLE 1.7. Contributions of Various Food Groups to the World Food Supply

	Calories (%)	Protein (%)
Cereals	49	43
Roots, tubers, pulses	10	10
Nuts, oils, vegetable fats	8	4
Sugar and sugar products	9	2
Vegetables and fruits	8	7
All animal products	16	34
Meat	(7)	(15)
Eggs	(1)	(2)
Fish	(1)	(5)
Milk	(5)	(11)
Other	(2)	(1)

Source: Adapted from several FAO World Food Surveys.

Besides the nutritional advantages, a major reason for human use of animals for food is that most countries have land areas unsuitable for growing cultivated crops. Approximately two-thirds of the world's agricultural land is permanent pasture, range, and meadow; of this, about 60% is unsuitable for producing cultivated crops that would be consumed directly by humans. This land, however, can produce roughage in the form of grass and other vegetation that is digestible by grazing ruminant animals, the most important of which are cattle and sheep (Fig. 1.3). These animals can harvest and convert the vegetation, which is for the most part undigestible by humans, to high-quality protein food. In the United States about 385 million acres of rangeland and forest, representing 44% of

FIGURE 1.3.
Animals produce food for humans by utilizing grass, crop residues, and other forages from land which cannot produce crops to be consumed directly by humans. (A) Sheep grazing a steep hillside. Courtesy of California Agriculture Magazine, University of California. (B) Cattle grazing a mountain valley in Switzerland. Courtesy of American Simmental Associations. (C) Cattle produce meat from the mountains and Plains areas of the western United States. Courtesy of American Simmental Association. (D) Sheep utilizing the crop residue remaining after havesting the corn grain. Courtesy of Winrock International.

the total land area, is used for grazing. Although this acreage now supports only about 40% of the total cattle population, it could carry twice this amount if developed and managed intensively.

Animal agriculture therefore does not compete with human use for production of most land used as permanent pasture, range, and meadow. On the contrary, the use of animals as intermediaries provides a means by which land that is otherwise unproductive for humans can be made productive (Fig. 1.4).

People today are concerned about energy, protein, population pressures (Fig. 1.5), and land resources as they relate to animal agriculture. Quantities of energy and protein present in foods from animals are smaller than quantities consumed by animals in their feed, because animals are inefficient in the ratio of nutrients used to nutrients produced. More acres of cropland are required per person for a diet high in foods from animals than for a diet including only plant products. As a consequence, animal agriculture has been criticized for wasting food and land resources that could otherwise be used to provide persons with inadequate diets. Consideration must be given to economic systems and consumer preferences to understand why agriculture perpetuates what critics perceive as resource-inefficient practices. These practices relate primarily to providing food-producing animals with feed that could be eaten by humans and using land resources to produce crops specifically for animals instead of producing crops that could be consumed by humans.

Hunger does exist, as it affects, in varying degrees, nearly 1 billion people in the world. This is almost 25% of the world's population, most of them in developing countries. However, hunger is being reduced, and the potential exists to make even more improvement. The book *Ending Hunger: An Idea Whose Time Has Come* states:

> Since 1900, seventy-five countries have ended hunger within their borders as a basic, society-wide issue. It is both important and heartening to note that there is no single prescribed way to achieve the end of the persistence of hunger in a society. Some countries focused on land reform, while others emphasized food subsidies, collectivized agriculture, or privately owned "family farms." For every country that saw a particular action as crucial to ending hunger, there is a country that ended hunger without it.

Ending Hunger quotes the National Academy of Sciences in its World Food and Nutrition Study, for which 1,500 scientists had been consulted: "If there is the political will in this country and abroad . . . it should be possible to overcome the worst aspects of widespread hunger and malnutrition within one generation."

Agriculture producers produce what consumers want to eat as reflected in the prices consumers can and are willing to pay. Eighty-five percent of the world's population desires food of animal origin in its diets, perhaps because foods of animal origin are considered more palatable than foods from plants. In more countries, as per-capita income rises, consumers tend to increase their consumption of meat and animal products, which are generally more expensive pound for pound than products derived from cereal grains (Fig. 1.6). Table 1.8 shows some comparative food prices for selected world capitals. The total of all food items varies from a low of $9.45, for Mexico City, to a high of $51.83, for Tokyo. Sirloin steak is the food item showing the widest price range—$0.83/lb in Mexico City versus $17.99/lb in Tokyo. Several factors besides supply and demand are reflected in the differences in these prices.

If many consumers in countries where animal products are consumed at a high rate

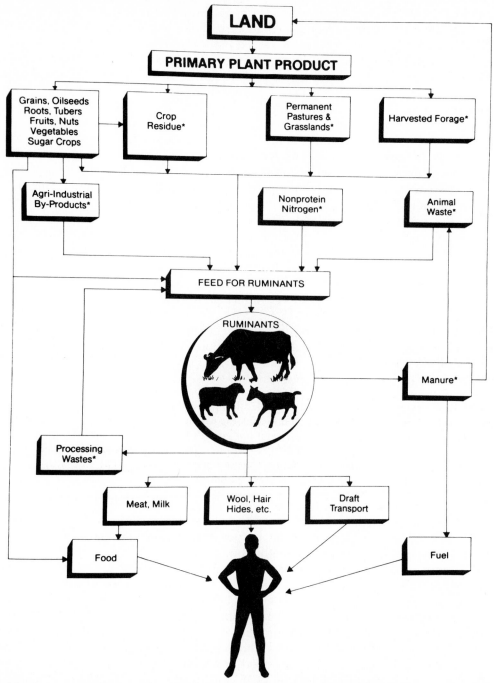

FIGURE 1.4.
A graphic illustration of land-plant-ruminant-animal-human relationship. Products marked with an asterisk are not normally consumed by humans, but ruminants convert many of these products into useful products for humans. Courtesy of Winrock International.

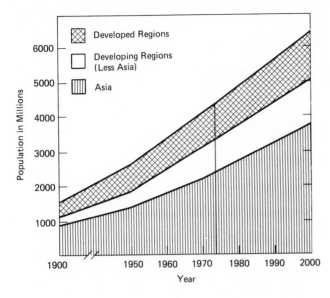

FIGURE 1.5.
Past, present, and projected world population, 1900–2000.

were to decide to eat only food of plant origin, consumption and price of foods from plants would increase, and consumption and prices of foods from animals would decrease. Agriculture would then adjust to produce greater quantities of food from plants and lesser quantities of food from animals. Ruminant animals can produce large amounts of meat without grain feeding. The amount of grain feeding in the future will be dictated by cost of grain and the price consumers are willing to pay for meat.

Some people in the United States advocate shifting from the consumption of foods from animals to foods from plants. They see this primarily as a moral issue, stating that

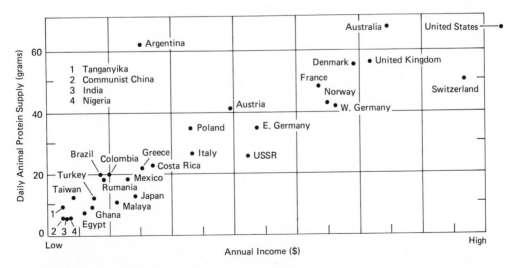

FIGURE 1.6.
Income and supply of animal protein per person from selected countries. Courtesy of USDA.

TABLE 1.8. Average Retail Food Prices in Selected World Capitals, November 1986 (in U.S. dollars per pound or units as indicated, converted at exchange rates for date shown)

Item	Buenos Aires	Mexico City	Paris	Pretoria	Rome	Stockholm	Tokyo	Washington DC
Steak, sirloin, boneless	$ 1.65	$0.83	$ 5.10	$ 2.61	$ 4.87	$ 8.71	$17.99	$ 4.59
Pork, roast, boneless	1.90	1.07	2.73	1.76	2.60	9.28	5.34	2.80
Broilers, whole	0.60	0.55	1.34	0.65	1.83	2.86	2.26	1.03
Eggs, large, dozen	0.88	0.54	0.85	0.75	1.29	2.30	1.74	0.77
Butter	1.30	1.61	2.06	0.90	2.32	2.31	3.65	2.48
Cheese, cheddar	2.66	1.35	2.75	1.88	3.73	4.19	3.05	3.20
Milk, whole, quart	0.26	0.23	0.73	0.41	0.85	0.69	1.25	0.49
Total, animal products	9.25	6.18	15.56	8.96	17.49	30.34	35.28	15.36
Total, market basket, Nov. 1986[a]	17.65	9.45	23.35	18.48	25.60	45.03	51.83	24.02
Total, market basket, Nov. 1985	13.86	9.88	23.46	12.20	20.07	36.59	48.87	22.79

[a]Also includes potatoes, apples, oranges, flour, rice, sugar, coffee, and cooking oil.
Source: FAS, USDA, *World Production and Trade, Weekly Roundup,* Dec. 1986.

it is unethical to let people elsewhere in the world starve when we could meet our own food needs by eating foods from plants rather than feeding plants to animals. We could then send the balance of plant-derived foods abroad. These people believe that grain is especially good to ship because grain can be shipped with comparative ease, because a surplus of grain exists in the United States, and because any surplus we have should be provided at no cost. Providing free food to other countries has met with limited success in the past. In some situations, it upsets their own agricultural production, and in many cases the food cannot be adequately distributed in the country because transportation and marketing systems are poor. There are strong feelings that the United States has a moral obligation to share its abundance with other people in the world, particularly those in developing countries. It appears that this can best be done by sharing our time and sharing our basic, but not necessarily advanced, technology. People need to have self-motivation to improve, need to be shown how to help themselves step by step, and need to fully develop the agricultural resources in their own countries.

Table 1.9 shows the prices of animal products and other popular foods in selected U.S. cities. In this survey, San Diego had the lowest deviation (-16%) from the national average for total market basket cost, and Honolulu had the highest ($+25.4\%$). The percent deviation of $+19.6$ for Washington, DC (not shown in Table 1.9), gives information that can be used in comparing Tables 1.8 and 1.9. Thus, it appears that certain U.S. cities (i.e., San Diego) would have total food costs similar to those in Pretoria and Buenos Aires.

The information reported in Tables 1.8 and 1.9 should be considered estimates when making comparisons. Some food items are not comparable in quality, and prices are for a short period of time. Certain food items may be daily specials or featured as loss leaders that day.

Increases in the efficiency of animal production in the United States have been remarkable during the past half century (Table 1.10). This increase in efficiency has occurred primarily because people had an incentive to progress under a free-enterprise system. They learned how to improve their standard of living by effectively using available resources. This progress has taken time under the environmental conditions that motivated people to be successful.

Citizens of the United States should not only share their technology with other countries but also share how this technology was developed (Fig. 1.7). These achievements have been built on knowledge developed through experience and research, the extension of knowledge to producers, and the development of an industry to provide transportation, processing, and marketing in addition to production. Dwindling dollars currently being spent to support agricultural research and extension of knowledge in the United States may not provide the technology needed for future food demands.

About 30% of the world human population and 32% of the ruminant animal population live in developed regions of the world, but ruminants of these same regions produce two-thirds of the world's meat and 80% of the world's milk. In developed regions, a higher percentage of animals are used as food producers, and these animals are higher in productivity on a per-animal basis than animals in developing regions. This is the primary reason for the higher level of human nutrition in developed countries of the world.

Possibly many so-called developing regions of the world can achieve levels of plant,

TABLE 1.9. Retail Prices of Selected Food Items in Selected U.S. Cities, June 19, 1986

Item	National Average	San Diego	Boston	Chicago	Atlanta	Denver	Dallas	Honolulu
Steak, sirloin bone in (lb)	$ 2.64	$ 1.98	$ 1.79	$ 1.79	$ 2.39	$ 2.89	$ 3.28	$ 2.89
Pork, loin chops	2.38	2.49	1.49	1.66	2.89	2.69	2.08	3.09
Broilers, whole	0.67	0.69	0.59	0.45	0.59	0.49	0.59	0.79
Eggs, large (doz)	0.72	0.79	0.89	0.61	0.57	0.69	0.59	1.19
Cheese (lb)	2.66	2.27	2.34	2.75	2.32	2.83	2.87	3.23
Milk, whole (qt)	0.60	0.49	0.48	0.57	0.72	0.61	0.56	0.70
Ice cream (gal)	3.06	2.80	2.60	3.20	2.20	3.00	4.00	4.40
Total animal products	12.73	11.51	10.18	11.03	11.68	13.20	13.97	16.29
Total market basket	47.58	39.97	43.26	45.12	45.87	48.12	50.08	59.64
Percent difference from national average		−16.0	−9.1	−5.2	−3.6	+1.1	+5.3	+25.4

[a]Includes vegetables and fruits in addition to animal products and other typical foods.
Source: National grocery survey reported in Rocky Mountain News, July 12, 1986.

14

TABLE 1.10. Improvement in Efficiency of Producing Foods of Animal Origin in the United States

Species and Measure of Productivity	1925 Value	1950 Value	1975 Value
Beef cattle			
Liveweight marketed per breeding female (lb)	220	310	482
Sheep			
Liveweight marketed per breeding female (lb)	60	90	130
Dairy cattle			
Milk marketed per breeding female (lb)	4,189	5,313	10,500
Swine			
Liveweight marketed per breeding female (lb)	1,600	2,430	2,850
Broiler chickens			
Age to market weight (weeks)	15.0	12.0	7.5
Feed per pound of gain (lb)	4.0	3.3	2.1
Liveweight at marketing (lb)	2.8	3.1	3.8
Turkeys			
Age to market weight (weeks)	34	24	19
Feed per pound of gain (lb)	5.5	4.5	3.1
Liveweight at marketing (lb)	13.0	18.6	18.4
Laying hens			
Eggs per hen per year (No.)	112	174	232
Feed per dozen eggs (lb)	8.0	5.8	4.2

Source: *Food from Animals,* CAST Report No. 82. March 1980, p. 13.

animal, and eventually human food productivity similar to those of developed regions (Fig. 1.8). Except perhaps in India, abundant world supplies of animal feed resources that do not compete with production of food for people are available to support expansion of animal populations and production. It has been estimated that through changes in resource allocation, an additional 8 billion acres of arable land (twice what is now being used) and 9.2 billion acres of permanent pasture and meadow (23% more than is now being used) could be put into production in the world. These estimates, plus the potential increase in productivity per acre and per animal in developed countries, demonstrate the magnitude of world food production potential. This potential cannot be realized, however, without proper government planning and increased incentive to individual producers.

In the long run, each nation must assume the responsibility of more of its own food supply by efficient production, barter, or purchase and by keeping future food production technology ahead of population increases and demand. Extensive untapped resources that can greatly enhance more food production exists throughout the world, including an ample supply of animal products. The greatest resource is each individual human being, who can, through self-motivation, become more productive and self-reliant.

CONTRIBUTIONS OF CLOTHING AND OTHER NONFOOD PRODUCTS

Products other than food from ruminants include wool, hair, hides, and pelts. Although synthetic materials have made some inroads into markets for these products, world wool

FIGURE 1.7.
Mechanization has increased plant and animal production in the United States. (A) In the late 1800s and early 1900s, a team of horses, man, and single moldboard plow could plow about 2 acres per day. Courtesy of Charles L. Benn, Iowa State University. (B) An early tractor in 1918 increased agricultural production on a per man basis. Courtesy of Michigan State University. (C) In the 1980s, one man with a powerful modern tractor pulling three plows, each with five moldboards can plow 110 acres in a 10-h day, accomplishing the work that once required 55 men and 110 horses. Courtesy of the USDA and CAST.

production has remained relatively stable over the past 15 years. It is important to note that in more than 100 countries, ruminant fibers are used in domestic production and cottage industries for clothing, bedding, housing, and carpets.

Annual production of animal wastes from ruminants contains millions of tons of nitro-

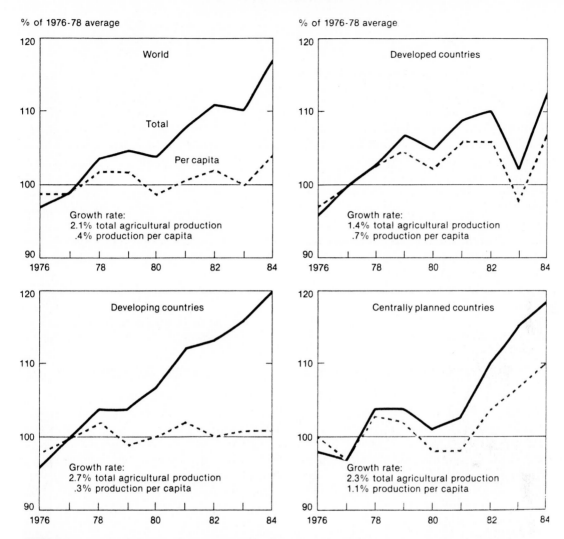

FIGURE 1.8.
Changes in world agricultural production. Developed countries include U.S., Canada, Europe, USSR, Japan, Republic of South Africa, Australia, and New Zealand. Developing countries include South and Central America, Africa (except Republic of South Africa), and Asia (except Japan and communist Asia). Centrally planned countries include China, Korea, Vietnam, Czechoslovakia, Hungary, Poland, and Rumania. Courtesy of the USDA.

gen, phosphorus, and potassium. The annual value of these wastes for fertilizer is estimated at $1 billion.

Inedible tallows and greases are animal by-products used primarily in soaps and animal feeds and as sources of fatty acids for lubricants and industrial use. Additional tallow and grease by-products are used in the manufacture of pharmaceuticals, candles, cosmetics, leather goods, woolen fabrics, and tin plating. The individual fatty acids can be used to produce synthetic rubber, food emulsifiers, plasticizers, floor waxes, candles, paints, varnishes, printing inks, and pharmaceuticals.

Gelatin is obtained from hides, skins, and bones and can be used in foods, films, and glues. Collagen, obtained primarily from hides, is used to make sausage casings.

CONTRIBUTION TO WORK AND POWER NEEDS

The early history of the developed world abounds with examples of the importance of animals as a source of work energy through draft work, packing, and human transport. In the United States during the 1920s, approximately 25 million horses and mules were used, primarily for draft purposes. The tractor has replaced all but a few of these draft animals. In parts of the developing world, however, animals provide as much as 99% of the power for agriculture even today.

In more than half the countries of the world, animals—mostly buffalo and cattle, but also horses, mules, camels, and llamas—are kept primarily for work and draft purposes (Fig. 1.9). About 20% of the world's human population depends largely or entirely on animals for moving goods. Animal draft power makes a significant contribution to the production of major foods (rice and other cereal grains) in several heavily populated areas of the world. India, for example, has more than 200 million cattle and buffalo, the largest number of any country. Although not slaughtered, because they are considered sacred, cattle of India contribute significantly to the food supply by providing work energy for fields and milk for people. It is estimated that India alone would have to spend more than $1 billion annually for gasoline to replace the animal energy it uses in agriculture.

ANIMALS FOR COMPANIONSHIP, RECREATION, AND ENTERTAINMENT

Estimates of the number of companion animals in the world are unavailable. There are an estimated 26 million family-owned dogs and 21 million family-owned cats in the United States. The pet food industry of the United States processes more than 3 million tons of cat and dog food annually valued at more than $1 billion. Many species of animals would qualify as companions where mankind derives pleasure from them. The contribution of animals as companions, especially to the young and elderly, is significant, even though it is difficult to quantify the emotional value. Animals used in rodeos, bull fighting, and other sport areas provide income for thousands of people and entertainment and recreation for millions (Fig. 1.10).

From the livestock that provide high-quality food to the pet that gives pleasure to its owner, domestic animals serve people in diverse and vital ways. This book is intended to discuss the contributions of livestock, a group of domestic animals whose contributions to human welfare have long been, and will continue to be, fundamentally important to meeting human needs.

ANIMAL CONTRIBUTIONS TO U.S. INCOME

The sale of livestock and livestock products generates a large amount of income to agricultural producers in the United States. In 1984 the sale of livestock and livestock products contributed nearly $73 billion to the U.S. economy. As these animals and products are processed and prepared for human consumption, they generate many billions of dollars more than their $73 billion sale price.

Figure 1.11 shows the cash receipts from livestock and livestock products for all 50 states, where 27 states have yearly receipts of more than $1 billion. Sales of livestock and

A

B

C

FIGURE 1.9.
Animals provide significant contributions to the draft and transportation needs of countries lacking mechanization in their agricultural technology. (A) Cattle used for draft in Honduras. Courtesy of Winrock International. (B) Water buffalos are used to cultivate many of the rice patties in the Far East. Courtesy of Dr. Budi S. Nara. (C) Camels being used to transfer fertilizer to farming areas in Ethiopia. FAO photo, courtesy of C. N. Coombes.

livestock products comprise nearly 50% of all farm cash receipts. Combined receipts from cattle and dairy products contribute the most to the total for livestock and livestock products. The top 10 states for each species or product are also identified in Table 1.11. There is another category, miscellaneous livestock and livestock products, in which several commodities are combined. For the United States, this miscellaneous category totals $2,158 million. In this category, Kentucky shows $580 million in receipts for horses and mules; Texas identifies $48 million in receipts for mohair.

CONTRIBUTION TO U.S. AND WORLD TRADE

Animal products are very important in U.S. trade relations with other countries. The U.S. livestock, poultry, and food service industries are dependent on exports and imports of animal products. Table 1.12 shows that tallow, grease, lard, hides, and skins are the major animal products exported. Beef and veal are the most important animal products imported.

A

B

C

FIGURE 1.10.
(A) The dog serves humans in a variety of ways. The guide dog leads the way for a visually-impaired man. Courtesy of Guide Dogs for the Blind, San Rafael, California. (B) Many people enjoy riding horses. Courtesy of *The Western Horseman.* (C) Horse racing is one of the highest attended spectator sports in the United States.

D

E

F

Courtesy of the Kentucky Derby. (D) Steer wrestling is exciting for the spectator but dangerous for the participants. Fast, well-trained horses and strong participants are needed to wrestle the steers to the ground in less than five seconds. Courtesy of Larry Thomas. (E) Calf roping is a favorite rodeo event. Courtesy of Larry Thomas. (F) Bull fighting originated in Spain centuries ago. It is popular in Mexico and several South American countries. Courtesy of José Rafael Cortes, editor of *La Nación* newspaper, San Cristobal, Venezuela.

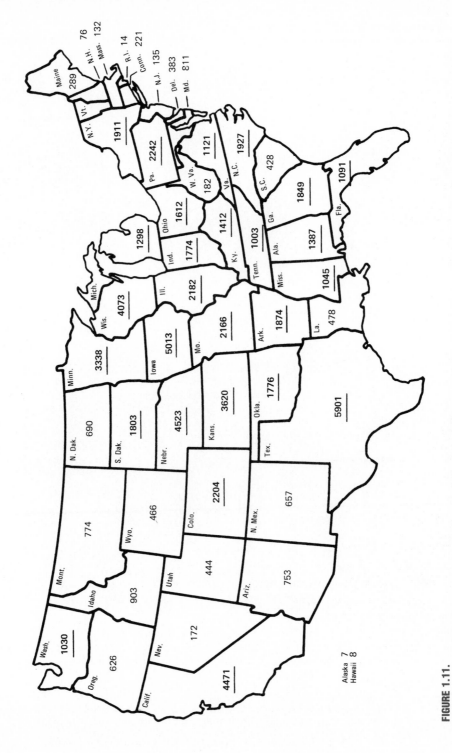

FIGURE 1.11.
Farm cash receipts, by states, for animals and animal products, 1984 (million dollars). States that have over 1 billion dollars in receipts have bold, underlined numbers. Adapted from the USDA data.

TABLE 1.11.   Leading States for Farm Cash Receipts, 1984

| Commodities | Rank | Value ($ millions) | Top 5 States ($ millions) | | | | |
			1	2	3	4	5
All commodities		141,835	Calif. 14,185	Texas 9,312	Nebr. 6,738	Ill. 6,738	Minn. 6,242
All livestock		72,739	Texas 5,901	Iowa 5,013	Nebr. 4,523	Calif. 4,471	Wisc. 4,073
Cattle and calves	1	30,601	Texas 3,666	Kans. 3,082	Colo. 1,757	Calif. 1,416	Okla. 1,388
Dairy products	2	17,661	Wisc. 3,008	Calif. 1,986	N.Y. 1,508	Pa. 1,306	Minn. 1,290
Hogs	5	9,691	Iowa 2,579	Ill. 1,080	Ind. 743	Minn. 727	Nebr. 635
Broilers	7	5,970	Ark. 1,000	Ga. 835	Ala. 687	N.C. 603	Miss. 448
Eggs	9	4,086	Calif. 409	Ga. 321	Ind. 291	Pa. 256	N.C. 239
Turkeys	14	1,655	N.C. 286	Minn. 230	Calif. 178	Ark. 123	Mo. 104
Wool	?	73	Texas 16	Calif. 8	Wyo. 7	Colo. 5	Mont. 5
Horses[a]	?	?	Ky. 580	—	—	—	—
Mohair[a]	?	?	Texas 48	—	—	—	—

[a]Other states list receipts from horses and mohair under category of miscellaneous livestock.
Source: USDA (Economic Indicators of the Farm Sector).

HUMAN HEALTH RESEARCH

Laboratory animals are commonly used to provide valuable information for improving human life. Larger, domestic farm animals are used less frequently, because purchase price and maintenance costs are high. Some examples of how farm animals have contributed to human health are noted.

Iowa State University has done an extensive study using swine as the model to evaluate how irradiation might increase the frequency of genetic defects. Colorado State University has studied the long-term effects of low levels of irradiation on the incidence of cancer in a beagle dog colony.

Miniature pigs (Yucatan swine whose ancestors came from the wilds of southern Mexico and central South America) are being used as laboratory animals, because their pulmonary, cardiac, dental, and even prenatal brain development closely resembles that of humans. The utilization of the miniature Yucatan pigs in biomedical research has in-

TABLE 1.12.   U.S. Exports and Imports of Major Animal Products, 1985

Commodity	Exports		Imports	
	Quantity	Value ($ millions)	Quantity	Value ($ millions)
Live animals (ex. poultry)	952 (thou head)	337.8	2,171.0 (thou head)	608.0
Cattle and calves	124.7 (thou head)	122.3	836.0 (thou head)	306.5
Hogs	18.2 (thou head)	7.9	1,226.6 (thou head)	127.8
Sheep and lambs	—	11.2	—	1.1
Horses, mules, burros	11.2 (thou head)	186.5	15.2 (thou head)	171.4
Baby chicks	26,470.5 (thou head)	52.4	—	—
Meats and meat products (ex. poultry)	941.3 (mil lb)	904.8	2,508.0 (mil lb)	2,225.8
Beef and veal	240.5 (mil lb)	467.2	1,489.6 (mil lb)	1,276.5
Pork	89.5 (mil lb)	76.0	932.6 (mil lb)	861.3
Lamb and mutton	5.0 (mil lb)	5.7	34.8 (mil lb)	33.6
Processed meats	—	7.7	—	6.7
Poultry	514.8 (mil lb)	249.2	—	98.1
Horse meat	43.6 (mil lb)	33.0	0.2 (mil lb)	1.0
Other Products				
Tallow, grease, and lard	2,960.5 (mil lb)	619.0	46.9 (mil lb)	14.5
Variety meats	550.7 (mil lb)	304.3	12.3 (mil lb)	6.4
Sausage casings	15.2 (mil lb)	14.4	29.4 (mil lb)	52.7
Hides and skins	26.0 (mil)[a]	1,295.2	49.1 (mil lb)[a]	245.7
Wool and mohair	10.3 (mil lb)	42.2	94.8 (mil lb)	145.2
Dairy products	820.6 (mil lb)	431.3	909.9 (mil lb)	772.9
Egg and egg products	61.8 (mil lb)	55.2	—	14.7
Feathers and down		23.3		
Bull semen		23.1		
Total for Animals and Animal Products		4,148.7		4,152.8
Total for Agricultural Products		29,025.4[b]		19,968.0
Total for all Products		206,925.0		343,553.0

[a]Cattle hides only.
[b]Grains and soybeans are most important.
Source: USDA, *Foreign Agricultural Trade of the United States.*

creased dramatically since 1978. Miniature pigs are considered ideal in studying human aging, disease resistance, and the effect of diet on diabetes and atherosclerosis (Fig. 1.12). In one research study, for example, it was discovered that genetically predisposed pigs became susceptible to diabetes after consuming a low-fiber diet with 40% of the calories coming from saturated fat. There is less cost in the miniature pigs, since their mature weights are approximately 120 lb compared to 600–800 lb for typical swine. A new breed of Yucatan swine, the "micropig," may prove even more useful as an experimental animal, as the mature weights range from 50 to 70 lb.

Cattle and sheep have been used to test artificial organs before these organs have been implanted into humans. Figure 1.13 shows some of the experimental animals used to test the Jarvik 7 artificial heart.

FIGURE 1.12.
Yucatan miniature pigs used as experimental animals to help solve human problems such as diabetes, obesity, and atherosclerosis. Courtesy of Colorado State University.

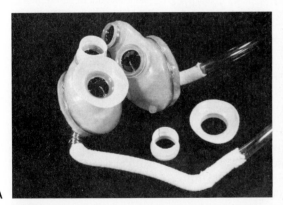

FIGURE 1.13.
(A) Jarvik 7 artificial heart was used experimentally in sheep and calves before it was used in humans. (B) Longest survival time in sheep with Jarvik 7 artificial heart was 289 days. (C) Dairy animal "Charley" had a natural heart for 90 days, a Jarvik 7 artificial heart for 74 days, then a natural heart transplant from a twin sister, until he was slaughtered at 39 months of age. Courtesy of Division of Artificial Organs, University of Utah.

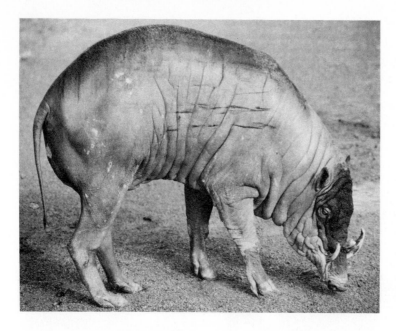

FIGURE 1.14.
The babirusa, a pig-like animal unique to a few islands in eastern Indonesia. Courtesy of Phillip Coffey, Jersey Wildlife Preservation Trust.

OTHER ANIMALS

This book gives major attention to cattle, sheep, swine, horses, poultry, and goats. Several other useful species are briefly mentioned in this chapter. In addition, there are other animals that provide useful products in specific areas of the world. There are also several domesticated animals about which little is known and undomesticated animals that have potential for long-term world agricultural development.

Some of the domesticated animals, relatives to cattle, are the banteng of Indonesia and the mithan of India, Burma, and Bangladesh. Wild bovines include the kouprey in Thailand and the gaur in India and Southeast Asia.

Among the undomesticated Asian pigs is the babirusa of eastern Indonesia (Fig. 1.14). The stomach of these unusual animals has an extra sac, suggesting they may have the ability to break down cellulose. These "ruminant pigs" browse leaves, a behavior more similar to deer than to pigs.

Several of these rarely known animals have disease resistance that might be incorporated into breeding programs of our more familiar animals. A question is raised as the potential of the unfamiliar animals is reviewed. Are we overlooking animals that could benefit mankind?

SELECTED REFERENCES

Baldwin, R. L. 1980. *Animals, Feed, Food and People: An Analysis of the Role of Animals in Food Production.* Boulder, CO: Westview Press.
Cravens, W. W. 1981. Plants and animals as protein sources. *J. Anim. Sci.* 53:817.
Devendra, C. 1980. Potential of sheep and goats in less developed countries. *J. Anim. Sci.* 51:461.

Ending Hunger: An Idea Whose Time Has Come. 1985. Sparks, NV: Praeger.

FAO Production Yearbook, Vol. 38. 1985. Rome: FAO.

Fitzhugh, H. A., Hodgson, H. J., Scoville, O. J., Nguyen, T. D., and Byerly, T. C. 1978. *The Role of Ruminants in Support of Man.* Morrilton, AR: Winrock International.

Food from Animals: Quantity, Quality and Safety. 1980. Council for Agricultural Science and Technology (CAST). Report No. 82.

Hodgson, H. J. 1979. Role of the dairy cow in world food production. *J. Dairy Sci.* 62:343.

McDowell, R. E. 1978. Contributions of animals to human welfare. *N.Y. Food Life Sci.* 11:15.

National Academy of Science. 1978. *Plant and Animal Products in the U.S. Food System.* Washington: National Academy Press.

National Research Council. 1983. *Little-Known Asian Animals with a Promising Economic Future.* Washington: National Academy Press.

Reid, J. T., White, O. D., Anrique, R., and Fortin, A. 1980. Nutritional energetics of livestock: Some present boundaries of knowledge and future research needs. *J. Anim. Sci.* 51:1393.

Willham, R. L. 1985. *The Legacy of the Stockman.* Morrilton, AR: Winrock International.

CHAPTER 2

Red Meat Products

Red meat products come primarily from cattle, swine, sheep, goats, and, to a lesser extent, horses and other animals. Poultry meat, sometimes called "white meat," is discussed in Chapter 5.

Red meats (referred to as meat from this point) are named according to their source: **beef** is typically from cattle over a year of age; **veal** is from young calves (veal carcasses are distinguished from beef by their grayish pink color of the lean); **pork** is from swine; **mutton** is from mature sheep; **lamb** is from young sheep; **chevon** is from goats, but more commonly is called goat meat.

PRODUCTION

World meat supply approaches 250 billion pounds with the United States, China, and the USSR leading all other countries (Table 2.1). Buffalo meat comes from the true buffalo, and it should not be confused with bison in the United States. The true buffalo has no hump; thus the name buffalo belongs only to water buffalo in Asia and the African Buffalo.

TABLE 2.1. World Red Meat Supply, 1985

Product	Bil lb[a]	Leading Countries (bil lb)
Pork	128	China (33.2); U.S. (14.8); USSR (13.1); West Germany (7.1)
Beef and veal	101	U.S. (24.0); USSR (15.6); Argentina (5.6); Brazil (5.0)
Mutton and lamb	14	USSR (1.8); Australia (1.0); China (0.7); U.K. (0.6)
Goat meat	5	China (0.7); Pakistan (0.5); Nigeria (0.3); Turkey (0.2)
Buffalo meat	2	Pakistan (0.5); Egypt (0.3); India (0.3); China (0.2)
Horsemeat	1	U.S. (0.2); Mexico (0.1); Argentina (0.1); Italy (0.1)
World total	251	U.S. (39.9); China (36.6); USSR (31.1); France (9.5)

[a]Carcass weight basis.
Source: 1985 FAO Production Yearbook.

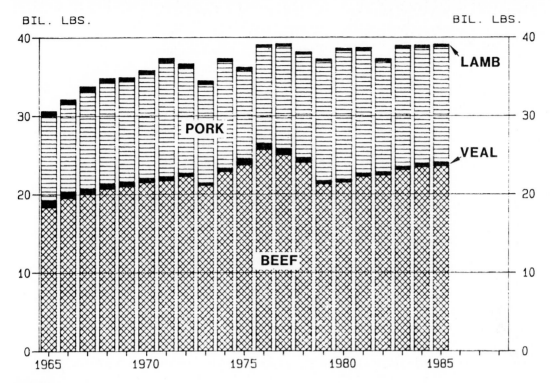

BIL. LBS.   BIL. LBS.

FIGURE 2.1.
Red meat produced in U.S. (carcass weight basis). Courtesy of Western Livestock Marketing Information Project.

Red meat production in the United States is shown in Fig. 2.1. Beef and pork comprise most of the annual production of nearly 40 billion pounds. The total production has been relatively stable over the past several years.

Figures 2.2, 2.3, and 2.4 show where cattle, hogs, and sheep are slaughtered in the United States. Texas, Kansas, Nebraska, Iowa, and Colorado are the leading states in cattle slaughter. In 1984, Iowa slaughtered nearly one-fourth of the nation's hogs, and Illinois, Minnesota, Michigan, Nebraska, Kentucky, Ohio, Indiana, Missouri, and South Dakota slaughtered an additional 40%. Colorado, California, Iowa, Texas, and South Dakota are the leading lamb-slaughtering states.

The United States has an annual supply of horsemeat of approximately 200 million pounds (Table 2.1). Although most of the horsemeat goes into pet food, several thousand pounds are shipped to Europe for human consumption. In 1980, approximately 15–20 slaughter plants in the United States specialized in slaughtering horses, shipping the meat to France, Belgium, the Netherlands, Switzerland, and Italy.

Meat animals are slaughtered in packing plants located near areas of live animal production. This reduces the transportation costs as carcasses, wholesale cuts, and retail cuts

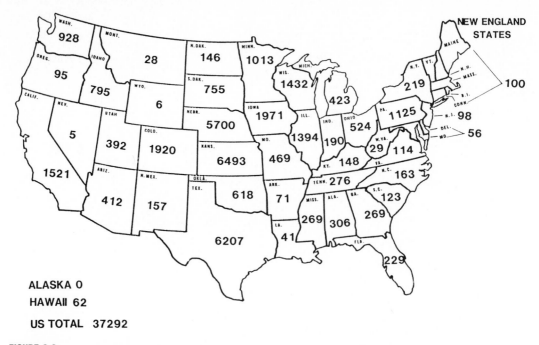

FIGURE 2.2.
Commercial cattle slaughter, 1986, 1000 head. Courtesy of Western Livestock Marketing Information Project.

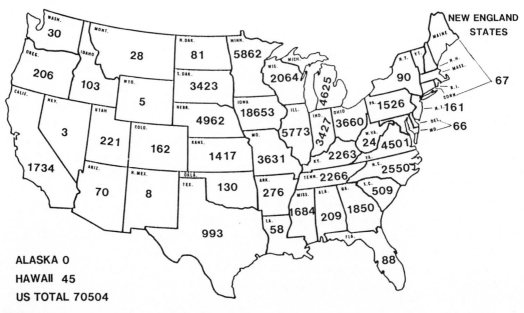

FIGURE 2.3.
Commercial hog slaughter, 1986, 1000 head. Courtesy of Western Livestock Marketing Information Project.

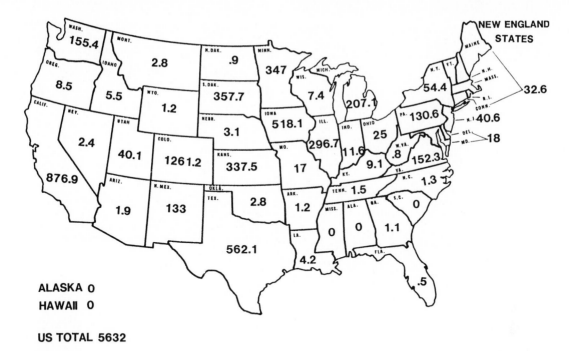

FIGURE 2.4.
Commercial sheep and lamb slaughter, 1986, 1000 head. Courtesy of Western Livestock Marketing Information Project.

can be transported at less cost than live animals. The top 10 packers are identified in Table 2.2.

PRODUCTS

Animals are sent to packing plants, where they are slaughtered. During the initial slaughter stage, the animals are made unconscious by using carbon dioxide gas or stunning (electrical or mechanical). The jugular vein and/or carotid artery is then cut to drain the blood from the animal. After bleeding, the hides are removed from cattle and sheep. Hogs are scalded to remove the hair, but the skin is usually left on the carcass. There is an increasing number of packers (especially small plants) that are skinning hogs, as it is more energy-efficient than leaving the skin on.

After the hide or hair is removed, the internal organs are separated from the carcass. Those parts removed from the carcass are sometimes referred to as the **"drop,"** viscera, **offal,** or **by-products.** Typically this is the head, hide, hair, shanks (lower parts of legs and feet), and internal organs.

Dressing percentage (sometimes referred to as **yield**) is the relation of hot carcass weight to live weight. It is calculated as follows:

$$\text{Dressing percentage} = \frac{\text{Hot carcass weight}}{\text{Live weight}} \times 100$$

TABLE 2.2. Leading Meat Packers and Processors (1985 net sales of more than $1 billion)

Company Location (parent firm)	Net Sales ($ billions)	No. Plants	Product				
			Beef	Pork	Lamb	Veal	Poultry
1. IBP Inc. Dakota City, NE (Occidental Petroleum)	6.50	14	x	x			
2. ConAgra, Inc. Omaha (Meat and Poultry Oper.)	3.20	39	x	x	x	x	x
3. Swift Indep. Pack. Co. Chicago (Valley View Holding Co.)	2.89	14	x	x	x		
4. Excel Corp. Wichita, KS (Cargill)	2.50	7	x				
5. Oscar Mayer Foods Co. Madison, WI (General Foods/Phillip Morris)	2.00	19	x	x			x
6. Sara Lee Corp. Chicago (Meat Group)	1.80	24	x	x			x
7. John Morrell & Co. Northfield, IL (United Brands Co.)	1.70	5	x	x	x		
8. Swift-Eckrich, Inc. Oak Brook, IL (Beatrice)	1.69	24	x	x			x
9. Wilson Foods Corp. Oklahoma City	1.53	9		x			
10. Geo. A. Hormel & Co. Austin, MN	1.50	17	x	x			
11. Monfort of Colo., Inc. Greeley, CO	1.46	4	x		x		

Source: Meat Industry, July 1986.

TABLE 2.3.   U.S. Livestock Slaughter and Meat Production, 1985

Species/ Class	Meat Products	Number Head (millions)	Average Live Weight (lb)	Total Carcass Weight (mil lb)	Average Carcass Weight (lb)	Average Dressing Percent
Cattle	Beef	36.3	1,065	23,548	649	60
Calves[a]	Veal	3.4	250	499	147	60
Hogs	Pork	84.5	240	14,721	174	73
Sheep and lambs	Mutton and lamb	6.1	110	352	59	50

[a]Part of these are classified as calf carcasses, which are intermediate between veal and beef carcasses.
Source: USDA.

Table 2.3 shows the average live slaughter weights, carcass weights, and dressing percentages for cattle, sheep, and hogs. For practical purposes, the average dressing percent for hogs is 70%; cattle, 60%; and sheep, 50%, although dressing percent can vary several percentage points within each species. The primary factors affecting dressing percent are fill (contents of digestive tract), fatness, muscling, weight of hide, and, in sheep, weight of wool.

Beef and pork carcasses are split down the center of the backbone (sheep carcasses are not split), giving two sides approximately equal in weight. After the carcasses are washed, they are put in a cooler to chill. The cooler temperature of approximately 35°F removes the heat from the carcass and keeps the meat from spoiling. Carcasses can be stored in coolers for several weeks; however, this usually does not occur in large packing plants, as profitability is dependent on slaughtering live animals one day and shipping carcasses or products out the next day or two. Smaller packers may keep carcasses in the coolers for 2 weeks. This allows the carcasses to age. This **aging** process increases meat tenderness owing to enzyme activity in the connective tissue.

In the past, most of the meat was shipped in carcass form, but today many packing plants process the carcasses into wholesale or primal cuts and even into retail cuts (Table 2.4; Figs. 2.5–2.8). This latter process involves shipping meat in boxes; thus the terms **"boxed beef"** and **"boxed lamb."**

TABLE 2.4.   Wholesale Cut of Beef, Veal, Pork, and Lamb

Beef	Veal	Pork	Lamb
Round	Leg/round	Leg/ham	Leg
Sirloin	Sirloin	Loin	Loin
Short loin	Loin	Blade shoulder	Rib
Rib	Rib	Jowl	Shoulder
Chuck	Shoulder	Arm shoulder	Neck
Foreshank	Foreshank	Spareribs	Foreshank
Brisket	Breast	Side	Breast
Short plate			
Flank			

Source: The National Live Stock and Meat Board.

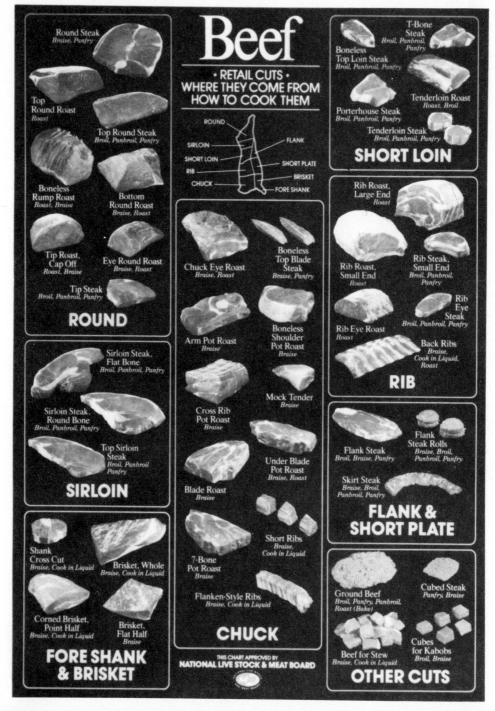

FIGURE 2.5.
Wholesale and retail cuts of beef with recommended types of cooking. Courtesy of The National Live Stock and Meat Board.

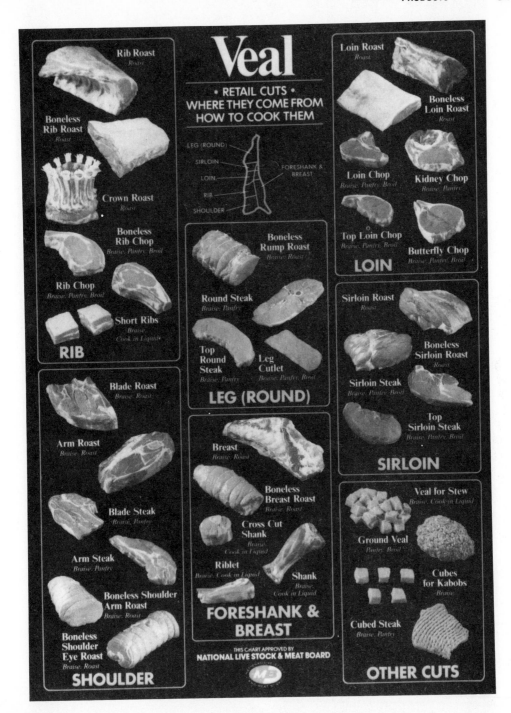

FIGURE 2.6.
Wholesale and retail cuts of veal with recommended type of cooking. Courtesy of The National Live Stock and Meat Board.

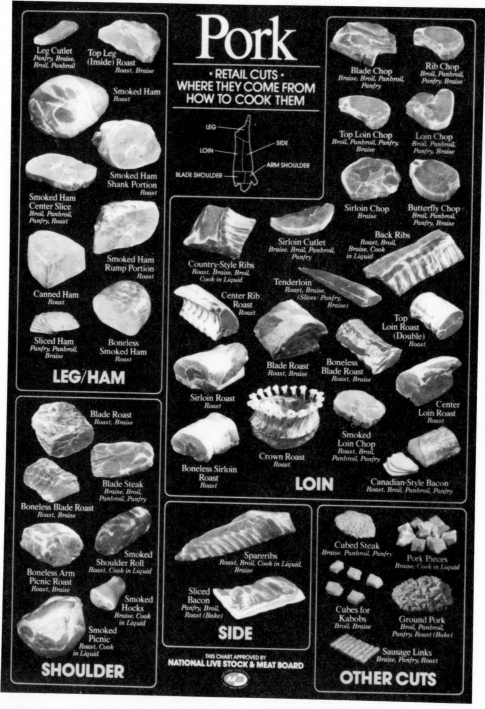

FIGURE 2.7.
Wholesale and retail cuts of pork with recommended type of cooking. Courtesy of The National Live Stock and Meat Board.

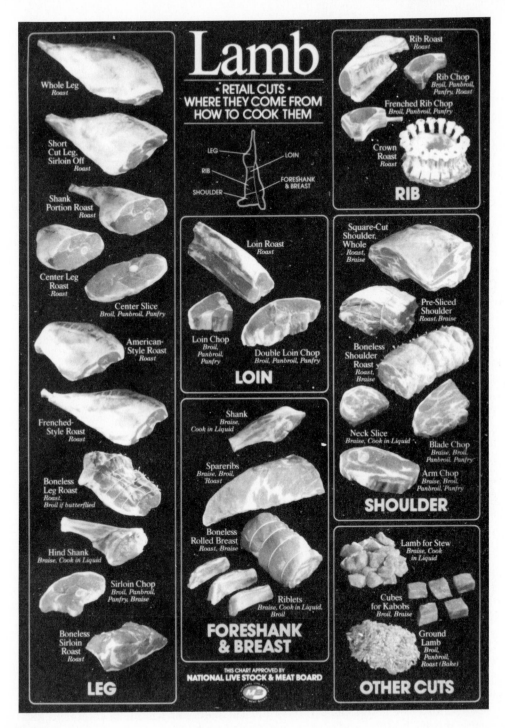

Lamb
·RETAIL CUTS·
WHERE THEY COME FROM
HOW TO COOK THEM

LEG
LOIN
RIB
FORESHANK & BREAST
SHOULDER

LEG

Whole Leg
Roast

Short Cut Leg, Sirloin Off
Roast

Shank Portion Roast
Roast

Center Leg Roast
Roast

Center Slice
Broil, Panbroil, Panfry

American-Style Roast
Roast

Frenched-Style Roast
Roast

Boneless Leg Roast
Roast, Broil if butterflied

Hind Shank
Braise, Cook in Liquid

Sirloin Chop
Broil, Panbroil, Panfry, Braise

Boneless Sirloin Roast
Roast

LOIN

Loin Roast
Roast

Loin Chop
Broil, Panbroil, Panfry

Double Loin Chop
Broil, Panbroil, Panfry

FORESHANK & BREAST

Shank
Braise, Cook in Liquid

Spareribs
Braise, Broil, Roast

Boneless Rolled Breast
Roast, Braise

Riblets
Braise, Cook in Liquid, Broil

THIS CHART APPROVED BY
NATIONAL LIVE STOCK & MEAT BOARD

RIB

Rib Roast
Roast

Rib Chop
Broil, Panbroil, Panfry, Roast

Frenched Rib Chop
Broil, Panbroil, Panfry

Crown Roast
Roast

SHOULDER

Square-Cut Shoulder, Whole
Roast, Braise

Pre-Sliced Shoulder
Roast, Braise

Boneless Shoulder Roast
Roast, Braise

Neck Slice
Braise, Cook in Liquid

Blade Chop
Braise, Broil, Panbroil, Panfry

Arm Chop
Braise, Broil, Panbroil, Panfry

OTHER CUTS

Lamb for Stew
Braise, Cook in Liquid

Cubes for Kabobs
Broil, Braise

Ground Lamb
Broil, Panbroil, Roast (Bake)

FIGURE 2.8.
Wholesale and retail cuts of lamb with recommended types of cooking. Courtesy of The National Live Stock and Meat Board.

Most red meats are marketed as fresh products. Some are cured, smoked, processed, or canned. Table 2.5 identifies how various quantities of major meat products are prepared for sale.

KOSHER MEATS

All **kosher meat** comes from animals that have split hooves, chew their cud, and have been slaughtered in a manner prescribed by the Torah (Orthodox Jewish law). Meat from undesirable animals or animals not properly slaughtered or with imperfections are called nonkosher. Red meats, poultry, and all other foods are termed kosher or nonkosher. Foods that are kosher bear the endorsement symbol Ⓤ.

Kosher slaughter, inspection, and supervision are performed only by trained individuals approved by rabbinic authority. The trachea and esophagus of the animal must be cut in swift continuous strokes with a perfectly smooth, sharp knife causing instant death and minimum pain to the animal. Various internal organs are inspected for defects and adhesions.

Only the forequarters are identified as kosher, as the hindquarters contain the forbidden sciatic nerve, which is imbedded deep in the muscle tissue and is difficult to remove. Hindquarters are sold as nonkosher meats.

The Torah forbids the eating of blood, including blood of arteries, veins, and muscle tissue. For this reason, certain arteries and veins must be removed from cattle; neck veins of poultry must be severed to allow the free flow of blood. Blood is extracted from meat by broiling and salting. There are specific steps required in proper broiling and salting to endorse the meat as kosher.

There are approximately 2 million Orthodox Jews in the United States who keep kosher. Estimates on quantities of kosher meat consumed are not available.

COMPOSITION

Physical and chemical composition are the two major ways of classifying meat composition. Physical composition can be observed visually; chemical composition is determined by chemical analysis.

Physical Composition

The major physical components of meat are lean (muscle), fat, bone (Fig. 2.9), and connective tissue. The proportion of fat, lean, and bone changes from birth to slaughter time. Chapter 17 discusses these compositional changes in more detail. Connective tissue, which to a large extent determines meat tenderness, exists in several different forms and locations. For example, tendons are composed of connective tissue (collagen) which attaches muscle to bone. Other collagenous connective tissues hold muscle bundles together and provide the covering to each muscle fiber. Figure 2.9 shows the physical structure of muscle: **myofibrils** are component parts of muscle fibers; muscle fibers combined together comprise a muscle bundle; and groups of muscle bundles are components of muscle systems.

TABLE 2.5. **Meat Products Processed and Canned Under Federal Inspection, 1984 (major items selected as list of alternative products is extensive)**

Item	Million Pounds
Fresh/frozen products	
Pork cuts	10,706
Pork boning	3,689
Beef cuts	9,737
Beef boning	8,731
Hamburger/ground beef	3,618
Other fresh/frozen	1,014
Cured	
Pork	4,023
Beef	285
Other meats	19
Smoked, dried, or cooked	
Ham	1,759
Bacon	1,704
Beef	368
Other meats	330
Sausage	
Fresh pork	878
Fresh beef	11
Fresh other	256
Franks and wieners	1,405
Bologna	688
Liver sausage and meat loaves	1,048
Sliced products	
Bacon	1,601
Ham	369
Convenience foods	
Pizza	579
Pies	215
Dinners	285
Entrees	459
Fats and oils	
Lard	1,402
Edible tallow	1,613
Other fat compounds	1,021
Canned meats (total)	2,060
Hams	155
Chili con carne	310
Pasta meat products	435
Luncheon meat	183

Source: USDA.

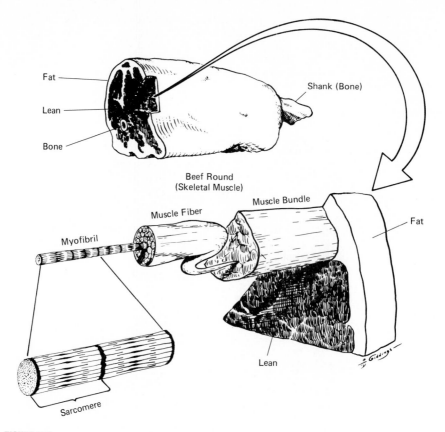

FIGURE 2.9.
The fundamental structure of meat and muscle in the beef carcass. Drawn by Dennis Giddings.

Chemical Composition

Since muscle or lean meat is the primary carcass component consumed, only its chemical composition will be discussed. Chemical composition is important, because it primarily determines the nutritive value.

Muscle consists of approximately 65–75% water, 15–20% protein, 2–12% fat, and 1% minerals (ash). As the animal is fattened, water and protein decrease and fat percent increases.

Fat-soluble vitamins (A, D, E, and K) are contained in the fat component of meat. Most B vitamins (water-soluble) are abundant in meat.

The major protein in muscle is actomyosin, which is a globulin consisting of two proteins, actin and myosin. Most of the other nitrogenous extracts in meats are relatively unimportant nutritionally. However, these other extracts provide aroma and flavor in meat, which stimulate the flow of gastric juices.

Simple carbohydrates in muscle are less than 1%. Glucose and glycogen are concentrated in the liver. They are not too important nutritionally, but they have an important effect on meat quality, affecting muscle color and water-holding capacity.

HEALTH CONSIDERATIONS

Nutritive Value

Heightened health awareness in the United States has made today's consumers more astute food buyers than ever before. Consumers look for foods to meet their requirements for a nutritious, balanced diet as well as convenience, versatility, economy, and ease of preparation. Processed meat products have long been among Americans' favorite foods. Consumers today choose from more than 200 styles, including more than 50 cold cuts, dozens of sausages, and a wide variety of hams and bacon. Nutrient-rich red meats—beef, pork, lamb, and veal—are primary ingredients in these products. Meat products in all delicious and unique varieties contain the same nutrients as the fresh meats from which they are made (Table 2.6). More than 40 nutrients are needed by the body for good health. These must be obtained from a variety of foods; no single food contains all nutrients.

Meat is nutrient-dense and is among the most valuable sources of important vitamins and minerals in today's diet. **Nutrient density,** a new measurement of food value, compares the amount of essential nutrients to the number of calories in food.

Red meats are abundant sources of iron, a mineral necessary to build and maintain blood hemoglobin, which carries oxygen to body cells. Heme iron, one form of iron found in red meat, is used by our bodies most effectively, and meat enhances the absorption of nonheme iron from meat and other food sources. Red meat is also a rich source of zinc,

TABLE 2.6.   Nutrient Contents of Red Meat (3 oz) and their Contribution to the RDA (recommended daily allowances)[a]

Nutrient	Beef		Pork		Lamb	
	Amount	RDA %	Amount	RDA %	Amount	RDA %
Calories	192	10	198	10	176	9
Protein	25 g	56	23 g	51	24 g	54
Fat[b]	9.4 g	14	11.1 g	17	8.1 g	12
Cholesterol[b]	73 mg	24	79 mg	26	78 mg	26
Sodium	57 mg	2	59 mg	2	71 mg	2
Iron	2.6 mg	14	1.1 mg	6	1.7 mg	10
Zinc	5.9 mg	39	3.0 mg	20	4.5 mg	30
Thiamin	0.08 mg	6	0.59 mg	39	0.09 mg	6
Niacin	3.4 mg	17	4.3 mg	22	5.3 mg	27
B_{12}	2.4 mg	79	0.7 mg	24	2.2 mg	37

[a]Based on 2,000-calorie diet and RDA for women, age 23–51.
[b]The American Heart Association recommends that fat contribute not more than 30% of total daily calories and that cholesterol be limited to 300 mg/day.
Source: USDA.

an essential trace element that contributes to tissue development, growth, and wound healing.

Red meats contain a rich supply of the B vitamins. Pork is a particularly rich source of thiamine. Thiamine is essential to convert carbohydrates into energy. Vitamin B_{12} occurs only in animal products such as meat, fish, poultry, and milk and is essential to protect nerve cells and for the formation of blood cells in bone marrow. Niacin, riboflavin, and vitamin B_6 are also found in significant amounts in red meats.

Finally, red meats are an excellent source of protein, a fundamental component of all living cells. Fresh and processed meats contain high-quality or complete protein with all essential amino acids in good quantity.

Human Disease Risk

Heart disease and cancer are two of the most dreaded human diseases. Over the past several years, millions of dollars have been utilized in research attempting to identify factors causing these diseases. The factors causing heart disease and cancer appear to be many and complex. This has allowed only limited progress in identifying cures for the diseases. Medical scientists do not agree on the major factors causing heart disease and cancer.

Coronary heart disease (CHD) is the leading cause of death in the United States. After several decades of increase, the death rate from CHD in this country reached a peak in the mid-1960s and has been declining ever since. This remarkable phenomenon has not been explained. However, changes in life-style, increased physical activity, and a new level of dietary consciousness have evolved during the same period.

Most CHD results from atherosclerosis, a process by which a complex, lipid-containing lesion develops in the arteries, eventually causing blockage of the blood supply to the heart or the brain. The most important risk factors from CHD are (1) genetics—a family history of CHD, especially premature CHD; (2) cigarette smoking; (3) high blood pressure (hypertension); (4) high blood cholesterol concentration (hypercholesterolemia); and (5) obesity.

Each person has a unique set of risk factors determined by environment and genetic susceptibility. Thus, treatment of individuals to alter CHD risk needs to be designed according to each person's risk.

Since meat contains fat and cholesterol, it has been implicated as a cause of hypercholesterolemia and thus a risk for CHD. During the past several years, there has been strong encouragement, primarily through commercial advertising, to lower blood cholesterol by consuming more polyunsaturated fats (vegetables) and decreasing animal fat consumption. Research has shown that dietary changes favoring more consumption of polyunsaturated fats has lowered blood cholesterol but has not changed total body cholesterol. Evidently, more of the blood cholesterol is transferred to soft tissues such as the arteries. Figure 2.10 shows fat and oil consumption by vegetable and animal origin; vegetable oils have replaced some of the animal fat in the diet, although total fat and oil consumption has increased since 1967.

The relationship between dietary and blood cholesterol has not been adequately clarified. Research data indicate that cholesterol intake is one of the least important dietary components that influence blood cholesterol. There is little scientific evidence indicating

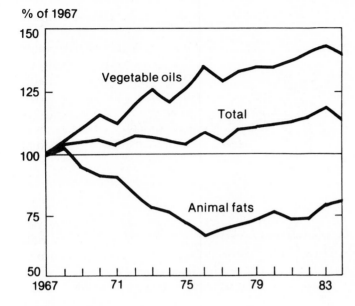

FIGURE 2.10.
Per capita consumption of fat and oils (animal fats include butter). Courtesy of the USDA.

that for most persons dietary changes will reduce the incidence of CHD except for the obese.

Many consumers are becoming more health conscious, and they desire to reduce calorie and saturated-fat intake. Numerous food items are marketed as "light" or "lite," which implies that the calorie content is reduced.

Red-meat marketing is also being directed to these consumers. "Light" beef is low-fat beef; i.e., it contains less fat, both external (subcutaneous) and internal (marbling). "Light" beef is at least 25% less fatty than typical USDA choice beef, or at least 90% fat-free on a retail cut basis (precooked). "Light" beef is also lower in calories and saturated fat than USDA choice beef. The lower fat content of "light" beef also leads to slight reductions in cholesterol content of the beef cut, because cholesterol is more highly associated with fat cells than muscle cells.

The biggest limitation to leaner and lower grading beef is that it has less tolerance for improper cooking methods than USDA choice grade beef. For example, leaner beef may have a higher moisture loss and be less tender when subjected to high cooking temperatures for a longer cooking time than is needed.

Most scientists agree that it is premature to issue broad dietary advice on cancer prevention. There is mounting evidence that total caloric intake may be far more important than individual nutrients, like fat. Energy expenditure (exercise) may also be important. Research supports moderate and balanced diets for decreasing the risk of cancer. This work may take some attention away from fat per se (and fat sources, including meat). However, fat is a concentrated source of calories, so attention will focus on fat to avoid an excessive caloric intake.

In recent years, several retail markets initiated advertisements to show that meat is not high in calories (Fig. 2.11). Additional nutritional information on meat has been made available through The National Live Stock and Meat Board's "Meat Nutri-Facts" program.

FIGURE 2.11.
Several major supermarkets have advertised the calorie content of lean pork, lamb, and beef. Courtesy of The National Live Stock and Meat Board.

CONSUMPTION

Meat is consumed for eating satisfaction and nutrient content (Figs. 2.12–2.14); however, there are food products other than meat that are lower cost per unit of protein. Visual displays, as shown in Fig. 2.15, appear at the retail meat counters. They show that a cooked, trimmed, 3-oz serving of meat is relatively low in calories, fat, and cholesterol while providing an abundant supply of other essential nutrients. However, consumers have a preference for the way in which the protein is packaged—particularly if it is more palatable in texture, flavor, and aroma. Meat has these preferred palatability characteristics.

Consumer acceptance of red meat and poultry is shown in Fig. 2.16. Poultry is included here to demonstrate that over the past 15 years the total annual per-capita meat consumption has not deviated greatly from 200 lb. However, the proportions of the different meats consumed have changed.

Table 2.7 shows the per-capita consumption (disappearance) of red meat on a carcass and retail basis. These data do not reflect the amount consumed, as there are fat trim, inedible bone, cooking losses, and plate waste. Consumption data accounting for most of these losses are shown in Table 2.8.

FIGURE 2.12.
A thick, tender and juicy steak is the most highly preferred beef product. Courtesy of The National Live Stock and Meat Board.

FIGURE 2.13.
A favorite pork dish is smoked, boneless ham. Courtesy of The National Live Stock and Meat Board.

FIGURE 2.14.
Processed beef and pork are popular meat items for sandwich lovers. Courtesy of The National Live Stock and Meat Board.

TABLE 2.7. Per-Capita Disappearance of Red Meat (lb)

Year	Beef		Veal		Pork		Lamb and Mutton		Total Red Meat	
	Carcass	Retail	Carcass	Retail	Carcass	Retail	Carcass	Retail	Carcass	Retail
1965	99.5	73.6	5.2	4.3	58.7	54.7	3.7	3.3	167.1	135.9
1970	113.5	84.0	2.9	2.4	66.4	62.3	3.2	2.9	186.0	151.6
1975	118.8	87.9	4.1	3.4	54.8	50.7	2.0	1.8	179.7	143.7
1980	103.4	76.5	1.8	1.5	73.5	68.3	1.5	1.4	180.2	147.7
1985	106.6	78.9	2.2	1.8	66.0	62.0	1.6	1.4	176.4	144.1

Source: USDA.

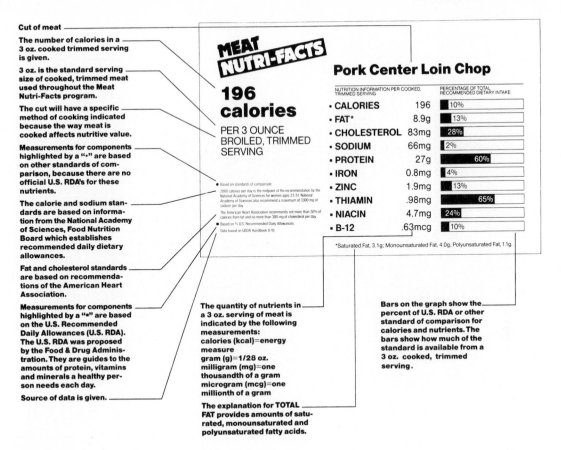

Cut of meat

The number of calories in a 3 oz. cooked trimmed serving is given.

3 oz. is the standard serving size of cooked, trimmed meat used throughout the Meat Nutri-Facts program.

The cut will have a specific method of cooking indicated because the way meat is cooked affects nutritive value.

Measurements for components highlighted by a "•" are based on other standards of comparison, because there are no official U.S. RDA's for these nutrients.

The calorie and sodium standards are based on information from the National Academy of Sciences, Food Nutrition Board which establishes recommended daily dietary allowances.

Fat and cholesterol standards are based on recommendations of the American Heart Association.

Measurements for components highlighted by a "▪" are based on the U.S. Recommended Daily Allowances (U.S. RDA). The U.S. RDA was proposed by the Food & Drug Administration. They are guides to the amounts of protein, vitamins and minerals a healthy person needs each day.

Source of data is given.

MEAT NUTRI-FACTS

196 calories

PER 3 OUNCE BROILED, TRIMMED SERVING

• Based on standards of comparison

▪ 2000 calories per day is the midpoint of the recommendation by the National Academy of Sciences for women ages 23-51. National Academy of Sciences also recommend a maximum of 3300 mg of sodium per day.

The American Heart Association recommends not more than 30% of calories from fat and no more than 300 mg of cholesterol per day.

▪ Based on % U.S. Recommended Daily Allowances.

Data based on USDA Handbook 8-10.

Pork Center Loin Chop

NUTRITION INFORMATION PER COOKED, TRIMMED SERVING		PERCENTAGE OF TOTAL RECOMMENDED DIETARY INTAKE
• CALORIES	196	10%
• FAT*	8.9g	13%
• CHOLESTEROL	83mg	28%
• SODIUM	66mg	2%
• PROTEIN	27g	60%
▪ IRON	0.8mg	4%
▪ ZINC	1.9mg	13%
▪ THIAMIN	.98mg	65%
▪ NIACIN	4.7mg	24%
▪ B-12	.63mcg	10%

*Saturated Fat, 3.1g; Monounsaturated Fat, 4.0g; Polyunsaturated Fat, 1.1g.

The quantity of nutrients in a 3 oz. serving of meat is indicated by the following measurements: calories (kcal)=energy measure gram (g)=1/28 oz. milligram (mg)=one thousandth of a gram microgram (mcg)=one millionth of a gram

The explanation for TOTAL FAT provides amounts of saturated, monounsaturated and polyunsaturated fatty acids.

Bars on the graph show the percent of U.S. RDA or other standard of comparison for calories and nutrients. The bars show how much of the standard is available from a 3 oz. cooked, trimmed serving.

FIGURE 2.15.
An explanation of the Meat Nutri-Facts display for a pork center loin chop. Courtesy of The National Live Stock and Meat Board.

TABLE 2.8. Red Meat Consumption (cooking, bone, and fat losses removed), 1984

Product	Daily (oz)	Yearly (lb)
Beef	2.00	45.6
Fresh pork	0.30	6.8
Veal	0.04	0.9
Lamb	0.03	0.7
Processed meat (total)	1.60	36.5
Pork (79% of total)	(1.26)	(28.7)
Beef (20% of total)	(0.30)	(7.3)
Lamb (1% of total)	(0.02)	(0.4)

Source: The National Live Stock and Meat Board.

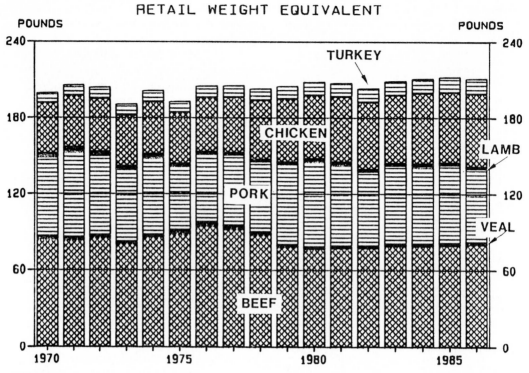

FIGURE 2.16.
Per capita red meat and poultry consumption. Courtesy of The Western Live Stock Marketing Information Project.

Data on consumption and disappearance of meat and other food items can be misleading. What people say they purchase and eat does not determine the amount actually consumed. Dr. Rathje (University of Arizona) has evaluated garbage in Tucson and other cities to determine what people eat by what they throw away. Over 10 years of research data show that (1) people throw away approximately 15% of all the solid food they purchase, and (2) consumers in middle and low income brackets consistently overestimate meat consumption, whereas those in higher income brackets underestimate it.

Per-capita disposable income has had a marked effect on the amount of money spent on meat and the amount of product consumed (Table 2.9). Some market analysts suggest this trend may not continue owing to diet and health issues, which create negative images for red meats.

Table 2.9 shows that per-capita disposable income has risen more rapidly than per-capita red meat expenditures. This has been particularly true from 1979 to 1984. This change reflects either a decreased preference for red meats or a preference for other goods and services as income has increased above a certain level.

TABLE 2.9.   Per-Capita Income and Red-Meat Expenditures

| Year | Disposable Per-Capita Income | Per-Capita Red Meat Expenditures | Expenditures for Red Meat as Percent of | |
			Disposable Per-Capita Income	Total Food Expenditures
1965	$ 2,448	$101.99	4.2	23.0
1970	3,390	136.00	4.0	23.2
1975	5,075	209.29	4.1	24.4
1980	8,032	287.93	3.6	21.7
1986	12,304	288.65	2.3	15.9

Source: U.S. Departments of Agriculture and Commerce.

With the advent of the fast-food outlets, hamburger consumption increased rapidly until the mid-1970s. From 1975 to 1985, a decrease in per-capita hamburger consumption has occurred. Table 2.10 shows the pattern of hamburger consumption compared to the amount of fresh and processed product.

MARKETING

Figure 2.17 shows channels of meat production and processing prior to reaching the retail meat outlets and eventually the consumers. On hog farms, most pigs are raised to slaughter weight. Cattle and sheep are sold by producers to feedlots for added weight gain prior to slaughter.

Packers and processors purchase the animals through several different marketing channels (Table 2.11). Direct purchases can occur at the packing plant, the feedlot, or a buying station near where animals are produced. Terminal markets are large livestock collection centers where an independent organization serves as a selling agent for the

TABLE 2.10.   Hamburger and Beef Consumption and Prices

| Year | Per Capita Consumption (lb retail) | | | | Prices ($ lb) | |
	Hamburger	Processed Beef	Beef Cuts	Total Beef[a]	Hamburger	Choice Beef
1971	16.7	9.3	57.4	83.4	0.68	1.08
1975	23.5	13.2	51.2	87.9	0.88	1.55
1980	18.0	10.1	48.4	76.5	1.36	2.38
1986	19.4	10.9	49.5	79.8	1.23	2.31

[a] Retail weight basis.
Source: Meat Facts, 1985; USDA.

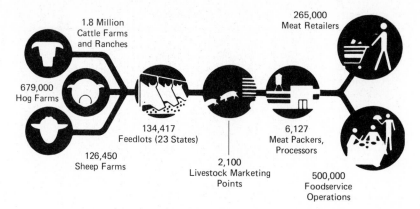

FIGURE 2.17.
Channels of meat production and processing in reaching the retail segment. Courtesy of The National Live Stock and Meat Board.

livestock owner. Oklahoma City, Omaha, and Sioux City, Iowa, are examples of some of the larger, terminal markets, most of which are located in the midsection of the United States.

Approximately 2,000 auctions, sometimes called sale barns, are located throughout the United States. The auction company has pens, sale ring, and an auctioneer. Auctioneers sell the livestock to the highest bidder among the buyers who attend the auction. Producers pay a marketing fee to the auction company for the sale service.

Most cattle, sheep, and swine are purchased on a live weight basis, with the buyer estimating the value of the carcass and other products. Some animals are purchased on the basis of their carcass weight and the desirability of the carcass produced. For cattle and swine there has been an increased number of purchases on a carcass basis during the past several years (Table 2.12). This method of marketing is sometimes called "grade and yield," where the seller is paid on a carcass merit basis.

The Packers and Stockyards Act, passed in 1921 and continually updated, is enforced

TABLE 2.11. Market Outlets Used by U.S. Packers in Purchasing Red-Meat Animals for Slaughter, 1984

Animals	Total (million head)	Direct Purchases[a] (%)	Terminal Markets[b] (%)	Auction Markets[c] (%)
Steers and heifers	26.4	90.2	5.7	4.1
Cows and bulls	8.1	42.0	9.9	48.1
Calves	2.8	50.2	2.9	46.9
Hogs	82.3	82.8	10.3	6.8
Sheep and lambs	6.5	78.9	8.9	12.2

[a]Direct purchases at packing plants, buying stations, country points, and feedlots.
[b]Terminals have more than one market agency selling on commission; auctions have only one.
[c]Animals sold by auctioneer to highest bidder.
Source: USDA (Packers and Stockyards' Statistical Resume).

TABLE 2.12. Livestock Purchased by Packers on Carcass Grade and Weight Basis

Year	Cattle		Calves		Hogs		Sheep and Lambs	
	No. Head (million)	Percent of Total Purchases	No. Head (million)	Percent of Total Purchases	No. Head (million)	Percent of Total Purchases	No. Head (million)	Percent of Total Purchases
1975	8.6	24.4	0.40	8.9	6.0	8.7	0.82	10.5
1980	8.4	27.4	0.47	21.1	10.2	11.0	1.54	28.3
1984	10.8	31.2	0.86	30.5	11.8	14.4	1.38	21.2

Source: USDA (Packers and Stockyards' Statistical Resume).

TABLE 2.13.   Consumer Attitudes Affecting Meat Consumption, 1985

Consumer Segments	Percent of Total Population
Meat lovers	10
Creative cooks	17
Price-driven	23
Active life-style	26
Health-oriented	24

by the USDA. This Act provides for uniform and fair marketing practices. Federal supervision is given at market outlets, including packing plants, for fee and yardage charges, testing of scales, commission charges, and other marketing transactions. Carcass grade and weight sales are covered as well as live-animal sales. Marketing protection is provided to both the buyer and the seller.

Consumer attitudes and preferences for meat have been changing during the 1980s. These attitudes have been surveyed, and some are identified in Table 2.13. This information shows that the market is segmented and that different products are needed for each segment. Other market surveys vividly demonstrate that many consumers have a preference for leanness which, in the minds of these consumers, means a low level of trimmable fat. Consumers are demanding more convenient (ease of preparation) meat products, as the numbers of two-income families and single parents have increased dramatically since 1975. There were more (50% vs. 33%) active, health-oriented consumers in 1985 than 2–3 years previously.

Advertising dollars spent by the livestock industry have been low in recent years. Even though there has been an increase in media advertising, the meat industry's total advertising has been small compared to other industries (Table 2.14).

TABLE 2.14.   Leading Advertisers Compared to Meat Industry

Industry	Spent in 1985 ($ millions)
Procter and Gamble	872
General Motors	764
R.J. Reynolds	678
AT&T	563
K Mart	554
McDonalds	480
Dairy	100
Poultry	30[a]
Beef	6[a]
Swine	3[a]
Sheep	less than 1

[a] Estimates for 1987 are poultry (75 mil), beef (22 mil), swine (7 mil).

TABLE 2.15. Consumer Expenditures for Meat Products, Poultry, and Fish in Grocery Stores, 1983

Item	Millions of Dollars	Percent of Food and Beverages	Percent of Total
Fresh meat	26,690	14.4	11.1
Beef	21,073	11.4	8.7
Veal	649	0.4	0.3
Lamb	1,039	0.6	0.4
Pork	3,929	2.1	1.6
Processed meats	13,155	7.1	5.5
Cured hams and picnics	2,288	1.2	0.9
Packaged bacon	2,290	1.2	1.0
Frankfurters	1,497	0.8	0.6
Sausage products	2,835	1.5	1.2
Cold cuts and other	3,118	1.7	1.3
Frozen meat	263	0.1	0.1
Canned meat	1,495	0.8	0.6
Total meat	41,603	22.5	17.3
Poultry	7,312	3.9	3.0
Fish and seafood	4,626	2.5	1.9
Total meat, poultry, and seafood	53,541	28.9	22.2
Total grocery store sales	241,020	—	—
Food and beverages	185,221	—	76.8
Nonfood and general merchandise	55,799	—	23.2

Source: *Supermarket Business* magazine and *1985 Meat Facts*.

Table 2.15 shows the dollars generated by red-meat products through the grocery stores and the large supermarket chains. Total meat represents 22% of total retail sales.

SELECTED REFERENCES

Publications

American Meat Institute. 1985. *Meat Facts. A Statistical Summary About America's Largest Food Industry.* Washington: American Meat Institute.

Boggs, D. L., and Merkel, R. A. 1979. *Live Animal, Carcass Evaluation and Selection Manual.* Dubuque, IA: Kendall-Hunt.

Diet and Coronary Heart Disease. 1985. *CAST Report No. 107.*

Facts About Beef; Facts About Pork; Facts About Lamb. Chicago: The National Live Stock and Meat Board.

Forest, J. C., Aberle, E. D., Hedrick, H. B., and Merkel, R. A. 1975. *Principles of Meat Science.* San Francisco: W. H. Freeman.

Horsemeat. *Livestock*, Sept. 1980.

Kashruth. Handbook for Home and School. 1972. New York: Union of Orthodox Jewish Congregations of America.

Romans, J. R., Jones, K. W., Costello, W. J., Carlson, C. W., and Ziegler, P. T. 1985. *The Meat We Eat*, 12th Ed. Danville, IL: Interstate Printers and Publishers.

Visuals

Meat Cut Identification (slide set). Vocational Education Productions, California Polytechnical State University, San Luis Obispo, CA 93407.

Lamb Carcass Evaluation Multi-Media Kit. Vocational Education Productions, California Polytechnic State University, San Luis Obispo, CA 93407.

Marketing Challenges Facing the Beef Industry. A 10-minute slide presentation produced by the Beef Industry Council, especially for industry audiences. Loan copies available from Beef Industry Council (444 N. Michigan Ave., Chicago, IL 60611) or State Beef Councils.

Meat Identification (slide set), *Meat Evaluation Handbook,* and *The Meat Board Guide to Identifying Meat Cuts (17-203).* The National Live Stock and Meat Board, 444 N. Michigan Ave., Chicago, IL 60611.

CHAPTER 3

By-Products of Meat Animals

By-products are products of considerably less value than the major product. In the United States, meat animals produce meat as the major product; hides, fat, bones, and internal organs are considered by-products. In other countries, primary products may be draft (work), milk, hides, and skins, with meat considered a by-product when old, useless animals are slaughtered.

There are numerous by-products resulting from animal slaughter and processing of meat, milk, and eggs. Animals dying during their productive life are also processed into several by-products; however, these by-products are not used in human consumption.

By-products are commonly classified into two categories based on human consumption: edible and inedible (Fig. 3.1; Table 3.1). Some by-products not considered edible by

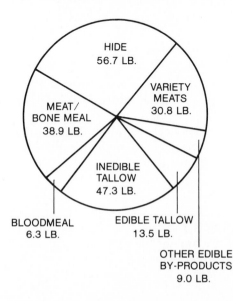

FIGURE 3.1.
Edible and inedible by-products from a 1050 lb. steer. Courtesy of the USDA.

TABLE 3.1.   Edible and Inedible By-Products from Cattle and Swine

	1,100-lb Steer		230-lb Hog	
Product	Expected Yield Per Head (lb)	Market Value Per Cwt (Oct. 1986)	Expected Yield Per Head (lb)	Market Value Per Cwt (3/86)
Edible				
Select liver	13.00	$ 16.25	2.50	$14.50
Heart	3.75	20.25	0.50	18.50
Cheek meat	4.00	77.00	0.88	59.50
Head meat	1.91	58.25	0.25	50.00
Salivary glands	1.25	10.65	—	16.50
Feet	—	—	5.00	
Lips	1.00	21.25	0.25	
Tail	2.00	73.00	0.25	
Spleen	1.25	8.40	—	
Tongue	3.23	91.75	0.75	45.00
Weasand meat[a]	0.43	130.00	0.12	
Tripe (stomach)	8.50	14.00	—	
Stomach	—	—	1.50	17.00
Sweetbread (thymus)	0.49	115.00		
Ears	—	—	0.62	22.00
Kidneys	—	—	0.04	14.00
Inedible				
Lung (standard trim)	5.00	6.60	1.00	
Gullet	0.95	6.75	0.62	
Ox gall (raw)	0.30	90.00	—	
Rendered fat	60.00	8.00	—	
Meat & bone meal				
50% protein	32.00	8.75		
Dried blood				
80% protein	6.50	14.50	7.00 (raw)	
Hide or hair	64.00	74.54	2.00	
Manure	50.00	—	4.00	

[a] Weasand is the muscular lining that surrounds the esophagus from larnyx to first stomach.
Source: Adapted from several sources.

humans are edible to other animals. By-products that are processed into animal feeds are good examples.

Figure 3.2 shows the major by-products obtained from a steer in addition to the retail beef products. Average total value of cattle by-products ranges between $0.05 and $0.07 per pound of liveweight, with the hide comprising the largest part of the total value.

EDIBLE BY-PRODUCTS

Variety meats are edible products originating from organs and body parts other than the carcass. Liver, heart, tongue, **tripe,** and **sweetbread** are among the more typical variety meats. Tripe comes from the lining of the stomach; sweetbread is the thymus gland.

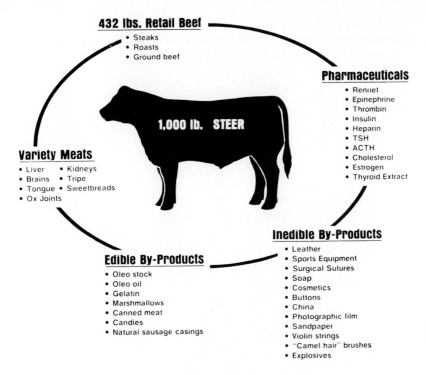

432 lbs. Retail Beef
- Steaks
- Roasts
- Ground beef

1,000 lb. STEER

Pharmaceuticals
- Rennet
- Epinephrine
- Thrombin
- Insulin
- Heparin
- TSH
- ACTH
- Cholesterol
- Estrogen
- Thyroid Extract

Variety Meats
- Liver
- Brains
- Tongue
- Ox Joints
- Kidneys
- Tripe
- Sweetbreads

Inedible By-Products
- Leather
- Sports Equipment
- Surgical Sutures
- Soap
- Cosmetics
- Buttons
- China
- Photographic film
- Sandpaper
- Violin strings
- "Camel hair" brushes
- Explosives

Edible By-Products
- Oleo stock
- Oleo oil
- Gelatin
- Marshmallows
- Canned meat
- Candies
- Natural sausage casings

FIGURE 3.2.
In addition to the retail product of beef, there are numerous by-products that come from beef. Courtesy of The National Live Stock and Meat Board.

Table 3.2 shows the U.S. per-capita disappearance of variety meats.

An average 1,100-lb slaughter steer produces approximately 34 lb of variety meats. Since the U.S. per-capita disappearance is 9 lb, surplus variety meats are exported into countries that have a preference for them (Table 3.3).

Other edible products are fats used to produce lard and tallow. These products are

TABLE 3.2. Per-Capita Disappearance of Variety Meats in the United States

Year	Pounds
1960	10.9
1965	10.4
1970	11.2
1975	10.2
1980	9.5
1984	9.1

Source: Meat Facts (AMF) and USDA.

TABLE 3.3.   U.S. Exports of Variety Meats, 1984

Country	Millions of Pounds	Millions of Dollars
France	87	46
Mexico	84	32
Japan	76	77
Egypt	52	24
United Kingdom	42	15

Source: 1985 Meat Facts.

eventually used in shortenings, margarine, pastries, candy, and other food items. In 1984, over 2 billion pounds of edible tallow and lard was produced in the United States.

INEDIBLE BY-PRODUCTS

Tallow, hides (skins), and inedible organs are the higher-valued inedible by-products. Not all skins are inedible, as some pork skins are processed into consumable food items. Life-saving and life-supporting pharmaceuticals originate from animal by-products, even though many of the same pharmaceuticals are made synthetically. Table 3.4 shows many of these pharmaceuticals and their animal by-product source.

Table 3.5 lists other inedible by-products and their uses. Animals contribute a large number of useful products. The many uses of inedible fats are discussed later in this chapter.

Hides and skins are valuable as by-products or as major products on a worldwide basis. Countries that are most important in hide and skin production are shown in Table 3.6.

Cattle and buffalo hides comprise 80% of farm animal hides and skins produced in the world. The United States has a major contribution in the world hide and skin market, as two-thirds of the 2,360 million pounds of cattle hides produced are exported to other countries (Chapter 1). The United States produces very few goat skins but ranks 14th in sheepskins, with an annual production of approximately 46 million pounds. Pork skin usually remains on the carcass, although some pigs are skinned.

The U.S. hide export market yields approximately $1 billion on a yearly basis. **Hides, skins,** and **pelts** are made into useful leather products through the tanning process (Table 3.7). The surplus hides are exported primarily to Japan, Korea, Mexico, Taiwan, and Romania.

The general term "hide" refers to a beef hide weighing more than 30 lb. Those weighing less than 30 lb are called skins. Skins come from smaller animals, such as pigs, sheep, goats, and small wild animals. Those skins from sheep are usually called sheep pelts with the wool left on. Hides, skins, and pelts are classified according to (a) the species, (b) weight, (c) size and placement of brand, and (d) type of packer producing them.

Value of hides may be reduced by branding, by nicking the hide while skinning, or by **warbles** (larvae of heel flies). Warbles emerge from the back region of cattle in the spring, making holes in the hide. Sheep hides may be damaged by outgrowths of grasses in the production of seeds (called beards), which can penetrate the skin, and by external parasites called **keds.**

TABLE 3.4. Selected Pharmaceuticals from Red Meat Animals—Their Source and Utilization

Pharmaceutical	Source	Utilization
Amfetin	Amniotic fluid	Reduces postoperative pain and nausea and enhances intestinal peristalsis
Cholesterol	Nervous system	Male sex hormone synthesis
Chymotrypsin	Pancreas	Removes dead tissue; for treatment of localized inflammation and swelling
Corticosteriods	Adrenal gland	Treatment for shock, Addison's disease
Corticotrophin (ACTH)	Pituitary gland	Diagnostic assessment of adrenal gland function; treatment of psoriasis, allergies, mononucleosis, and leukemia
Cortisone	Adrenal gland	Treatment for shock, arthritis, and asthma
Desoxycholic acid	Bile	Used in synthesis of cortisone for asthma and arthritis
Epinephrine	Adrenal gland	Relief of hay fever, asthma, and other allergies; heart stimulation
Fibrinolysin	Blood	Dead tissue removal; wound-cleansing agent; healing of skin from ulcers or burns
Glucagon	Pancreas	Counteracts insulin shock; treatment of some psychiatric disorders
Heparin	Intestines	Natural anticoagulant used to thin blood, retards clotting, especially during organ implants
Heparin	Lungs	Anticoagulant; prevention of gangrene
Hyaluranidase	Testes	Enzyme that aids drug penetration into cells
Insulin	Pancreas	Treatment of diabetes for 10 million Americans (primarily synthetic insulin is used)
Liver extracts	Liver	Treatment of anemia
Mucin	Stomach	Treat ulcers
Norepinephrine	Adrenal gland	Shrinks blood vessels reducing blood flow and slowing heart beat
Ovarian hormone	Ovaries	To treat painful menstruation and prevent abortion
Ox bile extract	Liver	Treatment of indigestion, constipation, and bile tract disorders
Parathyroid hormone	Parathyroid gland	Treatment of human parathyroid deficiency
Plasmin	Blood	Digests fibrin in blood clots—used to treat patients with heart attacks
Rennet	Stomach	Assists infants in digesting milk; cheese making
Thrombin	Blood	Assists in blood coagulation; treatment of wounds; skin grafting
Thyroid extract	Thyroid gland	Treatment of cretinism
Thyrotropin (TSH)	Pituitary gland	Stimulates functions of thyroid gland
Thyroxin	Thyroid	Treatment of thyroid deficiency
Vasopressin	Pituitary gland	Control of renal function

Source: Adapted from several sources.

TABLE 3.5. Other Inedible By-Products and Their Uses

By-Product	Use
Hog heart valves	They replace injured or weakened human heart valves. Since the first operation in 1971, more than 35,000 hog heart valves have been implanted
Pig skins	Used in treating massive human burns. These skins help prepare the patient for permanent skin grafting
Gelatin (from skin)	Coatings for pills and capsules
Brains	Cholesterol for an emulsifier in cosmetics
Blood	Sticking agent for insecticides; a leather finish; plywood adhesive; fabric printing and dyeing
Hides and skins	Many leather goods from coats, handbags, shoes, to sporting goods (see Table 3.7)
Bones	Animal feed, glue, buttons, china, and novelties
Gallstones	Shipped to Orient for use as ornaments in necklaces and pendants
Hair	Paint and other brushes; insulation; padding in upholstery, carpet padding, filters
Meat scraps and blood	Animal feed
Inedible fats	Animal feed, fatty acids (see Fig. 3.7)
Wool (pulled from pelts)	Clothing, blankets, lanolin
Poultry feathers	Animal feed, arrows, decorations, bedding, brushes

Leather tanning and finishing is a $1.5 billion industry in the United States, employing over 20,000 people in 330 establishments and accounting for $250 million in exports. About 160 plants process raw hides or skins directly into tanned leather. Tanning activities are concentrated in the Northeast, Midwest, Middle Atlantic states, and California. Some tanneries are relatively small concerns specializing in the manufacture of a particular kind of leather, whereas others employ several hundred people and produce a variety of leathers.

Every year, U.S. tanneries convert many millions of raw hides and skins into leather. The tanners' value added by manufacture constitutes over $500 million annually. Their

TABLE 3.6. Leading Countries in Fresh Hide and Skin Production

Cattle and Buffalo Hides		Sheepskins		Goatskins	
Country	Millions of Pounds	Country	Millions of Pounds	Country	Millions of Pounds
U.S.	2,360	USSR	277	India	162
India	1,798	New Zealand	275	China	130
USSR	1,684	Australia	267	Pakistan	75
Argentina	812	China	172	Nigeria	47
Brazil	682	Turkey	146	Bangladesh	32
World total	14,117	World total	2,554	World total	855

Source: FAO 1985 Production Yearbook.

TABLE 3.7. Leather Uses Related to Types of Hides and Skins

Skin Origin	Use
Cow and steer	Shoe and boot uppers, soles, insoles, linings; patent leather; garments; work gloves; waist belts; luggage and cases; upholstery; transmission belting; sporting goods; packings
Calf	Shoe uppers; slippers; handbags and billfolds; hat sweatbands; bookbindings
Sheep and lamb	Grain and suede garments; shoe linings; slippers; dress and work gloves; hat sweatbands; bookbindings; novelties
Goat and kid	Shoe uppers, linings; dress gloves; garments; handbags
Pig	Shoe suede uppers; dress and work gloves; billfolds; fancy leather goods
Horse	Shoe uppers; straps; sporting goods

Source: Leather Facts, New England Tanners Club.

product—leather—serves in turn as a raw material for the shoe and leather goods industries that provide jobs for over 200,000 people. Table 3.8 shows the most important hides and skins used in leather production. These amounts represent the total of both domestic and imported hides, skins, and pelts.

Table 3.8 shows that less than half of U.S. cattle hides are converted to leather domestically, most being exported for tanning. Some sheepskins are imported, as the domestic supply does not meet the demand. Pigskins are used primarily as food, but interest in leather production is increasing.

Just as meat is perishable, so too are hides and skins. If not cleaned and treated to prevent putrefaction, they begin to decompose and lose leather-making substance within hours after removal from the carcass. Hides are commonly treated (cured) by adding salt as the principal curing agent. The salt solution penetrates the hide in about 12 hours, then the hides are bundled for shipment (Fig. 3.3).

THE RENDERING INDUSTRY

A major focal point of by-products is the rendering industry, which recycles offal, fat, bone, meat scraps, and entire animal carcasses. The sources of these raw materials are packing and processing plants, butcher shops, restaurants, supermarkets, farmers, and

TABLE 3.8. Animal Slaughter and Leather Production, U.S., 1983

Species	Head Slaughtered (million)	Leather Production (million)
Cattle	36	15 hides
Sheep	7	9 skins
Pig	79	4 skins

Source: Leather Facts, New England Tanners Club.

FIGURE 3.3.
Beef hide room where hides are cured, tied, and stored in preparation for shipping. Courtesy of Iowa Beef Processors, Inc.

ranchers. Some large packing and processing plants have their own rendering plants integrated with their other operations.

Renderers have a regular pickup service which amounts to approximately 81 million pounds of animal material daily. This pickup service is essential to public health, as it reduces a major garbage disposal problem. Also, since animal by-products have value, consumers can buy meat at a lower price than would otherwise be possible.

Rendering of Red Meat Animal By-products

Animal fat and animal protein are the primary products of the renderer's art. Originally, animal fats went almost entirely into soap and candles. Today, from the same basic material, renderers produce many grades of tallow and semiliquid feed fat. The major uses of rendered fat are for animal feeds, fatty-acid production, and soap manufacture (Fig. 3.4).

Fatty acids are used in the manufacture of many products, such as plastic consumer items, cosmetics, lubricants, paints, deodorants, polishes, cleaners, calking compounds, asphalt tile, printing inks, and others. The fatty-acid industry has had a tremendous growth since 1950.

The rendered animal proteins are processed into several high-protein (more than 50% protein) feed supplements among which are meat and bone meals and blood meals. The supplements are more commonly fed to young monogastric animals—swine, poultry, and pet animals. These animals require a high-quality protein, particularly the amino acid lysine, which is typical of animal protein supplements.

Blood, not processed into blood meal, is used to produce products used in the pet food industry. Such products are blood protein (fresh, frozen whole blood) and blood cell protein (frozen, fresh, dewatered blood).

Rendering of Poultry By-products

Within a few hours after poultry is slaughtered, the offal is at the rendering plant. There it is cooked in steam-jacketed tanks at temperatures sufficient to destroy all pathogenic organisms. After cooking for several hours, the material is passed through presses which

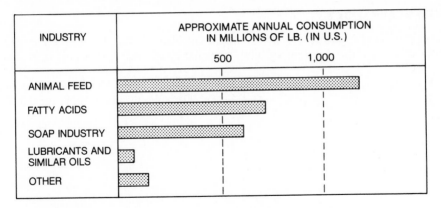

INDUSTRY	APPROXIMATE ANNUAL CONSUMPTION IN MILLIONS OF LB. (IN U.S.)	
	500	1,000
ANIMAL FEED		
FATTY ACIDS		
SOAP INDUSTRY		
LUBRICANTS AND SIMILAR OILS		
OTHER		

FIGURE 3.4.
The major uses of animal fat. Courtesy of The National Renderers Association.

remove most of the fat (poultry fat), while the remaining material becomes poultry by-product meal. While it is still hot, the poultry fat is piped to sterile tanks where appropriate antioxidants or "stabilizers" are added, and thence to tank cars or drums for shipment to feed manufacturers or other uses as shown in Fig. 3.4.

Poultry by-product meal (PBPM) is high in protein (approximately 58%) and contains 12–14% fat. PBPM consists of ground dry-rendered clean parts of the carcass of slaughtered poultry such as heads, feet, undeveloped eggs, and intestines, exclusive of feathers, except in such trace amounts as might occur unavoidably in good factory practice. It should not contain more than 16% ash and not more than 4% acid ash.

EXPORT MARKET

By-products from the livestock and poultry industries are an extremely important part of the export market of livestock products. Hides and skins, fats and oils, and variety meat comprise more than 50% of $3.5 billion export market in 1984 (Fig. 3.5).

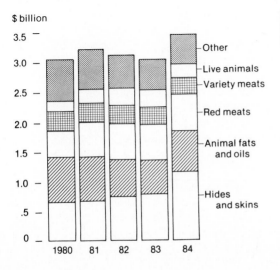

FIGURE 3.5.
U.S. exports of livestock products. Courtesy of the USDA.

SELECTED REFERENCES

Hog Is Man's Best Friend. Meat; By-products. Chicago: The National Live Stock and Meat Board.

Renderer Recycling for a Better Tomorrow. 1981. National Renderers Association, 2250 East Devon Ave., Des Plaines, IL 60018.

Romans, J. R., Jones K. W., Costello, W. S., Carlson, C. W., and Ziegler, P. T. 1985. Chapter 10: "Packing House By-Products," in *The Meat We Eat.* Danville, IL: Interstate Printers and Publishers.

CHAPTER 4

Milk and Milk Products

The world's population obtains most of its milk and milk products from cows, water buffalo, goats, and sheep (Fig. 4.1). Horses, donkeys, reindeer, yaks, camels, and sows contribute a smaller amount to the total milk supply for humans. Milk, with its well-balanced assortment of nutrients, is sometimes called "nature's most nearly perfect food." While milk is an excellent food product in many ways, it is not perfect. Nor is any other food. Milk and numerous milk products (e.g., cheese, butter, ice cream, and cottage cheese) are major components of the human diet in many countries.

Changes in dietary preferences and milk marketing are some of the challenges facing the dairy industry. Health considerations and new products are changing the consumption patterns of milk and milk products. Milk surpluses in the United States and the world result in an abundance of consumer products, but the surpluses represent economic challenges to milk producers.

This chapter will focus primarily on milk as human food, but the importance of milk in nourishing suckling farm mammals should not be overlooked. When the term "milk" is used, reference will be to milk from dairy cows unless otherwise specified.

MILK PRODUCTION

Table 4.1 shows the most important sources of milk for humans in the world and the United States. Total world milk production has increased during the past 25 years but not at the same rate as world human population (49% vs. 53%). Dairy cows produce over 90% of the world fluid milk supply (Fig. 4.2). During the past 30 years, yield of buffalo and sheep milk has been increasing while goat milk production has been decreasing. Some milk from dairy goats in the United States is marketed through supermarkets and other outlets similar to cow's milk (Fig. 4.3). However, most milk from goats is produced and consumed by individuals or families who raise a few goats.

Leading countries for milk production are also shown in Table 4.1. The Soviet Union produces more milk than the United States (214 bil lb vs. 135 bil lb). However, it is

FIGURE 4.1.
(A) Milk plays an important role in human nutrition throughout the world. Courtesy of Heifer Project International. (B) Milking native sheep in the desert of Iran. Courtesy of R. E. McDowell, Cornell University. (C) Goats being milked in Mexico. Courtesy of Winrock International. (D) Milking dairy cows in a modern United States dairy. Courtesy of Colorado State University.

interesting to note the difference in cow numbers and productivity per cow. The 44 million cows in the Soviet Union have an average annual production per cow of 4,965 lb, whereas the 11 million cows in the United States have an average yearly production of more than 13,000 lb per cow *(1986 Milk Facts)*.

The national dairy herd in the United States has decreased from over 20 million cows in 1956 to less than 11 million in 1986. Yet total milk production has remained approximately the same. Reduction in cow numbers has been offset by an increased production

TABLE 4.1.   Major Sources of World Milk Production, 1984

Species	Total Production (bil lb)	Leading Countries (bil lb)
Cows	988	USSR (214); U.S. (135); France (73); West Germany (58)
Buffalo	70	India (46); Pakistan (15); China (3)
Sheep	18	Turkey (2.9); France (2.3); Iran (1.6); Italy (1.4)
Goats	16	India (2.1); Turkey (1.4); Bangladesh (1.1)
All others	52	
Total world	1,144	

Source: 1984 FAO Production Yearbook.

per cow—from 5,800 lb in 1956 to 13,000 lb in 1986. The 12,500 lb per cow comes to 1,450 gal (1 gal = 8.62 lb), which would supply the annual per capita fluid milk consumption for nearly 50 people in the United States. However, since part of the milk is made into other dairy products (Fig. 4.4), the milk from one cow provides milk and manufactured products (cheese, frozen dairy products, etc.) for approximately 22 people.

FIGURE 4.2.
Dairy cows have the unique ability to convert feedstuffs into milk. Milk is an important source of nutrients to humans throughout the world. Courtesy of Dairy Council, Inc.

FIGURE 4.3.
Milk, cheese, and yogurt are among the most highly preferred milk products. Courtesy of American Dairy Association.

MILK COMPOSITION

Milk is a colloidal suspension of solids in liquid. Fluid whole milk is approximately 88% water, 8.6% **solids-not-fat (SNF)** and 3–4% milk fat. The SNF is the total solids minus the milk fat. It contains protein, lactose, and minerals.

Even though milk is a liquid, its 12% total solids is similar to the solids content of many solid foods. Differences in milk composition for several species of animals are given in Chapter 18.

The first milk a female produces after the young is born is called **colostrum.** True colostrum is milk obtained during the first milking. For the next several milkings, the milk is called transitional milk and is not legally salable until the 11th milking. Colostrum differs greatly in composition from milk. Colostrum is higher in protein, minerals, and milk fat, but it contains less lactose than milk. The most remarkable difference between colostrum and milk produced later is the extremely high immunoglobulin content of colostrum. These protein compounds accumulate in the mammary gland and are the antibodies that are transferred to the suckling young through milk. People seldom use colostrum from animals because of its unpleasant appearance and odor, although a few may use it for pudding. Additionally, the high globulin content causes a precipitation of proteins when colostrum is pasteurized.

Milk contains a high nutrient density—more than 100 milk components have been identified. **Nutrient density** refers to the concentration of major nutrients in relation to the caloric value of the food. In other words, milk contains important amounts of several nutrients while being relatively low in calories. The major components of milk solids can be grouped into nutrient categories including proteins, fats, carbohydrates, minerals, and vitamins.

Milk Fat

In whole milk, the approximate 3–4% milk fat is a mixture of lipids existing as microscopic globules suspended in the milk. The fat contributes about 48% of the total calories in whole milk. Fat-soluble vitamins (A, D, E, and K) are normal components in milk fat. Milk fat contains most of the flavor components of milk, so when milk fat is decreased, there may be a concurrent reduction in flavor.

FIGURE 4.4.
Milk products from dairy goats. Courtesy of George F. W. Haenlein, University of Delaware.

Approximately 500 different fatty acids and fatty-acid derivatives from 2 to 26 carbon atoms long have been identified in milk. Milk fat from ruminant animals contains both short-chain and long-chain fatty acids.

Carbohydrates

Lactose, the predominant carbohydrate in milk, is synthesized in the mammary gland. Approximately 4.8% of cow's milk is lactose. It accounts for approximately 54% of the SNF content in milk. Lactose is only about one-sixth as sweet as sucrose, and it is less soluble in water than other sugars. It contributes about 30% of the total calories in milk. Milk is the sole source of lactose in nature.

Proteins

Milk contains approximately 3.3% protein. Protein accounts for about 38% of total SNF and about 22% of the calories of whole milk. The proteins of milk are of high quality. They contain varying amounts of all amino acids required by humans. A surplus of the amino acid lysine offsets the low lysine content of vegetable proteins and particularly cereals.

Casein, a protein found only in milk, is approximately 82% of the total milk protein. Whey proteins, primarily lactalbumin and lactoglobulin, constitute the remaining 18%. Immunoglobulins, the antibody proteins in colostral milk, were previously discussed.

Vitamins

All vitamins essential in human nutrition are found in milk. Fat-soluable vitamins are in the milk fat portion of milk, and water-soluble vitamins are in the nonfat portion. Milk is usually fortified with vitamin D during processing.

Milk fat from Jersey and Guernsey cows has a rich, yellow color due to carotene (a precursor of vitamin A), which is yellow. Milk fat from Holstein cows has a pale, yellow color, and goat milk fat is white. Carotene can be split into two molecules of vitamin A. Milk fat of Holstein cows is pale yellow, because most of the carotene has been split, and the milk fat of goats is white, because all of the carotene has been changed to colorless vitamin A.

Water-soluble vitamins (C and B) are relatively constant in milk and are not greatly influenced by vitamin content of the cow's ration. The B vitamins are produced by rumen microorganisms, and vitamin C is formed by healthy epithelial tissue in most animals (excluding man, primates, and guinea pigs).

Minerals

Milk is a rich source of calcium for the human diet and a reasonably good source of phosphorus and zinc. Calcium and vitamin D are needed in combination to contribute to bone growth in young humans and to prevent **osteoporosis** in adults, particularly women. Milk is not a good source of iron, and the iodine content of milk varies with the iodine content of the animal's feed.

Table 4.2 and Fig. 4.5 shows how the milk supply is used to produce various milk products. More than 80% of the milk produced is marketed as fluid milk, cream, cheese, and butter.

The amount of milk needed to produce each product depends primarily on the milk fat content of the milk. Table 4.3 gives the approximate amount of milk used to produce each product.

TABLE 4.2.   Milk and Milk Products Resulting from the 1985 Milk Supply

Whole-Milk Product	Total Production (mil lb)	Percent of Total
Fluid milk and cream	58.2	39
Cheese	41.7	28
Butter	24.4	16
Frozen dairy products	12.9	9
Evaporated and condensed milk	2.3	1
Uses on farms where produced	2.5	2
Other uses	8.2	5

Source: 1986 Milk Facts.

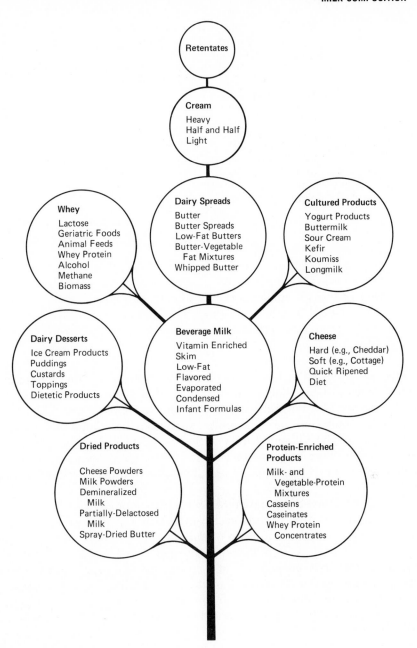

FIGURE 4.5.
The modern dairy tree showing the many products and by-products of milk. Courtesy of *J. Dairy Sci.* 64:1005.

TABLE 4.3.   Approximate Amount of Milk Used to Produce Several Selected Dairy Products

Dairy Product (lb)	Whole Milk (lb)
Butter	21.2
Whole-milk cheese	10.0
Evaporated milk	2.1
Condensed milk	7.4
Ice cream (1 gal)	12.0
Cottage cheese	6.2 (skim)
Nonfat dry milk	11.0

Source: USDA.

Fluid Milk

Eighty-five percent of the 135 million pounds of milk produced in the United States is grade A quality milk. Grade A milk in excess of fluid milk demand is processed into other milk products. How fluid milk typically moves from farm to consumer is shown in Fig. 4.6.

Depending on the milk fat content and the processing at the plant, the fresh fluid product is labeled whole milk, low-fat milk or skim milk. In further discussion in this chapter, "milk" will refer to whole milk. Whole milk is defined as a "lacteal secretion," and when it is packaged for beverage use it must contain not less that 3.25% milk fat and not less than 8.25% milk SNF. Low-fat milk usually has had some or most milk fat removed to have one of the following milk fat levels: 0.5, 1.0, 1.5, or 2.0% with not less than 8.25% milk SNF. Skim milk has had most of the milk fat removed (it contains less than 0.5% milk fat). It must contain at least 8.25% SNF and may be fortified with nonfat solids to 10.25%.

Most fluid milk consumed in the United States is **homogenized** to prevent milk fat from separating from the liquid portion and rising to the top. Homogenization is a physical process resulting in a stable emulsion of milk fat. A "cream line" does not appear in homogenized milk, nor does it form butter if churned.

Fat globules in raw milk average about 6 μm in diameter. Homogenization reduces the fat globule size to less than 2 μm in diameter. To visualize how small the fat globules in homogenized milk are, note that 25,000 μm equal approximately 1 in. Homogenized milk will likely deteriorate by becoming rancid more rapidly than nonhomogenized milk. Rancidity of homogenized milk is forestalled by pasteurizing milk prior to or immediately following homogenization, thus destroying the action of lipolytic enzymes.

Evaporated and Condensed Milk

Evaporated milk is made by preheating to stabilize proteins and removing about 60% of the water. It is sealed in the container and then heat-treated to sterilize its contents. Milk fat and SNF of evaporated milk must be at least 7.5 and 25%, respectively. Evaporated

Milk from Farm to Family

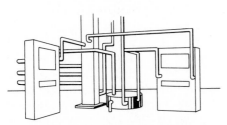

1 Cows are milked twice a day with sanitary milking machines.

2 The milk is pumped from the milking machine into the refrigerated farm tank.

3 Milk is sampled and checked before it is pumped into the refrigerated truck.

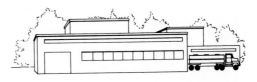

4 Each tank truck rushes cold milk from several farms to dairy plants.

5 Milk is processed and packaged in sterilized equipment.

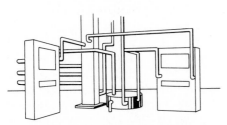

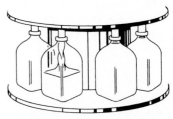

6 The milk is pasteurized in thermostatically controlled equipment, then homogenized (some is not homogenized).

7 Automatic machines package cold pasteurized milk in paper, glass or plastic containers.

8 And put the containers into cases. Conveyors rush the milk, ready for delivery, to a refrigerated room.

9 Refrigerated trucks bring the milk.

10 To stores and homes.

FIGURE 4.6.
The processes involved in making milk available to consumers.

skim milk must have at least 20% SNF and not more than 0.5% milk fat. Evaporated milk requires no refrigeration until opened. Once the can is opened, refrigeration is necessary to avoid spoilage.

Concentrated, or condensed, milk has similar milk fat (7.5%) and SNF (25.5%) requirements as evaporated milk. Concentrated milk also has water removed, but, unlike evaporated milk, it is not subjected to further heat treatment to prevent spoilage. Most concentrated milk is sold bulk for industry use. A common form of concentrated milk on grocery shelves is sweetened (sugar added) condensed milk; it is used for candy and other confections.

Dry Milk

This product is prepared by removing water from milk, low-fat milk, or skim milk. All of these products are to contain no more than 5% moisture by weight. Dry milk can be stored for long periods of time if it is sealed in an atmosphere of nonoxidizing gas such as nitrogen or carbon dioxide. The spray-dried or foam-dried product can be reconstituted easily in warm or cold water with agitation.

Fermented Dairy Products

Buttermilk, yogurt, sour cream, and other cultured milk products are produced under rigidly controlled conditions of sanitation, inoculation, incubation, acidification, and/or temperature. The word "cultured" appearing in a product name indicates the addition of appropriate bacteria cultures to the fluid dairy product and subsequent fermentation. Selected cultures of bacteria convert lactose into lactic acid, which produces a tart flavor. The specific bacterial cultures used and other controls in the fermentation process determine whether buttermilk, sour cream, yogurt, or cheese will be the end product.

Today buttermilk is a cultured product rather than the by-product of churning cream into butter as was the procedure in years past. At the correct stage of acid and flavor development, the buttermilk is stirred gently to break the curd that has formed. It is then cooled to stop the fermentation process.

The word "acidified" in the product name or production process indicates that the food was produced by souring milk or cream with or without the addition of microbial organisms.

Cottage cheese dry curd is soft, unripened cheese; it is produced by culturing or direct acidification. Finished dry curd contains less than 0.5% milk fat and not more than 80% moisture. Cottage cheese is prepared for market by mixing cottage cheese dry curd with a pasteurized creaming mixture called dressing. The finished product contains not less than 4% milk fat and not over 80% moisture. Low-fat cottage cheese is made by the same process; however, the milk fat range is 0.5–2%.

Sour cream is generally a lactic acid fermentation, although rennet extract is often added in small quantities to produce a thicker-bodied product. Federal standards provide that cultured sour cream contain not less than 18% milk fat. When stored for more than 3–4 weeks, sour cream may develop a bitter flavor as a result of continued bacterial proteolytic enzyme activity.

Yogurt, as a liquid or gel, can be manufactured from fresh whole, low-fat, or skim milk that is heated before fermentation. Federal standards specify that "yogurt" contain not less than 3.25% milk fat, "low-fat yogurt" may contain 0.5–2% milk fat, and "nonfat yogurt" may not contain more than 0.5% milk fat before any bulky flavors are added.

Today three main types of yogurt are produced: (1) flavored, containing no fruit; (2) flavored, containing fruit (fruit may be at the bottom or blended in); (3) unflavored yogurt.

Cream

Cream is a liquid milk product, high in fat, separated from milk. Federal standards require that cream contain not less than 18% milk fat.

Several cream products are marketed (Fig. 4.7). Half-and-half is a mixture of milk and cream containing not less than 10.5% but less than 18% milk fat. Light cream (coffee or table cream) contains not less than 18% but less than 30% milk fat. Light whipping cream or whipping cream contains not less than 30% but less than 36% milk fat. Heavy cream or heavy whipping cream contains not less than 36% milk fat. Dry cream is produced by removal of water only from pasteurized milk and/or cream. It contains not less than 40% but less than 75% milk fat and not more than 5% moisture.

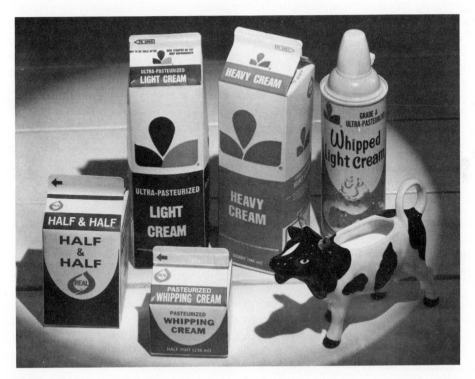

FIGURE 4.7.
There is a range of cream products to meet consumer needs. Courtesy of American Dairy Association.

Butter

It is known from food remnants in vessels found in early tombs that the Egyptians cooked with butter and cheese. For many centuries, butter making and cheese making were the only means of preserving milk and cream.

Butter churns are the oldest dairy equipment, dating from prehistoric days when nomads carried milk in a type of pouch made from an animal's stomach. Slung on the back of a horse or camel, the pouch bounced as the animal moved, churning the cream or milk into butter.

Wooden or crockery dasher-type churns used in great-grandmother's day are today's attic treasures. By jouncing a long-handled, wooden dasher up and down in the deep churn, the cream was sloshed around until butter particles separated from the remaining liquid, called buttermilk.

Butter is the dairy product made exclusively from milk or cream, or both, which contains not less than 80% by weight of milk fat. Federal standards establish U.S. grades for butter based on flavor, color, and salt characteristics.

In modern butter making, fresh sweet milk is weighed, tested for milk fat content, and checked for quality. The cream is then separated by centrifugation to contain 30–35% milk fat for batch-type churning or 40–45% milk fat for continuous churning.

Continuous butter making operations, which can produce 1,800–11,000 lb an hour, are an industry trend. New continuous butter making processes employ several different scientific principles to form butter.

Cheese

There are more than 400 different kinds of cheese that can be made (Fig. 4.8). The cheeses have more than 2,000 names, since the same cheese may have two or more different names. For example, cheddar cheese is one of the American-type cheeses in the United States, along with Colby, washed or stirred curd, and Monterey.

Most cheeses and cheese products are classified into one of four main groups: (1) soft, (2) semisoft, (3) hard, and (4) very hard. Classification is based on moisture content in the cheese. Thus the body and texture of cheeses range from soft, unripened cheese (such as cottage cheese with 80% moisture) to very hard, grated, shaker cheeses such as Parmesan and Romano. These latter, ripened cheeses have 32–34% moisture.

Cheese flavor varies from bland cottage cheese to pungent Roquefort and Limburger. Modern microbiology makes it possible to add specific species and strains of microorganisms required to produce the desired product. For example, a fungus (*Penicillium roqueforti*) is used to produce Roquefort cheese.

The most dramatic increase in cheese production in the United States has occurred with Italian varieties. The phenomenal growth of the pizza industry has caused an increased production and importation of mozzarella cheese. In 1980, 17% of the cheese (excluding cottage cheese varieties) produced in the United States was mozzarella, the largest production (44%) was of cheddar cheese.

Cheese making involves a biochemical process called coagulation, or curdling. First the milk is heated, then a liquid starter culture is added. Bacteria from the culture form acids and turn the milk sour. At a later time, rennet is added to thicken the milk. Rennet,

FIGURE 4.8.
There are numerous kinds of cheeses to satisfy the varying consumer preferences. Courtesy of American Dairy Association.

obtained from the stomachs of young calves, contains the enzyme rennin. Other enzymes of microbial origin have been used successfully to replace a declining supply of rennin. After stirring this mixture, a custardlike substance called curd is formed. Then the liquid part (whey) is removed. The cheese-making process reduces 100 lb of milk to 8–16 lb of cheese.

In cheese plants, disposal of whey is a problem. It is currently estimated that 35 billion pounds of whey containing more than 4 billion pounds of solids is produced. Slightly more than half the whey is used to produce human food or animal feeds. The remaining whey poses waste disposal problems. There continues to be an increasing industrial use of whey for production of food, feed, fertilizer, alcohol, and insulation.

Table 4.4 identifies the leading countries in yearly cheese production and the most important cheese-producing states in the United States.

Ice Cream

Although they come in many variations, ice cream and a group of similarly frozen foods are made in a similar way and have many of the same ingredients. Ice cream, frozen custard, French ice cream, and French custard ice cream are frozen dairy products highest in milk fat and milk solids. Ice cream may contain egg yolk solids. If egg yolks are in excess of 1.4% by weight, the product is called frozen custard, French ice cream, or French custard ice cream.

TABLE 4.4.   **Leading Cheese-Producing Countries and States (1982)**

Country	Total Production (bil lb)	State	Total Production (mil lb)
U.S.	4.9	Wisconsin	1,644
USSR	3.3	Minnesota	567
France	2.6	New York	348
West Germany	1.8	California	248
Italy	1.3	Iowa	206

Source: USDA.

Ice milk contains less milk fat, protein, and total solids than ice cream. Ice milk usually has more sugar than does ice cream. Soft ice milk and soft ice cream are soft and ready to eat when drawn from the freezer. About three-fourths of the soft-serve products are ice milks. Some of the soft-serve frozen dairy foods contain vegetable fats that have replaced all or most of the milk fat.

Frozen yogurt is relatively new, and no federal standards exist for it. Frozen yogurt has less milk fat and higher acidity than ice cream and less sugar than sherbet.

Sherbet is low in both milk fat and milk solids. It has more sugar than ice cream. The tartness of fruit sherbet comes from the added fruit and fruit acid. Nonfruit sherbet is flavored with such ingredients as spices, coffee, or chocolate.

Water ices are nondairy frozen foods. They contain neither milk ingredients nor egg yolk. Ices are made similarly to sherbet.

Mellorine differs from ice cream in that milk fat may be replaced partially by a vegetable fat, another animal fat, or both.

Eggnog

Eggnog contains milk products, egg yolk, egg white, and a nutritive carbohydrate sweetener. In addition, eggnog may contain salt, flavoring, and color additives. Federal standards specify that eggnog shall contain not less than 6% milk fat and 8.25% SNF.

Imitation Dairy Products

The FDA established regulations in 1973 to differentiate between imitation and substitute products. It defined an imitation product as one that looks like, tastes like, and is intended to replace the traditional counterpart but is nutritionally inferior to it. A substitute product resembles the traditional food but also meets the FDA's definition of nutritional equivalency.

Imitation milks usually contain ingredients such as water, corn syrup solids, sugar, vegetable fats, and protein from soybean, sodium caseinate, or other sources. Although imitation fluid milk may not contain dairy products per se, it may contain derivatives of milk such as whey, lactose, casein, salts of casein, and milk proteins, other than casein.

The dairy industry has developed a program for identification of real dairy foods. A "REAL" Seal (observe in Fig. 4.7) on a carton or package identifies milk or other dairy foods made from U.S.-produced milk that meets federal and/or state standards. This seal assures consumers that the food is not an imitation or substitute.

HEALTH CONSIDERATIONS

Nutritive Value of Milk

Milk and other dairy products make a significant contribution to the nation's supply of dietary nutrients. Particularly noteworthy are the relatively large percentages of calcium, phosphorus, protein, and B vitamins (Table 4.5).

Human milk is regarded as the best source of nourishment for infants. Cow's milk for infant feeding is modified to meet the nutrient and physical requirements of infants. Cow's milk is heated, homogenized, or acidified so the nutrients can be utilized efficiently by infants. Sugar is usually added to cow's milk for infant feeding to make milk more nearly like human milk.

Milk is low in iron; therefore young animals consuming nothing but milk may develop anemia. Baby pigs produced in confinement need a supplemental source of iron, usually given as an injection. Babies typically receive supplemental iron, and young children who consume large amounts of milk at the expense of meat should be given supplemental iron.

Milk and milk products are excellent sources of nutrients to meet the dietary require-

TABLE 4.5. Contribution of Dairy Foods (excluding butter) to Nutrients Consumed in the United States (1985)

Nutrients	Percent
Energy	10
Protein	21
Fat	11
Carbohydrate	6
Cholesterol	14
Minerals	
Calcium	76
Phosphorus	36
Magnesium	20
Iron	2
Zinc	20
Vitamins	
Ascorbic acid	3
Thiamin	9
Riboflavin	35
Vitamin B_6	11
Vitamin B_{12}	20
Vitamin A	10

Source: USDA.

ments of young children, adolescents, and adults. With aging, there is an increased prevalence of osteoporosis, a problem of loss of bone mass. Since osteoporosis involves the loss of bone matrix and minerals, a diet generous in protein, calcium, vitamin D, and fluoride is recommended to overcome the problem. Thus, milk, owing to its content of calcium and other nutrients, is an important food for this segment of the population. Furthermore, for the elderly as well as for the infant and young child, milk is an efficient source of nutrients readily tolerated by a sometimes weakened digestive system.

Wholesomeness

Milk is among the most perishable of all foods owing to its excellent nutrition composition and its fluid form. As it comes from the cow, milk provides an ideal medium for bacterial growth. Properly processed milk can be kept for 10–14 days under refrigeration. With ultra-high-temperature (UHT) processing, milk can be kept for several weeks at room temperature.

Protecting the quality of milk is a responsibility shared by public health officials, the dairy industry, and consumers. The U.S. Food and Drug Administration describes milk as "one of the best controlled, inspected and monitored of all food commodities."

The most important safeguards from a health standpoint are requirements that bacterial counts be low and that milk be pasteurized (Fig. 4.9). Milk sold through commercial outlets is certified to be from herds that are tested and found to be free from brucellosis and tuberculosis. The consumer who buys milk and milk products can be assured of obtaining a safe, desirable, wholesome food. There are definite health risks in consuming

FIGURE 4.9.
Milk samples are checked for undesirable bacteria. Courtesy of Dairy Council, Inc.

FIGURE 4.10.
The large milk trucks move the milk from dairy farms to processing plants. The tanks are washed and sanitized after each delivery. Courtesy of Dairy Council, Inc.

milk from a cow or goat where the milk has not been pasteurized. This concern is important where families or individuals have purchased the animal with an unknown health background or bought milk that has not been processed.

Milking Procedures

Milk is taken from the cow by a sanitized milking machine, then transported through sanitized pipes into holding tanks. When withdrawn from the cow, milk is at the cow's body temperature of about 100°F (38°C). It flows into a refrigerated tank, where it is rapidly cooled to 45°F (7°C) or less. Cold temperature maintains the high quality of the milk while it is held for delivery.

Tank truck drivers who pick up milk inspect it to see if it is cold and has an acceptable aroma. They take a milk sample which will be tested later at the processing plant. Then the cold milk is pumped from the refrigerated farm tank through a sanitized hose into the insulated tank on the truck (Fig. 4.10).

After the milk sample passes several tests at the processing plant, it is pumped through sanitized pipes into the processing plant's refrigerated or insulated holding tanks.

Milk is pasteurized at the processing plant. Pasteurization is a process of exposing milk to a temperature that destroys all pathogenic bacteria but neither reduces the nutritional value of milk nor causes it to curdle. Milk is most commonly pasteurized at 161°F (71.5°C) for 15 sec. Ultrapasteurized milk and UHT processed milk are heated to 280°F (138°C) for at least 2 sec; this sterilizes milk and increases shelf life.

As it flows out of the pasteurizer, most milk is homogenized by being pumped through a series of valves under pressure. The milk fat is broken up into particles too small to coalesce. As a result, they remain suspended throughout the milk rather than separating to the top as a layer of cream.

The pasteurized, ultrapasteurized, or UHT processed milk is cooled rapidly to below 45°F (7°C). UHT milk is packaged into presterilized containers and aseptically sealed. Since bacteria cannot enter the UHT milk, it can be kept unrefrigerated for at least 3 months. However, once the container is opened, the UHT milk picks up organisms from the air. Then the UHT milk must be handled and stored like any other fluid milk product.

Milk Intolerance

Milk is the main dietary source of carbohydrate lactose. In the 1960s, the potential problem of lactose intolerance was emphasized when reports revealed low levels of the enzyme lactase in digestive tracts of 70% of black and 10% of white persons in the United States. Symptoms include bloating, abdominal cramps, nausea, and diarrhea. Worldwide, lactose intolerance is relatively high among nonwhite populations. However, results of many studies have disclosed the difference between lactose intolerance and milk intolerance. While a large segment of certain populations may be diagnosed as lactose-intolerant, most of these individuals, once adapted, can tolerate the amount of lactose contained in typical servings of dairy foods.

For the few individuals who are truly milk-intolerant, suitable alternatives are available. These include consumption of recommended amounts of milk in smaller but more frequent servings throughout the day, most cheeses, many fermented and culture-containing dairy products (e.g., yogurt), lactose-hydrolyzed milk, and some dairy products containing up to 75% less lactose.

Milk protein allergy may be considered as another form of milk intolerance. An allergic reaction to the protein component of cow's milk may occur in a few infants who experience allergies early in life. The incidence is probably 1% or less in the infant and child population in industrialized countries, although a range of 0.3–7.5% has been reported. The condition is usually outgrown by 2 years of age, beyond which time true allergic reactions to milk in the general population are rare.

Human Disease Risk

In the 1950s, scientists hypothesized an association between increased dietary intakes of cholesterol and increased risk of atherosclerosis and coronary heart disease (CHD). In 1980, the U.S. Government issued dietary guidelines for Americans; one guideline recommended not eating excessive fat, especially saturated fat and cholesterol.

In line with these dietary recommendations, dairy foods have been singled out as being undesirable because of their cholesterol and saturated fatty-acid content. Dairy foods, however, are not particularly high in either cholesterol or saturated fatty acids. Some research demonstrates that whole milk, skim milk, and yogurt exhibit a cholesterol-lowering effect. Further evidence that dairy foods have little or no effect in CHD has been increasing while animal fat consumption has been decreasing.

The whole question of dietary change as a measure to reduce the incidence of CHD is unresolved. Many scientific investigators and professional organizations believe that the role of dietary cholesterol and saturated fatty acids in the prevention, mitigation, and cure of CHD has been overemphasized.

CONSUMPTION

Population growth by the year 2000 will no doubt cause an increase in consumption of dairy products; however, per-capita consumption is expected to decrease (*J. Dairy Sci.* 64:959).

Table 4.6 shows the per-capita consumption of dairy products in several selected countries. Finland and Ireland are highest in fluid milk consumption, and New Zealand leads in butter consumption. France, Israel, and West Germany are the countries with the highest per-capita cheese consumption.

Table 4.7 shows the per-capita sales of milk and milk products in the United States in 1965, 1975, and 1985. There is a noticeable consumer preference toward low-fat products. Figure 4.11 reflects the changes in consumer preferences during the past 10 years with yogurt, low-fat milk, sour cream (and dips), and cheese showing the most significant increases. In 1984, consumer expenditures for dairy products were $48 billion—12.8% of all food expenditures.

MARKETING

World

Only a small percentage of the world's milk products production enters world trade: butter (4%), cheese (4%), skim milk powder (18%), and casein (80%). The European Eco-

TABLE 4.6. Yearly Per-Capita Consumption of Selected Dairy Products in Selected Countries, 1985

Leading Countries and U.S., USSR[a]	Per-Capita Consumption (lb)		
	Milk	Butter	Cheese
Finland	421	24.3	
Ireland	399	20.3	
Austria	356		
Sweden	354		32.4
Norway	464		
East Germany		35.7	
Denmark		24.7	
New Zealand		26.9	
Greece			46.7
France			45.9
Italy			35.7
U.S.	221	5.3	27.7
USSR	218	14.1	6.4

[a]U.S. and USSR data shown for comparison to leading countries.
Source: USDA.

TABLE 4.7. Yearly Per-Capita Sales of Dairy Products in the United States (lb)

Product	1965	1975	1985
Plain whole milk	236.5	173.2	116.3
Lowfat milk	10.9	54.9	85.0
Skim milk	12.6	11.9	13.0
Flavored milk and drinks	8.8	10.0	10.7
Buttermilk	6.1	4.8	4.3
Yogurt	0.3	2.1	3.9
Eggnog	0.3	0.4	0.5
Half-and-half	4.0	2.5	3.1
Light cream	0.9	0.4	0.3
Heavy cream	0.8	0.6	0.9
Sour cream and dips	1.0	1.7	2.3
Butter	5.8	4.4	3.8
American cheese	5.7	7.9	9.5
Other cheese	3.4	6.1	10.1
Cottage cheese	4.7	4.6	4.0
Dry milk	5.1	3.2	1.8
Evaporated milk	15.8	8.7	7.6
Ice cream	18.8	18.6	18.1
Ice milk	6.6	7.6	6.9

Source: USDA.

nomic Community (EEC) and New Zealand account for about 80% of total world exports of dairy products. The EEC, Canada, and the United States have surpluses of skim milk powder in part because of price stabilization programs and because production costs exceed world market prices. Exports of skim milk powder go principally to countries with a milk deficit and are used in a variety of recombined products. Much of New Zealand and Australian production of casein is exported to the United States.

Nutritionally, the world may need milk, but there is no real market for its present surpluses of dairy products. Stocks (surpluses at the end of 1985) were 1,600 million tons of butter (mostly in EEC), 1,450 million tons of cheese (75% in EEC and the U.S.), and 1,100 million tons of nonfat dry milk (nearly half in the U.S.).

United States

Most milk produced on dairy farms goes to plants and dealers for processing. More than 50% of the fluid milk is marketed by supermarkets, primarily in plastic gallon containers.

Prices

Cooperative milk marketing associations, located near large population centers, give producers more bargaining power, as they control approximately 90% of the milk produced. The cooperatives hire professionals to market the milk to prospective purchasers.

There are classes and grades of milk that determine price. Grade A (fluid or market

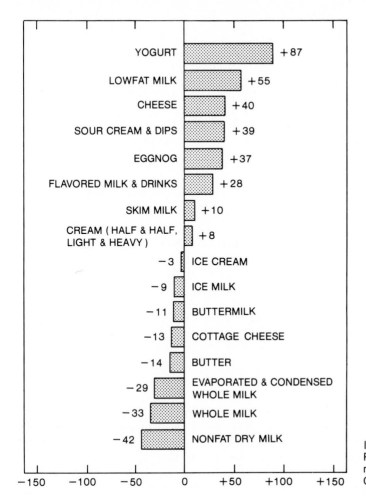

FIGURE 4.11.
Percent change in per capita sales of milk and milk products, 1975–85. Courtesy of Milk Industry Foundation.

milk) and grade B (manufacturing milk) are determined by the sanitary and microbial quality of the milk. Class I, the highest price class, is grade A milk for fluid use. Classes II and III are for milk used in manufactured products. Class II includes milk for cottage cheese, cream, and frozen desserts; class III is milk used for butter and cheese. Class II is grade A milk, and class III is surplus grade A or grade B milk. Producers receive a "blend" price based on the proportion of milk used in each price class.

Forty-four federal "milk-marketing orders" and 14 states establish the prices that processors must pay dairy farmers for about 95% of the fluid milk and fluid milk products consumed in the United States. In some states, milk control commissions not only determine what farmers are to be paid but the price that stores can charge customers. Federal milk orders are to promote orderly marketing conditions for dairy producers and assure consumers an adequate supply of milk.

In 1985 the average price of class I milk, received by dairy producers, was $13.88 per

Billion pounds

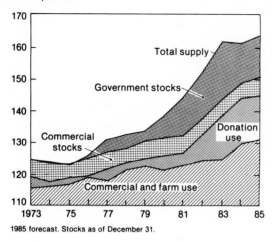

1985 forecast. Stocks as of December 31.

FIGURE 4.12.
Milk supply, use, and stocks. Courtesy of the USDA.

hundredweight. Total farm receipt from the sale of milk was $18.1 billion, which represented 12.7% of total cash farm receipts.

Surpluses

Since 1975, there has been a marked increase in milk supply in the United States. Each dairy cow produces 20% more milk than the average cow in 1975 (Fig. 4.12). Increased milk production, in addition to government support prices and purchases, has resulted in a surplus of milk and milk products. The excess production and government storage are the major problems facing the dairy industry.

SELECTED REFERENCES

Dairy Producer Highlights. 1986. National Milk Producers Federation, 1840 Wilson Blvd., Arlington, VA 22201.

Dunkley, W. L., and Pelissier, C. L. 1981. Relationship of United States dairy industry to dairying internationally. *J. Dairy Sci.* 64:695.

Fallert, R. F. 1981. Milk pricing—Past, present, the 1980's. *J. Dairy Sci.* 64:1105.

Hanman, G. E., and Blakeslee, B. 1981. Dairy producer and marketing cooperatives—Past, present and future. *J. Dairy Sci.* 64:1113.

Hedrick, T. I., Harmon, L. G., Chandan, R. C., and Seiberling, D. 1981. Dairy products industry in 2006. *J. Dairy Sci.* 64:959.

Kosikowski, F. V. 1981. Dairy foods of the world—Evolution, expansion and innovation. *J. Dairy Sci.* 64:996.

Larson, B. L. (editor). 1985. *Lactation.* Ames: Iowa State University Press.

Lowenstein, M., Speck, S. J., Barnhart, H. M., and Frank, J. F. 1980. Research on goat milk products: A review. *J. Dairy Sci.* 63:1629.

Milk Facts. 1986. Milk Industry Foundation, 888 Sixteenth St. N.W., Washington, DC 20006.

Newer Knowledge of Cheese and Other Cheese Products. 1983. National Dairy Council, 6300 N. River Road, Rosemont, IL 60018.

Newer Knowledge of Milk and Other Fluid Dairy Products. 1983. National Dairy Council, 6300 N. River Road, Rosemont, IL 60018.

Product Information Sheets (Butter and Cream; Milk; and Ice Cream). 1983. National Dairy Council, 6300 N. River Road, Rosemont, IL 60018.

Sellars, R. L. 1981. Fermented dairy foods. *J. Dairy Sci.* 64:1070.

Speckmann, E. W. L., Brink, M. F., and McBean, L. D. 1981. Dairy foods in nutrition and health. *J. Dairy Sci.* 64:1008.

Tobias, J., and Muck, G. A. 1981. Ice cream and frozen desserts. *J. Dairy Sci.* 64:1077.

CHAPTER 5 ▮▮▮▮▮▮▮▮▮▮▮

Poultry and Egg Products

Poultry meat and eggs are nutritious and relatively inexpensive animal products used by humans throughout the world. Feathers, down, livers, and other **offal** are additional useful products and by-products obtained from poultry.

Chickens and turkeys dominate the world poultry industry. However, in parts of Asia, ducks are commercially more important than broilers (young chickens), and in areas of Europe there are more geese than turkeys.

Application of genetics, nutrition, and disease control, along with sound business practices, has advanced the commercial poultry industry where eggs and meat can be produced very efficiently. The industry has developed two specialized and different types of chickens—one for meat production and one for egg production. Broiler lines are superior in efficient, economical meat production but have a lower egg-producing ability than the egg production lines. On the other hand, the egg-type birds produce large numbers of eggs very efficiently. They have been bred to mature at light weights and they are therefore slower-growing and inefficient meat producers.

POULTRY MEAT AND EGG PRODUCTION

Table 5.1 shows the poultry and egg production for the leading countries of the world. The United States is the leading country in poultry meat production (16.4 bil lb, of which 2.6 bil lb is turkey) and ranks third, after the Soviet Union and China, in egg production. The 8.9 billion pounds of eggs produced in the United States would convert to approximately 5.9 billion dozen eggs, since a dozen eggs averages approximately 24 oz in weight.

Broiler (or fryer) chickens and turkeys provide most of the world's production and consumption of poultry meat. Roaster chickens, mature laying hens (fowl), ducks, geese, pigeons, and guinea hens are consumed in smaller quantities, although some of these are very important food sources in some areas of the world. The more common kinds of poultry used for meat production in the United States are shown in Table 5.2.

Poultry are slaughtered and processed in large plants owned by totally integrated

TABLE 5.1.   World Poultry and Egg Production, 1985

Poultry Meat		Hen Eggs	
Country	Billions of Pounds	Country	Billions of Pounds
U.S.	17.2	USSR	9.5
USSR	6.1	China	9.4
China	3.9	U.S.	8.9
Brazil	3.4	Japan	4.7
Japan	3.1	France	2.0
France	2.8	India	1.8
Italy	2.3	Brazil	1.7
World total	68.1	World total	66.5

Source: 1985 FAO Production Yearbook.

poultry companies. Some packing and processing companies that slaughter and process red meat animals also slaughter and process poultry. These companies were identified in Chapter 2 (Table 2.2). The large companies that slaughter and process only poultry are identified in Table 5.3.

COMPOSITION

Meat

Table 5.2 shows chickens yield approximately 78% of their liveweight as carcass weight whereas the dressing percent for turkeys is 83%. The edible raw meat from carcasses of chickens and turkeys is approximately 67 and 78%, respectively.

The gross chemical contents of chicken and turkeys are shown in Table 5.4. Dark meat is higher in calories, lower in protein, and higher in cholesterol than light meat for both

TABLE 5.2.   Age and Weights of Various Kinds of Poultry Used for Meat Production

Classification	Typical Age (mo)	Average Slaughter Wt (lb)	Average Dressed Wt (lb)	Dressing (%)
Young chickens				
Broiler (fryer)	1.5–2	4–5	3–4	78
Rock Cornish	1.0	2	1.5	75
Roaster	3–5	6–8	5	75
Fowl mature (stewing)	19	5	3.8	75
Turkeys	4–6	10–25	8–20	83
Ducks	<2	7	4–5	70
Geese	6	11	8–9	75

Source: USDA and various others.

TABLE 5.3.   Large Companies That Slaughter and Process Only Poultry (1985 net sales of more than $750 million)

Company Location (parent company)	Net Sales ($ millions)	No. Plants
Perdue Farms Salisbury, MD	822	7
Holly Farms Poultry Wilkesboro, NC (The Federal Co.)	803	8
Tyson Foods Springdale, AR	750	16

chicken and turkey. If the skin remains on poultry, it usually adds cholesterol and total fat.

Eggs

An egg is spherical in shape with one end being rather blunt and larger than the other, smaller, more pointed end. Figure 5.1 shows a cross section of a hen's egg. The mineralized shell surrounds the contents of the egg. Immediately inside the shell are two membranes; one is attached to the shell itself, and the other tightly encloses the content of the egg. The air cell usually forms between the membranes in the blunt end of an egg shortly after it is laid, when the contents of the egg cool and contract.

The "egg white," or albumen, exists in four distinct layers. One of these layers surrounds the yolk and keeps the yolk in the center of the egg. The spiral movement of the developing egg in the magnum causes the mucin fibers of the albumen to draw together

TABLE 5.4.   Composition of Chicken and Turkey Per 100 g, Edible Portion

	Water (g)	Food Energy (calories)	Protein (g)	Total Fat (g)	Cholesterol (mg)	Ash (g)
Young chicken Light meat (no skin)	74.9	114	23.2	1.6	58	0.98
Dark meat (no skin)	76.0	125	20.1	4.3	80	0.94
Turkey (hen) Light meat (no skin)	73.6	116	23.6	1.7	58	1.0
Dark meat (no skin)	74.0	130	20.1	4.9	62	0.95

Source: USDA, Agricultural Handbook No. 8-5.

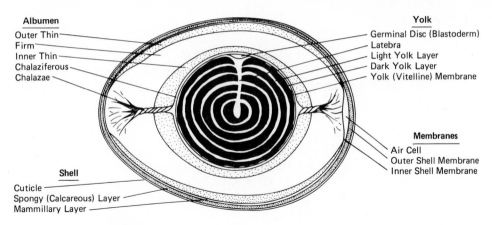

Albumen
Outer Thin
Firm
Inner Thin
Chalaziferous
Chalazae

Yolk
Germinal Disc (Blastoderm)
Latebra
Light Yolk Layer
Dark Yolk Layer
Yolk (Vitelline) Membrane

Membranes
Air Cell
Outer Shell Membrane
Inner Shell Membrane

Shell
Cuticle
Spongy (Calcareous) Layer
Mammillary Layer

FIGURE 5.1.
Longitudinal section of a hen's egg. Courtesy of the USDA.

into strands that form the chalaziferous layer and chalazae. The twisting and drawing together of these mucin strands tends to squeeze out the thin albumen to form an inner thin albumen layer.

The yellowish-colored yolk is in the center of the egg, its contents surrounded by a thin, transparent membrane called the vitelline membrane.

Egg weight varies from 1.5 to 2.5 oz, with the average egg weighing approximately 2 oz. The component parts of an egg are shell and membranes (11%), albumen (58%), and yolk (31%). The gross composition of an egg is shown in Table 5.5. The mineral content of the shell is approximately 94% calcium carbonate. The yolk has a relatively high fat and protein percent, although there is more total protein in the albumen.

The weight classes and grades of eggs are discussed in Chapter 7. Egg formation is presented in Chapter 9.

TABLE 5.5. Chemical Composition of the Egg, Including the Shell

	Percent of Total	Water (%)	Protein (%)	Fat (%)	Ash (%)
Whole egg	100	65.5	11.8	11.0	11.7
White	58	88.0	11.0	0.2	0.8
Yolk	31	48.0	17.5	32.5	2.0
	Percent of Total	Calcium Carbonate (%)	Magnesium Carbonate (%)	Calcium Phosphate (%)	Organic Matter (%)
Shell and shell membranes	11	94.0	1.0	1.0	4.0

Source: USDA.

POULTRY PRODUCTS

Meat

More than one-half of all broilers leave the processing plant as cut-up chicken (Table 5.6) or selected parts rather than whole birds (compared to less than 30% a decade ago).

Approximately 50% of the broilers, fowl (mature chickens), and turkeys are "further processed" beyond the "ready-to-cook" carcass or parts stage. The meat is separated from the bone and formed into products (e.g., turkey roll fingers and nuggets), cut and diced (e.g., for preparation of salads, casseroles), canned, or ground. Some of the cured and smoked ground products are frankfurters, bologna, pastrami, turkey ham, and salami. More than 1 out of 8 lb of all broilers is processed into such products as chicken patties, nuggets, battered and cooked chicken, and hot dogs. Even more of the broilers are expected to be processed in future years.

Eggs

Eggs are unique, prepackaged food products that are ready to cook in their natural state. Eggs find their way into the human diet in numerous ways. They are most popular as a breakfast entree; however, they are commonly used in sandwiches, salads, beverages, desserts, and several main dishes.

Shell eggs can be further processed into pasteurized liquid eggs and dried eggs and other egg products. Pasteurized liquid egg can be blast-frozen and sold to bakeries and other food processing plants as an ingredient in other fabricated food products or transported to users as a liquid product in refrigerated tank trucks. The dried egg products are sold to food manufacturers to produce products such as cake mixes, candy, and pasta.

The initial step in making egg products is breaking the shell and separating the yolks, whites, and shells. This is done in most egg-breaking plants by equipment completely automated to remove eggs from egg filler flats, wash and sanitize the shells, break the eggs for individual inspection of the contents by an operator, and separate the yolks from the whites. Processing equipment can also be used to produce various white and yolk products or mixtures of them. Present automatic systems are capable of handling 18,000–50,000 eggs per hour (Fig. 5.2).

TABLE 5.6. The Retail Parts and Cooked Edible Meat of a Broiler-Fryer Chicken

Part	Retail Parts (% of carcass)	Percent Edible[a]
Breast	28	63
Thighs	18	60
Drumsticks	17	53
Wings	14	30
Back and neck	18	27
Giblets	5	100
Total	100	53

[a]Net average yield of meat is 67% without neck and giblets.
Source: USDA and National Broiler Council.

A

B

C

D

FIGURE 5.2.
Automated egg handling equipment. (A) Overall view.
(B) Automatic egg washer. (C) Eggs on the flash-can-
dling area of the in-feed conveyor. (D) In-line scales.
Eggs of different sizes are weighed and ejected at dif-
ferent points on the line. Courtesy of the USDA, Egg
Grading Manual.

Some breeds of chickens lay white-shelled eggs, and other breeds produce brown-shelled eggs. The brown shell color comes from a reddish-brown pigment (ooporhyrin) derived from hemoglobin in the blood. Although there are shell color preferences by consumers in certain parts of the United States, there are no significant nutritional differences related to egg shell color.

Feathers and Down

In some areas of the world, ducks and geese are raised primarily for feathers and down. Down is a small, soft feather found beneath the outer feathers of ducks and geese. Feathers and down provide stuffing for pillows, quilts, and upholstery. They are also used in the manufacture of hats, clothing (outdoor wear), sleeping bags, fishing flies, and brushes.

The duck and goose industry in the United States is too small to meet the manufacturing demand for down. Thus, raw feathers and down are primarily imported (approximately 90% of the product used) from China, France, Switzerland, Poland, and several other European and Asian countries.

Breeder geese are often used for down production, as the older birds produce the best down. Four ducks or three geese will produce 1 lb of feather-and-down mixture, of which 15–25% is down. Down and feathers are separated by machine, washed, and dried.

Other Products and By-products

Goose livers, from force-fed geese, are considered a gourmet product in European countries. These enlarged fatty livers, weighing as much as 2 lb each, are produced by force-feeding corn to geese three times a day for 4–8 weeks. The liver is used to make a flavored paste called paté de foie gras (for hors d'oeuvres or sandwiches).

Poultry by-products include hydrolized feather meal and poultry meat and bone scraps. These by-products are occasionally included in livestock and poultry rations as protein sources (refer to Chapter 2 for more discussion on these by-products).

HEALTH CONSIDERATIONS

Nutritive Value of Poultry Meal

The white meat from fowl is approximately 33% protein, and dark meat is about 28% protein. This protein is easily digested and is of high quality, containing all the essential amino acids. The fat content of poultry meat is lower than that found in many other meats. Poultry meat is also an excellent source of vitamin A, thiamin, riboflavin, and niacin. The nutritive value of poultry and eggs to the U.S. population is shown in Table 5.7.

Eggs

Since the egg (Fig. 5.3) contains many essential nutrients, it is recognized as one of the important foods in the major food groups recommended in *Dietary Guidelines for Ameri-*

TABLE 5.7. Contribution of Poultry and Eggs to Nutrient Supplies Available to U.S. Civilian Consumption, 1985

Nutrients	Meat, Poultry,[a] and Fish (%)	Eggs (%)
Food energy	20	2
Protein	43	4
Fat	33	2
Carbohydrate	0.1	0.1
Cholesterol	40	40
Minerals		
Calcium	4	2
Phosphorus	28	4
Iron	29	4
Magnesium	15	1
Vitamins		
Vitamin A	17	2
Thiamin	26	1
Riboflavin	22	4
Niacin	47	0
Vitamin B_6	40	2
Vitamin B_{12}	72	6
Ascorbic acid	2	0

[a]Poultry comprises approximately 30% of the per-capita retail weight consumption of meat, poultry, and fish.
Source: USDA.

cans. These guidelines were developed by the U.S. Department of Agriculture to use in planning diets contributing to good health.

Eggs have a high nutrient density in that they provide excellent protein and a wide range of vitamins and minerals and have a low calorie count. The abundance of readily digestible protein, containing large amounts of essential amino acids, makes eggs a least-cost source of high-quality protein.

FIGURE 5.3.
"The incredible, edible egg." Courtesy of the American Egg Board.

TABLE 5.8. Estimated Nutrient Values for a Large Egg (based on 60.9-g shell weight with 55.1-g total liquid whole egg, 38.4-g white, and 16.7-g yolk)

Nutrients and Units (proximate)	Whole	White	Yolk
Solids (g)	13.47	4.6	8.81
Calories (kcal)	84	19	64
Protein (N × 6.25) (g)	6.60	3.88	2.74
Total lipids (g)	6.00	—	5.80
Cholesterol (mg)	264	—	258
Ash (g)	0.55	0.26	0.29

Source: Poultry Sci. 58:131.

Table 5.8 lists the nutrients and amounts of those nutrients found in eggs. When the values of these nutrients are compared with the dietary allowances recommended by the Food and Nutrition Board of the National Research Council (1980), it is found that two large eggs (a usual serving) supply an impressive percentage of essential nutrients—an average of 10–30% for adults (Table 5.8). Eggs are an especially rich source of high-quality protein, vitamins (A, D, E, folic acid, riboflavin, B_{12}, and pantothenic acid), and minerals (phosphorus, iodine, iron, and zinc). Some consumers believe that the dark yellow yolk is of higher nutrient value than the paler yolks, but there is no evidence to support this belief.

Table 5.9 lists the U.S. recommended daily allowances (RDA) supplied by two large eggs.

Wholesomeness

In the United States, over 90% of the poultry meat is inspected by the USDA for wholesomeness. USDA inspectors give assurance that the meat is free of disease and that it has been processed under sanitary conditions. The inspector gives an antemortem inspection of the poultry before they are slaughtered and a postmortem inspection of the carcasses and viscera.

Human Disease Risk

Eggs are relatively high in cholesterol (whole egg, 264 mg; yolk, 258 mg). Eggs have been included in the diet–coronary heart disease controversy, since high blood cholesterol is one of the several risk factors associated with CHD.

Cholesterol, a fatlike substance, is found in blood and nerve tissues and, in fact, in every cell of the body. Its exact function is not fully understood, but cholesterol is known to play a number of important physiological roles.

In addition to being present in all food products of animal origin, cholesterol is synthesized (made) and metabolized (used or excreted) daily in our bodies in amounts far greater than usually consumed in the diet.

TABLE 5.9. **U.S. Recommended Daily Allowances (RDA) in Relation to Nutrient Content of Large Eggs**

Nutrient	RDA	Percent RDA, Two Eggs
Protein	45.0 g	30
Vitamins		
A	5,000.0 IU	10
D	400.0 IU	15
E	30.0 IU	6
C	60.0 mg	
Folic acid	0.4 mg	15
Thiamin	1.5 mg	6
Riboflavin	1.7 mg	20
Niacin	20.0 mg	
B_6	2.0 mg	6
B_{12}	6.0 μg	15
Biotin	0.3 mg	8
Pantothenic acid	10.0 mg	15
Minerals		
Calcium	1.0 g	6
Phosphorus	1.0 g	20
Iodine	150.0 μg	35
Iron	18.0 mg	10
Magnesium	400.0 mg	4
Copper	2.0 mg	4
Zinc	15.0 mg	10

Source: Cotterill and Glauert (1981).

High serum cholesterol and CHD are statistically correlated, but CHD does occur in persons having normal or low blood cholesterol. CHD also occurs among vegetarians.

The level of cholesterol in the blood, according to the best scientific knowledge, is not markedly influenced in most people when moderate amounts of cholesterol are present in the foods eaten. It is possible that a high serum cholesterol level is a symptom and not a cause of CHD. There are several known disorders where a high level of cholesterol is a side effect.

CONSUMPTION

Meat

The per-capita consumption of chicken and turkey in the United States is over 70 lb (Fig. 5.4), of which turkey is 12.1 lb. Poultry consumption is approximately one-third of the total per-capita consumption of red meat and poultry consumption (Fig. 5.5). In recent years, poultry consumption has increased at the expense of red meat consumption. This is due, in part, to the price differential existing between the various meats (Fig. 5.6).

The dollars expended for broilers and turkey is shown in Table 5.10. The percentage of income spent for these two meats has decreased slightly from 1980 to 1984, even though the total dollars spent has increased.

Pounds of poultry

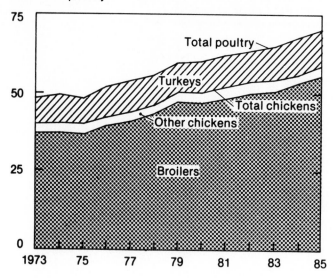

FIGURE 5.4.
Per capita consumption of poultry in the United States. Courtesy of the USDA.

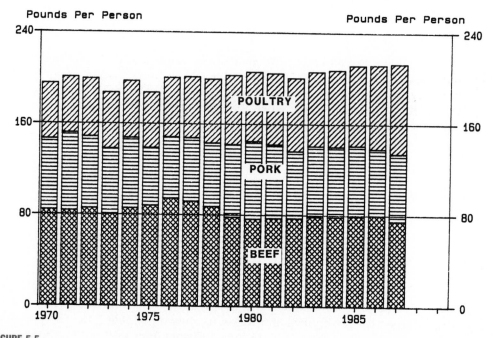

FIGURE 5.5.
Per capita meat consumption, retail weight equivalent, 1970–1986. Courtesy of The Western Livestock Marketing Information Project.

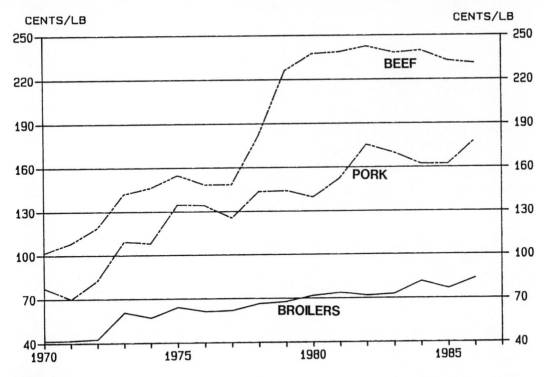

FIGURE 5.6.
Quarterly retail price for broilers, pork, and beef. Courtesy of The Western Livestock Marketing Information Project.

Heavy hens, the type used to produce hatching eggs for broiler production, are sometimes available in retail stores as "stewing hens" or "baking hens." These account for about 10% of the poultry meat consumed by Americans each year.

Nutritive Value of Eggs

Per capita consumption of eggs is approximately 254 eggs per year. Most of these are consumed as shell eggs (85%) rather than processed eggs (Fig. 5.7). The per capita consumption is nearly the number of eggs a hen lays per year.

TABLE 5.10.   Expenditures Per Person For Poultry

	Broilers		Turkey		Total Poultry	
Year	Dollars	Percent[a]	Dollars	Percent	Dollars	Percent
1980	35.96	0.45	9.32	0.12	45.28	0.56
1986	47.30	0.38	14.40	0.12	61.70	0.50

[a]Percent of per capita disposable income.
Source: USDA.

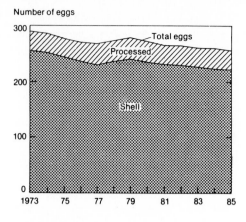

Number of eggs

FIGURE 5.7.
Per capita egg consumption in the United States. Courtesy of the USDA.

MARKETING

In the United States, large amounts of poultry meat are processed in commercial facilities and then transported under refrigeration to local and world markets. Broilers are the primary poultry meat exported, with more than 400 million pounds exported each year (Fig. 5.8). More than 26 million pounds of turkey (valued at $16.4 million) is exported from the United States, primarily to West Germany.

Chicken meat is usually marketed fresh in the United States, whereas turkeys are marketed frozen. However, some turkeys are sold fresh year round in some markets. Freshly frozen poultry meat can be marketed on a year-round basis through the use of frozen storage and environmentally controlled production units.

Most of the broilers are sold to consumers through retail stores (Fig. 5.9). Fresh chicken can reach the retail market counter the day following slaughter. "Sell-by" date on each package is usually 7 days from processing and is the last date recommended for sale of

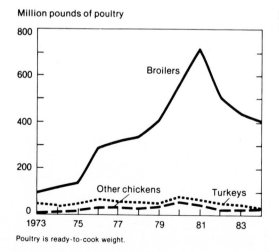

Million pounds of poultry

Poultry is ready-to-cook weight.

FIGURE 5.8.
U.S. exports of poultry meat. Courtesy of the USDA.

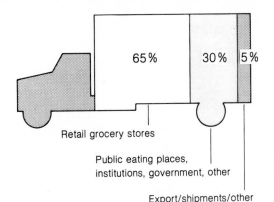

Retail grocery stores

Public eating places,
institutions, government, other

Export/shipments/other

FIGURE 5.9.
Market distribution of more than 12.5 billion pounds of
ready-to-cook chicken. Courtesy of the National Broiler
Council.

chicken. However, with proper refrigeration (28–32°F), shelf life can be extended up to 3 days longer. Some processors guarantee 14 days of shelf life for their broilers.

Almost all commercially available chicken and turkey meat is inspected for wholesomeness by federal or state government employees. About two-thirds of the chickens are graded by the USDA for quality.

Over 40% of the broilers are marketed to consumers under the producer's brand names. In certain markets, such as New York, over 90% of the broilers are sold with brand names. Poultry brands identify the single firm that raised, processed, and marketed the bird, not just the firm that processed the retail product, as is the case of red meats. The four largest vertically integrated poultry firms represented 34% of the broiler market in 1983, compared with only 17% in 1973.

Chicken and turkeys are frequently featured by supermarket stores as loss leaders. Stores knowingly take a loss on the poultry to attract customers to their store where additional items purchased usually more than cover the losses. The practice of using loss leaders is sometimes adopted to stay competitive with other stores which first initiate the practice.

Before and during the 1950s, the marketing of turkey was highly seasonal, as 90% of the turkeys were consumed during the last quarter of the year for the Thanksgiving and Christmas holidays. Today only about 40% of turkey sales are made during that period. Current trends indicate that turkey is being consumed more on a year-round basis and has become a meat staple.

SELECTED REFERENCES

Publications

A Scientist Speaks About Fowl. 1981. American Egg Board, 1460 Renaissance, Park Ridge, IL 60068.

A Scientist Speaks About Egg Nutrition. 1982. American Egg Board, 1460 Renaissance, Park Ridge, IL 60068.

A Scientist Speaks About Egg Products. 1981. American Egg Board, 1460 Renaissance, Park Ridge, IL 60068.

Cotterill, O. J., and Glauert, J. L. 1981. Shell egg nutritional labeling. *Poultry Tribune* 87:16.

Egg Grading Manual. USDA Agricultural Handbook No. 75. Revised April, 1983.

Eggcyclopedia. 1981. American Egg Board, 1460 Renaissance, Park Ridge, IL 60068.

Moreng, R. E., and Avens, J. S. 1985. *Poultry Science and Production.* Reston, VA: Reston Publishing.

Mountney, G. J. 1976, *Poultry Products Technology.* Westport, CT: Avi.

Stadelman, W. J., and Cotterill, O. J. 1986. *Egg Science and Technology.* Westport, CT: Avi.

Visuals

Egg Quality film (loan). Information Division, AMS, USDA, Washington, DC 20250.

Egg Quality slides. Photography Division, GPA, USDA, Washington, DC 20250.

Turkey Product slides. National Turkey Federation, 11319 Sunset Hills Rd., Reston, VA 22090.

Processing Chicken Broilers (slides and tape—80 slides; 13 min) and *Egg Breaking Operations* (slides and tape—80 slides; 12 min). Poultry Science Dept., Ohio State University, 674 W. Lane Ave., Columbus, OH 43210.

The Incredible Edible Egg on Video. Northwest Egg Producers Coop. Assoc., P.O. Box 1038, Olympia, WA 98507.

CHAPTER 6

Wool and Mohair

Skins of common mammals have a covering to which various terms are applied depending on the nature of the growth. For example, cattle, pigs, horses, and dairy goats have hair; sheep have wool; mink and nonangora rabbits have fur; angora rabbits have angora; and angora goats have mohair. Fibers that grow from the skin of animals gives protection from abrasions to the skin and help keep animals warm.

Hair from most mammals has little commercial value (it is used mostly in padding and cushions). Fur of mink and nonangora rabbits is either naturally beautiful or can be dyed to give attractive colors; therefore, it has considerable value as fur. Fur is composed of fine short fibers and relatively long, coarse guard hairs in contrast to the skin covering of cattle, horses, and pigs, which is composed entirely of guard hairs.

Furs made from rabbits, mink, foxes, and bison (American buffalo) may be classed as status clothing, because they are usually costly and attractive in appearance. White rabbit furs can be dyed any color (including the pastels), but colored furs cannot. Because colored furs cannot be dyed, color **mutations** have been important in the mink business because they provide a wide array of fur colors. At present, the mink industry in the United States is suffering economically because of competition from furs made in other countries and high costs of producing mink.

Hides from young market lambs are often marketed with wool left on to be processed into heavy winter lambskin coats. Lambs from the **Karakul** breed of fat-tail sheep are used to produce Persian lamb skins. The black, tight curls of fiber are more like fur than wool. Persian lamb skins have been produced in the United States, but larger numbers are currently produced in Afghanistan, Iran, and the Soviet Union.

Hides from sheep with longer wool on the pelts also may be marketed intact for processing to produce ornamental rug pieces. The long wool may be dyed. The long wool may also be loosened and removed from the hides after slaughter and sold separately from the hides.

GROWTH OF HAIR, WOOL, AND MOHAIR

Hair or wool fiber grows from a **follicle** located in the outer layers of the skin. Growth occurs at the base of the follicle, where there is a supply of blood, and cells produced are pushed outward. The cells die after they are removed from the blood supply, because they can no longer obtain nutrients or eliminate wastes. A schematic drawing of a wool follicle is presented in Fig. 6.1. The wool fiber has an inner core, the cortex; and an outer sheath, the cuticle. The cuticle causes the fibers to cling together. The intermingling of wool fibers is known as **felting.** The felting of wool is advantageous in that wool fibers can be entangled to make **woolens,** but it is also responsible for the shrinkage that occurs when wool becomes wet.

Wool fibers have waves called **crimp** (Fig. 6.2). Crimp is caused by the presence of hard and soft cellular material in the cortex. The soft cortex is more elastic and is on the outer side of the crimp.

Hair does not exhibit any crimp, because all the cells in the cortex are hard. Also, the inner core of hair is not solid, as is the inner core of most wool fibers. Some fibers in wool of a low quality are large, lack crimp, and do not have a solid inner core. These fibers are called **kemp,** and they reduce the value of the fleece.

Mohair fiber follicles develop in groups consisting of three primary follicles each. Secondary follicles develop later. Mohair fibers have very little crimp (less than 1 crimp for each inch of length). The mohair cortex is composed largely of so-called *ortho* cells, in contrast to the cortex of wool, which is composed of both *ortho* and *para* cells. Mohair has a different scale structure from wool, but it does have scales and will felt. Kemp and **medullated fibers** reduce the value of wool and mohair fleeces, because such fibers do not dye well, and they show in apparel made from fleeces containing such fibers.

Skin of young lambs is made up of two layers; the surface skin (outer skin) is the epidermis, and the underlying skin is the dermis (Fig. 6.1). Between these two layers is

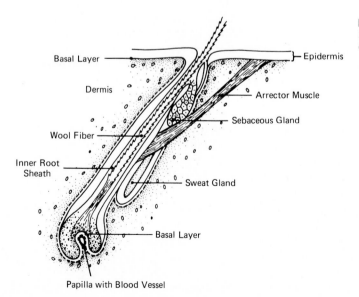

FIGURE 6.1.
Schematic drawing of a wool follicle.
Courtesy of R. W. Henderson.

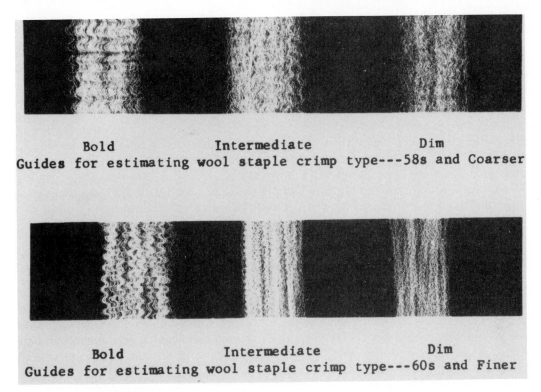

Bold   Intermediate   Dim
Guides for estimating wool staple crimp type---58s and Coarser

Bold   Intermediate   Dim
Guides for estimating wool staple crimp type---60s and Finer

FIGURE 6.2.
Several expressions of crimp in coarser and finer wool. Courtesy of the USDA.

the basal layer. Development of wool follicles occurs in skin of the fetus. The basal layer thickens, and areas that are to become wool follicles push down into the dermis. Two glands develop—the sebaceous gland that secretes sebum, a greasy secretion; and the sweat gland, which produces and secretes sweat. The downward growth of the basal layer rests on the papilla, which has a supply of blood. The wool fiber grows toward the skin's surface, and as it becomes removed from the blood supply, the cells die.

Two major types of follicles are produced. The first type to develop are known as primary follicles. They appear on the poll and face of the fetus by 35–50 days' gestation and over all other parts of the body by 60 days. They are arranged in groups of three. All primary follicles are fully developed and producing fiber at birth.

A second type of follicle, associated with the primary follicles, develops later and is known as secondary follicles. These follicles have an incomplete set of accessory structures—usually the sweat gland and arrector muscle are absent. The primary and associated secondary follicles are grouped into follicle bundles.

Since all the primary follicles are formed prior to birth of the lamb, the density of primary follicles per square inch of skin declines rapidly during the first 4 months of postnatal life and then declines gradually until the lamb approaches maturity.

Secondary follicles show little activity the first week of a lamb's life, after which there is a burst of activity of secondary follicles. Activity is maximum between 1 and 3 weeks of age with marked reduction after the third week.

Adverse prenatal environmental conditions can affect the number of secondary follicles initiated in development. Also, early postnatal influence could affect secondary follicle fiber production.

Growth of the wool fiber takes place in the root bulb, which is located in the follicle. Permanent dimensions of the fiber are determined in the area of the bulb, and there are no changes in growth characteristics of the fiber after it is formed. Defective portions of a fiber are caused by a reduction in the size of the fiber that creates a weakened area when sheep are under stressed conditions—i.e., poor nutrition and high body temperatures. Such fibers are likely to break under pressure.

Two types of undesirable fibers are kemp and medullated fibers. Both are hollow and brittle, and both are produced by primary follicles rather than secondary follicles, as is the case with true wool.

FACTORS AFFECTING THE VALUE OF WOOL

The two most important factors under control of the producer that affect the amount of wool produced by sheep are nutrition and breeding. The amount of feed available to the sheep (energy intake) and the percentage of protein in the diet influence wool production. Wool production is decreased when sheep are fed diets having less than 8% protein. When the diet contains more than 8% protein, the amount of energy consumed is the determining factor in wool production.

Improving wool production per animal through breeding involves selection for increased clean fleece weight, **staple length, fineness** and uniformity of length and fineness throughout the fleece. Much progress can be made by performance-testing rams for quantity and quality of wool production, but for overall improvement of the flock, lamb production and quality must be included (Fig. 6.3).

Sheep and angora goat producers are interested in maximizing net income from their sheep and goat operations. Increasing net income from wool and mohair can be accomplished by selecting to improve wool and mohair production and grade. The values of wool and mohair that are produced can be enhanced by giving attention to the following items:

1. The sheep or goats should be shorn when the wool or mohair is dry.
2. The portions of clips that are loaded with dung (called **tags**) should be sorted from fleeces and sacked separately.
3. Wool or mohair should be sacked by wool grades so that when **core samples** are taken from a sack of otherwise good wool, a few fleeces of low-grade wool (or mohair) won't cause the entire sack of wool or mohair to be placed in a low grade.
4. Wool should be shorn without making many double clips.
5. Fleeces should be tied with paper twine after they are properly folded with the clipped side out.
6. A lanolin-base paint should be used for branding animals. This type of paint is scourable.

FIGURE 6.3.
The ewe has been bred primarily for wool production with her lamb being a secondary product. Her heavy fleece will soon be shorn from her body. Courtesy of Woolknit Associates, Inc.

7. Fleeces from black-faced sheep should be packed separately from other wool. Also, black fleeces should be packed separately. Black fibers do not take on light-colored dyes and consequently stand out in a garment made from black and white fibers dyed a light pastel color.
8. All tags, sweepings, and wool or mohair from dead animals should be packed separately.
9. Hay that has been baled with wire rather than twine should be fed to sheep and goats. Pieces of twine will get into the wool or mohair and markedly lower the value.
10. Environmental stresses should be avoided. Being without feed or water for several days or having a high fever can cause a break or weak zone in the wool fibers.
11. Wool containing coarse fibers, called **kemp,** should be avoided, or kempy fleeces should be packed separately. Cloth and wool containing these coarse fibers are highly objectionable and thus are low in price.

Figure 6.4 shows undesirable fleeces, several of which have been identified. These undesirable fleeces have a lower value than fleeces containing white fibers and minimal foreign material. Some of the additional fleeces in Fig. 6.4 not previously discussed are the following:

Burry—contains vegetable matter such as grass seeds and prickly seeds which adhere tenaciously to wool
Chaffy—contains vegetable matter such as hay, straw, and other plant material
Cotted—wool fibers are matted or entangled
Dead—wool pulled from sheep that have died but have not been slaughtered
Murrain—wool obtained from decomposed sheep

National Wool Act

The Congress of the United States in 1954 established the National Wool Act, because it was recognized that domestic wool was an essential and strategic commodity. An incentive payment program was developed to encourage expansion of sheep production. This

FIGURE 6.4.
Several types of undesirable fleeces are identified. Each of these should be sorted and packed separately from the more desirable fleeces. Courtesy of the USDA.

Act provides that the price to encourage increased wool production be set each year. A formula determines a price to be paid for the weight of unshorn lambs marketed. There was also provided a means for wool growers to vote a referendum which, if passed, would fund the American Sheep Producers Council so that advertising and promotion could be financed. The money available in the incentive program is limited to a portion of funds derived from duty on imported wool. Thus, this incentive payment program is not financed from tax dollars.

Wool Incentive Payment

The best method of illustrating how the incentive payment works is to present actual cases in 1985 of two producers. The 1985 support price was $1.73 per pound. Formula for figuring incentive:

$$\begin{array}{l}
\$1.73 \text{ support price} \\
-\underline{0.61} \text{ (predicted) national average sale price for U.S.} \\
1.12 \text{ difference}
\end{array}$$

Then divide national average into this difference to obtain the "factor":

$$0.61\overline{)1.1200} = 1.84$$

1.84 is then the "factor" used to multiply the individual price received by each grower.

Example: $ .67 price received by producer A
                                    × 1.84
                                    $1.23

This makes $1.23 per pound support level received (in addition to the 67¢), which makes this producer's total income per pound of wool $1.90.

                                    $ .60 price received by producer B
                                    × 1.84
                                    $1.10

This makes $1.10 per pound support level received in addition to the 60¢ sale price, or a total of $1.70 per pound of wool for this producer's income.

If average price is below the support price, everyone receives payment. If average price is above the support price, no payments are made. If support price has been reached for payment, a producer may receive a percentage or support price incentive even if the wool was sold at prices above the incentive level; i.e., if the level is $1.50 and the wool sold for $1.75, the producer would still get incentive payment. Therefore the terminology "incentive" to produce better wool is used.

CLASSES AND GRADES OF WOOL

The value of wool is determined by its class and grade. Staple length determines the class and fineness of fibers determine the grade. The grades of wool are given by any one of three methods, but all three of the methods are concerned with a value that indicates the fineness (diameter) of the wool fibers. The three methods of reporting wool grades are these:

1. The American grade, or blood (old terminology), which is based on the theoretical amount of fine wool breeding (Merino and Rambouillet) represented in the sheep pro- ducing the wool (Fig. 6.5). The terms fine, $\frac{1}{2}$ blood, $\frac{3}{8}$ blood, $\frac{1}{4}$ blood, and low-$\frac{1}{4}$ blood are used to describe typical fiber diameter. Wool classified as fine or $\frac{5}{8}$ blood would have small fiber diameters, while $\frac{1}{4}$ blood wool would be coarse.
2. The spinning count system refers to the number of "hanks" of yarn, each 560 yards long, which can be spun from 1 lb of **wool top.** These grades range from 80 (fine) to 36 (coarse). Thus a grade of 50 by the spinning count method means that 28,000 yards (560 yd × 50 = 28,000 yd) of yarn can be spun from 1 lb of wool top.
3. The micron diameter method is based on the average of actual measurements of sev- eral wool fibers. This is the most accurate method of determining the grade of wool. A micron is $\frac{1}{25,400}$ of an inch.

Wool samples from each of the major grades are shown in Figs. 6.5 and 6.6. The systems of grading, along with the breeds from which the grades of wool come, are presented in Table 6.1. The Merino and Rambouillet breeds are considered fine wool breeds with the Cotswold, Lincoln, and Romney identified as coarse wool breeds. In Table 6.1, breeds listed between the fine and coarse wool breeds are called medium wool breeds.

Classes of wool are staple, French combing, and clothing; however, fibers of finer

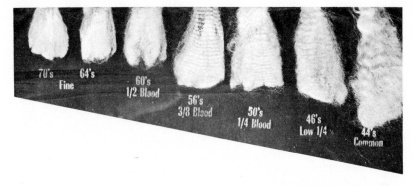

FIGURE 6.5.
Wool samples of the major grades of wool based on spinning count and the blood system. Courtesy of J. H. Landers.

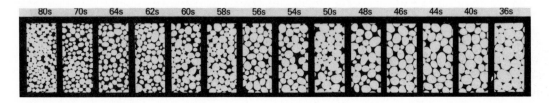

FIGURE 6.6.
Cross section of magnified wool fibers demonstrating the wool grades based on fiber diameter (micron diameter). Courtesy of the USDA.

TABLE 6.1.   Market Grades of Wool and Breeds of Sheep that Produce These Grades of Wool

USDA Grades Based on Spinning Count	USDA Grades Based on Blood System	Grades Based on Micron Diameter	Breeds According to Average Grades
80s	Fine	17.70–19.14	Merino
70s	Fine	19.15–20.59	Rambouillet
64s	Fine	20.60–22.04	Targhee, Southdown
62s	½ blood	22.05–23.49	Corriedale, Columbia
50s to 60s	¼ blood to ½ blood	23.50–30.99	Panama, Romeldale
48s to 56s	Low-¼ blood to ½ blood	26.40–32.69	Shropshire, Hampshire, Suffolk, Oxford, Dorset, Cheviot
46s to 48s	Common to low ¼ blood	32.69–38.09	Romney
36s to 40s	Common and braid	36.20–40.20	Cotswold
36s to 46s	Common and braid	32.70–40.20	Lincoln

grades that do not meet the length requirement may go into a higher class than they would if they were not fine. Much of the staple length wool is combed into **worsted**-type yarns for making garments of a hard finish that hold a press well.

The fineness of wool fibers from a sheep depends on the area of the body from which the fibers came. The body of a sheep can be divided into seven areas. Area 1 is the shoulder, and area 2 is the neck, where wool grows longer and finer than average wool from the same sheep. Area 3 is the back. It has long, average fineness, but it has been exposed to the most weathering. Area 4 is the side. It has long, average fineness, and it constitutes most of the wool. Area 5 is the tags. It is long, coarse, dirty, and stained wool. Area 6 is the britch. It is long, coarser-than-average wool that tends to be medullated. Area 7 is the belly. It has short, fine, matted wool that may be very heavy in vegetable matter. Preference is given to a fleece that has uniformity of length and fineness in all areas of the sheep's body.

PRODUCTION OF WOOL AND MOHAIR

Production of wool can be expressed as **greasy wool** or **scoured** (clean) **wool**. Greasy wool is wool or the fleece shorn once each year from sheep (Fig. 6.7). Scoured wool originates in the initial wool processing stage. Scouring is the washing and rinsing of wool to remove grease, dirt, and other impurities.

Table 6.2 shows the production of greasy and scoured wool for several countries of the world. The world grease wool production is approximately 6.3 billion pounds with the United States producing approximately 10 million pounds. Scoured or clean wool production represents 50–60% of the greasy wool produced.

Growth of wool varies greatly between sheep breeds. Fine-wool sheep grow 4–6 inches of wool per year, $\frac{1}{4}$ to $\frac{1}{2}$ blood breeds of sheep grow 5–10 inches of wool per year, and common to braid wool breeds grow 9–18 inches of wool per year. Mohair growth is approximately 12 inches per year.

Weights of **fleeces** also vary greatly among the breeds of sheep. Even though fine-wool breeds have fleeces of short length, weight of their fleeces varies from 9 to 12 lb; $\frac{1}{4}$

FIGURE 6.7.
Sheep being shown for their yearly production of wool. Courtesy of the National Wool Growers (USDA photo).

FIGURE 6.8.
Angora goats with a full growth of mohair. Courtesy of
Woolknit Associates, Inc.

FIGURE 6.9.
Mohair, kid-mohair, and cashmere,
strengthened by different proportions of
synthetic fibers, are popular yarns for making attractive garments that "breathe" and
reduce perspiration. Courtesy of George
F. W. Haenlein, University of Delaware.

TABLE 6.2. **World Production of Greasy and Scoured Wool Production, 1985**

Country	Greasy Wool (mil lb)	Scoured Wool (mil lb)
Australia	1,797	1,064
USSR	975	585
New Zealand	827	617
China	441	280
Argentina	353	213
South Africa	232	115
Uruguay	156	110
U.K.	127	89
U.S.	117	63
World total	6,632	4,010

Source: 1985 FAO Production Yearbook.

to ½ blood breeds of sheep produce fleeces weighing 7–15 lb; and coarse wool (common and braid) breeds produce fleeces weighing 9–16 lb.

Annual world mohair production is approximately 59 million pounds of which the United States produces between 29 and 30 million pounds or about half the world production. Texas produces about 97% of the mohair that is produced in the United States (Fig. 6.8). Mohair growth is about 1–2 inches per month; therefore, goats are usually shorn twice per year, with length of clip averaging 6–12 inches.

USES OF WOOL AND MOHAIR

Fibers from sheep, angora rabbits, and goats (Fig. 6.9) are used in making cloth and carpets. The consumption of wool in the United States is small when compared to cotton and numerous manufactured fibers (Fig. 6.10).

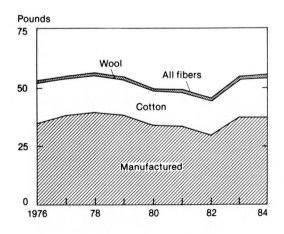

FIGURE 6.10.
United States per capita consumption of fibers. All fibers do not include flax and silk. Manufactured fibers include synthetic fibers such as nylon, dacron, and orlon. Courtesy of the USDA.

FIGURE 6.11.
The sewn-in Woolmark label (pure wool) and Woolblend label (wool blend) give assurance of quality-tested fabrics. Courtesy of the Wool Bureau, Inc.

Cloth made from wool has both highly desirable and undesirable qualities. Wool has a pleasant warming quality and can absorb considerable quantities of moisture while still providing warmth. Wool is also highly resistant to fire. Wool has a tendency to shrink when it becomes wet, however, and cloth from wool causes some people to itch.

Researchers have contributed greatly to the alteration of wool so that fabrics or garments will not shrink when washed. A process called WURLAN treatment has been developed which makes woolen garments machine-washable. In this treatment the wool fibers are coated with a very thin layer of resin which adds only 1% to the weight of the wool. The resin does not alter wool in any significant way except to prevent the fibers from absorbing water. There is an increasing trend to blend wool with other fibers. This improves aesthetics and dyability and provides materials with more durability and comfort.

Only fabrics that have passed the quality tests demanded by the International Wool Secretariat and its U.S. Branch, The Wool Bureau, Inc., are eligible to carry the Woolmark or Woolblend labels (Fig. 6.11).

Woolen garments are as old as the Stone Age, yet as new as today. Wool takes many forms: wovens, knits, piles, felts, each the result of a different process. It's a long way from a bale of raw wool to a beautiful bolt of fabric and ultimately to beautiful, versatile garments (Fig. 6.12).

SELECTED REFERENCES

Publications

Pohle, E. M., Keller, H. R., Ray, H. D., Lineberry, C. T., and Reals, H. C. 1977. Physical properties of grease mohair and related mill products. (Spring and fall clips.) Marketing Research Report 954. Washington: Agricultural Marketing Service, USDA.

Scott, George E. 1982. *The Sheepman's Production Handbook*. Denver: Abegg.

Shelton, M., and Lupton, C. 1986. Where there's wool, there's a way. *Science of Food and Agriculture*, November, p. 6.

Terrill, C. E. 1971. Mohair. In *Encyclopedia of Science and Technology*, 3d Ed. New York: McGraw-Hill.

Visuals

From Fleece to Fashion. Pendleton Woolen Mills, P.O. Box 1691, Portland, OR 97207.

A

B

C

FIGURE 6.12.
Wool sweaters and wool knit suits offer comfort and fashion to the most discerning customers. Courtesy of Woolknit Associates, Inc. (A) Wool sweaters in landscape patterns by Panache and Alps. Knitted in America. (B) The wool suit by Castleberry is the well-dressed delight. Knitted in America. (C) Pattern play in all-American wool sweaters by Pendleton. Knitted in America.

CHAPTER 7

Market Classes and Grades of Livestock, Poultry, and Eggs

The production and movement of more than 80 billion pounds of highly perishable livestock and poultry products to the consumer in the United States is accomplished through a vast and complex marketing system. Marketing is the physical movement, transformation, and pricing of goods and services, and a number of buyers and sellers working to convey livestock and livestock products from the point of production to the point of consumption. Producers need to understand marketing if they are to produce products preferred by the consumer, decide intelligently between various marketing alternatives, understand how animals and products are priced, and eventually raise productive animals profitably.

Market classes and grades have been established to segregate animals, carcasses and products into uniform groups based on preferences of buyers and sellers. The USDA has established extensive classes and grades to make the marketing process simpler and more easily communicated. Use of USDA grades is voluntary. Some packers have their own private grades, and these grades are often used in combination with the USDA grades. An understanding of market classes and grades helps producers recognize quantity and quality of the products they are supplying to consumers.

MARKET CLASSES AND GRADES OF RED MEAT ANIMALS

Slaughter Cattle

Slaughter cattle are separated into classes based primarily on age and sex. Age of the animal has a significant effect on tenderness, with younger animals typically producing more tender meat than older animals. Age classifications for meat from cattle are veal, calf, and beef. Veal is from young calves, 1–3 months of age, with carcasses weighing less than 150 lb. Calf is from animals ranging in age from 3 months to 10 months and carcass weights between 150 and 300 lb. Beef comes from more mature cattle, over 12

TABLE 7.1.   Official USDA Grade Standards for Live Slaughter Cattle and Their Carcasses

Class or Kind	Quality Grades (highest to lowest)	Yield Grades (highest to lowest)
Beef		
Steer and heifer	Prime, choice, select, standard, commercial, utility, cutter, canner	1, 2, 3, 4, 5
Cow	Choice, select, standard, commercial, utility, cutter, canner	1, 2, 3, 4, 5
Bullock	Prime, choice, select, standard, utility	1, 2, 3, 4, 5
Bull	(No designated quality grades)	1, 2, 3, 4, 5
Veal	Prime, choice, select, standard, utility	Not applicable
Calf	Prime, choice, select, standard, utility	Not applicable

months of age, having carcass weights higher than 300 lb. Classes and grades established by the USDA for cattle are based on sex, quality grade, and yield grade, all of which are used in the classification of both live cattle and their carcasses (Table 7.1).

The sex classes for cattle are **heifer, cow, steer, bull,** and **bullock.** Occasionally the sex class of stag is used by the livestock industry to refer to males that have been castrated after their secondary sex characteristics have developed. Sex classes separate cattle and carcasses into more uniform carcass weights, tenderness groups, and how carcasses are processed. **Quality grades** are intended to measure certain consumer palatability characteristics, and yield grades measure amount of fat, lean, and bone in the carcass. Slaughter steers representing some of the eight quality grades and five yield grades are shown in Figs. 7.1 and 7.2.

Quality grades are based primarily on two factors: (a) **maturity** (physiological age) of the carcass, and (b) amount of **marbling.** Maturity is determined primarily by observing bone and cartilage structures. For example, soft, porous, red bones with a maximum amount of pearly-white cartilage characterize A maturity, whereas very little cartilage and hard, flinty bones characterize C, D, and E maturities. As maturity in carcasses increases from A to E, the meat becomes less tender.

Marbling is intramuscular fat or flecks of fat within the lean, and it is evaluated at the exposed rib eye muscle between the 12th and 13th ribs. Nine degrees of marbling, ranging from abundant to practically devoid, are designated in the USDA marbling standards. Various combinations of marbling and maturity that identify the carcass quality grades are shown in Fig. 7.3. Figure 7.4 shows rib eye sections representing several different quality grades and several different degrees of marbling. Note that C maturity starts at 42 months of age. Slaughter cows more than 42 months of age will grade commercial, utility, cutter, or canner regardless of amount of marbling.

Yield grades, sometimes referred to as cuttability grades, measure the quantity of boneless, closely trimmed retail cuts (BCTRC) from the major wholesale cuts of beef (round, loin, rib, and check). BCTRC should not be confused with the total percentage of retail cuts (including hamburger) from a beef carcass. A numerical scale of 1 through 5 is used to rate yield grade, with 1 denoting the highest percentage of BCTRC. Although market-

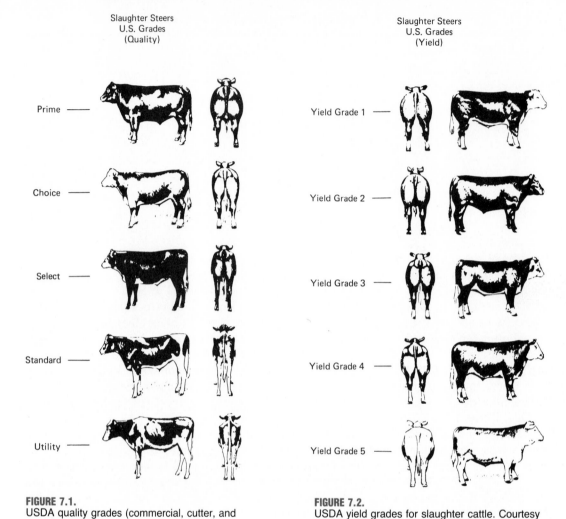

Slaughter Steers
U.S. Grades
(Quality)

Prime

Choice

Select

Standard

Utility

Slaughter Steers
U.S. Grades
(Yield)

Yield Grade 1

Yield Grade 2

Yield Grade 3

Yield Grade 4

Yield Grade 5

FIGURE 7.1.
USDA quality grades (commercial, cutter, and canner omitted). Courtesy of the USDA.

FIGURE 7.2.
USDA yield grades for slaughter cattle. Courtesy of the USDA.

ing communications generally quote yield grades in whole numbers, these yield grades are often shown in tenths in research data and information on carcass contests.

Packers can elect to have carcasses quality graded, but if they are quality graded, they must be yield graded. Table 7.2 shows the yield grades and their respective percentage of BCTRC. As an example, a carcass with a yield grade of 3.0 has a BCTRC of 50%; a 700-lb carcass with this yield grade would produce 350 lb of boneless, closely trimmed retail cuts from the round, loin, rib, and chuck.

Yield grades are determined from the following four carcass characteristics: (a) amount of fat measured in tenths of inches over the rib eye muscle, also known as the *longissimus dorsi* (Fig. 7.5); (b) kidney, pelvic, and heart fat (usually estimated as percent of carcass weight); (c) area of the rib eye muscle, measured in square inches (Fig. 7.6); and (d) hot

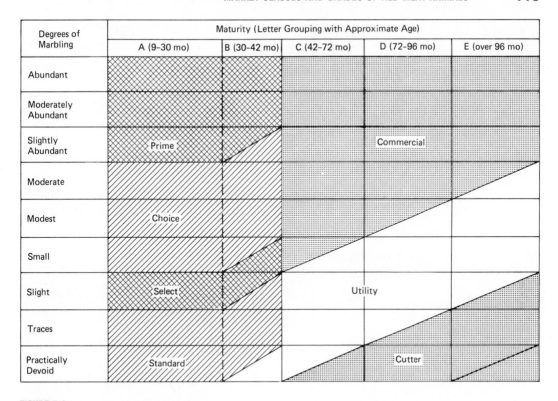

FIGURE 7.3.
Marbling and the maturity of the carcass are the major factors that determine the quality grade of the beef carcass. Courtesy of the USDA.

TABLE 7.2.   Beef Carcass Yield Grades and the Yield of BCTRC (Boneless, Closely Trimmed Retail Cuts) from the Round, Loin, Rib, and Chuck

Yield Grade	BCTRC (%)	Yield Grade	BCTRC (%)	Yield Grade	BCTRC (%)
1.0	54.6	2.8	50.5	4.6	46.4
1.1	54.2	3.0	50.0	4.8	45.9
1.4	53.7	3.2	49.6	5.0	45.4
1.6	53.3	3.4	49.1	5.2	45.0
1.8	52.8	3.6	48.7	5.4	44.5
2.0	52.3	3.8	48.2	5.6	44.1
2.2	51.9	4.0	47.7	5.8	43.6
2.4	51.4	4.2	47.3	—	—
2.6	51.0	4.4	46.8	—	—

Courtesy of the USDA.

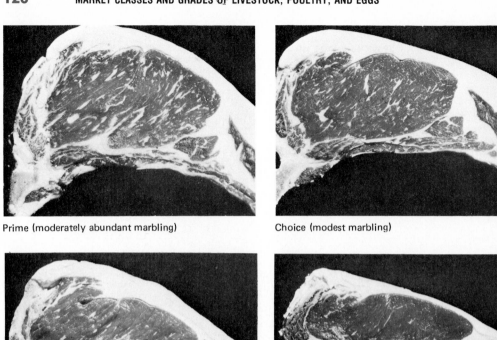

Prime (moderately abundant marbling)

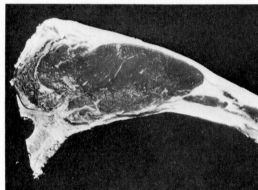

Choice (modest marbling)

Select (slight marbling)

Standard (traces of marbling)

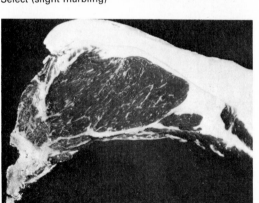

Commercial (moderately abundant marbling; however,
E maturity put it into this grade) See Figure 5-3.

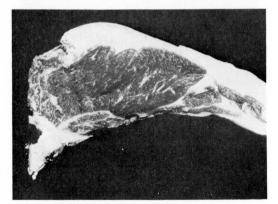

Utility (modest marbling)

FIGURE 7.4.
Exposed ribeye muscles (between twelfth and thirteenth ribs) showing various degrees of marbling of several beef carcass quality grades. Courtesy of The National Live Stock and Meat Board.

¾ the length of the longissimus dorsi muscle

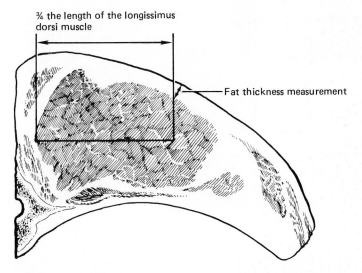

Fat thickness measurement

FIGURE 7.5.
Location of the fat measurement over the ribeye (longissimus dorsi muscle). Courtesy of Dennis Giddings.

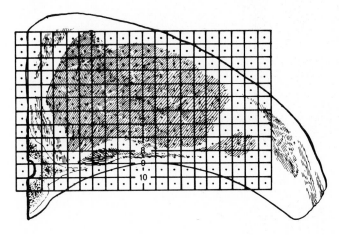

FIGURE 7.6.
Plastic grid is placed over the ribeye muscle to measure the area. Each square represents 0.1 square inch. Courtesy of Dennis Giddings.

carcass weight. The latter measurement reflects amount of intramuscular fat. Generally, as the carcass increases in weight, amount of fat between the muscles increases also.

Measures of fatness in beef carcasses have the greatest effect in determining yield grade. A tentative yield grade can be determined by estimating or measuring the fat of the rib eye muscle. Figure 7.7 shows the five yield grades with varying amounts of fat over the rib eye muscle and the area of the rib eye.

Quality grading and yield grading of beef carcasses by packers are voluntary. Approximately 50% of the 23 billion pounds of carcass beef produced each year is both quality- and yield-graded. Most of the graded beef is slaughter steers and heifers from feedlots. The distribution of quality grades and yield grades is shown in Fig. 7.8.

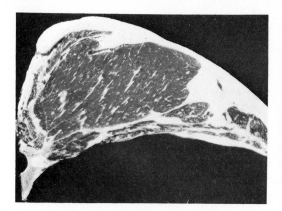

Yield Grade 1
(Fat 0.2 in. ribeye area 13.9 sq in.)

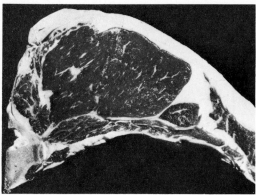

Yield Grade 2
(Fat 0.4 in. ribeye area 12.3 sq in.)

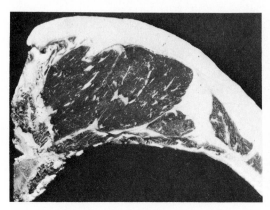

Yield Grade 3
(Fat 0.6 in., ribeye area 11.8 sq in.)

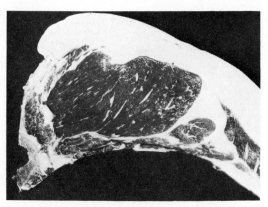

Yield Grade 4
(Fat 0.9 in., ribeye area 10.5 sq in.)

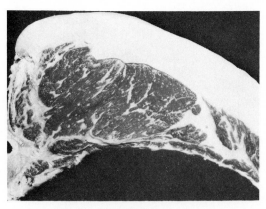

Yield Grade 5
(Fat 1.1 in., ribeye area 10.9 sq in.)

FIGURE 7.7.
The five yield grades of beef shown at twelfth and thirteenth ribs. Courtesy of The National Live Stock and Meat Board.

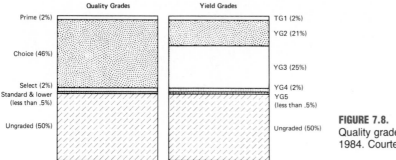

FIGURE 7.8.
Quality grades and yield grades of beef, 1984. Courtesy of the USDA.

Feeder Cattle

The revised 1979 USDA feeder grades for cattle are intended to predict feedlot weight gain and the slaughter weight end point of cattle fed to a desirable fat-to-lean composition. The two criteria used to determine feeder grade are frame size and thickness. The three measures of frame size and thickness are shown in Figs. 7.9 and 7.10. Some examples of feeder cattle grade terminology would be large No. 1, medium No. 2, and small No. 2. Feeder cattle are given a USDA grade of inferior if the cattle are unhealthy or double-muscled. These cattle would not gain satisfactorily in the feedlot.

Although frame size and ability to gain weight in the feedlot are apparently related in the sense that large-framed cattle usually gain fastest, frame size appears to predict more accurately carcass composition or yield grade at different slaughter weights than gaining ability. The USDA feeder grade specifications identify the different liveweights from the three frame sizes when they reach the same thickness of fat over the rib eye (Table 7.3).

Slaughter Swine

Sex classes of swine are **barrow** and **gilt, sow, boar,** and **stag.** Boars and sows are older breeding animals, whereas gilts are younger females that have not produced any young. The barrow is the male pig castrated early in life, and the stag is the male pig castrated after it has developed certain boarlike characteristics. Because of relationships between

TABLE 7.3. Slaughter Weights of Large-, Medium-, and Small-Frame Slaughter Cattle at 0.50 Inches of Fat

	Slaughter Weight	
	---	---
Frame Size	Steers (lb)	Heifers (lb)
Large	>1,200	>1,000
Medium	1,000–1,200	850–1,000
Small	<1,000	<850

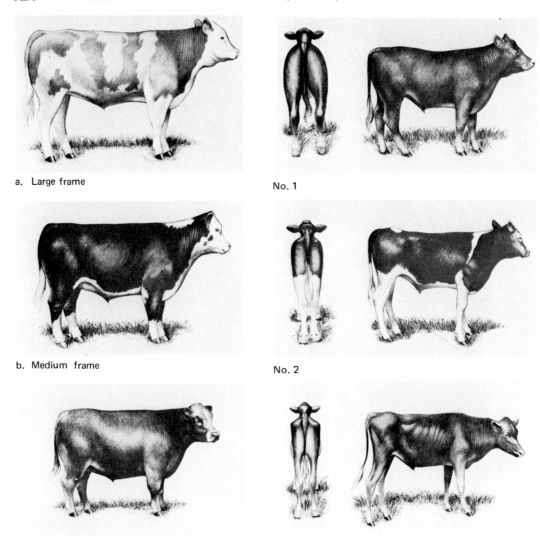

a. Large frame

No. 1

b. Medium frame

No. 2

c. Small frame

No. 3

FIGURE 7.9.
The three frame sizes of the USDA feeder grades
for cattle. Courtesy of the USDA.

FIGURE 7.10.
The three thickness standards of the USDA feeder
grades for cattle. Courtesy of the USDA.

sex and sex condition and the acceptability of prepared meats to the consumer, separate
grade standards have been developed for barrow and gilt carcasses and for sow carcasses.
There are no official grades standards for boar and stag carcasses.

The grade for barrow and gilt carcasses, which are not separated according to sex, are
based on two general criteria: (a) quality characteristics of the lean, and (b) expected
combined yields of the four lean cuts (ham, loin, blade (Boston) shoulder, and picnic
shoulder). There are only two quality grades for lean in a pork carcass: acceptable and

unacceptable. Quality of the lean is assessed by observing the exposed surface of a cut muscle, usually between the 10th and 11th ribs. Acceptable lean is gray-pink in color, is fine, and has fine marbling. Carcasses that have unacceptable lean quality (too dark or too pale, soft, or watery) or bellies too thin for suitable bacon production are graded U.S. Utility, as are soft and oily carcasses. Carcasses with acceptable lean quality are graded U.S. No. 1, U.S. No. 2, U.S. No. 3, or U.S. No. 4. These grades are based on expected yields of the four lean cuts as shown in Table 7.4.

Carcasses differ in their yields of the four lean cuts primarily because of differences in amount of fatness and muscling in relation to skeletal size. Back fat thickness and an evaluation of muscling are the two factors used to determine the numerical yield grade (Table 7.5).

Figure 7.11 shows muscling scores that are used to determine grades.

Feeder Pig Grades

The USDA grades of feeder pigs are U.S. No. 1, 2, 3, 4, and utility (Fig. 7.12). These grades correspond to USDA grades for market swine to be slaughtered at 200–260 lb.

TABLE 7.4. Expected Yields of the Four Lean Cuts, Based on Percent of Chilled Carcass Weight

Grade	Four Lean Cuts (%)
U.S. No. 1	53.0 and higher
U.S. No. 2	50.0 to 52.9
U.S. No. 3	47.0 to 49.0
U.S. No. 4	<47.0

TABLE 7.5. Preliminary Grade Based on Back Fat Thickness Over the Last Rib (assumes average muscle thickness)

Preliminary Grade[a]	Back Fat Thickness (inches)
U.S. No. 1	<1.00
U.S. No. 2	1.00–1.24
U.S. No. 3	1.25–1.49
U.S. No. 4	1.50 and over[b]

[a]Swine with thick muscling qualify for next higher grade; those with thin muscling are downgraded to next lower grade.
[b]Animals with an estimated last-rib back fat thickness of 1.75 inches or over have to be U.S. No. 4 and cannot be graded U.S. No. 3, even with thick muscling.
Source: USDA.

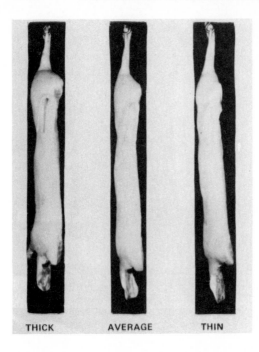

THICK   AVERAGE   THIN

FIGURE 7.11.
Three degrees of muscling in pork carcasses (thick and average represent the minimum accepted for each of the two degrees). Courtesy of the USDA.

Utility grade is for unthrifty or unhealthy feeder pigs. Thus, feeder pig grades combine an evaluation for thriftiness and slaughter potential.

Slaughter Sheep

Slaughter sheep are classified by their sex and maturity. Live sheep and their carcasses are also graded for quality grades and yield grades (Table 7.6).

Lamb carcasses, ranging from approximately 2 to 14 months of age, always have the characteristic break joint on one of their shanks following the removal of their front legs (Fig. 7.13). Mutton carcasses are distinguished from lamb carcasses by the appearance of the spool joint instead of the break joint (Fig. 7.13). The break joint ossifies as the sheep matures. Yearling mutton carcasses ranging from 12 to 25 months of age usually have the

TABLE 7.6.   USDA Maturity Groups, Sex Classes, and Grades of Slaughter Sheep

Maturity Group	Sex Class	Quality Grade (highest to lowest)	Yield Grade (highest to lowest)
Lambs	Ewe, wether, or ram	Prime, Choice, Good, Utility	1, 2, 3, 4, 5
Yearling mutton	Ewe, wether, or ram	Prime, Choice, Good, Utility	1, 2, 3, 4, 5
Mutton	Ewe, wether, or ram	Choice, Good, Utility, Cull	1, 2, 3, 4, 5

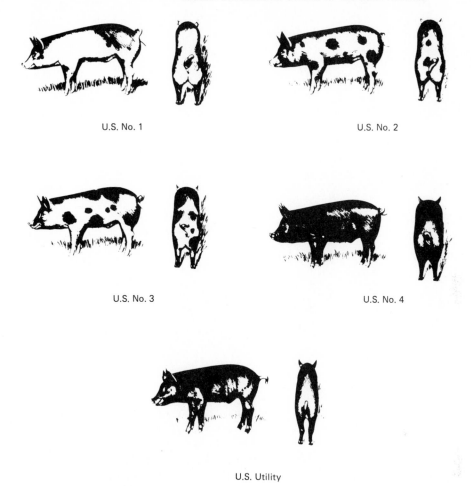

U.S. No. 1

U.S. No. 2

U.S. No. 3

U.S. No. 4

U.S. Utility

FIGURE 7.12.
The USDA grades for feeder pigs. Courtesy of the USDA.

spool joint present but may occasionally have a break joint. Yearling mutton is also distinguished from lamb and mutton by color of the lean (intermediate between pinkish red of lamb and dark red of mutton) and shape of rib bones. Most U.S. consumers prefer lamb to mutton, because it has a milder flavor and is more tender.

Quality grades are determined from a composite evaluation of conformation, maturity, **flank streaking,** and **flank firmness** and fullness. Conformation is an assessment of overall thickness of muscling in the lamb carcass. Maturity of lamb carcasses is determined by bone color and shape and muscle color. Flank streaking (steaks of fat within the flank muscle) predicts marbling since lamb carcasses are not usually ribbed to expose marbling in the rib eye muscle. Flank firmness and fullness are determined by taking hold of the flank muscle with the hand. Most lamb carcasses grade either prime or choice, with few carcasses grading in the lower grades.

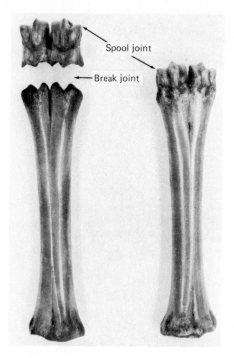

Spool joint

Break joint

FIGURE 7.13.
Break and spool joints. The cannon bone on the left exhibits the typical break joint in which the foot and pastern are removed at the cartilaginous junction. In the cannon bone on the right, the cartilaginous junction has ossified, making it necessary for the foot and pastern to be removed at the spool joint. Courtesy of the USDA.

Lamb carcasses may be quality-graded, yield-graded, or both. Yield grades estimate boneless, closely trimmed retail cuts from the leg, loin, rack, and shoulder. The approximate percent retail cuts for selected yield grades are shown in Table 7.7.

Yield grades are determined from a composite evaluation of leg conformation score, percent kidney and pelvic fat, and fat thickness. Leg score is subjectively evaluated on a numerical scale from 1 to 15 by assessing thickness, fullness, and plumpness of muscling in the hind leg. Figure 7.14 shows examples of two different leg scores. Amount of kidney and pelvic fat is evaluated subjectively and expressed as a percent of the carcass weight. Fat thickness, in tenths of inches, is measured over the center of the rib eye muscle between the 12th and 13th ribs. This measurement of amount of external fat is the most important yield grade factor, since it is a good indicator of amount of fat trimmed

TABLE 7.7.   Lamb Carcass Yield Grades and the Percent Retail Cuts

Yield Grade	Percent Retail Cuts	Yield Grade	Percent Retail Cuts
1.0	49.0	3.5	44.6
1.5	48.2	4.0	43.6
2.0	47.2	4.5	42.8
2.5	46.3	5.0	41.8
3.0	45.4	5.5	41.0

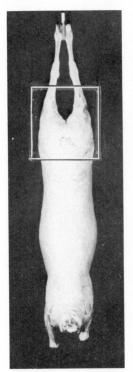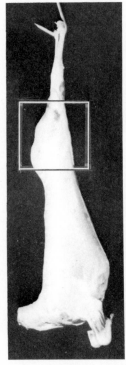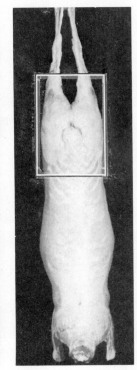

Leg score = Average Prime (score of 14)   Leg score = Low Choice (score of 10)

FIGURE 7.14.
Comparison of two different lamb leg scores used as a partial determining factor for yield grade. Boxes define the area evaluated for thickness, fullness, and plumpness of muscling in the hind leg. Courtesy of The National Live Stock and Meat Board.

in making retail cuts. Figure 7.15 shows cross sections of lamb carcasses representing the five yield grades and amount of fat thickness over the rib eye for each yield grade.

Feeder Lamb Grades

Choice and prime slaughter lambs are produced in relatively large numbers, fed on grass and their mother's milk. Lambs weighing less than 100 lb at weaning are considered feeder lambs. They require additional feeding to produce a more desirable carcass.

There are no official USDA grades for feeder lambs. However, some market reports and industry terminology may refer to prime, choice, good, utility, and cull feeder lambs. These grades are not widely used at the present time, since most lambs, correctly finished for slaughter, are slaughter-graded as choice regardless of their breed or body shape. Most feeder lambs are classified by weight rather than grade. "Lightweight" feeder lambs typically weigh 60–75 lb; "medium weight," 75–85 lb; and "heavy," more than 85 lb. Lambs may also be classified as natives (produced in midwestern and eastern farm states) or westerns (produced in the western U.S.).

Fat .05 in.
Yield Grade 1.7

Fat .10 in.
Yield Grade 2.4

Fat .25 in.
Yield Grade 3.6

Fat .35 in.
Yield Grade 4.6

Fat .45 in.
Yield Grade 5.5

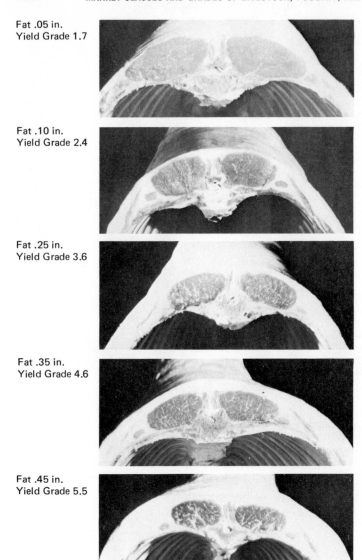

FIGURE 7.15.
The five yield grades of lamb showing the progressive increases in the amount of fat over the loin eye at the twelfth and thirteenth ribs. Courtesy of The National Live Stock and Meat Board.

POULTRY PRODUCTS

Poultry Meat

The class of poultry must be displayed on the package label or a tag on the wing of the bird. The age class indicates tenderness, as meat from younger birds if more tender than that from older birds. Table 7.8 shows the age classification labels for poultry meat.

Meat from the young age classes is usually prepared by broiling, frying, roasting, or barbecuing. The mature, less tender poultry meat is best prepared by baking, stewing, fabricating, or including it in other prepared dishes.

TABLE 7.8. Labels Used to Identify Poultry Age Classes

Type	Young	Mature
Chicken	Young chicken Rock Cornish game hen Broiler Fryer Roaster Capon	Mature chicken Old chicken Hen Stewing chicken Fowl
Turkeys	Young turkeys Fryer-roaster Young hen Young tom	Mature turkey Yearling turkey Old turkey
Ducks	Duckling Young duckling Broiler duckling Fryer duckling Roaster duckling	Mature Old

The grades are "U.S. grade A," "U.S. grade B," and "U.S. grade C" for each of the classes, with A being the highest grade (Fig. 7.16). Carcasses of A quality are free of deformities that detract from their appearance or that affect normal distribution of flesh. They have a well-developed covering of flesh and a well-developed layer of fat in the skin. They are free of pinfeathers and diminutive feathers, exposed flesh on the breast and legs, and broken bones. They have no more than one disjointed bone, and they are practically free of discolorations of the skin and flesh and defects resulting from handling, freezing, or storage. Carcasses of B quality may have moderate deformities. They have a moderate covering of flesh, sufficient fat in the skin to prevent a distinct appearance of the flesh through the skin, and no more than an occasional pinfeather or diminutive feather. They may have moderate areas of exposed flesh and discoloration of the skin and flesh. They may have disjointed parts but no broken bones, and they may have moderate defects resulting from handling, freezing, or storage.

Eggs

The grading of shell eggs involves classifying individual eggs according to established standards. Eggs are graded by sorting them into groups, each group having similar weight and quality characteristics. Table 7.9 shows how eggs are classified according to a size and weight relationship.

The USDA quality standards used to grade individual shell eggs are as follows:

Exterior Quality Factors
Cleanliness of shell
Soundness of shell (cracks, checks, and texture)
Shape

Interior Quality Factors
Albumen thickness
Condition of yolk
Size and condition of air cell
Abnormalities (e.g., blood spots, meat spots)

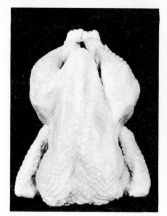
Broiler or fryer, A quality

Broiler or fryer, B quality

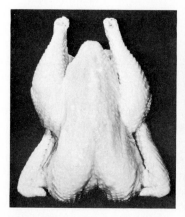
Hen or stewing chicken, A quality

Hen or stewing chicken, B quality

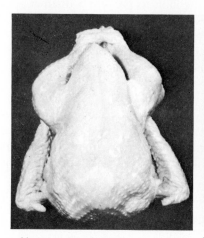
Young turkey, A quality

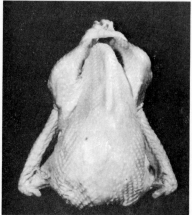
Young turkey, B quality

FIGURE 7.16.
Classes and grades of ready-to-cook poultry. Courtesy of the USDA.

TABLE 7.9. **Consumer Weight Classes of Eggs, Minimum Net Weight Per Dozen**

Size	Ounces (per dozen)
Jumbo	30
Extra Large	27
Large	24
Medium	21
Small	18
Peewee	15

Source: USDA.

Exterior quality factors are apparent from external observation; interior quality factors involve an assessment of contents of the egg. The latter is accomplished through a process called **candling** (visually appraising the eggs while light is shown through them).

Although shell color is not a factor in the U.S. standards and grades, eggs are usually sorted by color and sold either as "whites" or "browns." Eggs sell better when sorted by color and packed separately. Contrary to popular opinion, there are no differences between similar-quality brown and white shelled eggs other than color.

The U.S. standards for quality of individual shell eggs are applicable only to eggs from domestic chickens; these standards are summarized in Table 7.10. Consumer egg grades are U.S. grade AA, U.S. grade A, and U.S. grade B. Figure 7.17 shows selected shields for communicating the quality grade.

FIGURE 7.17.
United States Department of Agriculture grade shields showing two egg quality grades. Courtesy of the USDA.

SELECTED REFERENCES

Publications

Boggs, D. L., and Merkel, R. A. 1984. *Live Animal Carcass Evaluation and Selection Manual.* Dubuque, IA: Kendall/Hunt.

Foods from Animals: Quantity, Quality, and Safety. Council for Agricultural Science and Technology (CAST). Report No. 82, March 1980.

Egg Grading Manual. USDA Agriculture Handbook No. 75, April 1983.

Facts about U.S. Standards for Grades of Feeder Cattle. USDA Agricultural Marketing Service AMS-586, April 1980.

McCoy, J. H. 1979. *Livestock and Marketing.* Westport, CT: AVI.

Meat Evaluation Handbook. 1977. Chicago: The National Live Stock and Meat Board.

TABLE 7.10. Summary of U.S. Standards for Quality of Individual Shell Eggs

Quality Factor	Specifications for Each Quality Factor				
	AA Quality	A Quality	B Quality	Dirty	Check
Shell	Clean, unbroken Practically normal	Clean, unbroken, practically normal	Clean to slightly stained[a]; unbroken; abnormal	Unbroken Adhering dirt or foreign material, prominant stains, moderate stained areas in excess of B quality.	Broken or cracked shell but membranes intact, not leaking[c]
Air cell	1/8 in. or less in depth; unlimited movement and free or bubbly	3/16 in. or less in depth; unlimited movement and free or bubbly	Over 3/16 in. in depth; unlimited movement and free or bubbly		
White	Clear, firm	Clear, reasonably firm	Weak and watery; small blood and meat spots present[b]		
Yolk	Outline slightly defined; practically free from defects	Outline fairly well defined; practically free from defects	Outline plainly visible; enlarged and flattened; clearly visible germ development, but no blood; other serious defects		

[a]Moderately stained areas permitted (1/32 of surface if localized, or 1/16 if scattered).
[b]If they are small (aggregating not more than 1/8 in. in diameter).
[c]Leaker has broken or cracked shell and membranes, and contents are leaking or free to leak.
Source: USDA Egg Grading Manual.

Official United States Standards for Grades of Slaughter Swine (Jan. 14, 1985); Grade of Pork Carcasses (Jan. 14, 1985); Grades of Veal and Calf Carcasses (Oct. 6, 1980); Grades of Lamb, Yearling Mutton and Mutton Carcasses (Oct. 17, 1982); Grades of Carcass Beef and Slaughter Cattle (Mar. 12, 1975). Washington: USDA, Agricultural Marketing Service.

Visuals

Grading Eggs for Quality (sound filmstrip). Vocational Education Productions, California Polytechnic State University, San Luis Obispo, CA 93407.

Egg Grades . . . A Matter of Quality (16-mm film, 12 min). National Audio Visual Center, Order Section, 8700 Edgeworth Dr., Capitol Heights, MD 20743-3701.

United States Standards of Poultry and Poultry Parts (three-part slide series: Part 1—whole fryer; part 2—fryer parts; part 3—turkey). 1986. Learning Resources Center, VPI, Blacksburg, VA 24061.

U.S. Standards for Quality of Individual Shell Eggs (photos illustrating interior and exterior grade factors; chart, 15×22 in.). 1984. Superintendent of Documents, U.S. Government Printing Office, Washington, DC 20402.

CHAPTER 8

Visual Evaluation
of Slaughter
Red Meat Animals

Most red meat animals, ready for slaughter, are purchased at market time based on visual appraisal of their apparent carcass merit. Livestock producers attempt to combine high productivity and visual acceptability in these slaughter animals in anticipation of high market prices and positive net returns.

Productivity of breeding and slaughter meat animals is best identified by using meaningful performance records and effective visual appraisal. Opinions as to relationship of form and function in red meat animals (cattle, sheep, swine) differ widely among producers, who continually discuss the value of visual appraisal and body measurements in breeding programs, in determining market grades, and in defining the so-called "ideal types." It is important to separate true relationships from opinion in the area of animal form and function, which continues to be a controversial topic.

"Type" is defined as an ideal or standard of perfection combining all the characteristics that contribute to an animal's usefulness for a specific purpose. "Conformation" implies the same general meaning as type and refers to form and shape of the animal. Both type and conformation describe an animal according to its external form and shape that can be evaluated visually or measured more objectively with a tape or some other device.

Red meat animals have three productive stages: (a) breeding (reproduction), (b) feeder (growth), and (c) slaughter (carcass or product). It has been well demonstrated that performance records are much more effective than visual appraisal in improving reproduction and growth stages. Visual appraisal does have importance in these stages, primarily in identifying reproductive and skeletal soundness and health status. Visual appraisal has some importance in evaluating productive differences in the carcass stage of productivity. Both performance records (such as back fat probes and sonoray readings for meatiness) and visual appraisal are important in evaluating slaughter animals. Visual appraisal of slaughter red meat animals can be used effectively to predict primarily carcass composition (fat, lean, and bone) when relatively large differences exist.

EXTERNAL BODY PARTS

Effective communication in many phases of the livestock industry requires a knowledge of the external body parts of the animal. The locations of the major body parts for swine, cattle, and sheep are shown in Figs. 8.1–8.3.

LOCATION OF THE WHOLESALE CUTS IN THE LIVE ANIMAL

The next step for effective visual appraisal, after becoming familiar with the external parts of the animal, is to understand where major meat cuts are located in the live animal. Wholesale and retail cuts of the carcasses of beef, sheep, and swine were identified in Chapter 2. A carcass is evaluated after the animal has been slaughtered, eviscerated, and, in the case of beef and swine, split into two halves. The lamb carcass remains as a whole carcass. Furthermore, when it is evaluated, the carcass is hanging by the hind leg from the rail on the packing plant. This makes it difficult to perceive how the carcass would appear as part of the live animal standing on all four legs. Figure 8.4, which shows the location of the wholesale cuts on the live animal, assists in correlating carcass evaluation to live-animal evaluation.

VISUAL PERSPECTIVE OF CARCASS COMPOSITION OF THE LIVE ANIMAL

The carcass is composed of fat, lean (red meat), and bone. The meat industry goal is to produce large amounts of highly palatable lean and minimal amounts of fat and bone.

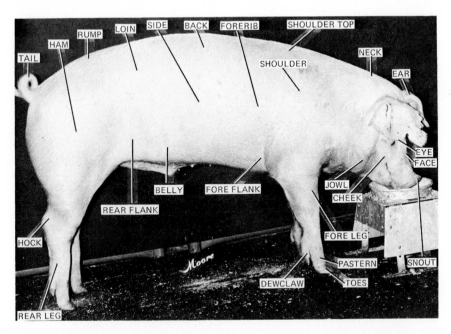

FIGURE 8.1.
The external parts of swine. Courtesy of the Chester White Record Association.

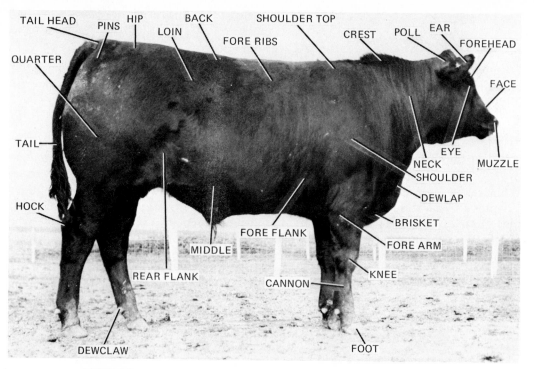

FIGURE 8.2.
The external parts of beef cattle. Courtesy of the American Angus Association.

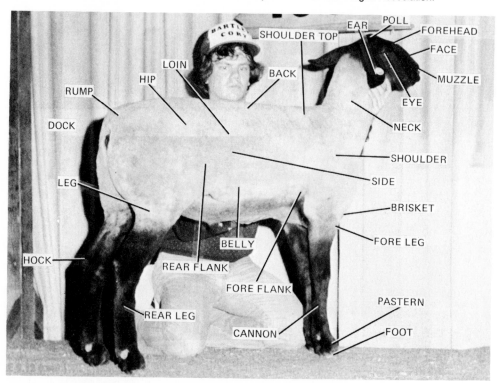

FIGURE 8.3.
The external parts of sheep. Courtesy of *Sheep Breeder and Sheepman* magazine.

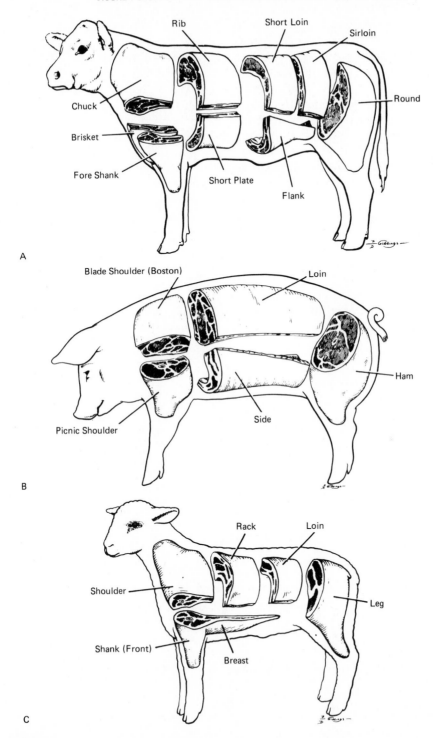

FIGURE 8.4.
Location of the wholesale cuts on the live steer, pig, and lamb. Courtesy of Dennis Giddings.

These composition differences are reflected in the yield grades of beef and lamb and the lean percentage of swine discussed in Chapter 7. Effective visual appraisal of carcass composition requires knowing the body areas of the live animal where fat deposits and muscle growth occur. Plates B and C show the fat and lean composition at several cross sections of a yield grade 2 steer and a yield grade 5 steer. The ribbon on the frozen carcasses shows the location of the cross sections. Cross section 1 takes off the bulge to the round; cross section 2 is in front of the hip bone down through the flank, cross-section 3 is at the twelfth and thirteenth rib, and cross-section 4 is at the point of the shoulder down through the brisket. Note the contrasts in fat (18% vs. 44%) and lean (66% vs. 43%) percentages. Both steers graded choice and were slaughtered at approximately 1,100 lb. The conformation characteristics of the two steers, which are primarily influenced by fat deposits and some muscling differences, are contrasted in Figs. 8.5 and 8.6. The conformation of these two steers is also contrasted to the conformation of a slaughter steer that is underfinished and thinly muscled (Fig. 8.7). All body parts referred to in these figures can be identified in Figs. 8.2 and 8.4, with the exception of the twist. The twist is the distance from the top of the tail to where the hind legs separate as observed from a rear view.

Plates D and E show the cross sections of a yield grade 1 lamb and a yield grade 5 lamb. Cross section 4 is cut through the shoulder area, cross section 5 is at the back, and cross section 6 is through the leg area. Note that many fat deposits and muscle areas are similar to those in Fig. 8.5.

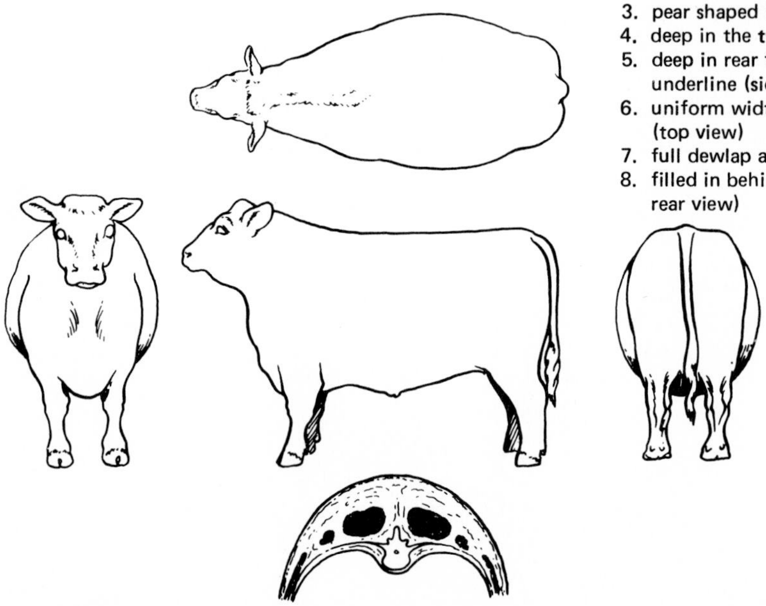

Conformation characteristics

1. short, deep body (side view)
2. flat, wide top (rear view)
3. pear shaped (rear view)
4. deep in the **twist** (rear view)
5. deep in rear flank which makes a straight underline (side view)
6. uniform width or wider in middle of back (top view)
7. full dewlap and brisket (front view)
8. filled in behind the shoulders (side and rear view)

FIGURE 8.5.
Conformation characteristics of slaughter steer typical of yield grade 5. Courtesy of Dennis Giddings.

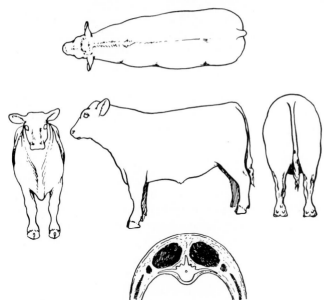

Conformation characteristics

1. relatively long body with moderate body depth (side view)
2. well turned (curved) top (rear view)
3. wide through center of round (rear view)
4. trim in the **twist** (rear view)
5. higher in rear flank than foreflank (side view)
6. wider in the rump than in the middle of the back (top view)
7. trim in dewlap and brisket (front view)
8. slightly dished behind the shoulders (side and rear view)

FIGURE 8.6.
Conformation charactertistics of slaughter steer typical of yield grade 2. Courtesy of Dennis Giddings.

Underfinished

1. narrow, pleated brisket (front view)
2. relatively shallow body (side view)
3. prominent hip and rib bones (side and rear view)
4. markedly dished behind the shoulders (side and rear view)

Thinly muscled

1. narrow through center of round (rear view)
2. flat and narrow forearm (front and side view)

FIGURE 8.7.
Conformation characteristics of slaughter steer that is underfinished (contrast with Figure 8.5) and thinly muscled (contrast with Figure 8.6). Courtesy of Dennis Giddings.

Plates F and G show the cross section of a U.S. No. 1 pig and a U.S. No. 4 pig. Percent lean cuts is used in pork carcass evaluation instead of yield grades. Although the terminology is different, they measure basically the same things. Percent lean cuts are measured by obtaining the total weight of the trimmed boston shoulder, picnic shoulder, loin, and ham (Fig. 8.4) and dividing the total by carcass weight. The lean cuts percentage for the No. 1 would be 53% or more, as contrasted with less than 47% for the U.S. No. 4. The ribbon on the frozen carcass marked the location of the cross sections. Cross section 4 is through the shoulder area, 5 is through the middle of the back, and 6 is through the ham area.

Even though the size and shape of slaughter cattle, sheep, and swine are different, these three species are remarkably similar in muscle structure and fat deposit areas. Therefore, what an individual can learn from visual evaluation of one species can be applied to another species. Regardless of species, an animal that shows a square appearance over the top of its back and appears blocky and deep from a side view has a large accumulation of fat. Fat accumulates first in flank areas, brisket, dewlap, and throat (jowl); between the hind legs; and over the edge of the loin. Fat also fills in behind the shoulders and gives the animal a smooth appearance. Movement of the shoulder blade can be observed when lean cattle and swine walk. The wool covering of sheep can easily camouflage the fat covering. The amount of fat in sheep can be determined by pressing the fingers of the closed hand lightly over the last two ribs and over the spinal processes of the vertebrae. Sheep that have a thick padding of fat in these areas will produce poor yield grading carcasses.

Slaughter red meat animals having an oval turn to the top of their back, while having thickness through the center part of their hind legs (as viewed from the rear), have a high proportion of lean to fat. It is important that slaughter animals have an adequate amount of fat, because thin animals typically do not produce a highly palatable consumer product.

Accuracy in visually appraising slaughter red meat animals is obtained by making visual estimates of yield grades and percent lean cuts and their component parts, and then comparing the visual estimates with the carcass measurements. Accurate visual appraisal can be used as one tool in producing red meat animals with a more desirable carcass composition of lean to fat.

SELECTED REFERENCES

Publications

Beeson, M. W., Hunsley, R. E., and Nordby, J. E. 1970. *Livestock Judging and Evaluation.* Danville, IL: Interstate.

Boggs, D. L., and Merkel, R. A. 1984. *Live Animal Carcass Evaluation and Selection Manual.* Dubuque, IA: Kendall/Hunt.

Crouse, J. E., Dikeman, M. E., and Allen, D. M. 1974. Prediction of beef carcass composition and quality by live-animal traits. *Journal of Animal Science* 38:264.

Kauffman, R. G., Grummer, R. H., Smith, R. E., Long, R. A., and Shook, G. 1973. Does live-animal and carcass shape influence gross composition? *Journal of Animal Science* 37:1112.

Visuals

The Livestock Judging Slide Sets (Beef, Sheep and Swine). Vocational Education Production, California Polytechnic State University, San Luis Obispo, CA 93407.

Steer Judging (video tape No. 15); *Sheep Judging* (video tape No. 16); *Swine Judging* (video tape No. 40). Illinois Film Center, 1325 South Oak St., Champaign, IL 61820.

COLOR PLATE SECTION

CARCASS COMPOSITION

The proper ratio of fat to lean in an animal carcass is a controversial issue that has see-sawed back and forth in the minds of producers, feeders, and packers since the industry began. At the turn of the century, fat-type animals such as the lard hog, baby beef, and fat lamb were the cornerposts of that era's livestock production system. Today, however, vegetable oils have replaced lard, lean beef has replaced the fatted calf, and even the lamb has been stretched longer and leaner through modern production techniques. Today's consumer demands lean meat, and today's grading system reflects those demands.

By visually appraising a hanging carcass, one can quickly identify the relationship of lean to fat. Appraisal of the live animal, however, can be a more difficult and discriminating challenge. The following set of photographs graphically demonstrates a lean/fat ratio both on individual animals and as a comparison between different yield grades. The pictures are courtesy of Iowa State University and Colorado State University, where the entire carcasses were frozen. This process allowed technicians to remove different layers of the carcass such as the hide, fat, and muscling and to section entire carcasses for easy comparison purposes. The pictures are organized by color plates, with individual captions for each photo presented here by plate and photo number.

Plate A compares body conformation of different yield grades in both live animals and their carcasses.

1. Thickness of muscling in two medium-framed feeder steers. The steer on the left has more thickness of lean meat in the round than the steer on the right.
2 & 3. Pounds of fat trimmed from one-half of the body of a yield grade 4 steer and a yield grade 2 steer. Black stripes are hide and fat which remain at key locations on the body.
4. Rear view showing how the fat increases in thickness from the middle of the back to the edge of the loin.
5. Two yearling Hereford bulls of approximately the same weight. Breeding cattle can be visually appraised for fat and lean compositional differences. The bull on the left would sire slaughter steers having yield grade 4 or 5 carcasses, while the bull on the right would sire yield grade 1 or 2 steers. This assumes that the bulls would be bred to similar frame-size cows, and the steers slaughtered at approximately 1,150 pounds. Compare to plates B and C.

Plate B & C compare fat to lean composition on a yield grade 2 steer (plate B) and a yield grade 5 steer (plate C).

1. Live animal—side view
2. Numbered ribbons indicate where following cross sections were made

3, 4, 5, & 6. Cross sections from rump, hip, mid-carcass, and shoulder
7. The percentages of fat, lean, and bone found in this carcass.

Plate D & E compare the fat to lean composition of lamb carcasses on yield grade 1 lamb (plate D) and yield grade 5 lamb (plate E).

1. Side view
2. Rear view
3. Numbered ribbons indicate where following cross sections were made
4, 5, & 6. Shows cross sections from shoulder, mid-carcass and rump

Plate F & G compare fat to lean composition of a U.S. No. 1 pig (plate F) with a U.S. No. 4 pig (plate G).

1. Live animal—side view
2. Rear view
3. Numbered ribbons indicate where following cross sections were made
4, 5, & 6. Shows cross sections from shoulder, mid-carcass and rump

Plate H shows the composition and appearance of a steer carcass at various stages of removal of hide, fat, and muscle.

1. Live steer with hair clipped from one side
2. Frozen steer with hide removed and fat exposed
3. Fat removed from one-half of steer's body
4. Rear view with fat removed from the left side. Note cod and twist fat on the right side.
5. Skeleton of the beef animal after all the muscle has been removed.

1

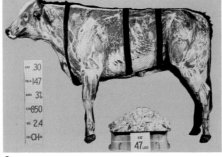

2

3

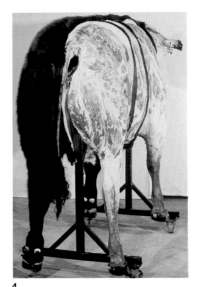

5

4

PLATE A.
Courtesy of Colorado State University.

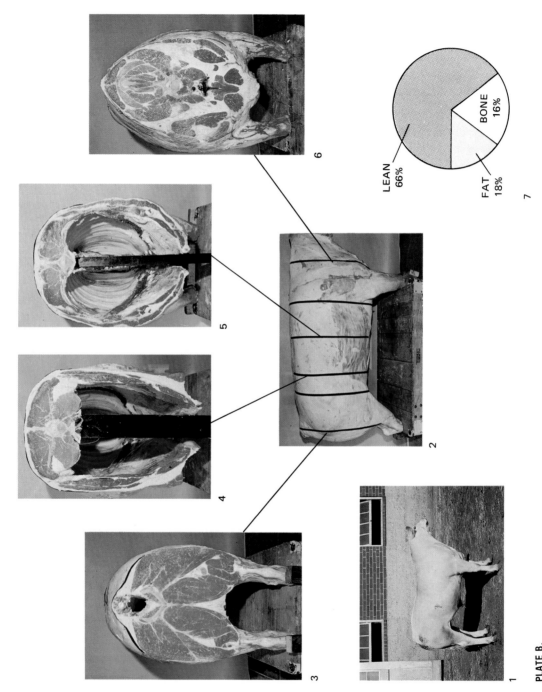

LEAN 66%

BONE 16%

FAT 18%

PLATE B.
Courtesy of Iowa State University.

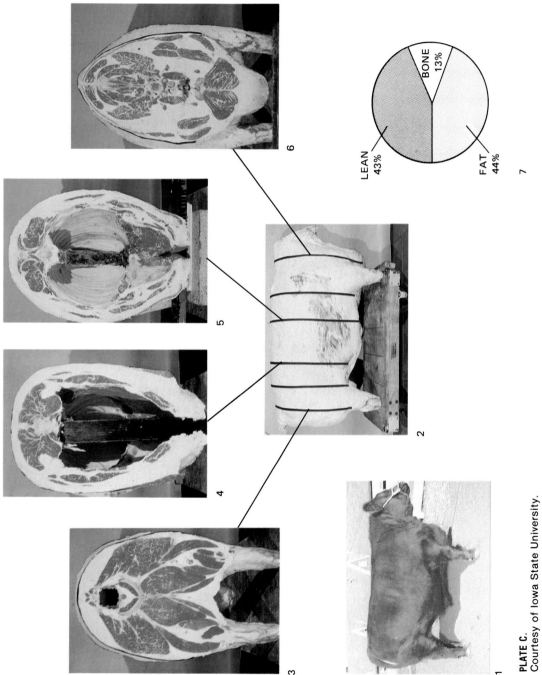

LEAN 43%

BONE 13%

FAT 44%

7

6

5

4

3

2

1

PLATE C.
Courtesy of Iowa State University.

1

2

4

5

6

3

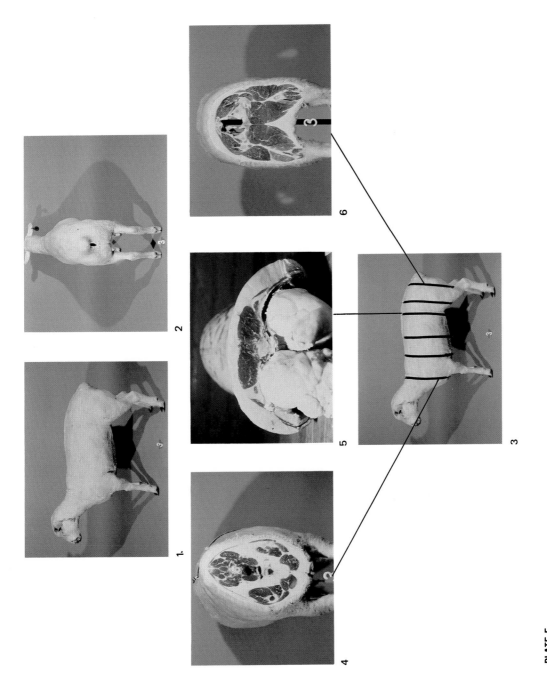

PLATE E.
Courtesy of Iowa State University.

PLATE F.
Courtesy of Iowa State University.

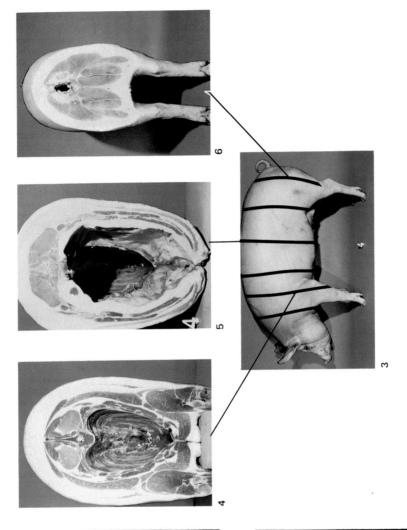

PLATE G.
Courtesy of Iowa State University.

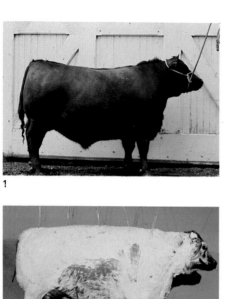

1

2

3

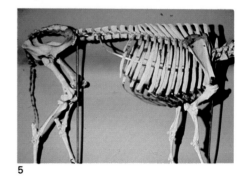

4

5

PLATE H.
Courtesy of Iowa State University.

CHAPTER 9

Reproduction

Reproductive efficiency in farm animals, as measured by number of calves or lambs per 100 breeding females or numbers of pigs per litter, is a trait of great economic importance in farm animal production. It is essential to understand the reproductive process in the creation of new animal life, because it is a focal point of overall animal productivity. Producers who manage animals for high reproductive rates must understand the production of viable sex cells, estrous cycles, mating, pregnancy, and birth.

FEMALE ORGANS OF REPRODUCTION AND THEIR FUNCTION

Figures 9.1 and 9.2 show the reproductive organs of the cow and sow. The female reproductive anatomy of the various farm animal species is similar, although there are a few obvious differences.

The organs of reproduction of the typical female farm mammal include a pair of ovaries, which are suspended by ligaments just back of the kidneys, and a pair of open-ended tubes, the oviducts (also called the Fallopian tubes), which lead directly into the uterus (womb). The uterus itself has two horns, or branches, that in farm mammals merge together at the lower part into one structure so that the lower opening, or exit, from the uterus is a canal. This canal is called the cervix. Its surface is fairly smooth in the mare and the sow, but is folded in the cow and ewe. The cervix opens into the vagina, a relatively large canal or passageway that leads posteriorly to the external parts, which are the vulva and clitoris. The urinary bladder empties into the vagina through the urethral opening.

Ovaries

Ovaries produce ova (female sex cells, also called eggs) and the female sex hormones, estrogen and progesterone. Each ovum (Fig. 9.3) develops in a recently formed follicle within the ovary (Fig. 9.4). Some tiny follicles develop and ultimately attain maximum

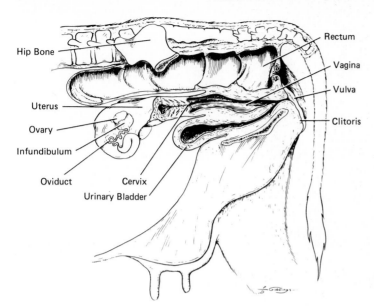

Hip Bone

Uterus

Ovary

Infundibulum

Oviduct   Cervix

Urinary Bladder

Rectum

Vagina

Vulva

Clitoris

FIGURE 9.1.
Reproductive organs of the cow.
Courtesy of Dennis Giddings.

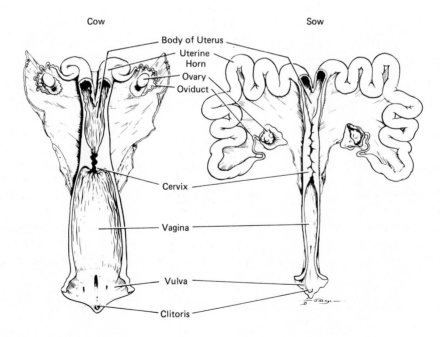

Cow   Sow

Body of Uterus

Uterine
Horn

Ovary

Oviduct

Cervix

Vagina

Vulva

Clitoris

FIGURE 9.2.
A dorsal view of the reproductive organs of the cow and the sow. The most noticeable difference is the longer uterine horns of the sow compared to the cow. Courtesy of Dennis Giddings.

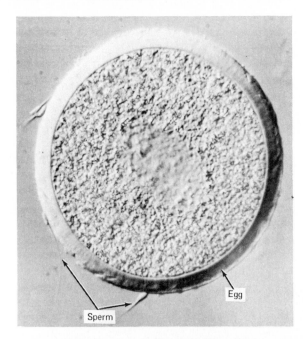

FIGURE 9.3.
Bull sperm and cow egg, each magnified 300 ×.
Note the comparative sizes. Each is a single cell
and contains one-half the chromosome number
typical of other body cells. Courtesy of Colorado
State University.

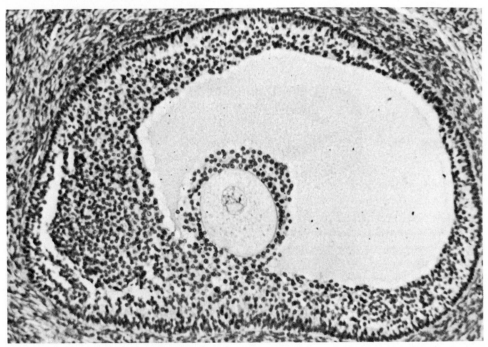

FIGURE 9.4.
The large structure, outlined with a circle of dark cells, is a follicle located on a cow's ovary (magnified 265 ×)
The smaller circle, near the center, is the egg. The large light-gray area is the fluid that fills the follicle. When the
follicle ruptures, the fluid will wash the egg into the oviduct. Courtesy of Colorado State University.

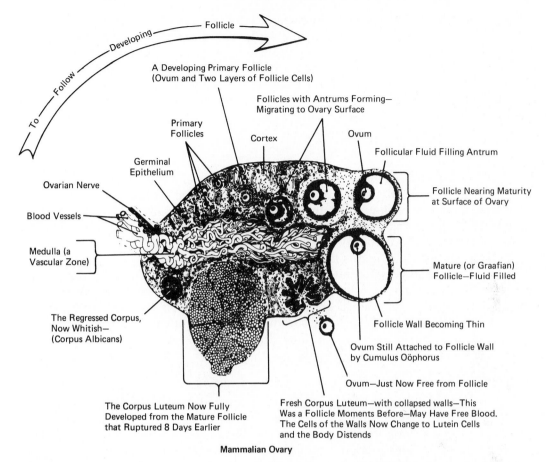

Mammalian Ovary

FIGURE 9.5.
A cross section of the bovine ovary showing how a follicle develops to full size and then ruptures, allowing the egg to escape. The follicle then becomes a "yellow body" (corpus luteum), which is actually orange-colored in cattle. The corpus luteum degenerates in time and disappears. Of course, many follicles cease development, stop growing, and disappear without ever reaching the mature stage. From Bone, J. F., Animal Anatomy and Physiology, 4th ed., Corvallis: Oregon State University Book Stores, © 1975.

size, about 20–35 mm in diameter, after having migrated from deep in the ovary to the surface of the ovary. These growing follicles produce estrogens. These mature (Graafian) follicles rupture, thus freeing the ovum (ovulation). Many of the tiny follicles grow to various stages, cease growth, deteriorate, and are absorbed (Fig. 9.5).

After the ovum escapes from the mature follicle, cells of the follicle change into a corpus luteum, or "yellow body." The corpus luteum produces progesterone, which becomes a vitally important hormone if pregnancy is maintained.

Oviducts

Immediately after ovulation, the ova are caught by the infundibulum of the oviduct. The ova are tiny, 200 μm or less in diameter (200 μm = $\frac{1}{5}$ mm), which is approximately the size of a dot made by a sharp pencil. Sperm are transported through the uterus into the oviduct after the female is inseminated (naturally or artificially). Therefore the oviducts are the sites where ova and sperm meet and where fertilization takes place. After fertilization, 3–5 days is required in cows and ewes and probably about the same amount of time in other farm animals for the ova to travel down the remaining two-thirds of the oviduct. From the oviduct, the newly developing embryos pass to the uterus and soon attach to it.

Uterus

The uterus varies in shape from the type that has long, slender left and right horns, as in the sow, to the type that is primarily a fused body with short horns, as in the mare. In the sow, the embryos develop in the uterine horn; in the mare, the embryo develops in the body of the uterus. Each surviving embryo develops into a fetus and remains in the uterus until parturition (birth).

The lower outlet of the uterus is the cervix, an organ composed primarily of connective tissue that constitutes a formidable gateway between the uterus and the vagina. Like the rest of the reproductive tract, the cervix is lined with mucosal cells. These cells make significant changes as the animal goes from one estrous cycle to another and during pregnancy. The cervical passage changes from one that is tightly closed or sealed in pregnancy to a relatively open, very moist canal at the height of estrus.

Vagina

The vagina serves as the organ of copulation at mating and as the birth canal at parturition. Its mucosal surface changes during the estrous cycle from very moist when the animal is ready for mating to almost dry, even sticky, between periods of heat. The tract from the urinary bladder joins the posterior ventral vagina; from this juncture to the exterior vulva, the vagina serves the dual role of a passageway for the reproductive and urinary systems.

Clitoris

A highly sensitive organ, the clitoris is located ventrally and at the lower tip of the vagina. The clitoris is the homologue of the penis in the male (i.e., it came from the same embryonic source as the penis). Some research indicates that clitoral stimulation or massage, following AI in cattle, will increase the chance of conception, but it has not been verified.

Reproduction in Poultry Females

The hen differs from farm mammals in that the young are not suckled. The egg is layed outside the body, and there are no well-defined estrous cycles or pregnancy. Since eggs are an important source of human food, hens have been selected and managed to lay eggs consistently throughout the year.

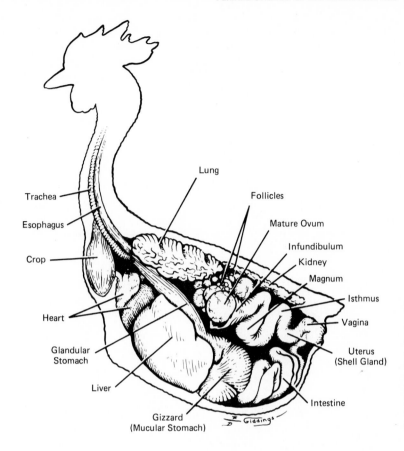

FIGURE 9.6.
Reproductive organs of the hen in relation to other body organs. The single ovary and oviduct are on the hen's left side; an underdeveloped ovary and oviduct are sometimes found on the right side, having degenerated in the developing embryo. Courtesy of Dennis Giddings.

The anatomy of the reproductive tract of the hen is shown in Figs. 9.6 and 9.7. At hatching time, the female chick has two ovaries and two oviducts. The right ovary and oviduct do not develop. Therefore the sexually mature hen has only a well-developed left ovary and oviduct. The ovary appears as a cluster of tiny gray eggs or yolks in front of the left kidney and attached to the back of the hen. The ovary is fully formed, although very small, when the chick is hatched. It contains approximately 3,600–4,000 miniature ova. As the hen reaches sexual maturity, some of the ova develop into mature yolks (yellow part of the laid egg). The remaining ova are variable in size from the nearly mature to those of microscopic size.

The oviduct is a long, glandular tube leading from the ovary to the cloaca (common opening for reproductive and digestive tracts). The oviduct is divided into five parts: The infundibulum (3–4 in. long), which receives the yolk; the magnum (approximately 15 in. long), which secretes the thick albumen, or white of the egg; the isthmus (about 4 in. long), which adds the shell membranes; the uterus (approximately 4 in. long), or shell gland, which secretes the thin white, the shell, and the shell pigment; and the vagina (about 2 in. long).

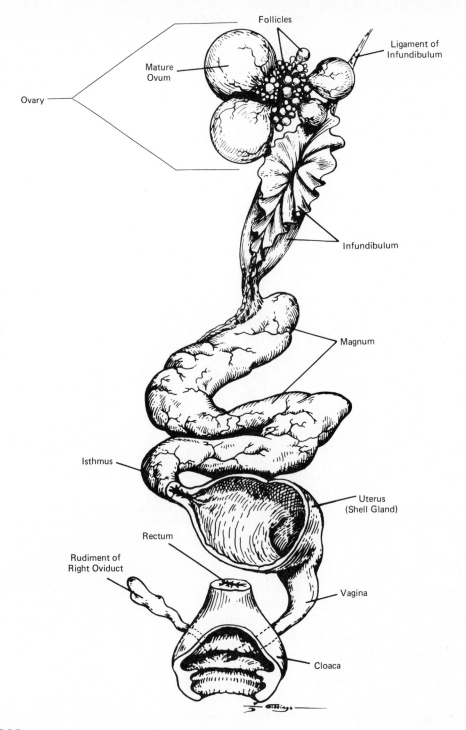

FIGURE 9.7.
Reproductive organs of the hen. Sections of the uterus and cloaca are cut away to better view internal structure.
Courtesy of Dennis Giddings.

Ovulation is the release of a mature yolk (ovum) from the ovary. When ovulation occurs, the infundibulum engulfs the yolk and starts it on its way through the 25- to 27-inch oviduct. The yolk moves by peristaltic action through the infundibulum into the magnum area in about 15 min.

During the 3-hour passage through the magnum, more than 50% of the albumen is added to the yolk. The developing egg passes through the isthmus in about $1\frac{1}{4}$ hours. Here water and mineral salts and the two shell membranes are added. In the 21-hour stay in the uterus, the remainder of the albumen is added, followed by the addition of shell and shell pigment. Moving finally into the vagina, the fully formed egg enters the cloaca and is laid. The entire time from ovulation to laying is usually slightly more than 24 hours. About 30 min after a hen has laid an egg, she releases another yolk into the infundibulum, and it will likewise travel the length of the oviduct.

After the fertilized egg is incubated for 21 days, the chick is hatched. The egg is biologically structured to support the growth and life processes of the developing chick embryo during incubation and for 3–4 days after the chick is hatched.

There are several egg abnormalities which occur because of factors affecting ovulation and the developmental process. Double-yolked eggs result when two yolks are released about the same time or when one yolk is lost into the body cavity for a day and is picked up by the funnel when the next day's yolk is released. Yolkless eggs are usually formed about a bit of tissue that is sloughed off the ovary or oviduct. This tissue stimulates the secreting glands of the oviduct and a yolkless egg results. The abnormality of an egg within an egg is due to reversal of direction of an egg by the wall of the oviduct. One day's egg is added to the next day's egg, and shell is formed around both. Soft-shelled eggs generally occur when an egg is laid prematurely and insufficient time in the uterus prevents the deposit of the shell. Thin-shelled eggs may be caused by dietary deficiencies, heredity, or disease. Glassy- and chalky-shelled eggs are caused by malfunctions of the uterus of the laying bird. Glassy eggs are less porous and will not hatch but may retain their quality.

MALE ORGANS OF REPRODUCTION AND THEIR FUNCTION

Figures 9.8 and 9.9 show the reproductive organs of the bull and boar. The organs of reproduction of a typical male farm mammal include two testicles, which are held in the scrotum. Male sex cells (called sperm or spermatozoa) are formed in the tiny seminiferous tubules of the testicles. The sperm from each testicle then pass through very small tubes into the epididymis, which is a highly coiled tube that is held in a covering on the exterior of the testicle. Each epididymal tube leads to a larger tube, the vas deferens (also called the ductus deferens). The two vasa deferentia converge at the upper end of the urethral canal, close to where the urinary bladder opens into the urethra. The urethra is the large canal that leads through the penis to the outside of the body. The penis has a triple role. It serves as a passageway for semen and urine and it is the organ of copulation.

The left and right parts of the seminal vesicles, which lie against the urinary bladder, consist of glandular tissue that secretes a substance into the urethra which supplies nutrients for the sperm. The prostate gland contains 12 or more tubes, each of which empties into the urethra. Another gland, the bulbourethral (Cowper's) gland, which also empties its secretions into the urethral canal, is posterior to (behind) the prostate.

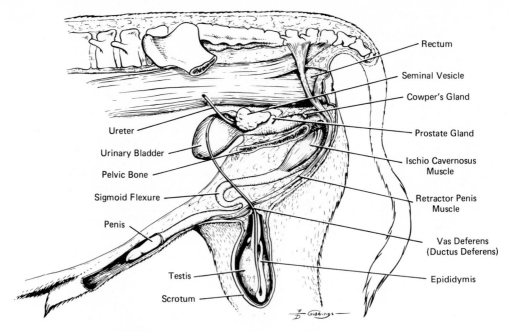

FIGURE 9.8.
Reproductive organs of the bull. Courtesy of Dennis Giddings.

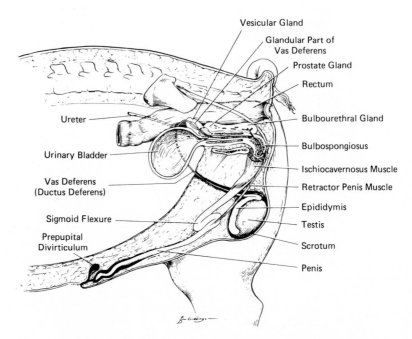

FIGURE 9.9.
Reproductive organs of the boar. Courtesy of Dennis Giddings.

Testicles

The testicles produce (a) the sperm cells that fertilize the ova of the female, and (b) a hormone called testosterone that conditions the male so that his appearance and behavior are masculine. Details of the structure of the spermatozoa of the bull are shown in Figs. 9.10–9.12.

If both testicles are removed (as is done in castration), the individual loses his sperm factory and is sterile. Also, without testosterone his masculine appearance is not apparent, and he approaches the status of a neuter—an individual whose appearance is somewhere between that of a male and that of a female. A steer does not have the crest, or powerful neck, of the bull. The bull has heavier, more muscular shoulders and a deeper voice than a counterpart steer. If a bull calf is castrated, the reproduction organs such as the vas deferens, seminal vesicles, and prostate and bulbourethral glands all but cease further development. If castration is done in a mature bull, the remaining genital organs tend to shrink in size and in function.

Within each testicle, sperm cells are generated in the seminiferous tubules, and tes-

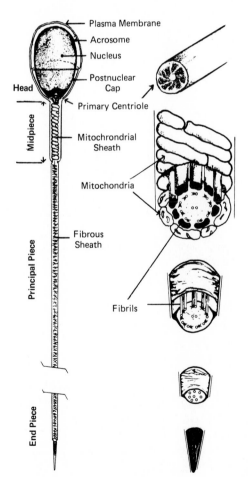

FIGURE 9.10.
A diagrammatic sketch of structure of bull sperm. Courtesy of Dr. Arthur S. H. Wu, Oregon State University.

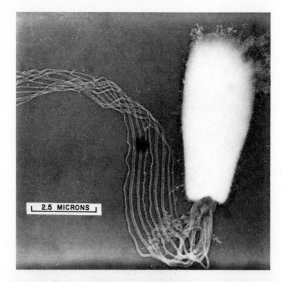

2.5 MICRONS

FIGURE 9.11.
A spermatozoa (sperm cell) from a bull (7000 ×)
viewed through the electron microscope after treatment with 0.15 N NaOH at 25° C for 16 h. Note how
the covering membrane of the neck region of the
sperm has been removed, exposing the nine fibrils of
the axial filament. Three filaments are larger than the
rest. From Wu, A. S. H., and McKenzie, F. F. "Micro-
structure of Spermatozoa After Denudation as Revealed by Electron Microscope," *Journal of Animal
Science* 14(4):1151–66, 1955.

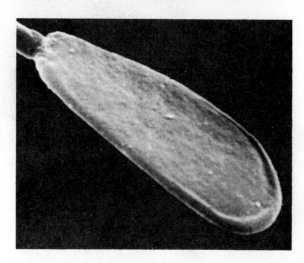

FIGURE 9.12.
Bull sperm (12,000 ×) viewed with the scanning
electron microscope showing the depth of the
sperm head covered by the raised acrosome.
Courtesy of Dr. Arthur S. H. Wu, Oregon State
University.

tosterone is apparently produced in the cells between the tubules, called Leydig cells or
interstitial cells (Fig. 9.13).

Epididymis

The epididymis is the storage site for sperm cells, which enter it from the testicle to
mature. In passing through this very long tube (30–35 m in the bull, longer in the boar
and stallion), the sperm acquire the capacity to fertilize ova. Sperm taken from the part
of the epididymis nearest the testicle and inseminated into females are not likely to be
able to fertilize ova, whereas sperm taken near the vas deferens have the capacity to
fertilize.

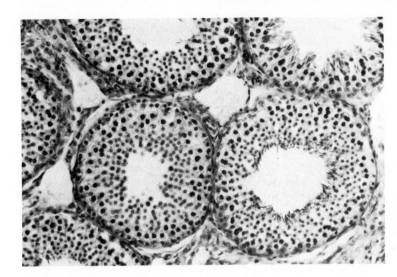

FIGURE 9.13
A cross section through the seminiferous tubules of the testis of the bull (magnified 240 ×). The tubule in the lower right-hand corner demonstrates the more advanced stages of spermatogenesis as the spermatids are formed near the lumen (opening) of the tubule. Courtesy of Colorado State University.

In the sexually mature male animal, sperm reside in the epididymis in large numbers. In time, the sperm mature, then degenerate, and are absorbed in the part of the epididymis farthest from the testicle unless they have been moved on into the vas deferens and have been ejaculated.

Scrotum

The scrotum is a two-lobed sac that contains and protects the two testicles. It also regulates temperature of the testicles, maintaining them at a temperature lower than body temperature—1.6–3.9°C lower in the bull and 5–7°C lower in the ram and goat. When the environmental temperature is low, the tunica dartos muscle of the scrotum contracts, pulling the testicles toward the body and its warmth; when the environmental temperature is high, this muscle relaxes, permitting the testicles to drop away from the body and its warmth. This heat-regulating mechanism of the scrotum begins at about the time of puberty. Hormone function precedes puberty by 40–60 days.

When the environmental temperature is elevated such that the testicles cannot cool sufficiently, the formation of sperm is impeded, and a temporary condition of lowered fertility results. Providing shade, keeping the males in the shade during the heat of the day, even providing air conditioning, are ways to manage and prevent this temporary sterility.

Vas Deferens

The vas deferens is essentially a transportation tube that carries the sperm-containing fluid from each epididymis to the urethra. The vasa deferentia join the urethra near its origin as the urethra leaves the urinary bladder. In the mature bull, the vas deferens is about 3 mm in diameter except in its upper end, where it widens into a reservoir, or ampulla, about 10–17 cm long and 1 cm wide.

Under the excitement of anticipated mating, the secretion loaded with spermatozoa from each epididymis is propelled into each vas deferens and accumulates in the ampulla of the deferent duct. This brief accumulation of semen in the ampulla is an essential part of sexual arousal. The sperm reside briefly in the ampula until the moment of ejaculation, when the contents of each ampulla are pressed out into the urethra, and then through the urethra and the penis en route to their deposition in the female tract.

The ampulla is found in the bull, stallion, goat, and ram—species that ejaculate rapidly. It is not present in the boar or dog, animals in which ejaculation normally takes several minutes (8–12 min is typical in swine). In such animals, sperm in numbers travel all the way from the epididymis through the entire length of the vas deferens and the urethra. On close observation of the boar at the time of mating, one can see the muscles over the scrotum quivering rhythmically as some of the contents of each epididymis are propelled into the vasa deferentia and on into the urethra. This slow ejaculation of the boar contrasts to the sudden expulsion of the contents of the ampulla of the vas deferens at the height of the mating reaction, or orgasm, in the bull, stallion, ram, and goat.

Urethra

The urethra is a large, muscular canal extending from the urinary bladder. The urethra runs posteriorly through the pelvic girdle and curves downward and forward through the full length of the penis. Very near the junction of the bladder and urethra, tubes from the seminal vesicles and tubes from the prostate gland join this large canal. The bulbourethral gland joins the urethra at the posterior floor of the pelvis.

Accessory Sex Glands

The seminal vesicles and prostate and bulbourethral glands are known as the accessory sex glands. Their primary functions are to add volume and nutrition to the sperm-rich fluid coming from the epididymis. Semen consists of two components: the sperm and the fluids secreted by the accessory sex glands. The semen characteristics of some farm animals are shown in Table 9.1.

TABLE 9.1.   Semen Characteristics of Several Male Animals

Animal	Semen Characteristics			
	Volume per Ejaculate (ml)	Composition of Ejaculate	Sperm Concentration per ml $\times 10^9$	Total Sperm per Ejaculate $\times 10^9$
Bull (cattle)	3–10	Single fraction	0.8–1.2	4–18
Ram (sheep)	0.5–2.0	Single fraction	2–3	1–4
Boar (swine)	150–250	Fractionated	0.2–0.3	30–60
Stallion (horse)	40–100	Fractionated	0.15–0.40	8–50
Buck (goat)	0.5–2.5	—	2.0–3.5	1–8
Dog (dog)	1.0–5.0	—	2–7	4–14
Buck (rabbit)	0.5–6.5	—	0.3–1.0	1.5–6.5
Tom (turkey)	0.1–0.7	—	8–30	1–20
Cock (chicken)	0.1–1.5	—	0.4–1.5	0.05–2.0

Seminal Vesicles

The seminal vesicles are sizable organs that lie over the neck of the bladder and on either side of the pelvic urethra. They open into the urethra near the openings of the vasa deferentia. Seminal vesicles are prominent in the bull, stallion, ram, and goat but are absent in the dog and cat. In the bull, the seminal vesicles are about 10 cm long and 2.5 cm in diameter. The boar's seminal vesicles are large, thin-walled, pyramid-shaped glands in which the apex points posteriorly. The combined weight of the two glands in the boar varies from 150 to 850 g and the contents from 38 g to more than 500 g.

Seminal vesicles contribute nutrients for the sperm including ascorbic acid, citric acid, inorganic phosphorus, acid-soluble phosphoris, and fructose.

Prostate Gland

The prostate gland lies near the neck of the bladder. In the bull, the main part of the prostate is about 37 mm across and 12 mm in diameter. Additional prostate tissue is scattered among the muscles of the pelvic urethra. The prostate empties into the urethra through several small openings in the muscular wall of the urethra. It supplies antagglutin and minerals to the ejaculate.

Bulbourethral Glands

The bulbourethral glands (sometimes referred to as Cowper's glands) are located on either side of the pelvic urethra, just posterior to the urethra-penis where the urethra-penis dips downward in its curve. Except for the boar, they are small in farm animals. In the bull, they are about 25×12 mm, but in the boar, they are comparably larger (15×3–5 cm). The secretion from the bulbourethral glands is thick and viscous, very slippery (lubricating), and whitish in color. It has a high sialoprotein content. This sialoprotein is involved in the formation of the gelatinous fraction of the semen in the boar.

Penis

The penis is the organ of copulation. It provides a passageway for semen and urine. It is a muscular organ characterized especially by its spongy, erectile tissue that fills with blood under considerable pressure during periods of sexual arousal, making the penis rigid and erect.

The penis of the bull is about 1 m in length and 3 cm in diameter, tapering to the free end, or glans penis. In the bull, boar, and ram, the penis is S-shaped when relaxed. This S curve, or sigmoid flexure, becomes straight when the penis is erect. The S curve is restored after copulation, when the relaxing penis is drawn back into its sheath by a pair of retractor penis muscles. The stallion penis has no S curve; it is enlarged by engorgement of blood in the erectile tissues.

The free end of the penis is termed the glans penis. The opening in the ram penis is at the end of a hairlike appendage that extends about 20–30 mm beyond the larger penis proper. This appendage also becomes erect and during ejaculation whirls in a circular fashion, depositing semen in the anterior vagina. It does not regularly penetrate the ewe's cervix, as some investigators claim it does. Only a small portion of the penis of the bull, boar, ram, and goat extends beyond its sheath during erection. The full extension awaits the thrust after entry into the vagina has been made. The stallion and ass usually extend the penis completely before entry into the vagina.

It is emphasized here that all these accessory male sex organs depend on testosterone for their tone and normal function. This dependence is especially apparent when the testicles are taken away (at castration); the usefulness of the accessory sex organs is then diminished or even terminated.

Reproduction in Male Poultry

The reproductive tract of male poultry is shown in Fig. 9.14. There are several differences when compared to the reproductive tracts of the farm mammals previously described. The testes of male poultry are contained in the body cavity. Each vas deferens opens into small papillae, which are located in the cloacal wall. The male fowl has no penis but does have a rudimentary organ of copulation. The sperm are transferred from the papillae to the rudimentary copulatory organ which transfers the sperm to the oviduct of the hen during the mating process. The sperm are stored in primary sperm-host glands located in the oviduct. These sperm are then released on a daily basis and transported to secondary storage glands in the infundibulum. Fertilization occurs in the infundibulum. Sperm stored in the oviduct are capable of fertilizing the eggs for 30 days in turkeys and 10 days in chickens.

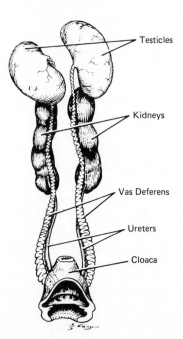

Testicles

Kidneys

Vas Deferens

Ureters

Cloaca

FIGURE 9.14.
Male poultry reproductive tract (ventral view). Courtesy of Dennis Giddings.

WHAT MAKES TESTICLES AND OVARIES OPERATE

Testicles

Testicles produce their hormones under stimuli coming to them from the anterior pituitary gland (AP) situated at the base of the brain. The anterior pituitary elaborates two hormones important to male reproductive performance. Luteinizing hormone (LH) and follicle-stimulating hormone (FSH) are known as gonadotropic hormones because they stimulate the gonads (ovary and testicle). LH produces its effect on the interstitial tissue (Leydig cells) of the testicle, causing the tissue to produce the male hormone, testosterone. FSH stimulates cells in the seminiferous tubules to nourish the developing spermatozoa.

Some species that respond to changes in length of daylight exhibit more seasonal fluctuation than others in reproductive activities. These influences of daylight are on the neurophysiological mechanism in the brain. The hypothalamus is the floor and part of the wall of the third ventricle of the brain and secretes releasing factors through blood vessels that affect the anterior pituitary and its production of FSH and LH (see Figs. 17.11 and 18.3 in Chapters 17 and 18).

Ovaries

Ovarian hormones in the sexually mature female owe the rhythmicity on their production to hormones that originate in the anterior pituitary body and to the interplay between gonad-stimulating hormones produced there and ovarian hormones whose level and potency vary as the estrous cycle progresses. FSH (the same hormone as FSH in the male) circulates through the bloodstream and affects the responsive follicle cells of the ovary, which respond by secreting the so-called estrogens or estrus-producing hormones, estradiol and estrone. When the amount of estradiol and estrone in the blood reaches a sufficient level, the pituitary is caused to reduce its production of FSH. With this drop in FSH production, the production of estradiol and estrone subsides. When the estrus-producing hormones have been lost from the body and their depressing effect on the anterior pituitary has been spent, the anterior pituitary again steps up its production of FSH and the cycle is repeated.

Luteal cells, the successors of the follicle cells, secrete a hormone called progesterone, which plays a role in the female reproductive cycle. Progesterone helps cause the occurrence and recurrence of the desire to mate, which is called **estrus** or **heat.**

Estrus is the period of time when the female will accept the male for breeding purposes. The female of each species exhibits some behavior patterns that demonstrate she is in heat (see Chapter 34). For example, a mare in estrus, when "teased" with the presence of a stallion (Fig. 9.15), will not avoid or kick him. The mare in heat will stand solidly, sometimes squatting and urinating when approached by the stallion.

Synchronized with the estrous cycle is the important and essential phenomenon of ovulation. Ovulation occurs in the cow after estrus. It occurs in the sow, ewe, goat, and mare toward the latter part of, but nevertheless during, estrus. These species all ovulate spontaneously—i.e., ovulation takes place whether copulation occurs or not. By contrast,

FIGURE 9.15.
Estrus is determined in the mare by "teasing" her with the presence of a stallion. Courtesy of Colorado State University.

copulation (or some such stimulation) is necessary to trigger ovulation in such animals as the rabbit, cat, ferret, and mink, which are considered "induced ovulators." Ovulation in these animals takes place at a fairly consistent time after mating.

Ovulation is controlled by hormones. The follicle of the ovary grows, matures, fills with fluid, and softens a few hours before rupturing. The follicle ruptures owing to a sudden release of LH rather than bursting as a result of pressure inside.

The FSH of the anterior pituitary accounts for the increase in size of the ovarian follicle and for the increased amount of estradiol and estrone, which are products of the follicle cells. LH, another hormone from the anterior pituitary, alters the follicle cells and granulosa cells of the ovary, changing them into luteal cells. These luteal cells are in turn stimulated by LH from the anterior pituitary to produce progesterone.

In the course of 7–10 days, what was formerly an egg-containing follicle develops into the corpus luteum, a luteal body of about the same size and shape as the mature follicle. The luteal cells of the corpus luteum produce a sufficient quantity of progesterone to depress FSH secretion from the anterior pituitary until the luteal cells reach maximum development (in nonpregnant animals), cease their development, and (in 3 weeks) lose their potency and disappear.

If pregnancy occurs, the corpus luteum continues to function, persisting in its progesterone production and preventing further estrous cycles. Thus, no more heat occurs until after pregnancy has terminated. The process of follicle development, ovulation, and corpus luteum development and regression is shown in Fig. 9.5. Table 9.2 shows the length of estrus, estrous cycles, and time of ovulation for the different farm animals.

The changing length of daylight is a potent factor that influences the estrous cycle, the onset of pregnancy, and the seasonal fluctuations in male fertility. The amount of daylight acts both directly and indirectly on the animal. It acts directly on the central nervous system by influencing the secretion of hormones and indirectly by affecting plant growth, thus altering the level of quality of nutrition available. In cattle, increasing length of day is associated with increased reproductive activity in both males and females. In sheep, the breeding season reaches its height in the autumn, as the hours of daylight

TABLE 9.2. Duration and Frequency of Heat and Time of Ovulation

Animal	Duration of Heat		Length of Cycle (days)		Approximate Time of Ovulation
	Average	Range	Average	Range	
Heifer, cow (cattle)	12 h	6–27 h	21	19–23	30 h after beginning of heat
Ewe (sheep)	30 h	20–42 h	17	14–19	26 h after beginning of heat
Mare (horse)	6 days	1–37 days	21	10–37	1 day before the end of heat
Gilt, sow (swine)	44 h	1½–4 days	21	19–23	30–38 h after beginning of heat
Doe (goat)	39 h	20–80 h	17	12–27	On second day of heat
Doe (rabbit)[a]	Constant estrus				8–10 h after mating
Queen (cat)	5 days	4–7	10	8–14	24 h after mating
Bitch (dog)[a,b]	9 days	4–13	—	—	24–48 h after heat begins
Jill (mink)	2 days	Seasonal breeding (March)			40–50 h after mating

[a]The dog and rabbit may exhibit a pseudo or false pregnancy after mating has occurred.
[b]The dog has no cycle. There are generally two heats per year—in the fall and spring.

shorten. Of course, individuals of both sexes vary in their intensity and level of fertility. When selection has resulted in improvements in the traits associated with reproduction, individuals exhibit higher levels of fertility (i.e., more intense expression of estrus, occurrence of estrus over more months of the year, or occurrence of spermatogenesis at a high level over more months of the year) than unselected individuals.

PREGNANCY

When the sperm and the egg unite (fertilization), conception occurs, which is the beginning of the gestation period. The fertilized egg begins a series of cell divisions (Fig. 9.16). In farm mammals, the embryo migrates through the oviduct to the uterus in 3–4 days. By then it has developed to the 16- or 32-cell stage. The chorionic and amniotic membranes develop around this new embryo, and the chorion attaches to the uterus. The embryo (and later the fetus) obtain nutrients and discharges wastes through these membranes. This period of attachment (20–30 days in cattle and 14–21 days in swine) is critical. Unless the environment is sufficiently favorable, the embryo dies. Embryonic mortality causes a significant economic loss in farm animals, especially swine, in which multiple ovulations and embryos are typical of the species. It is vital that management protect the female and embryos early in pregnancy. High temperatures will cause embryonic death. The female should enter the breeding season in a thrifty, weight-gaining condition.

The embryonic stage in the life of an individual is defined as that period in which the

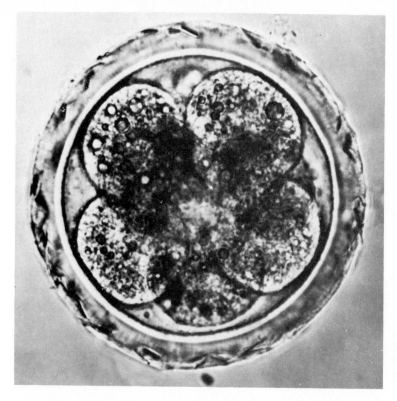

FIGURE 9.16.
A bovine embryo in the six-cell stage of development (magnified 620 ×). Courtesy of Colorado State University.

body parts differentiate to the extent that the essential organs are formed. This period lasts 45 days in cattle.

When the embryonic stage is completed, the young organism is called a fetus. The fetal period, which lasts until birth, is mainly a time of growth. The duration of pregnancy in cattle is included in Table 9.3. Length of pregnancy varies chiefly with the breed and age of the mother.

TABLE 9.3.   Gestation Length and Number of Offspring Born

Animal	Gestation Length (days)	Usual Number of Offspring Born
Cow (cattle)	285	1
Ewe (sheep)	147	1–3
Mare (horse)	336	1
Sow (swine)	114	6–14
Doe (goat)	150	2–3
Doe (rabbit)	31	4–8
Jill (mink)	50	4
Queen (cat)	52	4
Bitch (dog)	60	7

FIGURE 9.17.
Parturition in the ewe. (A) The water bag (a fluid-filled bag) becomes visible. (B) The head and front legs of the lamb appear. (C) The lamb is forced out by the uterine contractions of the ewe. (D) In a few minutes after birth, the lamb is nursing the ewe. From *Animal Agriculture,* 2nd ed., edited by H. H. Cole and W. N. Garrett, W. H. Freeman and Company, © 1980.

PARTURITION

Parturition (birth) marks the termination of pregnancy (Fig. 9.17). Extraembryonic membranes that had been formed around the embryo in early pregnancy are shed at this time and are known as afterbirth. These membranes are attached to the uterus during pregnancy by the organ known as the placenta, which is responsible for the transfer of nutrients and wastes between mother and fetus. The extraembryonic membranes and the placenta develop the capacity to produce hormones, especially estrogens and progesterone, in some farm animal species.

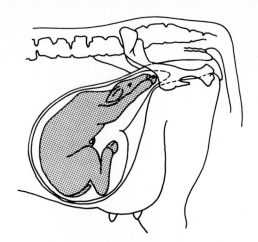

Normal Presentation

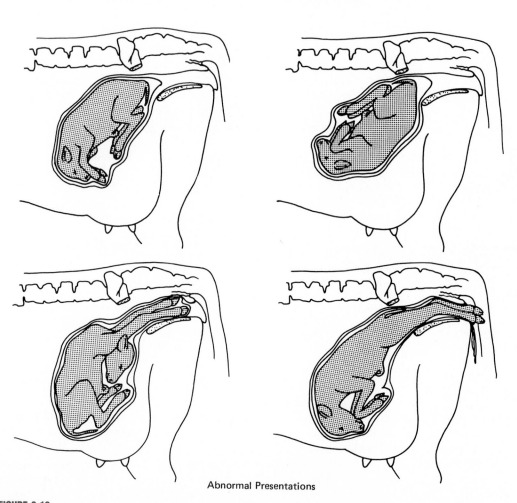

Abnormal Presentations

FIGURE 9.18.
Normal and some abnormal presentations of the calf at parturition. From Battaglia and Mayrose, *Handbook of Livestock Management Techniques,* Macmillan Publishing Company, New York, 1981, pp. 131, 134, 135.

The parturition process is initiated by release of the hormone cortisol from the fetal adrenal cortex. Progesterone levels decline whereas estrogen, prostaglaudin $F_{2\alpha}$, and oxytocin levels rise, resulting in uterine contractions.

Parturition is a synchronized process. The cervix, until now tightly closed, relaxes. Relaxation of the cervix, along with pressure generated by uterine muscles on the contents of the uterus, permits the passage of the fetus into the vagina and on to the exterior.

A

B

Figure 9.19.
Live offspring are born to each breeding female in the herd or flock when the intricate mechanisms of reproduction function properly. (A) Dairy cow and calf. Courtesy of Colorado State University. (B) Sow and litter of pigs. Courtesy of D. C. England, Oregon State University. (C) Mare and foal. Courtesy of *The Western Horseman.* (D) Ewe and lambs. Courtesy of Colorado State University. (E) Chicks in the process of hatching. Courtesy of *Poultry Digest.* (F) Beef cow and calf. Courtesy of the American Polled Hereford Association.

C

D

E

F

FIGURE 9.19 (cont.)

Another hormone, relaxin, is thought to aid in parturition. Relaxin, which originates in the corpus luteum or placenta, helps to relax cartilage and ligaments in the pelvic region to help parturition occur.

At the beginning of parturition, the offspring typically assumes a position that will offer the least resistance as it passes into the pelvic area and through the birth canal. Fetuses of the cow, mare, and ewe assume similar positions in which the front feet are extended with the head between them (Fig. 9.18). Fetal piglets do not orient themselves in any one direction, which does not appear to affect the ease of birth. Calves, lambs, or foals may occasionally present themselves in a number of abnormal positions (Fig. 9.18). In many of these situations, assistance needs to be given at parturition. Otherwise, the offspring may die or be born dead, and in some instances, the mother may die as well. An abnormally small pelvis opening or an abnormally large fetus can cause some mild to severe parturition problems.

Producers can manage their herds and flocks for high reproductive rates. Management decisions are most critical that cause males and females, selected for breeding, to reach puberty at early ages, have high conception rates, and have minimum difficulty at parturition. Keeping animals healthy, providing adequate levels of nutrition, selection of genetically superior animals, and producer attention to parturition are some of the more critical management inputs to minimize reproductive loss. Live offspring born to each breeding female is the key end point to successful farm animal reproduction (Fig. 9.19).

SELECTED REFERENCES

Publications

Battaglia, R. A., and Mayrose, V. B. 1981. *Handbook of Livestock Management Techniques.* New York: Macmillan.

Bone, J. F. 1982. *Animal Anatomy and Physiology.* Reston, VA: Reston Publishing.

Bearden, H. J., and Fuquay, J. W. 1984. *Applied Animal Reproduction.* Reston, VA: Reston Publishing.

Frandson, R. D. 1981. *Anatomy and Physiology of Farm Animals.* Philadelphia: Lea and Febiger.

Hafez, E. S. E. (editor). 1980. *Reproduction in Farm Animals.* Philadelphia: Lea and Febiger.

Pickett, B. W., Voss, J. L., Squires, E. L., and Amann, R. P. 1981. *Management of the Stallion for Maximum Reproductive Efficiency.* Fort Collins: Colorado State University Press, Animal Reprod. Lab. Gen. Series 1005.

Sorenson, A. M., Jr. 1979. *Animal Reproduction: Principles and Practices.* New York: McGraw-Hill.

Visuals

Embryo Development of the Chick (silent filmstrip). Vocational Education Productions, California Polytechnic State University, San Luis Obispo, CA 93407.

Heat Detection in Dairy Cows, Artificial Insemination, and *The Calving Process* (videotapes). Agricultural Products and Services, 2001 Killebrew Dr., Suite 333, Bloomington, MN 55420.

CHAPTER 10

Artificial Insemination, Estrous Synchronization, and Embryo Transfer

In the process of artificial insemination (AI), semen is deposited in the female reproductive tract by artificial techniques rather than by natural mating. AI was first successfully accomplished in the dog in 1780 and in horses and cattle in the early 1900s. AI techniques are also available for use in sheep, goats, swine, poultry, laboratory animals, and bees.

The primary advantage of AI is that it permits extensive use of outstanding sires to maximize genetic improvement. For example, a bull may sire 30–50 calves naturally per year over a productive lifetime of 3–8 years. In an AI program, a bull can produce 200–400 units of semen per ejaculate, with four ejaculates typically collected per week. If the semen is frozen and stored for later use, hundreds of thousands of calves can be produced by a single sire (one calf per 1.5 units of semen), and many of these offspring can be produced long after the sire is dead. AI also can be used to control reproductive diseases, and sires can be used that have been injured or are dangerous when used naturally.

SEMEN COLLECTING AND PROCESSING

There are several different methods of collecting semen. The most common method is the artificial vagina (Fig. 10.1), which is constructed to be similar to an actual vagina. The artificial vagina is commonly used to collect semen from bulls, stallions, rams, and buck goats and rabbits. The semen is collected by having the male mount an estrous female or training him to mount another animal or object (Figs. 10.2, 10.3). When the male mounts, his penis is directed into the artificial vagina by the person collecting the semen, and the semen accumulates in the collection tube. Semen from the boar and dog is not typically collected with an artificial vagina. It is collected by applying pressure with a gloved hand, after grasping the extended penis, as the animal mounts another animal or object.

Semen can also be collected by using an electroejaculator, in which a probe is inserted into the rectum and an electrical stimulation causes ejaculation. This is used most com-

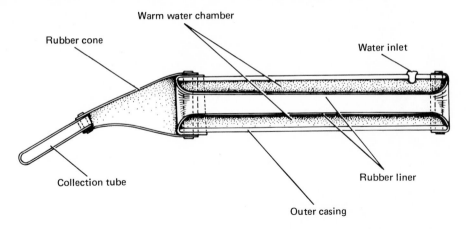

FIGURE 10.1.
Longitudinal section of an artificial vagina. Courtesy of Dennis Giddings.

FIGURE 10.2.
Using the artificial vagina to collect semen from the stallion. The handler of the stallion is holding the front leg to prevent the horse from striking the collector. Courtesy of Colorado State University.

FIGURE 10.3.
Collecting semen from a bull using the artificial vagina. Courtesy of Colorado State University.

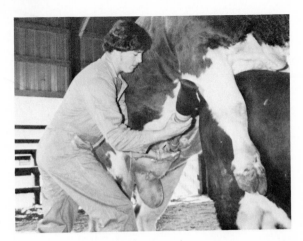

FIGURE 10.4.
Approximately 4 ml of bull semen in the collection tube attached to the artificial vagina. Note that the collection tube is immersed in water to control the temperature. The livability of the sperm is decreased when they are subjected to sudden temperature changes. Courtesy of Colorado State University.

monly in bulls and rams that are not easily trained to use the artificial vagina and from which semen is collected infrequently.

If semen is collected too frequently, the number of sperm per ejaculate decreases. Semen from a bull is typically collected twice a day for 2 days a week (Fig. 10.4). Semen from rams can be collected several times a day for several weeks, but bucks (goats) must be ejaculated less frequently. Boars and stallions give large numbers of sperm per ejacu-

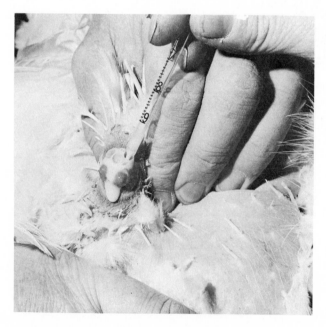

FIGURE 10.5.
Collection of semen from a turkey. Two individuals are usually involved in the collection process. One holds the turkey while the other individual manually stimulates semen into the cloaca while drawing semen into the collection tube. Courtesy of Lake, P.E., and Stewart, J. M., *Artificial Insemination in Poultry,* 1978, Ministry of Agriculture, Fisheries, and Food. No. 213, London: Her Majesty's Stationery Office, British Crown Copyright.

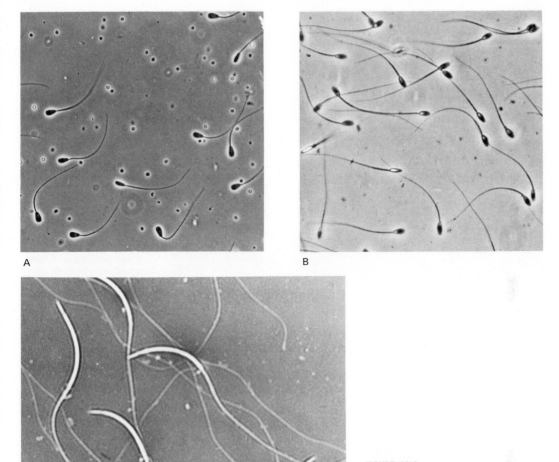

A

B

FIGURE 10.6.
(A) Normal bull semen. (B) Normal stallion semen. (C) Normal semen from domestic fowl as viewed under the microscope. Courtesy of Colorado State University (A and B); Dr. P. E. Lake and British Crown Copyright (C).

C

late, so semen from them is usually collected every other day at most. Semen is collected from tom turkeys 2–3 times per week. The simplest and most common technique for collecting sperm from toms is the abdominal massage. This technique usually requires two persons: the "collector" helps hold the bird and operates the semen collection apparatus while the "milker" stimulates a flow of semen by massaging the male's abdomen (Fig. 10.5).

After the semen is collected, it is evaluated for volume, sperm concentration, motility of the sperm, and sperm abnormalities (Fig. 10.6). The semen is usually mixed with an extender which dilutes the ejaculate to a greater volume. This greater volume allows a

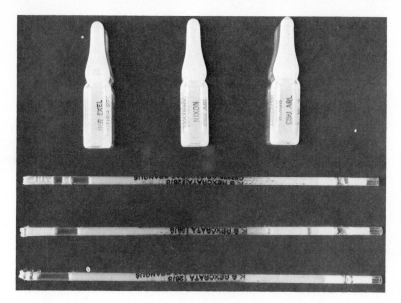

FIGURE 10.7.
Semen is typically stored and frozen in ampules or straws. Three ampules appear at the top of the figure, with three straws of bull semen below. The animal's name, registration number, and location of collection are printed on the ampules and straws. Courtesy of Colorado State University.

single ejaculate to be processed into several units of semen, where one unit of semen is used for each female inseminated. This extender is usually composed of nutrients such as milk and egg yolk, a citrate buffer, antibiotics, and glycerol. The amount of extender used is based on the projected number of viable sperm available in each unit of extended semen. For example, each unit of semen for insemination in cattle should contain 10 million motile, normal spermatozoa.

Some semen is used fresh; however, it can only be stored this way for a day or two. Most semen is frozen in liquid nitrogen and stored in ampules (glass vials), plastic straws (Fig. 10.7), or pellets. Bull semen can be stored in this manner for an indefinite period of time and retain its fertilization capacity. Most cattle are inseminated with frozen semen that has been thawed. The fertilizing capacity of frozen semen from boars, stallions, and rams is only modestly satisfactory; however, improvement has been noted in the past few years. Turkey semen cannot be frozen satisfactorily. For maximum fertilization capacity, it should be used within 30 min of collection. The semen cannot be extended, so 12 toms are usually kept for each 100 hens in the flock.

INSEMINATION OF THE FEMALE

Prior to insemination, the frozen semen is thawed. Thawed semen should not be refrozen and used again, because conception rates will be reduced.

High conception rates using AI depend on the female's cycling and ovulating, detecting estrus; using semen that has been properly collected, extended, and frozen; thawing and handling the semen satisfactorily at the time of insemination; insemination techniques; and avoiding extremes in stress and excitement to the animal being inseminated.

Detecting Estrus

Estrus must be detected accurately, because it signals time of ovulation and determines proper timing of insemination. The best indication of estrus is the condition called "standing heat" in which the female stands still when mounted by a male or another female.

Cows are typically checked for estrus twice daily, in the morning and evening. They are usually observed for 30 min to detect standing heat. Other observable signs are restlessness, attempting to mount other cows, and a clear, mucous discharge from the vagina. Some producers use sterilized bulls or hormone-treated cows as "heat checkers" in the herd. These animals are sometimes equipped with a chin marker that greases or paints marks on the back of the cow when she is mounted.

Estrus in sheep or goats is checked using sterilized males equipped with a brisket-marking harness. Gilts and sows in heat assume a rigid stance with ears erect when hands are placed firmly in their backs. The vulva is usually red and swollen. The presence of a boar and the resulting sounds and odors can help detect heat in swine as the females in estrus will attempt to locate a boar. Signs of estrus in the mare are elevation of the tail, contractions of the vulva (winking), spreading of the legs, and frequent urination.

Proper Timing of Insemination

The length of estrus and ovulation time are quite variable in farm animals. This variability poses difficulty in determining the best time for insemination. An additional challenge is that sperm are short-lived when put into the female reproductive tract. Also, estrus is sometimes expressed without ovulation occurring; sows, for example, typically show estrus 3–5 days after farrowing, but ovulation does not occur. Sows should not be bred at this time either artificially or naturally.

Insemination time should be as close to ovulation time as possible; otherwise sperm stay in the female reproductive tract too long and lose their fertilizing capacity. Cows found in estrus in the morning are usually inseminated the evening of the same day, and cows in heat in the evening are inseminated the following morning. Insemination, therefore, should occur toward the end or after estrus has been expressed in the cow. Ewes are usually inseminated in the second half of estrus, and goats are inseminated 10–12 h after the beginning of estrus. Sows ovulate from 30 to 38 h after the beginning of estrus, so insemination is recommended at the end of the first day or at the beginning of the second day of estrus. Sometimes sows are inseminated both days, which improves the conception rate.

Insemination of dairy cows occurs while the cow is standing in a stall or stanchion. Beef cows are penned and inseminated in a chute that restrains the animal. The most common insemination technique in cattle involves the inseminator having one arm in the rectum to manipulate the insemination tube through the cervix (Fig. 10.8). The insemination tube is passed just through the cervix, and the semen is deposited into the body of the uterus. The insemination procedure for sheep and goats is similar to that for cattle; however, a speculum (a tube approximately 1.5 in. in diameter and 6 in. long) allows the inseminator to observe the cervix in sheep and goats. The inseminating tube is passed

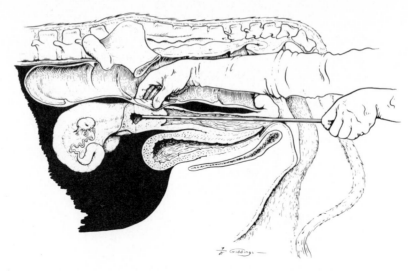

FIGURE 10.8.
Artificial insemination of the cow. Note that the insemination tube has been manipulated through the cervix. Inseminator's forefinger is used to determine when the insemination rod has entered into the uterus. Courtesy of Dennis Giddings.

through the speculum into the cervix, where the semen is deposited into the uterus or cervix.

The sow is usually inseminated without being restrained. The inseminating tube is easily directed into the cervix, because the vagina tapers into the cervix. The semen is expelled into the body of the uterus. The mare is hobbled or adequately restrained prior to insemination. The vulva area is washed, and the tail is wrapped or put into a plastic bag. The plastic-covered arm of the inseminator is inserted into the vagina, and the index finger is inserted into the cervix. The insemination tube is passed through the cervix, and the semen is deposited into the uterus.

The turkey hen is inseminated by first applying pressure to the abdominal area to cause eversion of the oviduct (Fig. 10.9). The insemination tube is inserted into the oviduct approximately 2 inches, and the semen is released. The first insemination is made when 5–10% of the flock have started laying eggs. A second insemination a week later assures a high level of fertility. Thereafter, insemination is done at 2-week intervals. Fertility in the turkey usually persists at a high level for 2–3 weeks after insemination. This is possible because sperm are stored in special glands of the hen, where they are nourished and retain their fertilizing capacity. Most of these glands are located near the junction of the uterus and vagina.

EXTENT OF ARTIFICIAL INSEMINATION

The number of farm animals inseminated each year is not well documented. It is estimated that in the United States more than 60% of the dairy cows (approximately 8 million head) and 5% of the beef cows (approximately 2 million head) are inseminated each year.

It is estimated that approximately 15,000 sows are artificially bred in the United States

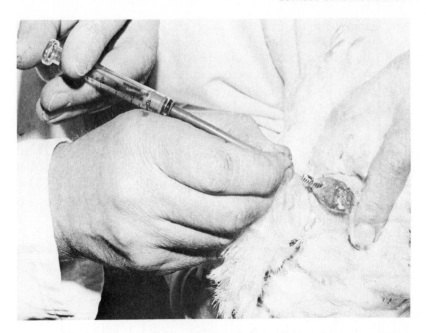

FIGURE 10.9.
Insemination of a hen turkey is accomplished by everting the opening of the oviduct through the opening of the cloaca. The tube is inserted one or two inches, pressure on the oviduct is relaxed, and the proper amount of semen is deposited as the syringe is slowly withdrawn. Courtesy of Lake, P. E, and Stewart, J. M, *Artificial Insemination in Poultry*, 1978, Ministry of Agriculture, Fisheries, and Food, No. 213, London: Her Majesty's Stationery Office, British Crown Copyright.

each year, which is less than 1% of the total sows bred in this country. Satisfactory techniques to freeze boar semen were accomplished in 1971, but frozen semen yields smaller litter sizes than fresh semen. More than 400,000 sows are artificially inseminated each year in certain European countries, which represent 20–30% of the sows bred in these countries.

AI in horses is still limited because of difficulty in providing extended semen storage. AI in sheep and goats in the United States is limited because herds and flock are dispersed over wide areas and the cost per unit of semen is high. Little AI is done in chickens; however, AI is rather extensive in turkeys. It is especially important in the broadbreasted turkey, which, because of the size of its breast, has difficulty mating naturally.

ESTROUS SYNCHRONIZATION

Estrous synchronization is controlling or manipulating the estrous cycle so that females in a herd or flock express estrus at approximately the same time. Estrous synchronization is a useful part of an AI program, because checking heat and breeding animals under range conditions is time-consuming and expensive. Also, estrous synchronization is a successful tool in making embryo transfer programs successful.

Considerable research has been done on estrous synchronization over the past few decades, where several different products have been evaluated. However, none of the products have been available commercially, because they have not been cleared by the

Food and Drug Administration. Also, some of the earlier research work demonstrated that conception rates were lowered when estrous synchronization products were used.

In 1979, a prostaglandin was cleared for use in cattle. Prostaglandins are naturally occurring fatty acids that appear to have important functions in several of the body systems. The prostaglandin that has a marked effect on the reproductive system is prostaglandin F_2 alpha ($PGF_{2\alpha}$).

It was pointed out in Chapter 9 that the corpus luteum (CL) controls the estrous cycle in the cow by secreting the hormone progesterone. Progesterone prevents the cow from expressing heat and ovulation. The prostaglandin destroys the CL, thus destroying the source of progesterone. About 3 days after the injection of prostaglandin, the cow will be in heat. For prostaglandin to be effective, the cow must have a functional CL. It is ineffective in heifers that have not reached puberty and in noncycling mature cows. Also, prostaglandin is ineffective if the CL is immature or has already started to regress. Prostaglandin is, then, only effective in heifers and cows that are in days 5–18 of their estrous cycle. Because of this relationship, prostaglandin is given on either a one-injection or a two-injection system.

One-Injection System

This system requires one prostaglandin injection and an 11-day AI breeding season. The first 5 days are a conventional AI program of heat detection and insemination. After 5 days, a calculation can be made to determine what percentage of the females in the herd have been in heat. If a smaller than expected percentage are cycling, the cost of the prostaglandin may not justify the anticipated benefit. If the decision is to proceed with the injection, the remaining animals not previously bred are injected on day 6. Then the AI breeding season continues 5 more days, for a total of 11 days.

Field trial results in herds where a high percentage of females cycling show a 50–60% pregnancy rate at the end of 11 days of AI breeding for those receiving prostaglandin versus 30–40% pregnancy rate for those not receiving it. The amount of prostaglandin required per calf that is produced by AI will be between 1.2 and 2.0 units.

Two-Injection System

In this system, all cows are injected with prostaglandin at two different times. Counting the first injection as day 1, the second injection is administered on day 11, 12, or 13. All cows capable of responding to the drug should be in heat during the first 5 days after the second injection of prostaglandin. Heat detection and insemination can occur each of these 5 days, or a fixed-time insemination can be performed 76–80 h after the second injection.

Field trial results in herds in which a high percentage of the females are cycling show a 35–55% pregnancy rate at the end of the fifth day or 76- to 80-h one-time insemination versus 10–12% in the cows not receiving prostaglandin. The two-injection system will require 4–7 units of prostaglandin per AI calf.

Prostaglandin is not a wonder drug. It will only work in well-managed herds where a high percentage of the females are cycling. Biologically, it has been well demonstrated that prostaglandin can synchronize estrus. However, producers must weigh the cost against the economic benefit. Caution should be exercised in administering prostaglandin to pregnant cows, as it may cause abortion. Estrous synchronization in cattle may not be advisable in areas where inadequate protection is given to young calves during severe blizzards. Also, excellent herd health programs must be utilized to prevent high losses from calf scours and other diseases that become more serious where large numbers of newborn calves are grouped together.

Another estrous synchronization product, Syncro-Mate-B, has been available for beef and dairy heifers since 1983. Heifers are injected with 5 mg of estradiol valerate and 3 mg of norgestomet and implanted with 6 mg norgestomet implant subcutaneously on top of the ear. Nine days later, the implant is removed. Most heifers will show estrus approximately 24–48 h after implant removal. Heifers can be inseminated 12 h after being detected in estrus or time inseminated 48–54 h after implant removal. Syncro-Mate-B will cause many noncycling heifers to show heat; however, conception rates in these heifers at this estrus are low (~20%). In cycling heifers treated with Syncro-Mate-B, conception rates are usually 40–60%.

In a sense, there are some natural occurrences of estrous synchronization. The weaning process in swine is an example, because the sow will typically show heat 3–8 days after the pigs are weaned. Estrus is suppressed through the suckling influence, so when the pigs are removed, the sow will show heat.

EMBRYO TRANSFER

Embryo transfer is sometimes referred to as ova transplant or embryo transplant. In this procedure, an embryo in its early stage of development is removed from its own mother's (the donor's) reproductive tract and transferred to another female's (the recipient's) reproductive tract. The first successful embryo transfer was accomplished in 1890. In the past several decades, successful embryo transfers have been reported in sheep, goats, swine, cattle, and horses.

FIGURE 10.10.
Ten Holstein embryo transfer calves resulting from one superovulation and transfer from the Holstein cow in the background. The ten recipient cows are shown on the left-hand side of the fence. Courtesy of Colorado State University.

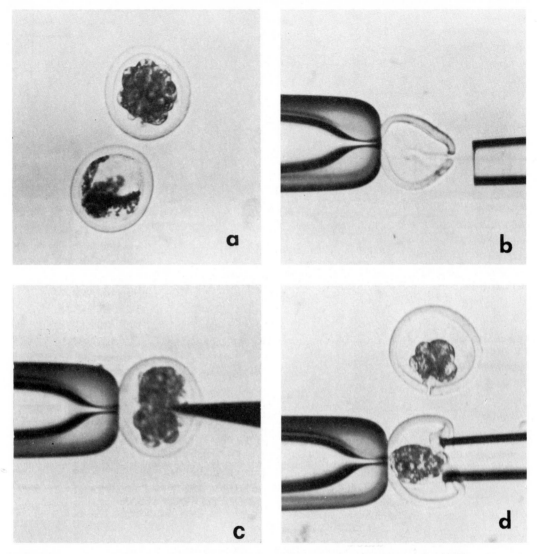

FIGURE 10.11.
The process of embryo splitting. (a) A morula (40 to 70 cells of the fertilized egg) and an unfertilized egg. (b) The unfertilized egg with the cellular contents sucked out of it. (c) The morula is divided into two groups of cells with the microsurgical blade. (d) One-half of the cells are left in the morula while the other one-half of the cells are placed inside the other unfertilized egg. The result is two genetically identical embryos ready for transfer. Courtesy of Williams et al. 1983 NAAB Conference and Beef AI and Embryo Transfer.

In recent years, commercial embryo transfer companies have been established in the United States and several other foreign countries. Most commercial work is done with beef and dairy cattle. It is estimated that approximately 65,000 pregnancies from embryo transfers in beef cattle and 35,000 pregnancies in dairy cattle occurred in North America in 1985. This compares to 20 done in 1972.

Superovulation is the production of a greater than normal number of eggs. Females

FIGURE 10.12.
"Question" and "Answer" were the world's first split-embryo foals resulting from nonsurgical embryo transfer.
Courtesy of Colorado State University.

that are donors of eggs for embryo transfer are injected with hormones to stimulate increased egg production. In cattle, the average donor produces approximately six transferable embryos per superovulation treatment; however, a range of 0–25 embryos can be expected (Fig. 10.10). This procedure gives embryo transfer its greatest advantage—i.e., increasing the number of offspring that a superior female can produce. The key to justification of embryo transfer is identification of genetically superior females. Embryo transfer is usually confined to seed stock herds, where genetically superior females can be more easily identified and where high costs can be justified.

Until recently, embryo transfer had to be done surgically, but nonsurgical techniques are now used. The procedure after superovulation is to breed the donor 12 and 24 h after she comes into heat with two doses of semen each time. The fertilized eggs are recovered about a week later by flushing the uterus with a buffered solution. This is essentially the reverse of the AI process and does not require surgery. The embryos are located and evaluated using a microscope and loaded into an AI rod. Nonsurgical transfer of an embryo into a recipient is done with the AI rod. Conception rates are usually higher if the transfer is done surgically. Embryos are usually transferred shortly after being collected, so recipient females need to be in the same stage of the estrous cycle (within 24 h) for successful transfer to occur. Estrous synchronization of females is necessary, or large numbers of females must be kept for this purpose.

Embryos can now be frozen in liquid nitrogen and remain dormant for months or years. Conception rates are lower when frozen embryos are used than when fresh embryos are used, but research is bringing about improvement. In cattle, approximately 50–60% of the frozen embryos will be normal after thawing. Thirty to forty percent of these normal embryos will result in confirmed pregnancies in 60–90 days. This compares to a pregnancy rate of 55–65% from fresh embryos transferred the same day of collection.

Recent advances in embryo transfer research has permitted the embryo to be mechanically divided so that identical twins can be produced from a single embryo (Figs. 10.11, 10.12). Perhaps in the future an embryo of a desired mating and sex could be selected from an inventory of frozen embryos. This embryo would then be thawed and transferred nonsurgically in the cow when she comes into heat. Embryo splitting and embryo transfer are an area of biotechnology that continue to advance very rapidly.

SELECTED REFERENCES

Publications

Ernst, R. A., Ogasawara, F. X., Rooney, W. F., Schroeder, J. P., and Ferebee, D. C. 1970. *Artificial Insemination of Turkeys.* University of California Agric. Ext. Publ. AXT-338.

Herman, H. A. 1981. *Improving Cattle by the Millions. NAAB and the Development and Worldwide Application of Artificial Insemination.* Columbia: University of Missouri Press.

Kiessling, A. A., Hughes, W. H., and Blankevoort, M. R. 1986. Superovulation and embryo transfer in the dairy goat. *J. Am. Vet. Med. Assoc.* 188:829.

Lake, P. E., and Steward, J. M. 1978. *Artificial Insemination in Poultry.* Scotland Ministry of Agriculture, Fisheries, and Food No. 213.

Perry, E. J., ed. 1968. *The Artificial Insemination of Farm Animals.* New Brunswick, NJ: Rutgers University Press.

Pickett, B. W., and Back, D. G. 1973. *Procedures, Collection, Evaluation and Insemination of Stallion Semen.* Fort Collins: Colorado State University Exp. Sta., Animal Reprod. Lab., Gen. Series 935.

Salisbury, G. W., VanDemark, N. L., and Odge, J. R. 1978. *Physiology of Reproduction and Artificial Insemination of Cattle,* 2d Ed. San Francisco: W. H. Freeman.

Seidel, G. E., Jr. 1981. Superovulation and embryo transfer in cattle. *Science* 211:351.

Seidel, G. E., Jr., Seidel, S. M., and Bowen R. A. 1978. *Bovine Embryo Transfer Procedures.* Colorado State University Exp. Sta., Animal Reprod. Lab., Gen. Series 975.

Squires, E. L., Cook, V. M., and Voss, J. L. 1984. Collection and Transfer of equine embryos. Colorado State University Animal Reprod. Lab. Bull. No. 01.

Synchronization of Beef Cattle with Prostaglandin. 1979. DeForest, WI: American Breeders Service.

Visuals

Reproduction Kit (sound filmstrips) covering "Embryo Transfer of Beef and Dairy Cattle" and "Artificial Insemination of Beef and Dairy Cattle." Vocational Education Production, California Polytechnic State University, San Luis Obispo, CA 93407.

Swine Reproduction Series (sound filmstrips) covering "Fresh Semen Artificial Insemination" and "Frozen Semen Artificial Insemination." Vocational Education Productions, California Polytechnic State University, San Luis Obispo, CA 93407.

Equine A.I. (sound filmstrips) covering "Introduction to Equine AI," "Teasing and Rectal Palpation," "Semen Collection and Evaluation," "Broodmare Health Care," and "Inseminating and Diagnosing Pregnancy." Vocational Education Productions, California Polytechnic State University, San Luis Obispo, CA 93407.

Modern Cattle Breeding Techniques (sound filmstrips) covering "Artificial Insemination—An Overview" and "Embryo Transfer—The New Horizon." Prentice-Hall Media, 150 White Plains Road, Tarrytown, NY 10591.

Artificial Insemination in Poultry (slide-tape—57 slides; 10 min). Poultry Science Department, Ohio State University, 674 W. Lane Ave., Columbus, OH 43210.

CHAPTER 11

Genetics

Body tissues of animals and plants are composed of cells, whose structure can be observed microscopically. These cells, with certain exceptions, have an outer membrane, an internal cytoplasm, and a nucleus (see Chapter 17, Fig. 17.3). This nucleus contains rod-like bodies called **chromosomes** (Fig. 11.1). Body cells contain these chromosomes in pairs, and each chromosome contains genes, which are the functional units of inheritance. When cells divide to produce more body cells, the chromosomes replicate by a process called **mitosis** (Fig. 11.2). During mitosis, each member of each chromosome pair divides so that the two new cells formed (daughter cells) are identical to the original cell that divided.

Each species has a characteristic number of chromosomes (Table 11.1), which is maintained through meiosis and fertilization. For comparative purposes, humans have 46 pairs of chromosomes.

PRODUCTION OF GAMETES

The testicles of the male and the ovaries of the female produce cells that become gametes (sex cells) by a process called gametogenesis. The gametes produced by the testicles are called sperm; the gametes produced by the ovaries are called eggs, or ova. Specifically, the production of sex cells that will become sperm is called **spermatogenesis;** the production of ova, **oogenesis.** The unique type of cell division in which gametes (sperm or ova) are formed is called **meiosis.** Each newly formed gamete contains only one member of each of the chromosome pairs present in the body cells.

Let us examine gametogenesis in a theoretical species in which there are only two pairs of chromosomes in each body cell. Meiosis occurs in the primordial germ cells (cells capable of undergoing meiosis) located near the other wall of the **seminiferous tubules** of each testicle and near the surface of each ovary. The initial steps in meiosis are similar for the male and female. The chromosomes replicate themselves so that each chromosome is doubled. Then each pair of chromosomes comes together in extremely accurate pairing called synapsis. After chromosome replication and synapsis, the cell is called a

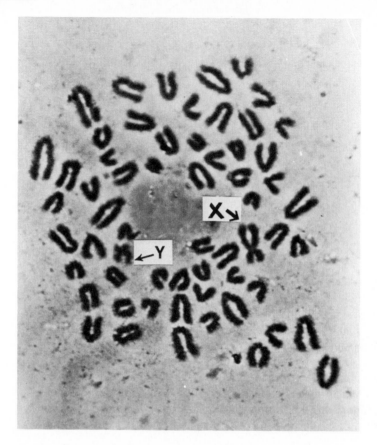

FIGURE 11.1.
The 30 pairs of chromosomes of a bull magnified several hundred times. Note the X and Y chromosomes. Courtesy of Nat M. Kieffer, Texas A&M University.

TABLE 11.1.   Pairs of Chromosomes in Livestock and Poultry

Species	Number of Pairs
Turkeys	41
Chickens	39
Horses	32
Cattle	30
Goats	30
Sheep	27
Swine	19

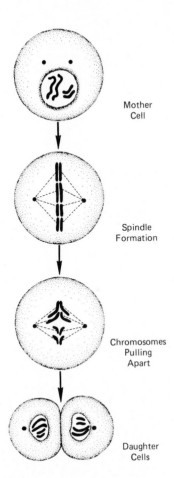

Mother
Cell

Spindle
Formation

Chromosomes
Pulling
Apart

Daughter
Cells

FIGURE 11.2.
Mitosis. Drawn by Dennis Giddings.

primary spermatocyte in the male and a primary oocyte in the female. Because subsequent differences exist between spermatogenesis and oogenesis, the two will be described separately.

SPERMATOGENESIS

The process of spermatogenesis in our theoretical species is shown in Fig. 11.3. The primary spermatocyte contains two pairs of chromosomes in synapsis. Each chromosome is doubled following replication. Thus, the primary spermatocyte contains two bodies or structures formed by four parts. By two rapid cell divisions, in which no further replication of the chromosomes occurs, four cells, each of which contains two chromosomes, are produced. The spermatid cells each lose much of their cytoplasm and develop a tail. This process, spermatogenesis, results in formation of sperm. Four sperm are produced from each primary spermatocyte. Whereas four chromosomes in two pairs are present in the primordial germ cell, only two chromosomes (one half of each chromosome pair) are present in each sperm. Thus, the number of chromosomes in the sperm has been reduced to half the number in the primordial germ cell.

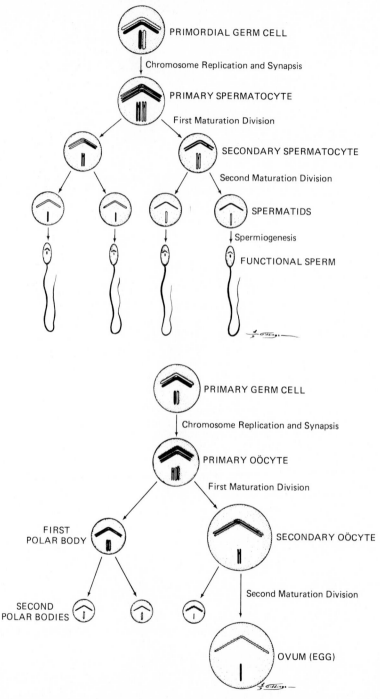

FIGURE 11.3.
Meiosis or reduction cell division occuring in the testicle and ovary (example with two pair of chromosomes).
Drawn by Dennis Giddings.

OOGENESIS

The process of **oogenesis** in our theoretical species is shown in Fig. 11.3. Like the primary spermatocyte of the male, the primary oocyte of the female contains a tetrad. The first maturation division produces one relatively large nutrient-containing cell, the secondary oocyte, and a smaller cell, the first polar body. Each of these two cells contains a dyad. The unequal distribution of nutrients that results from the first maturation division in the female serves to maximize the quantity of nutrients in one of the cells (the secondary oocyte) at the expense of the other (the first polar body). The second maturation division produces the egg (ovum) and the second polar body, each of which contains two chromosomes. The first polar body may also divide, but all polar bodies soon die and are reabsorbed. Note that the egg, like the sperm, contains only one chromosome of each pair that was present in the primordial germ cell. The gametes each contain one member of each pair of chromosomes that existed in each primordial germ cell.

FERTILIZATION

When a sperm and an egg of our theoretical species unite to start a new life, each contributes one chromosome to each pair of chromosomes in the fertilized egg, now called a zygote (Fig. 11.4). Fertilization is defined as the union of the sperm and the egg along with the establishment of the paired condition of the chromosomes. The zygote is termed **diploid** (diplo- means "double") because it has chromosomes in pairs, one member of each pair having come from the sire and one member having come from the dam. Ga-

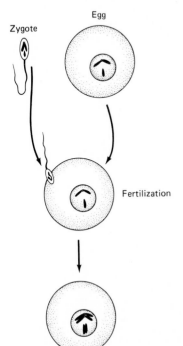

FIGURE 11.4.
Combining of chromosomes through fertilization (two pairs of genes used for simplification of example. Drawn by Dennis Giddings.

metes have only one member of each pair of chromosomes; therefore, gametes are termed **haploid** (haplo- means "half"). Gametogenesis thus reduces the number of chromosomes in a cell to half the diploid number. Fertilization reestablishes the normal diploid number.

DNA AND RNA

The two members of each typical pair of chromosomes in a cell are alike in size and shape and carry **genes** that affect the same hereditary characteristics (Fig. 11.5). Such chromosomes are said to be **homologous.** The genes are points of activity found in each of the chromosomes that govern the way in which traits develop. The genes form the coding system that directs enzyme and protein production. Thus, they control the development of traits.

Chromosomes in advanced organisms, such as poultry and livestock, are composed basically of a protein sheath surrounding **deoxyribonucleic acid** (DNA for short). Segments of DNA are the genes. DNA is itself composed of three parts: deoxyribose sugar, phosphate, and four nitrogenous bases. The combination of deoxyribose, phosphate, and one of the four bases is called a **nucleotide;** when many nucleotides are chemically bonded to one another, they form a strand that composes one-half of the DNA molecule. (A molecule formed by many repeating sections is called a polymer.) Two of these strands wind around each other in a double helix to form the DNA molecule.

The bases of DNA are the parts that hold the key to inheritance. The four bases are adenine (A), thymine (T), guanine (G), and cytosine (C). In the two strands of DNA, A is always complementary to (pairs with) T, and G is always complementary to C (Fig.

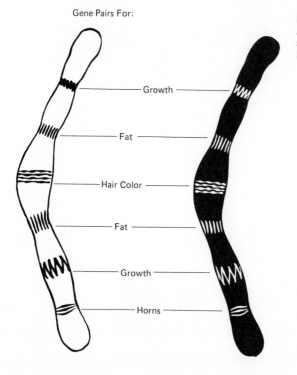

Gene Pairs For:

Growth

Fat

Hair Color

Fat

Growth

Horns

FIGURE 11.5.
A simplified example showing a pair of chromosomes containing several pairs of genes. Drawn by Dennis Giddings.

11.6). During meiosis and mitosis, the chromosomes are replicated by the unwinding and pulling apart of the DNA strands, and a new strand is formed alongside the old. The old strand served as a template, so that wherever an A occurs on the old strand, a T will be directly opposite it on the new, and wherever a C occurs on the old strand, a G will be placed on the new. Complementary bases pair with each other until two entire double-stranded molecules are formed where originally there was one.

Nearly all genes code for proteins, which are also polymers. It is important to realize that DNA and protein are both polymers. For each of the 26 amino acids of which proteins are made, there is at least one "triplet" sequence of three nucleotides. For example, two DNA triplets, TTC and TTT, code for the amino acid called lysine; four triplets, CGT, CGA, CGG, and CGC, all code for the amino acid called alanine. If we think of a protein molecule as a word, and amino acids as the letters of the word, each triplet sequence of DNA can be said to code for a letter of the word, and the entire encoded message, the series of base triplets, is the gene.

The processes by which the code is "read" and protein is synthesized are called **transcription** and translation. To understand these processes, another group of molecules, the **ribonucleic acids** (RNAs), must be introduced. There are three types of RNA; transfer RNA (tRNA), which identifies both an amino acid and a base triplet in mRNA; **messenger RNA (mRNA),** which carries the information codes for a particular protein; and ribosomal RNA (rRNA), which is essential for ribosome structure and function. All three RNAs are coded by the DNA template.

The first step in protein synthesis is that of transcription. Just as the DNA molecule serves as a template for self-replication using the pairing of specific bases, it can also serve as a template for the mRNA molecule. Messenger RNA is similar to DNA but is single-stranded rather than double-stranded and has the base uracil (U) in place of the base thymine found in DNA. It is shorter, coding for only one or a few proteins. The triplet sequence that codes for one amino acid in mRNA is called a codon. Through transcription, the encoded message held by the DNA molecule becomes transcribed onto the mRNA molecule. The mRNA then leaves the nucleus and travels to an organelle called a ribosome, where protein synthesis will actually take place. The ribosome is composed of rRNA and protein.

The second step in protein synthesis is the union of amino acids to their respective tRNA molecules. The tRNA molecules are coded by the DNA. They have a structure and contain an anticodon that is complementary to an mRNA codon. Each RNA unites with one amino acid. This union is very specific such that, as an example, lysine never links with the tRNA for alanine but only to the tRNA for lysine.

The mRNA attaches to a ribosome for translation of its message into protein. Each triplet codon on the mRNA (which is complementary to one on DNA) associates with a specific tRNA bearing its amino acid, using a base-pairing mechanism similar to that found in DNA replication and mRNA transcription. This matching of each tRNA with its specific mRNA triplets begins at one end of the mRNA and continues down its length until all the codons for the protein-forming amino acids are aligned in the proper order. The amino acids are each chemically bonded to each other by so-called peptide bonding as the mRNA moves through the ribosome, and the fully formed protein disassociates from the tRNA-mRNA complex and is ready to fulfill its role as a part of a cell or as an enzyme to direct metabolic processes. Figure 11.7 summarizes the steps of protein syn-

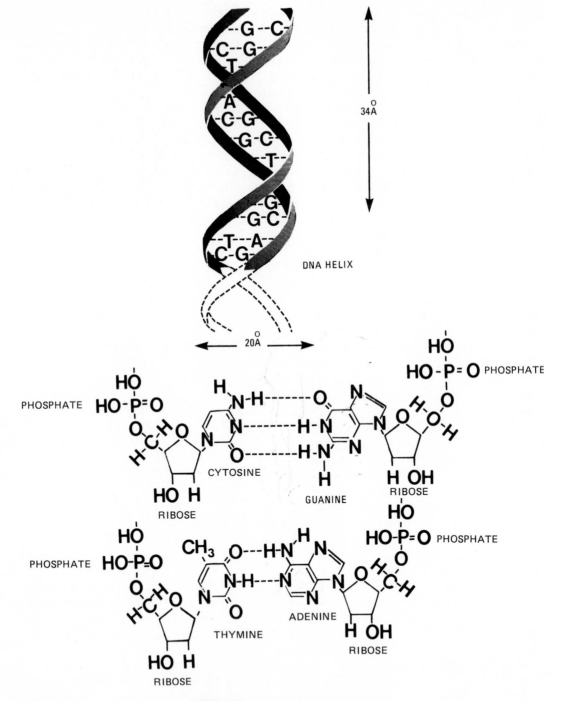

FIGURE 11.6.
DNA helix and structure of nucleotides.

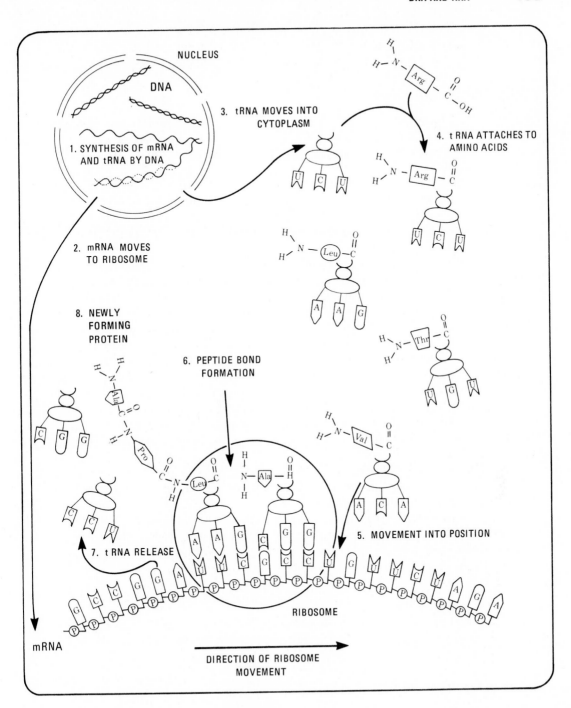

FIGURE 11.7.
Protein synthesis in the cell.

thesis. It depicts the amino acids arginine (arg), leucine (leu), threonine (thr), and valine (val) being moved into position to join a chain that already includes the amino acids alanine (ala), proline (pro), and leucine (leu).

GENES AND CHROMOSOMES

Because chromosomes are in pairs, genes are also in pairs. The location of a gene in a chromosome is called a locus (plural, loci). For each locus in one of the members of a pair of homologous chromosomes, a corresponding locus occurs in the other member of that chromosome pair. The transmission of genes from parents to offspring depends entirely on the transmission of chromosomes from parents to offspring.

A special pair of chromosomes, the so-called sex chromosomes (X and Y), exist as a pair in which one of the chromosomes does not correspond entirely to the other in terms of what loci are present. The Y chromosome is much shorter in length than the X chromosome.

The X and Y chromosomes determine the sex of animals. A female has two X chromosomes, and the male has an X chromosome and a Y chromosome. The female, being XX, can contribute only an X chromosome to the offspring. The male contributes either an X or a Y, thus determining the sex of the newborn.

The genes located at corresponding loci in **homologous chromosomes** may correspond to each other in the way that they control a trait, or they may contrast. If they correspond, the individual is said to be **homozygous** at that locus (homo- means "alike"; -zygous refers to the individual); if they differ, the individual is said to be **heterozygous** (hetero- means "different"). Those genes that occupy corresponding loci in homologous chromosomes but affect the same character in different ways are called **alleles.** Genes that are alike and thus affect the character developing in the same way are called "identical genes." Some authorities speak of alike genes in homologous chromosomes as identical alleles.

The geneticist usually illustrates the chromosomes as lines and indicates the genes by alphabetical letters. When the genes at corresponding loci on homologous chromosomes differ, one of the genes often overpowers, or dominates, the expression of the other. This allele is called **dominant.** The allele whose expression is prevented is called **recessive.** The dominant allele is symbolized by a capital letter. The recessive allele is symbolized by a lowercase letter. For example, in cattle, black hair color is dominant to red hair color, so we let B = black and b = red. Three combinations of genes are possible:

$$\dagger B \quad \dagger B \qquad \dagger B \quad \dagger b \qquad \dagger b \quad \dagger b$$

Both BB animals and bb animals are homozygous for the genes that determine hair color, but one is homozygous-dominant (BB) and the other is homozygous-recessive (bb). The animal that is Bb is heterozygous; it has allelic genes.

SIX FUNDAMENTAL TYPES OF MATING

With three kinds of individuals (homozygous-dominant, heterozygous, and homozygous-recessive) and one pair of genes considered, six types of matings are possible. Using the genes designated as B = black and b = red in cattle, the six mating possibilities are $BB \times BB$,

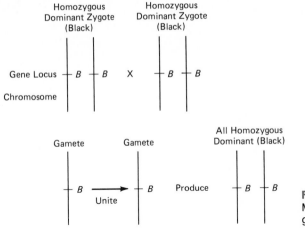

FIGURE 11.8.
Mating of homozygous dominant × homozygous dominant.

$BB \times Bb$, $BB \times bb$, $Bb \times Bb$, $Bb \times bb$, and $bb \times bb$. Keep in mind that the discussion based on these genes is applicable to any other pair of genes in any species.

Homozygous-Dominant × Homozygous-Dominant *(BB × BB).* When two homozygous-dominant individuals are mated, each of them can produce only one kind of gamete— the gamete carrying the dominant gene. In the particular example we are using, this gamete carries gene B. The union of gametes from two homozygous-dominant parents results in zygote that is homozygous-dominant $(B \times B = BB)$. Thus, homozygous-dominant parents produce only homozygous-dominant offspring (Fig. 11.8).

Homozygous-Dominant × Heterozygous *(BB × Bb).* The mating of a homozygous-dominant with a heterozygous individual results in an expected ratio of 1 homozygous-dominant:1 heterozygous. The homozygous-dominant parent produces only one kind of **gamete,** the one carrying the dominant gene (B). The heterozygous parent produces two kinds of gametes, one carrying the dominant gene (B) and one carrying the recessive gene (b). The latter two kinds of gametes are produced in approximately equal numbers. The chances that a gamete from the parent producing only the one kind of gamete (the one having the dominant gene) will unite with each of the two kinds of gametes produced by the heterozygous parent are equal; therefore, the number of homozygous-dominant offspring $(B \times B = BB)$ and heterozygous offspring $(B \times b = Bb)$ produced should be approximately equal (Fig. 11.9).

Homozygous-Dominant × Homozygous-Recessive *(BB × bb).* The homozygous-dominant individual can produce only gametes carrying the dominant gene (B). The recessive individual must be homozygous-recessive to show the recessive trait; therefore, it produces only gametes carrying the recessive gene (b). When the two kinds of gametes unite, all offspring produced receive both the dominant and the recessive gene $(B \times b = Bb)$ and are thus all heterozygous (Fig. 11.10).

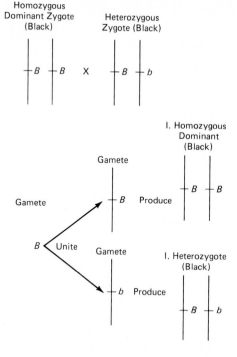

FIGURE 11.9.
Mating of homozygous dominant × heterozygote.

Heterozygous × Heterozygous *(Bb × Bb).* Each of the two heterozygous parents produces two kinds of gametes in approximately equal ratios: one kind of gamete carries the dominant gene *(B)*; the other kind carries the recessive gene *(b)*. The two kinds of gametes produced by one parent each have an equal chance of uniting with each of the two kinds of gametes produced by the other parent. Thus, four equal-chance unions of gametes are possible. If the gamete carrying the *B* gene from one parent unites with the gamete carrying the *B* gene from the other parent, the offspring produced are homozygous-dominant *(B × B = BB)*. If the gamete carrying the *B* gene from one parent unites with the

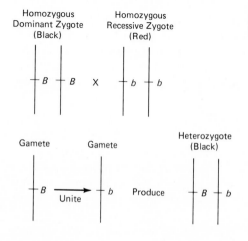

FIGURE 11.10.
Mating of homozygous dominant × homozygous recessive.

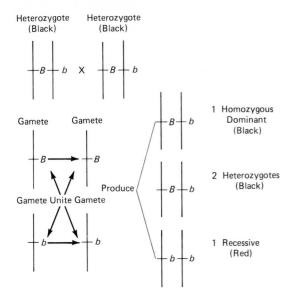

FIGURE 11.11.
Mating of heterozygote × heterozygote.

gamete carrying the *b* gene from the other parent, the offspring are heterozygous *(B × b = Bb)*. The latter combination of genes can occur in two ways: the *B* gene can come from the male parent and the *b* gene from the female parent, or the *B* gene can come from the female parent and the *b* gene from the male parent. When the gametes carrying the *b* gene from both parents unite, the offspring produced are recessive *(b × b = bb)*. The total expected ratio among the offspring of two heterozygous parents is 1*BB*:2*Bb*:1*bb* (Fig. 11.11). This 1:2:1 ratio is the genetic, or *genotypic,* ratio. The appearance of the offspring is in the ratio of 3 dominant:1 recessive because the *BB* and *Bb* animals are all black (dominant) and cannot be genetically distinguished from one another by looking at them. This 3:1 ratio, based on external appearance, is called the phenotypic ratio.

Heterozygous × Homozygous-Recessive *(Bb × bb).* The heterozygous individual produces two kinds of gametes, one carrying the dominant gene *(B)* and the other carrying the recessive gene *(b),* in approximately equal numbers. The recessive individual produces only the gametes carrying the recessive gene *(b).* There is an equal chance that the two kinds of gametes produced by the heterozygous parent will unite with the one kind of gamete produced by the recessive parent *(B × b = Bb; b × b = bb).* The offspring produced when these gametes unite thus occur in the expected ratio of 1 heterozygous:1 homozygous-recessive (Fig. 11.12).

Homozygous-Recessive × Homozygous-Recessive *(bb × bb).* The recessive individuals are homozygous; therefore, they can produce gametes carrying only the recessive gene *(b).* When these gametes unite *(b × b = bb),* all offspring produced will be recessive (Fig. 11.13). This example illustrates the principle that recessives, when mated together, breed true.

The knowledge of what results from each of the six fundamental types of matings

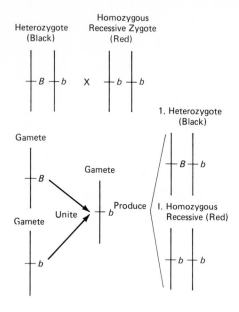

FIGURE 11.12.
Mating of heterozygote × homozygous recessive.

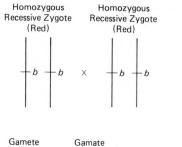

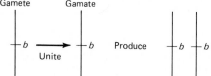

FIGURE 11.13.
Mating of homozygous recessive × homozygous recessive.

provides a background for understanding complex crosses. When more than one pair of genes is considered in a mating, one can understand the expected results, because one can combine each of the combinations of one pair of genes with each of the other combinations of one pair of genes to obtain the expected ratios.

PARENTAL AND FILIAL GENERATIONS

In genetics research, the dominant is often mated with the recessive, after which the offspring are intermated. If we use the same genes as before (B=black and b=red), the mating sequence is:

First parental (P_1) generation (or P_1 zygotes)	$BB \times bb$
P_1 gametes	$B, \quad b$
First filial (F_1) generation (or F_1 zygotes)	Bb
F_1 zygotes	$Bb \times Bb$
F_1 gametes	B, b, B, b
F_2 zygotes	$BB, Bb, Bb, \quad bb$

$$\underbrace{BB, Bb, Bb,}_{\text{3 black}} \quad \underset{\text{1 red}}{bb}$$

MORE THAN ONE PAIR OF GENES

Suppose that there are two pairs of genes to be considered, each pair independently affecting a particular trait. Let us consider two pairs of genes, one pair of which determines coat color in cattle, and the second of which determines whether the animal is hornless (polled) or horned. The genes are designated as follows: $B =$ black (dominant), $b =$ red (recessive), $P =$ polled (dominant), and $p =$ horned (recessive).

If a bull that is heterozygous for both traits *(BbPp)* is mated to cows that are also heterozygous for both traits *(BbPp),* one can determine the expected phenotypic and genotypic ratios. Results of crossing $Bb \times Bb$ (giving a ratio of 3 black:1 red in the offspring, or $\frac{3}{4}$ black and $\frac{1}{4}$ red) has previously been shown. Similarly, a cross of $Pp \times Pp$ gives a ratio of 3 polled:1 horned. Combined, the $BbPp \times BbPp$ produces the following phenotypes:

$$BbPp \times BbPp$$

$$3 \text{ black} \begin{cases} 3 \text{ polled} = 9 \text{ black polled} \\ \\ 1 \text{ horned} = 3 \text{ black horned} \end{cases}$$

$$1 \text{ red} \begin{cases} 3 \text{ polled} = 3 \text{ red polled} \\ \\ 1 \text{ horned} = 1 \text{ red horned} \end{cases}$$

Of the $\frac{3}{4}$ that are black, $\frac{3}{4}$ will be polled and $\frac{1}{4}$ will be horned. Therefore, $\frac{3}{4}$ of $\frac{3}{4}$ (or $\frac{9}{16}$) will be black and polled; and $\frac{1}{4}$ of $\frac{3}{4}$ (or $\frac{3}{16}$) will be black and horned. Similarly, of the $\frac{1}{4}$ that are red, $\frac{3}{4}$ will be polled and $\frac{1}{4}$ will be horned, and $\frac{3}{4}$ of $\frac{1}{4}$ (or $\frac{3}{16}$) will be red and polled and $\frac{1}{4}$ of $\frac{1}{4}$ (or $\frac{1}{16}$) will be red and horned. This distribution of phenotypes is more often expressed as a 9:3:3:1 ratio instead of a set of fractions:

$$1BB \begin{bmatrix} 1PP = 1BBPP \\ 2Pp = 2\,BBPp \\ 1pp = 1\,BBpp \end{bmatrix}$$

$$2Bb \begin{bmatrix} 1PP = 2BbPP \\ 2Pp = 4\,BbPp \\ 1\,pp = 2\,Bbpp \end{bmatrix}$$

$$1\,bb \begin{bmatrix} 1\,PP = 1\,bbPP \\ 2\,Pp = 2\,bbPp \\ 1\,pp = 1\,bbpp \end{bmatrix}$$

TABLE 11.2.   Genotypes and Phenotypes with Two Heterozygous Gene Pairs

Sperm	Eggs			
	BP	*Bp*	*bP*	*bp*
BP	*BBPP*[a] black, polled[b]	*BBPp* black, polled	*BbPP* black, polled	*BbPp* black, polled
Bp	*BBPp* black, polled	*BBpp* black, horned	*BbPp* black, polled	*Bbpp* black, horned
bP	*BbPP* black, polled	*BbPp* black, polled	*bbPP* red, polled	*bbPp* red, polled
bp	*BbPp* black, polled	*Bbpp* black, horned	*bbPp* red, polled	*bbpp* red, horned

[a]Genotype. Number of different genotypes: 1 *(BBPP)*, 2 *(BBPp)*, 2 *(BbPP)*, 4 *(BbPp)*, 1 *(BBpp)*, 1 *(bbPP)*, 2 *(Bbpp)*, 2 *(bbPp)*, 1 *(bbpp)*.
[b]Phenotype. Number of different phenotypes: 9 black, polled; 3 black, horned; 3 red, polled; 1 red, horned.

Table 11.2 is another method demonstrating the different phenotypes and genotypes that can be produced from a *BbBp* (bull) × *BbPp* (cow) mating. The four different sperm have opportunity to combine with each of the four different eggs. Nine different genotypes and four different phenotypes are possible.

GENE INTERACTIONS

A gene may interact with other genes in the same chromosome (linear interaction), with its corresponding gene in a homologous chromosome (allelic interaction), or with genes in nonhomologous chromosomes (epistatic interaction). In addition, genes interact with the cytoplasm and with the environment. The environmental factors with which genes interact are internal, such as hormones, and external, such as nutrition, temperature, and amount of light.

Linear Interactions

The order in which genes occur in the chromosome is significant, because a reversal or alteration of that order may produce an entirely different phenotype. It has been demonstrated in the common fruit fly *(Drosophila)* that different phenotypes can result after the arrangement of genes on chromosomes has been altered; this has not been demonstrated in farm animals, perhaps because little is known about the location of genes in their chromosomes.

Allelic Interactions

When contrasting genes occupy corresponding loci in homologous chromosomes, each gene exerts its influence on the trait, but the effects of each of the genes depend on the relationship of dominance and recessiveness. Allelic interactions may also be termed dominance interactions. When unlike genes occupy corresponding loci, one might show complete dominance over the other. In this situation, only the effect of the dominant gene is expressed. A good example is provided by cattle, in which the hornless (polled) condition is dominant to the horned condition. The heterozygous animal is polled and is phenotypically indistinguishable from the homozygous polled animal.

There may be lack of dominance, in which heterozygous animals show a phenotype that is different from either homozygous phenotype and is usually intermediate between them. A good example is observed in sheep, in which two alleles for ear length lack dominance to each other. Sheep that are *LL* have long ears, *Ll* have short ears, and *ll* are earless. Thus the heterozygous *Ll* sheep are different from both homozygotes and are intermediate between them in phenotype.

Sometimes heterozygous individuals are superior to either of the homozygotes. This condition, in which the heterozygotes show overdominance, is an example of a selective advantage for the heterozygous condition. "Overdominance" means that heterozygotes possess greater vigor or are more desirable in other ways (such as producing more milk or being more fertile) than either of the two homozygotes that produced the heterozygote. Because the heterozygotes of breed crosses are more vigorous than the straightbred parents, they are said to possess **heterosis.** This greater vigor or productivity of crossbreds is also said to be an expression of **hybrid vigor.**

The three types of dominance can be illustrated by lines to show the possible expression of a trait:

⊢ *AA, Aa*	⊢ *AA*	⊢ *Aa*
	⊢ *Aa*	⊢ *AA*
⊢ *aa*	⊢ *aa*	⊢ *aa*
Complete dominance	Lack of dominance	Overdominance

These illustrations show that when complete dominance exists, homozygous and heterozygous dominant individuals *(AA* and *Aa)* express the same phenotype: a phenotype that is quite different from the one expressed by the homozygous recessive *(aa).* With lack of dominance, the phenotypes of the two homozygotes *(AA* and *aa)* are different from each other, and the phenotype of the heterozygote *(Aa)* is intermediate between the two homozygote phenotypes. With overdominance, the phenotype of the overdominant heterozygote *(Aa)* is superior to either of the two homozygotes *(AA* and *aa),* and one of the homozygotes is superior to the other.

The effects of dominance on the expression of traits that are important in livestock production ("production traits," such as fertility, milk production, growth rate, feed conversion efficiency, carcass merit, and freedom from inherited defects) can be illustrated by bar graphs (Fig. 11.14). Figure 11.14A, which illustrates complete dominance, shows equal performance of the homozygous-dominant and heterozygous individuals, both of which are quite superior to the performance of the recessive individual.

With lack of dominance, one homozygote is superior, the heterozygote is intermediate, and the other homozygote is inferior (Fig. 11.14B). When traits are controlled by genes that show the effects of overdominance, the heterozygote is superior to either of the homozygous types (Fig. 11.14C).

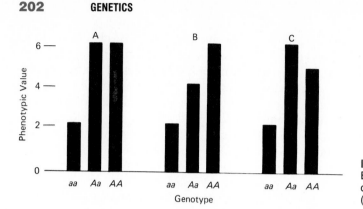

FIGURE 11.14.
Bar graphs illustrating: (A) complete dominance; (B) lack of dominance; (C) overdominance.

Epistatic Interactions

A gene or a pair of genes in one pair of chromosomes may alter or mask the expression of genes in other chromosomes. A gene that interacts with other genes that are not allelic to it is said to be **epistatic** to them. In other cases, a gene or one pair of genes in one pair of chromosomes may influence many other genes in many pairs of chromosomes. An example in which a pair of recessive genes alters the expression of many other genes is dwarfism in cattle. A dwarf animal may possess genes for rapid and economical gains and desirable carcass characteristics, but, because the animal is homozygous for dwarfism, none of these genes fully express themselves. We noted in previous discussions that when an animal that is heterozygous for two traits is mated to another animal that is heterozygous for two traits, offspring are produced in the phenotypic ratio of 9:3:3:1. The derivation of this ratio is illustrated in the following example, which uses the genes B = black, b = red, P = polled, and p = horned. When $BbPp$ animals are intermated the results are as follows:

$$BbPp \times BbPp$$

$$3 \text{ black} \begin{cases} 3 \text{ polled} = 9 \text{ black polled} \\ 1 \text{ horned} = 3 \text{ black horned} \end{cases}$$

$$1 \text{ red} \begin{cases} 3 \text{ polled} = 3 \text{ red polled} \\ 1 \text{ horned} = 1 \text{ red horned} \end{cases}$$

Let us explore some examples in which the expression of certain genes is altered by **epistatic** genes.

One Dominant Epistatic Gene. In cattle, B = black, b = red, Bs = blackish, and bs = nonblackish. Blackish (often called mahogany) is the dark color that is observed in Jersey and Ayrshire cattle. The gene for black, acting as a dominant epistatic gene, prevents the Bs and bs genes that are present from expressing themselves:

$$BbBsbs \times BbBsbs$$

$$3 \text{ black} \begin{cases} 3 \text{ blackish} \\ 1 \text{ nonblackish} \end{cases} = 12 \text{ black}$$

$$1 \text{ red} \begin{cases} 3 \text{ blackish} \\ 1 \text{ nonblackish} \end{cases} \begin{array}{l} = 3 \text{ blackish} \\ = 1 \text{ red} \end{array}$$

One of the groups of three has been placed along with the group of nine to give 12 black, because the black epistatic gene prevents the nine black individuals that carry the Bs gene from being phenotypically different from the three nonblackish individuals that have $bsbs$.

A Pair of Recessive Epistatic Genes. In rabbits, B=black, b=chocolate, C=color, and c=albino (white). An albino is white because there is no pigment in the skin or hair. It has pink eyes because there is no pigment in the eyes, and the blood vessels in the eyes therefore give them a pink appearance:

$$CcBb \times CcBb$$

$$3 \text{ colored} \begin{cases} 3 \text{ black} = 9 \text{ black} \\ 1 \text{ chocolate} = 3 \text{ chocolate} \end{cases}$$

$$1 \text{ albino} \begin{cases} 3 \text{ black} \\ \phantom{3 \text{ black}} = 4 \text{ albino} \\ 1 \text{ chocolate} \end{cases}$$

The effect of the homozygous combination of cc genes is to remove the pigment from those individuals that would otherwise be black and those that would otherwise be chocolate. Thus the cc genes act as a pair of recessive epistatic genes. One group of three and one individual have the same (albino) **phenotype**. Therefore, they are combined into a group of four albinos out of the 16.

Dominant and Recessive Epistatic Genes. In chickens, I=inhibition of color or white that is found in Leghorns, i=noninhibited (colored), W=colored, and w=white as is present in White Rock chickens. I is epistatic to W, and ww is epistatic to ii:

$IiWw \times IiWw$ (both of these heterozygous genotypes result in a white phenotype)

$$3 \text{ inhibited} \begin{cases} 3 \text{ colored} \\ 1 \text{ white} \end{cases} \Bigg\} 12 = 13 \text{ white}$$

$$1 \text{ noninhibited} \begin{cases} 1 \text{ white} \\ 3 \text{ colored} \end{cases} = 3 \text{ colored}$$

The group of nine, one group of three, and one individual are white. Therefore, they are combined to give 13 white. The resulting ratio is 13 white:3 colored.

Two Pairs of Recessive Epistatic Genes. In White Rock chickens, C=color, c=albino, W=color, and w=white. Homozygous cc prevents W from showing, and ww prevents C from showing:

$CcWw \times CcWw$ (both of these heterozygous genotypes result in a colored phenotype)

$$3 \text{ colored} \begin{cases} 3 \text{ colored} \quad 9 \quad = 9 \text{ colored} \\ 1 \text{ white} \quad\quad 3 \end{cases}$$

$$1 \text{ white} \begin{cases} 3 \text{ colored} \quad 3 \\ 1 \text{ white} \quad\quad 1 \end{cases} = 7 \text{ white}$$

Some WW and Ww chickens are albinos, and they and the albino White Rock ww have no pigment in the eyes. The CC or Cc chickens that are ww are true White Rocks, having pigment in the eyes.

INTERACTIONS BETWEEN GENES AND ENVIRONMENT

Genes interact with both external and internal environments. The external environment includes temperature, light, altitude, humidity, disease, and feed supply. Some breeds of cattle (Brahman) can withstand high temperatures better than others. Some breeds (Scottish Highland) can withstand the rigors of extreme cold better than others.

Perhaps the most important external environmental factor is feed supply. Some breeds or types of cattle can survive when feed is in short supply for considerable periods of time, and they may consume almost anything that can be eaten. Other breeds of cattle select only highly palatable feeds, and these animals have poor production when good feed is not available.

An important contributor to the status of the internal environment of animals is their hormonal condition. When the gene for blackish described previously is in the heterozygous condition *(Bsbs)*, its expression depends on the amount of androgens present. Thus, heterozygous bulls are dark mahogany color whereas heterozygous females are light mahogany. It has been shown that castration of heterozygous bulls will prevent them from being dark and that injections of testosterone into heterozygous females will cause them to become dark mahogany. Homozygous *(BsBs)* blackish animals are dark regardless of androgen levels; therefore, both males and females that are *BsBs* are dark. Both males and females that are homozygous recessive *(bsbs)* lack the gene *Bs*; therefore, both are red in color.

Allelic, epistatic, and environmental interactions all influence the genetic improvement that can be made through selection. When the external environment has a large effect on production traits, genetic improvement is quite low. For example, if animals in a population are maintained at different nutritional levels, those that are best fed obviously grow faster. In such a case, much of the difference in growth may be due to the nutritional status of the animals rather than to differences in their genetic composition. The producer has two alternatives: either standardize the environment so that it causes less variation among the animals, or maximize the expression of the production trait by improving the environment. The first approach is geared to increasing genetic improvement so that selection is more effective. Improvements made by this approach are permanent. The latter approach does not improve the genetic qualities of animals, but it allows animals to express their genetic potentials. The most sensible approach is to expose the breeding animals to an environment similar to one in which commercial animals are expected to perform economically.

SELECTED REFERENCES

Publications

Gardner, E. J., and Snustead, D. P. 1981. *Principles of Genetics*, 6th Ed. New York: John Wiley.

Hartl, D. L. 1980. *Principles of Population Genetics*. Sutherland, MA: Sinauer Associates.

Jones, W. E. 1982. *Genetics and Horse Breeding*. Philadelphia: Lea and Febiger.

Lasley, J. F. 1987. *Genetics of Livestock Improvement*. Englewood Cliffs, NJ: Prentice-Hall.

Petters, R. M. 1986. Recombinant DNA, gene transfer and the future of animal agriculture. *J. Anim. Sci.* 62:1759.

Van Vleck, L. D., Oltenacu, E., and Pollack, J. 1986. *Genetics for the Animal Sciences*. New York: W. H. Freeman.

Warwick, E. J., and Legates, J. E. 1979. *Breeding and Improvement of Farm Animals.* New York: McGraw-Hill.

Visuals

Cattle Abnormalities, (8861), 26 min (16-mm color film). Bureau of Audio-Visual Instruction, University of Wisconsin Extension Service, Box 2093, Madison, WI 53701.
Genetics of Coat Color of Horses (SL 501 slide set). Educational Aids, National 4-H Council, 7100 Connecticut Ave., Chevy Chase, MD 20815.

CHAPTER 12

Genetic Change Through Selection

CONTINUOUS VARIATION AND MANY PAIRS OF GENES

Most economically important traits in farm animals such as milk production, egg production, growth rate, and carcass composition are controlled by hundreds of pairs of genes; therefore, it is necessary to expand one's thinking beyond inheritance involving one and two pairs of genes. Consider even a simplified example of 20 pairs of heterozygous genes (1 gene pair on each pair of 20 chromosomes) affecting yearling weight in sheep, cattle, or horses. The estimated numbers of genetically different gametes (sperm or eggs) and genetic combinations are shown in Table 12.1. Remember that for one pair of heterozygous genes, there are three different genetic combinations (i.e., *AA*, *Aa*, and *aa*), and for two pairs of heterozygous genes, there are nine different genetic combinations (Table 11.2).

Most farm animals would likely have some gene pairs heterozygous and some homozygous, depending on the mating system being utilized. Table 12.2 shows the number of gametes and genotypes where eight pairs of genes are either heterozygous or homozygous and each gene pair is located on a different pair of chromosomes.

TABLE 12.1. Number of Gametes and Genetic Combinations with Varying Numbers of Heterozygous Gene Pairs

No. Pairs of Heterozygous Genes	No. of Genetically Different Sperm or Eggs	No. of Different Genetic Combinations (genotypes)
1	2	3
2	4	9
n	2^n	3^n
20	$2^n = 2^{20} = {\sim}1$ million	$3^n = 3^{20} = {\sim}3.5$ billion

TABLE 12.2.   Number of Gametes and Genetic Combinations with Eight Pair of Genes with Varying Amounts of Heterozygosity and Homozygosity

Genes in Sire: Genes in Dam:	*Aa* *aa*		*Bb* *Bb*		*Cc* *CC*		*Dd* *Dd*		*Ee* *Ee*		*FF* *FF*		*GG* *gg*		*Hh* *Hh*		Total
No. different sperm for sire	2	×	2	×	2	×	2	×	2	×	1	×	1	×	2	=	64
No. different eggs for dam	1	×	2	×	1	×	2	×	2	×	1	×	1	×	2	=	16
No. different genetic combinations possible in offspring	2	×	3	×	2	×	3	×	3	×	1	×	1	×	3	=	324

Many economically important traits in farm animals show continuous variation primarily because they are controlled by many pairs of genes. As these many genes express themselves, and the environment also influences these traits, producers usually observe and measure large differences in the performance of animals for any one trait. For example, if a large number of calves were weighed at weaning (~205 days of age) in a single herd, there would be considerable variation in the calves' weights. Distribution of weaning weights of the calves would be similar to examples shown in Fig. 12.1 or 12.2. The bell-shaped curve distribution demonstrates that most of the calves are near the average for all calves, with relatively few calves having extremely high or low weaning weights when compared at the same age.

Figure 12.2 shows the use of the statistical measurement of standard deviation (SD) which describes the variation or differences in a herd where the average weaning weight was 440 lb and the calculated standard deviation was 40 lb. Using herd average and

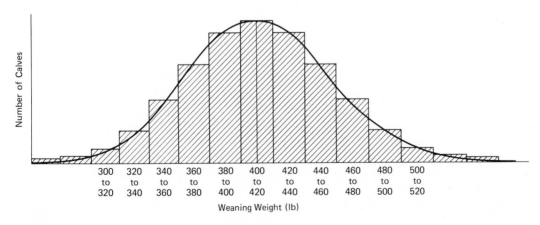

FIGURE 12.1.
Variation or differences in weaning weight in beef cattle. The variation shown by the bell-shaped curve could be representative of a breed or a large herd. The dark vertical line in the center is the average or the mean and, in this example, is 410 pounds.

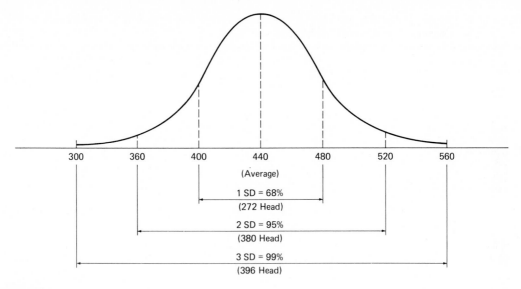

FIGURE 12.2.
A normal, bell-shaped curve for weaning weight showing the number of calves in the area under the curve. (Four hundred calves in the herd.)

standard deviation, the variation in weaning weight is shown in Fig. 12.2 and can be described as follows:

$$1 \text{ SD: } 440 \text{ lb} \pm 1 \text{ SD } (40 \text{ lb}) = 400\text{--}480 \text{ lb}$$
$$\text{(68\% of the calves are in this range)}$$
$$2 \text{ SD: } 440 \text{ lb} \pm 2 \text{ SD } (80 \text{ lb}) = 360\text{--}520 \text{ lb}$$
$$\text{(95\% of the calves are in this range)}$$
$$3 \text{ SD: } 440 \text{ lb} \pm 3 \text{ SD } (120 \text{ lb}) = 320\text{--}560 \text{ lb}$$
$$\text{(99\% of the calves are in this range)}$$

One percent of the calves (4 calves in a herd of 400) would be on either side of the 320- to 560-lb range. Most likely, two calves would be below 320 lb and two calves would weigh more than 560 lb.

Weaning weight is a phenotype, since the expression of this characteristic is determined by the genotype (genes received from the sire and the dam) and the environment to which the calf is exposed. The genetic part of the expression of weaning weight is obviously not simply inherited. There are many pairs of genes involved, and, at the present time, the individual pairs of genes cannot be identified similar to traits controlled by one to two pair of genes.

Consider the following formula:

$$phenotype = genotype \pm environment$$

The phenotype will more closely predict the genotype if producers expose their animals to a similar environment; however, the latter must be within economic reason. This resemblance between phenotype and genotype, where many pairs of genes are involved,

FIGURE 12.3.
Variation in belt pattern in Hampshire swine. Courtesy of *The Hampshire Herdsman.*

is predicted with the estimate of heritability. Use of **heritability,** along with selecting animals with superior phenotypes, is the primary method of making genetic improvement in traits controlled by many pairs of genes. The application of this method is discussed in more detail later in this chapter.

Traits that are influenced very little by the environment can also show considerable variation. An example would be the white belt, a breed-identifying characteristic in Hampshire swine (Fig. 12.3). Note that the white belt can be nonexistent in some Hampshires whereas others can be almost white. To be eligible for pedigree registration, Hampshire pigs must be black with the white belt entirely circling the body including both front legs and feet. Too much white can also limit registrations. It is difficult to select a herd of purebred Hampshires that will breed true for desired belt pattern. Apparently some of the genes for this trait exist in heterozygous or epistatic combinations.

In a beef herd, where the average weaning weight is 440 lb, or a dairy herd, where milk production averages 12,500 lb per cow, the managers of these herds want to increase the herd average. This assumes the increased production is economically feasible—that the economic benefits of improvement will cover the additional costs.

SELECTION

Selection is differential reproduction—preventing some animals from reproducing while allowing other animals to become parents of numerous offspring. In the latter situation, the selected parents should be genetically superior for the economically important traits. Factors affecting the rate of genetic improvement from selection include **selection differential, heritability,** and **generation interval.**

SELECTION DIFFERENTIAL

Selection differential is sometimes called **reach,** because it shows the superiority (or inferiority) of the selected animals compared to the herd average. To improve weaning weights in beef cattle, producers cull as many below-average-producing cows as economically feasible. Then replacement heifers and bulls would be selected that are above the herd average for weaning weight. For example, if the average weaning weight of the selected replacement heifers was 470 lb in a herd averaging 410 lb, then the selection differential for the heifers would be 60 lb. Part of this 60-lb difference is due to genetic differences, and the remaining part is due to differences caused by the environment.

HERITABILITY

A **heritability** estimate is a percentage figure that shows what part of the total phenotypic variation (phenotypic differences) is due to genetics. Heritability can also be defined as that portion of the selection differential that is passed on from parent to offspring.

Realized heritability is the portion actually obtained compared to what was attempted in selection. To illustrate realized heritability, let us suppose a farmer has a herd of pigs whose average postweaning gain is 1.80 lb/day. If the farmer selects from this original herd, a breeding herd whose members have an average gain of 2.3 lb/day, the farmer is selecting for an increased daily gain of 0.5 lb/day.

If the offspring of the selected animals gain 1.95 lb, then an increase of 0.15 lb (1.95 − 1.80) instead of 0.5 lb (2.3 − 1.8) has been obtained. To find the portion obtained of what was reached for in the selection, one divides 0.15 by 0.5, which gives 0.3. This figure is the realized heritability. If one multiplies 0.3 by 100%, the result is 30%, which is the percentage obtained of what was selected for in this generation.

Table 12.3 shows heritability estimates for several species of livestock. Traits having heritability estimates 40% and higher are considered highly heritable; 20–39% are classified as having medium heritability, and lowly heritable traits are below 20%.

PREDICTING GENETIC CHANGE

The rate of genetic change that can be made through selection can be estimated using the following equation:

$$\text{Genetic change per year} = \frac{\text{heritability} \times \text{selection differential}}{\text{generation interval}}$$

Consider the following example of selecting for weaning weight in beef cattle and predicting the genetic change. The herd average is 400 lb and bulls and heifers are selected from within the herd.

BULLS

Average of selected bulls	540 lb
Average of all bulls in herd	400 lb
Selection differential	140 lb
Heritability	0.30
Total genetic superiority	42 lb
Only half passed on (42 ÷ 2)	21 lb

HEIFERS

Average of selected heifers	460 lb
Average of all heifers in herd	400 lb
Selection differential	60 lb
Heritability	0.30
Total genetic superiority	18 lb
Only half passed on (18 ÷ 2)	9 lb

Fertilization combines the genetic superiority of both parents which in this example is 21 lb + 9 lb = 30 lb. The calves obtain half of their genes from each parent, therefore $\frac{1}{2}$ (42) + $\frac{1}{2}$ (18) equals 30 lb.

TABLE 12.3. Heritability Estimates for Selected Traits in Several Species of Farm Animals

Species Trait	Percent Heritability	Species Trait	Percent Heritability
Beef Cattle		Thoroughbred racing	
Calving interval	10	Log of earnings	50
Age at puberty	40	Time	15
Scrotal circumference	50	Pacer: best time	15
Birth weight	40	Trotter	
Weaning weight	30	Log earnings	40
Post weaning gain	45	Time	30
Yearling weight	40		
Yearling hip (frame size)	40	**Poultry**	
Mature weight	50	Age at sexual maturity	35
Carcass quality grade	40	Total egg production	25
Yield grade	30	Egg weight	40
Cancer eye susceptibility	30	Body weight	40
		Shank length	45
Dairy Cattle		Egg hatchability	10
Services per conception	5	Livability	10
Birth weight	50		
Milk production	25	**Sheep**	
Fat production	25	Number born	15
Protein	25	Birth weight	30
Solids-not-fat	25	Weaning weight	30
Type score	30	Postweaning gain	40
Teat placement	20	Mature weight	40
Mastitis (susceptibility)	10	Fleece weight	40
Milking speed	30	Face covering	55
Mature weight	35	Loin eye area	55
Excitability	25	Carcass fat thickness	50
		Weight of retail cuts	50
Goats			
Milk production	30	**Swine**	
Mohair production	20	Litter size	10
		Birth weight	5
Horses		Litter weaning weight	15
Withers height	45	Postweaning gain	30
Pulling power	25	Backfat probe	40
Riding performance		Carcass fat thickness	50
Jumping (earnings)	20	Loin eye area	45
Dressage (earnings)	20	Percent lean cuts	45
Cutting ability	5		

The selected heifers represent only approximately 20% of the total cow herd, so the 30 lb is for one generation. Since the heifer replacement rate in the cow herd is 20% it would take 5 years to replace the cow herd with selected heifers. Because of this **generation interval**, the genetic change per year from selection would be 30 lb $\div$ 5 = 6 lb/year.

Generation interval is defined as the average age of the parents when the offspring are born. This is calculated by adding the average age of all breeding females to the average age of all breeding males and dividing by 2. The generation interval is approximately 2 years in swine, 3–4 years in dairy cattle, and 5–6 years in beef cattle.

The previous example of 5 lb/year shows genetic change if selection was for only one trait. If selection is practiced for more than one trait, genetic change is $1/\sqrt{n}$ where n is the number of traits in the selection program. If four traits were in the selection program, the genetic change per trait would be $1/\sqrt{4} = \frac{1}{2}$. This means that only $\frac{1}{2}$ the progress would be made for any one trait compared to giving all the selection to one trait. This reduction in genetic change per trait should not discourage producers from multiple-trait selection, as herd income is dependent on several traits. Maximizing genetic progress in a single trait may not be economically feasible, and it may lower productivity in other economically important traits.

Sometimes traits are genetically correlated. A genetic correlation between two traits means that some of the same genes affect both traits. In swine, the genetic correlation between rate of gain and feed per pound of gain (from similar beginning weights to similar end weights) is negative. This is desirable from a genetic improvement standpoint, because animals that gain faster will require less feed (primarily less feed for maintenance). This relationship is also desirable because rate of gain is easily measured whereas feed efficiency is expensive to measure.

Yearling weight or mature weight, in cattle, is positively correlated with birth weight, which means that as yearling or mature weight increases, birth weight will also increase. This may pose a potential problem, since birth weight reflects calving difficulty and calf death loss to a large extent.

EVIDENCE OF GENETIC CHANGE

The previously shown examples of selection are theoretical. Does selection really work, or is the observed improvement in farm animals a result of improving only the environment? Let's evaluate several examples.

There have been marked visual, meat-to-bone ratio changes in the thick-breasted modern turkey selected from the narrow-breasted wild turkey, which is the only animal native to the United States. Estimates for genetic change for meat-to-bone ratio have been 0.5% per year for the modern turkey. Genetic selection and improving the environment, particularly through improved nutrition and health, have produced the modern turkey. Tom turkeys can produce 27 lb of liveweight (22 lb dressed weight) in 5 months; the wild turkey would weigh 10 lb in 6 months.

The modern turkey would likely not survive in the same environment as the wild turkey. The modern turkey needs an environment with intensive management. For example, turkeys are mated artificially because the heavy muscling in the breast prevents them from mating naturally. Selection has been effective in changing rate of growth and meatiness in turkeys, but it has necessitated a change in the environment for these birds

TABLE 12.4.   Height and Weight Differences in Mature Draft Horses, Light Horses, Ponies, and Miniature Horses

Horse Type	Approximate Height at Withers		Approximate Mature Weight (lb)
	Hands	Inches	
Draft	17	68	1,600
Light	15	60	1,100
Pony	13	52	700
Miniature	8	32	250

to be productive. Therefore, the improvement in turkeys has resulted from improving both the genetics and the environment.

There are tremendous size differences in horses. Draft horses have been selected for large body size and heavy muscling to perform work. The light horse is more moderate in size, and ponies are small. Miniature horses have genetic combinations that result in a very small size. The miniature horse is considered a novelty and a pet. Table 12.4 shows the height and weight variations in different types of horses. Apparently these different horse types have been produced from horses that originally were quite similar in size and weight. The extreme size differences in horses result primarily from genetic differences, as the size differences are apparent when the horses are given similar environmental opportunities.

One of the best documented examples of genetic change in farm animals comes from a back fat selection experiment in swine. Figure 12.4 shows the results of 13 generations of selection for high and low back fat thickness. Also, a line of pigs was maintained in which no selection for back fat thickness was practiced. This nonselected line would show the environmental changes in back fat thickness from year to year. In this experiment, back fat thickness was the only trait for which selection was practiced. Selection for and against back fat thickness was effective, since back fat thickness is a highly heritable trait (50%).

The back fat probe has been used to measure differences in back fat thickness in live pigs (Fig. 12.5). Use of the back fat probe has improved the effectiveness of selection, as back fat thickness can be identified in breeding animals without having to slaughter them.

The effectiveness of selection against excessive back fat thickness in swine is shown in Table 12.5. The reduction in lard production per hog has decreased dramatically, even though the average slaughter weight has increased from 1960 to 1983.

Selection for milk production in dairy cattle has an additional challenge in a genetic improvement program because the bull does not express the trait. This is an example of a **sex-limited** trait.

Genetic evaluation of a bull is based primarily on how his daughters' milk production compares to that of their **contemporaries.** Milk production is a moderately heritable trait (25%), and the genetic change over the past 25 years is most noteworthy, even though the trait is not expressed in the bull.

A

B

C

FIGURE 12.4.
Examples of 13 generations of selecting for high and low fatness in swine. (A) High-fat slaughter pig. (B) Low-fat slaughter pig. (C) Carcass loin sections (from left to right): high-fat (slaughter weight, 212 lb.; 3.25 in backfat); control pig—no selection (slaughter weight, 216 lb.; 2.10 in backfat); low-fat (214 lb. slaughter weight; 1.30 in backfat). Courtesy of the USDA.

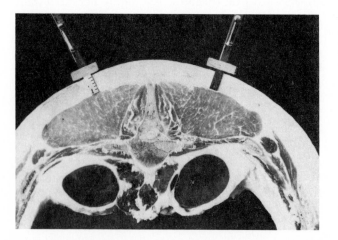

FIGURE 12.5.
A pork carcass cross section demonstrating how the steel ruler backfat probe measures backfat thickness in the live pig. Courtesy of Iowa State University.

TABLE 12.5.   Trends in U.S. Lard Production

Year	Total Hog Slaughter (mil head)	Total Lard Production (mil lb)	Average Slaughter Weight (lb)	Lard Production per Hog (lb)
1950	79.3	2,631		33.2
1960	84.1	2,562	236	30.4
1970	87.0	1,913	240	22.0
1980	97.2	1,217	242	12.5
1985	79.6	852	247	10.7

Source: USDA.

GENETIC IMPROVEMENT THROUGH ARTIFICIAL INSEMINATION

The primary reason for using artificial insemination is that genetically superior sires can be used more extensively. AI is used more in dairy cattle than any other species of farm animals. Using AI dairy bulls has resulted in higher milk production than using natural service bulls.

Figure 12.6 shows the rate of genetic improvement for yearling weight in beef cattle under different selection schemes. The most significant points are these:

1. Marked difference between herd A (no AI, males and females selected for yearling weight) and herd D (bulls obtained from a herd where no selection is practiced for yearling weight).

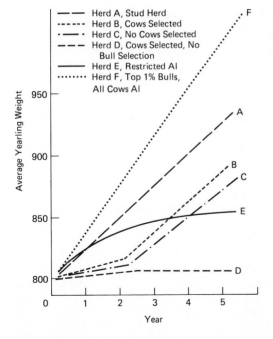

FIGURE 12.6.
Expected genetic change in different types of beef herds. Adapted from Magee. *Mich. State Univ. Qtr. Bul.* 48:4.

2. The small difference between herd B (selects bulls from herd A on yearling weight and selects for yearling weight in his females) and herd C (selects bulls from herd A but practices no selection for yearling weight on females). Effective sire selection will account for 80–90% of the genetic change in a herd.
3. Greatest change in herd F where the top 1% of the sons of the top 1% of the bulls were used to build generation upon generation of AI selection. The rate of improvement using AI in herd F is 1.5 times that of herd A with natural service. The primary reasons for the greater improvement in herd F is a larger selection differential.

SELECTION METHODS

The three methods of selection are (a) **tandem,** (b) **independent culling level,** and (c) **selection index** where the latter is based on an index of net merit.

Tandem is selection for one trait at a time. When the desired level is achieved in one trait, then selection is practiced for the second trait. The tandem method is rather ineffective if selection is for more than two traits or if the desirable aspect of one trait is associated with the undesirable aspect of another trait. The tandem method is not recommended, as it is the least effective of the three methods of selection.

Independent culling level establishes minimum culling levels for each trait in the selection program. Even though it is the second most effective type, it is the most prevalent one used. Independent culling level is most useful when the number of traits being considered is small and when only a small percentage of offspring is needed to replace the parents. For example, with poultry, 100 or more chicks might be produced from each hen and 1,500 or more chicks from each rooster. Here only 1 female out of 50 or 1 male out of 750 is needed to replace the parents. It is simple using this method to select from the upper 10% of the females and the upper 1% of the males.

Table 12.6 shows an example of using independent culling levels in selecting yearling bulls. Birth weight is correlated to calving ease. Weaning weight and postweaning gain measure growth. Puberty and semen production are measured with scrotal circumference. Any bull that does not meet the minimum or maximum level is culled. Birth weights are evaluated on a maximum level, since high birth weights result in increased calving difficulty.

The circled records in Table 12.6 indicate bulls that did not meet the independent culling levels. Bulls A, B, and D would be culled.

In the selection index method, each animal is rated numerically, in comparison to each other animal for each trait being considered. A simplified example of the selection index

TABLE 12.6. Independent Culling Level Selection in Yearling Bulls

Trait	Culling Level	Bull				
		A	B	C	D	E
Birth weight (lb)	85 (max)	(105)	82	85	83	80
Weaning weight (lb)	500 (min)	550	(495)	505	570	600
Postweaning gain (lb/day)	3.0 (min)	3.7	3.6	3.1	3.3	3.5
Scrotal circumference (cm)	30 (min)	34	37	31	(29)	35

TABLE 12.7.   Simplified Selection Index Method of Selecting Yearling Bulls Based on Total Score

Trait	Bulls				
	A	B	C	D	E
Birth weight	5	2	4	3	1
Weaning weight	3	5	4	2	1
Postweaning gain	1	2	5	4	3
Scrotal circumference	3	1	4	5	2
Total	12	10	17	14	7

method would be to rank each of the five bulls in Table 12.3 on a scale of 1 to 5 with 1 representing the best performance and 5 the poorest. Table 12.7 gives the ranks of the bulls based on the 1 to 5 rating scale.

In Table 12.7, the total score (sum of rankings for each of the four traits) is shown for each bull. The bull with the lowest total score would represent the best bull. The bull with the highest total score would be the poorest bull.

A comparison of Tables 12.6 and 12.7 identifies some strengths and weaknesses of independent culling level versus index method of selection. Bull E is evaluated as the best bull under both selection methods.

Bulls B and C identify major weaknesses of independent culling levels. Under the selection index method, bull B was the second-best bull; using independent culling levels, the bull was culled (and likely castrated) when weaned, at approximately 7 months of age. Bull B is superior in four traits but slightly below the culling level for weaning weight.

Bull C barely survives the four culling levels; however, he is ranked second to last in the index method. Therefore, mediocre animals may be selected using independent culling levels but culled using the index method.

An advantage of independent culling levels is that selection can occur during different productive stages during the animal's lifetime (i.e., at weaning). This is more cost-effective than the index method, where no culling should occur until records are recorded for all traits. A combination of the two methods may be most useful and cost effective.

The example in Table 12.7 is oversimplified. Equal emphasis was given to each of the four traits. A selection index could be computed for the traits in Table 12.7 by putting the heritabilities, economic values, genetic correlations, and variabilities of these four traits into a single formula. This process is rather complicated, as it requires the use of complex statistical methods. An example of such an index, used in boar selection, is given in Chapter 25.

BASIS FOR SELECTION

The primary bases of selection—(1) pedigree (names and records of relatives), (2) individual appearance or individual performance, and (3) progeny testing—are shown in Fig. 12.7.

Pedigree information is most useful before animals have expressed their own individ-

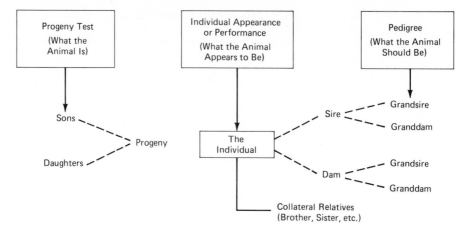

FIGURE 12.7.
The selection methods (how they tie together and a brief appraisal of each).

ual performance or their progeny's performance is known. Pedigree is also useful for assessing genetic abnormalities, assessing traits expressed much later in life (longevity), and selecting for traits expressed only in one sex. In evaluating pedigree information, only the information from the animal's closest relatives should be used because of the genetic relationship involved. For example, the genetic relationship of an animal to each parent is $\frac{1}{2}$ (50%), to each grandparent $\frac{1}{4}$ (25%), and to each great grandparent $\frac{1}{8}$ (12.5%).

Selection on individual appearance and performance should be practiced on traits that are economically important and have high heritabilities. One primary advantage of selection on individual performance is that it permits rapid generation turnover and thus shortens the general interval.

Progeny testing is more accurate than the other two methods, provided the progeny test is accurate and extensive. Progeny tests are particularly useful in selecting for carcass traits (where good carcass indicators are not available in the live animal), sex-limited traits such as milk production, and traits with low heritabilities.

Disadvantages of progeny testing are that too few animals can be progeny-tested, the longer generation interval required to obtain progeny information, and the decreased accuracy in poorly conducted tests. Progeny testing is usually limited to sires, because a sufficient number of offspring cannot be obtained on a dam to give an adequate progeny test.

In a proper progeny test, sires are assigned females at random after the dams have been classified by age, line of breeding, and known performance. A sire, bred to a selected group of females, can give biased information that does not accurately evaluate his breeding ability. An adequate progeny test involves a minimum number of progeny. The greater the numbers, the greater the validity of the test. Progeny test data will be biased and accuracy greatly reduced if dams and offspring are not fed and managed uniformly until all data are collected.

All three bases of selection can be used effectively in animal breeding programs. Outstanding young sires can best be identified by using individual performance and pedigree information on the sire, dam, and half-sibs. These outstanding young sires then need to prove themselves genetically based on a valid progeny testing program. Results of progeny tests will determine which sires to use extensively in later years.

SELECTED REFERENCES

Publications

Genetics and Goat Breeding (Symposia). 1982. Proceedings of the Third International Conference on Goat Production and Disease. Scottsdale, AZ: Dairy Goat Journal.

Herman, H. A. 1981. *Improving Cattle by the Millions. NAAB and the Development and Worldwide Application of Artificial Insemination.* Columbia: University of Missouri Press.

Hetzer, H. O., and Harvey, W. R. 1967. Selection for high and low fatness in swine. *J. Anim. Sci.* 26:1244.

Hintz, R. L. 1980. Genetics of performance in the horse. *J. Anim. Sci.* 51:582.

Lasley, J. F. 1987. *Genetics of Livestock Improvement.* Englewood Cliffs, NJ: Prentice-Hall.

Van Vleck, L. D., Oltenacu, E., and Pollack, J. 1986. *Genetics for the Animal Sciences.* New York: W. H. Freeman.

Warwick, E. J., and Legates, J. E. 1979. *Breeding and Improvement of Farm Animals.* New York: McGraw-Hill.

Visuals

Development of the Modern Chicken (79 slides and audio tape; 15 min) and *Development of the Modern Turkey* (76 slides and audio tape; 14 min). Poultry Science Department, Ohio State University, 674 W. Lane Ave., Columbus, OH 43210.

CHAPTER 13

Mating Systems

Purebred breeders (sometimes called seedstock producers) and commercial breeders (producers) are the two general classifications of animal breeders. **Purebred** livestock typically come from the pure breeds where the ancestry is recorded on a pedigree by a breed association (Fig. 13.1). Most commercial slaughter livestock are crossbreds, resulting from crossing two or more breeds or lines of breeding.

A knowledge of breeding or mating systems is important because producers can utilize these mating systems to preserve genetic superiority and utilize hybrid vigor. Genetic improvement can be optimized in most herds and flocks by utilizing a combination of selection and mating systems.

Mating systems are identified primarily by genetic relationship of the animals being mated. The two major systems of mating are **inbreeding** and **outbreeding.** Inbreeding is mating of animals more closely related than the average of the breed or population. Outbreeding is mating of animals not as closely related as the average of the population.

Since mating systems are based on the **relationship** of animals being mated, it is important to understand more detail about genetic relationship. Proper pedigree evaluation also involves understanding the genetic relationship between animals. Relationship is best described in knowing which genes two animals have in common and whether the genes in an animal or animals exist primarily in a heterozygous or homozygous condition. Figure 13.2 shows the mating systems and their relationship to a homozygous or heterozygous condition.

INBREEDING

Because inbreeding is the mating of related animals, the resulting inbred offspring have an increased homozygosity of gene pairs compared to noninbred animals in the same population (breed or herd). An example of inbreeding is shown in Fig. 13.3, where animal A has resulted from mating B (the sire) to C (the dam). Note that B and C have the same parents (D and E). B and C are genetically related because they have a brother–

CERTIFICATE OF REGISTRY   No.579621 F

National Suffolk Sheep Association

P. O. Box 324   Columbia, Missouri 65205

EWE Single

~~"Naxos"~~

Strahl 85-1	Sire Strahl Y414 N186304M	Sire Marshall 82-518 N170065M
		Dam Strahl S14 Enf N371105F
Ear Tag_____		"Vegas"
Year Born January 21, 1985		Sire Nelsh MG202 N180403M
Production Registry		
Index	Dam Strahl VG 30 N395647F	Dam Strahl BL160 N219401F

Breeder Maurice Strahl & Sons, Greenfield, Indiana

Owner (2nd) August 29, 1985, Burke & Nelsh Suffolks, Platte City, Missouri

DATE August 19, 1985 _____ Secretary _Betty Buellin_

SAMPLE CERTIFICATE OF REGISTRY – Not Valid for Registry
purposes

The above pedigree is on record in the NATIONAL SUFFOLK SHEEP ASSOCIATION.

FIGURE 13.1.
Pedigree of a Suffolk ewe, Strahl 85-1 (registration number 579621 F). Note in this pedigree that the sire, dam, and four grandparents are listed, with a registration number following each of their names. Courtesy of the National Suffolk Sheep Association.

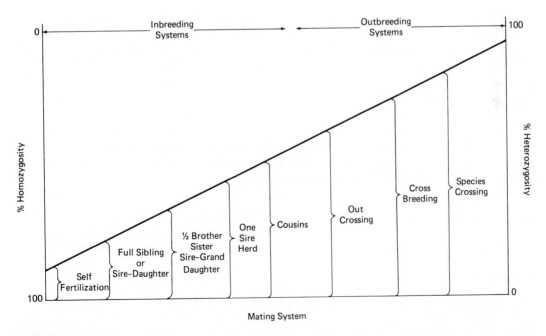

FIGURE 13.2.
Relationship of mating system to the amount of heterozygosity or homozygosity. Self-fertilization is currently not an available mating system in animals.

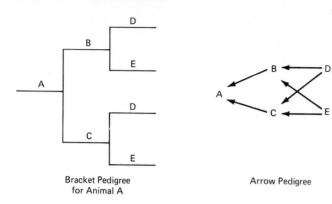

FIGURE 13.3.
Bracket pedigree and arrow pedigree showing animal A resulting from a full brother-sister mating.

sister relationship. B and C do not have the same identical genetic make-up because each has received a sample half of genes from each parent. The arrow in the arrow pedigree represents a sample half of genes from the parent to the offspring.

Animal A is inbred because it has resulted from the mating of related animals. Although the calculations are not shown, animal A has 25% more homozygous genes as compared with a noninbred animal in the same population.

Figure 13.4 shows another example of inbreeding where animal X has resulted from a sire–daughter mating. Sire A was mated to dam D, with D being sired by A. The increased homozygosity of animal X is 25%, which is the same as animal A in Fig. 13.3.

The different forms of inbreeding are:

1. intensive inbreeding—mating of closely related animals whose ancestors have been inbred for several generations.
2. linebreeding—a mild form of inbreeding where inbreeding is kept relatively low while maintaining a high genetic relationship to an ancestor or line of ancestors.

Intensive Inbreeding

Intensive inbreeding occurs by mating closely related animals for several generations. Figure 13.5 shows an example of intensive inbreeding. The increase in the homozygosity of animal H's genes is higher than 25% (compare with Fig. 13.4) because both the sire and grandsire of animal H were also inbred.

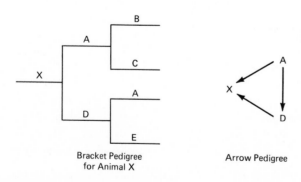

FIGURE 13.4.
Bracket pedigree and arrow pedigree showing animal X resulting from a sire-daughter mating.

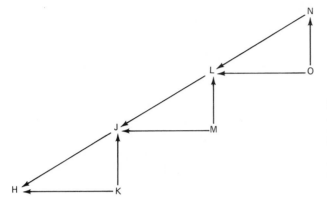

FIGURE 13.5.
An arrow pedigree showing animal H resulting from three generations of sire-daughter matings. Note that J is the sire of H and also the sire of K, the latter being the dam of H. Animal J has resulted from two previous, successive generations of sire-daughter matings.

There are numerous, genetically different inbred lines that can be produced in a given population such as a breed. The number of different, completely homozygous inbred lines is 2^n, where n is the number of heterozygous gene pairs. For example, with 2 pairs of heterozygous gene pairs, there are 2^2 or 4 different inbred lines that are possible. For example, with *BbPp* genes in a herd completely heterozygous (see Chapter 11), the different, completely homozygous lines that can result from inbreeding are *BBPP*, *BBpp*, *bbPP*, and *bbpp*. Each of these inbred lines is genetically pure and will breed true.

There have been research projects with swine and beef cattle in which inbred lines have been produced, anticipating that the crossing of these inbred lines would produce results similar to hybrid corn (Fig. 13.6). One continuing research project is at The San Juan Basin Research Center (SJBRC) at Hesperus, Colorado, where cattle have been inbred for more than 30 years. Figure 13.7 shows the arrow pedigree of an inbred bull (Royal 4160) produced in the Royal line of Hereford cattle. Note that College Royal Domino 3 is the sire of 10 animals in Royal 4160's pedigree, while Royal 3016 has sired 6 animals in the same pedigree. The increased homozygosity of this bull is 58%, which represents some of the most highly inbred cattle in the world.

Information obtained from inbreeding studies with farm animals demonstrates the following results and observations:

1. Increased inbreeding is usually detrimental to reproductive performance, preweaning and postweaning growth. Also, inbred animals are more susceptible to environmental stresses. Whereas 60–70% of the inbred lines show the detrimental effects to increased inbreeding, there are 30–40% of the lines that show no detrimental effect with some lines demonstrating improved productivity.
2. In a Colorado research beef herd, the inbred lines showed a yearly genetic increase of 2.6 lb in weaning weight over a 26-year period while the crosses of inbred lines made a 4.6 lb increase over the same time period. Heterosis is demonstrated in the line crosses, and the 4.6 lb increase is typical of what breeders might expect from using intense selection in an outbred herd.
3. Inbreeding quickly identifies some desirable genes and also undesirable genes, particularly the hidden, serious recessive genes.
4. Inbred animals with superior performance are most likely to have superior breeding

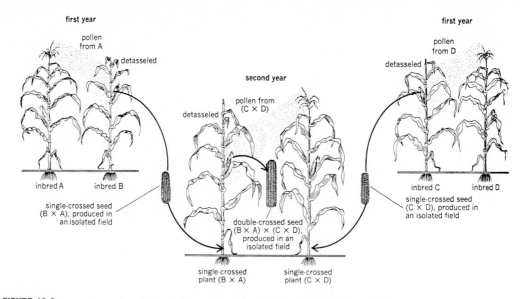

FIGURE 13.6.
Crossing of inbred lines of corn to eventually produce the double-crossing hybrid corn seed utilized widely today. Note the relative size of the corn ears from the inbred lines compared with the resulting hybrid cross. Courtesy of the USDA.

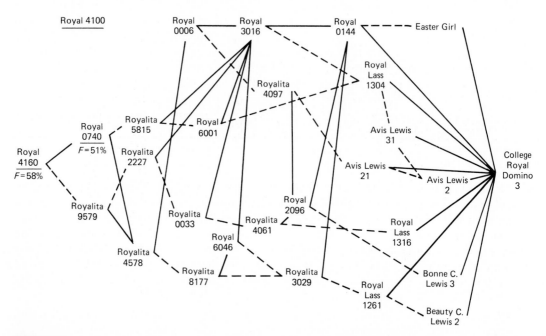

FIGURE 13.7.
Arrow pedigree of Royal 4160, which has an inbreeding coefficient of 58%. Solid lines represent the genetic contribution of the bulls, whereas the cow's contribution is represented by a broken line. Courtesy of the CSU Expt. Sta. (San Juan Basin Research Center) General Series 982.

values, which will result in more uniform progeny with high levels of genetically influenced productivity.

5. Crossing of inbred lines results in heterosis; however, in most cases heterosis compensates for inbreeding depression.

6. Crossing of inbred lines of animals has not yielded the same results as crossing inbred lines of corn. The reasons appear to be: (a) inbreeding animals is slower (can not self-fertilize as in corn); (b) easier and less costly to produce more inbred lines of corn; (c) inbred lines of animals are eliminated because of extremely poor reproductive performance and being less adapted to environmental stress.

7. There is merit in using some inbreeding in developing new lines of poultry and swine. This is discussed in more detail later in the chapter.

It is not logical for breeders to develop their own lines of highly inbred beef cattle. Inbreeding depression will usually affect the economics of the operation. Both purebred and commercial producers can take advantage of highly productive inbred lines by crossing these inbred bulls on unrelated cows.

Inbreeding such as sire–daughter matings are logical ways to test for undesirable recessive genes. Also, seedstock producers use inbreeding in well-planned linebreeding programs. Consideration should be given to implementing linebreeding when breeders have difficulty introducing sires from other herds that are genetically superior to those they are producing.

Linebreeding

Linebreeding is a mild form of inbreeding used to maintain a high genetic relationship to an outstanding ancestor, usually a sire. This mating system is best used by seedstock producers who have a high level of genetic superiority in their herd. They find it difficult to locate sires which are superior to the ones they are raising in their herds.

Occasionally a breeder may produce a sire with a superior combination of genes, where the sire consistently produces high-producing offspring. Some of these sires may not be out-produced by younger sires. This is observed in some dairy bulls where they remain competitively superior as long as they produce semen. These are the sires around which a linebreeding program can be built.

Figure 13.8 gives an example of linebreeding. Impressive, an outstanding quarter horse stallion, is linebred to his ancestor, Three Bars by three separate pathways. The inbreeding of Impressive is approximately 9%, whereas the genetic relationship of Impressive to Three Bars is approximately 44%. Inbreeding below 20% is considered low, whereas a genetic relationship is high when it approaches 50%.

Impressive has nearly the same genetic relationship as if Three Bars had been his sire (44% vs. 50%). Progeny of Impressive have produced outstanding records, particularly as halter point and working point winners in the showring.

OUTBREEDING

The different forms of outbreeding are:

1. species cross—crossing of animals of different species (e.g., horse to donkey or cattle to bison).

2. crossbreeding—mating of animals of different established breeds.
3. outcrossing—mating of unrelated animals within the same breed.
4. grading up—mating of purebred sires to commercial grade females and their female offspring for several generations. Grading up can involve some crossbreeding or it can be a type of outcrossing system.

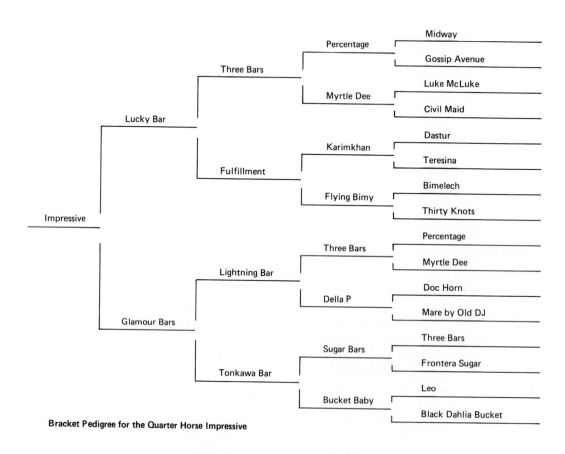

Bracket Pedigree for the Quarter Horse Impressive

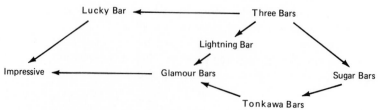

**Arrow Pedigree Showing the Genetic Pathways by which
Three Bars Contributes to the Inbreeding and Linebreeding of Impressive**

FIGURE 13.8.
Horse pedigree showing linebreeding. Courtesy of the American Quarter Horse Association.

Species Cross

A species designation is part of the zoological classification (Fig. 13.9), where taxonomists (a branch of zoology) classify animals on similarities in body structure. These differences and similarities in body structure translate to genetic differences and similarities. For example, some animals of different species, but the same genus, can be crossed to produce viable offspring. Animals of different genus cannot be successfully crossed because chromosome number and other genes are different. Therefore, a species cross is the widest possible kind of outbreeding that can be achieved.

One of the most common species cross is the **mule,** resulting from crossing the jack of the ass species and the mare of the horse species *(Equus asinus × Equus caballus).* Mules existed in large numbers as work animals before the advent of the tractor. The **hinny** is the reciprocal cross of the mule *(Equus caballus* stallion *× Equus asinas* **jennet**). The hinny never became as popular as the mule.

Mare mules are usually sterile, which gives verification to genetic differences between the ass and horse. There have been a few reports of fertile mare mules but most of these have not been documented.

Crossing of the zebu or humped cattle on European-type cattle *(Bos indicus × Bos taurus)* is very common in the southeastern part of the U. S. These crosses are more adaptable and productive in that hot, humid environment than either of the straight species.

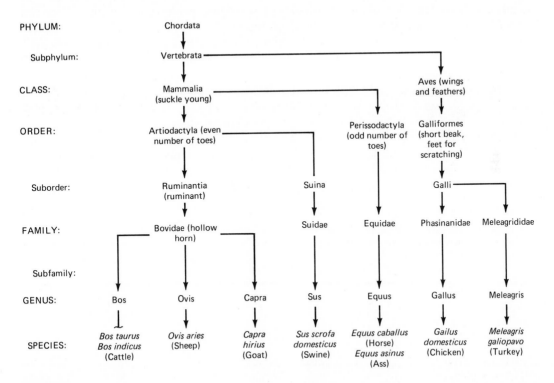

FIGURE 13.9.
Zoological classification that identifies the major species of farm animals. Adapted from Plimpton, R. F. and Stephens, J. F. 1979. *Animal and Science for Man: A Study Guide.* Minneapolis: Burgess Publishing Company.

Some authorities raise questions about *Bos indicus* and *Bos taurus* being separate species, and their crosses are usually referred to as crossbreds rather than species crosses.

Numerous crosses of American bison and cattle have been made. Some of these crosses have been designated as separate breeds called **Cattalo** or **Beefalo.** These crosses were intended to be more adaptable to harsh environments (cold and limited forage). Fertility problems have existed in some of these crosses, and these numbers are limited.

Sheep and goats have been crossed even though they have different genus classification. Fertilization occurs but embryos die in early gestation.

There are other species crosses that have occurred. Most species crosses have little commercial value, although insight into the evolutionary process is interesting. Recent advances in genetic engineering might make some genetic combinations between species more feasible. Gene splicing (inserting a gene or genes from one animal to another) has occurred between species. In the future, opportunity to combine desirable genes, both within a species and between species, could be a reality.

Crossbreeding

There are two primary reasons for using crossbreeding: (a) breed complementation and (b) heterosis. Breed complementation implies crossing breeds so their strengths and weaknesses complement one another. There is no one breed superior in all desired production characteristics; therefore, planned crossbreeding programs, using breed complementation, can significantly increase herd productivity.

Crossbreeding, if properly managed, allows for the effective use of heterosis (hybrid vigor), which has a marked effect on productivity in swine, poultry and beef cattle production. Heterosis is defined as the increase in productivity in the crossbred progeny over and above the average of breeds or lines that are crossed. An example of calculating heterosis is shown in Table 13.1. The calculated heterosis for calf crop percent in this example is 5%, whereas heterosis for weaning weight is 4%.

Crossbreeding is sometimes questioned when the crossbred performance is less than the parent breed. In Table 13.1, for example, the weaning weight of the crossbreds is 520 lb, while one parent (Breed B) is 540 lb. However, for calf crop percent, the crossbreds are 84%, which is higher than either Breed A or Breed B. The value of crossbreeding, in

TABLE 13.1.   Computation of Heterosis for Percent Calf Crop and Weaning Weights

	Calf Crop (%)	Weaning Weight (lb)	Lb Calf Weaned per Cow Exposed
Breed A	82	460	377
Breed B	78	540	421
Average of the two breeds (without heterosis)	80	500	399
Average of crossbreds (with heterosis)	84	520	447
Superiority of crossbreds over average of two breeds	4	20	
Percent heterosis	5% (4 ÷ 80)	4% (20 ÷ 500)	

TABLE 13.2.   Relationship of Heritability and Heterosis for Most Traits

Traits	Heritability	Heterosis
Reproduction	Low	High
Growth	Medium	Medium
Carcass	High	Low

this example, is best demonstrated by combining calf crop percent and weaning weight (calf crop percent × weaning weight = lb calf weaned per cow exposed). Note in Table 13.1 that lb of calf weaned per cow exposed is 447 lb for the crossbreds, whereas Breed A is 377 lb and Breed B is 421 lb.

The amount of heterosis expressed is related to the heritability of the trait. Table 13.2 shows that heterosis is highest for lowly heritable traits and lowest for highly heritable traits. These relationships are helpful to commercial producers as they can use selection and crossbreeding to enhance genetic improvement. Figure 13.10 shows the relative importance of selection and crossbreeding in an improvement program. This figure demonstrates that selecting genetically superior animals is more important than crossbreeding. However, using the two in combination gives the highest level of performance.

Crossbreeding is most commonly used in swine, beef cattle, and sheep. Little crossbreeding is done in dairy cattle because of the primary emphasis on one trait (milk production) and the superiority of the Holstein breed for that trait. Poultry breeders utilize heterosis primarily through crossing lines that have been developed from crossing breeds and inbreeding separate and distinct lines.

Figures 13.11, 13.12, and 13.13 show the crossbreeding systems most frequently used in swine, beef cattle, and sheep. More specific detail in using these crossbreeding systems for these species is given in Chapters 21, 25, and 27.

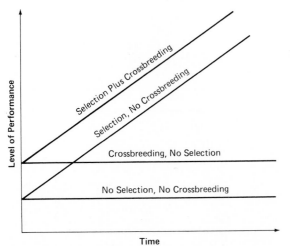

FIGURE 13.10.
Improvement in performance with various combinations of selection and crossbreeding.

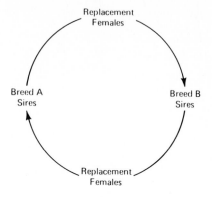

FIGURE 13.11.
Two-breed rotation cross. Females sired by breed A are mated to breed B sires and females sired by breed B are mated to breed A sires.

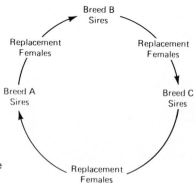

FIGURE 13.12.
Three-breed rotation cross. Females sired by a specific breed are bred to the breed of sire next in rotation.

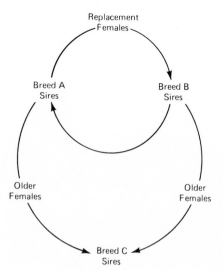

FIGURE 13.13.
Terminal (static) or modified-terminal crossbreeding system. It is terminal or static if all females in herd (A × B) are then crossed to breed C sires. All male and female offspring are sold. It is a modified-terminal system if part of females are bred to A and B sires to produce replacement females. The remainder of the females are terminally crossed to breed C sires.

Outcrossing

The most widely used breeding system for most species is outcrossing. As unrelated animals within the same breed are mated, the gene pairs are primarily heterozygous, although there is a slight increase in homozygosity over time. Homozygosity for several breeds has been estimated between 10 and 20%. This slight increase in homozygosity occurs because animals mated are somewhat related because they are members of the same breed.

Usefulness of outcrossing is primarily dependent on the effectiveness of selection (selection differential × heritability). The gene pairs stay primarily in a heterozygous condition as there is no attempt to maximize homozygosity or heterozygosity.

The continuous use of purebred sires of the same breed in a grade herd or flock is called **grading up.** In this situation, grading up is similar to outcrossing. Keeping replacement females from within the herd causes the inheritance of the purebred animals used to be substituted for that of the herd on which such purebred males are used. The percentage of inheritance of the desired purebred is 50% ($\frac{1}{2}$), 75% ($\frac{3}{4}$), 84.5% ($\frac{7}{8}$), and 94% ($\frac{15}{16}$) for four generations when grading up is practiced. The fourth generation resembles the purebred sires so closely in genetic composition that it approximates the purebred level.

The grading up system is useful in the breeding of cattle and horses, but it has little value in breeding sheep, swine, or poultry. High-producing purebred sheep, swine, and poultry breeding stock are available at reasonable prices; therefore, the breeder can buy them for less than he can produce them by grading up. The use of production-tested males that are above average in performance in a commercial herd can grade up the herd not only to a general purebred level but to a high level of production.

A recent use of grading up on a large scale has occurred with the introduction of many beef cattle breeds. Most of the introduction has been with males (bulls or semen) as females have been less available and more expensive because of numbers needed.

Imported bulls (or their semen) have been used on commercial cows or purebred cows of another breed. Grading up, as used here, is a type of crossbreeding, although the intent is not to maintain heterosis but to increase the frequency of genes from the introduced breed.

After several successive generations of mating the new breed to cows carrying a certain percentage of the new breed, the resulting offspring have been designated purebreds. In most breeds, this designation has been given when the calves had $\frac{7}{8}$ or $\frac{15}{16}$ of the genetic composition of the new breed. Figure 13.14 shows how these matings are made. It would require a minimum of 7 years to produce the first $\frac{15}{16}$ calves of the new breed.

FORMING NEW LINES OR BREEDS

New breeds have been formed and are currently being formed by crossing several breeds. These are sometimes given a general classification of **synthetic breeds** or **composite breeds.** In beef cattle, the Brangus, Barzona, Beefmaster and Santa Gertrudis breeds are composite (synthetic) breeds formed several years ago, whereas MARC I (crosses of Charolais, Brown Swiss, Limousin, Hereford, and Angus breeds) and RX3 (crosses of Red Angus, Hereford, and Red Holstein breeds) are examples of composites currently being developed. Columbia, Targhee, and Polypay are examples of synthetic breeds of sheep.

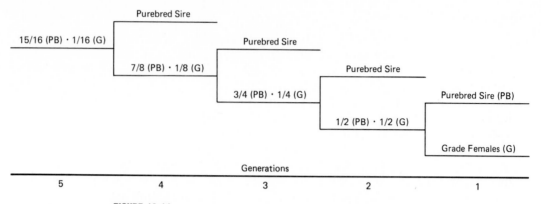

FIGURE 13.14.
Utilizing grading up to produce purebred offspring from a grade herd.

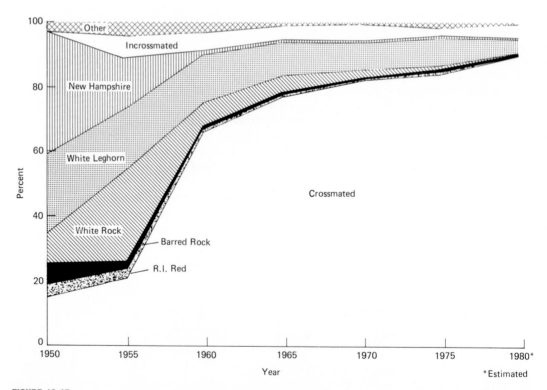

FIGURE 13.15.
In poultry breeding the specific breeds are losing their identity in the production of crossmated and incrossmated lines. Data is from over 30 million birds recorded in the National Poultry Improvement Plan. Courtesy of the USDA.

Crossing several swine breeds, associated with some inbreeding, has been used to develop the hybrid boars now being merchandized by several companies. Hybrid boars are used extensively in the swine industry.

In poultry, breeding for egg production differs from breeding for broiler production. Different traits are emphasized in the production of these two products. Both inbreeding and heterosis are utilized in the production of specific lines and strains of birds that are highly productive in the production of either eggs or broilers.

Figure 13.15 shows the change from poultry breeds to crossmated lines and strains of birds. The in-cross mated represents many of the synthetic egg-type strains or lines. These are primarily two- and three-way crosses of primarily Mediterranean breeds (e.g., White Leghorns). The crossmated chickens are the commercial broiler chickens that are primarily cornish-type males or White Plymouth females. The area of the chart depicting cross-mated shows the tremendous growth of breeding needed by the broiler industry.

SELECTED REFERENCES

Lasley, J. F. 1987. *Genetics of Livestock Improvements*. Englewood Cliffs, NJ: Prentice-Hall.
Warwick, E. J. and Legates, J. E. 1979. *Breeding and Improvement of Farm Animals*. New York: Mc-Graw-Hill.

CHAPTER 14

Nutrients and Their Functions

A **nutrient** is any feed constituent that functions in the support of life. There are many different feeds available to animals to provide these nutrients.

Most animal feeds are classified as **concentrates** and **roughages.** Concentrates include cereal grains (i.e., corn, wheat, barley, oats, and milo), oil meals (i.e., soybean meal, linseed meal, and cotton seed meal), molasses, and dried milk products. Concentrates are high in energy, low in fiber, and highly (80–90%) digestible. Roughages include legume hays, grass hays, and straws, which are by-products from the production of grass seed, and grain. Additional roughages are silage, stovers (dried corn, cane, or milo stalks and leaves with the grain portion removed), soilage (cut green feeds), and grazed forages. Roughages are less digestible than concentrates. Roughages are typically 50–65% digestible, but the digestibility of some straws is significantly lower.

NUTRIENTS

The six basic classes of nutrients (water, carbohydrates, fats, proteins, vitamins, and minerals) are found in varying amounts in animal feeds. Nutrients are composed of at least 20 of the more than 100 known chemical elements. These 20 elements and their chemical symbols are: calcium (Ca), carbon (C), chlorine (Cl), cobalt (Co), copper (Cu), fluorine (F), hydrogen (H), iodine (I), iron (Fe), magnesium (Mg), manganese (Mn), molybdenum (Mo), nitrogen (N), oxygen (O), phosphorus (P), potassium (K), selenium (Se), sodium (Na), sulphur (S), and zinc (Zn).

Water

Water contains hydrogen and oxygen. The terms water and moisture are used interchangeably. Typically, water refers to drinking water, whereas moisture is used in reference to the amount of water in a given feed or ration. The remainder of the feed, after accounting for moisture, is referred to as dry matter. Moisture is found in all feeds, ranging from 10% in air-dry feeds to more than 80% in fresh green forage. Livestock and poultry will consume several times more water than dry matter each day and will die from lack of water more quickly than from lack of any other nutrient. Water in feed is no more valuable than water from any other source. This is important to understand if proper assessment is to be made with feeds varying in their moisture content.

Water has important body functions because it enters into most of the metabolic reactions, assists in transporting other nutrients, helps maintain normal body temperature, and gives the body its physical shape, as water is the major component within cells.

Carbohydrates

Carbohydrates contain carbon, hydrogen, and oxygen in either simple or complex forms. The more simple carbohydrates, such as starch, supply the major energy source for cattle diets, particularly feedlot diets. The more complex carbohydrates, such as cellulose, are the major components of the cell walls of plants. These complex carbohydrates are not as easily digested as simple carbohydrates and they require host or microbial interaction for effective utilization.

Fats

Fats and oils, also referred to as lipids, contain carbon, hydrogen, and oxygen, although there is more carbon and hydrogen in proportion to oxygen than with carbohydrates. Fats are solid and oils are liquid at room temperature due generally to relative saturation. Fats contain 2.25 times more energy per pound than carbohydrates.

Most fats are comprised of three fatty acids attached to a glycerol backbone.

Example:

$$
\text{Glycerol + three fatty acids} \longrightarrow \text{a triglyceride + water}
$$

There are saturated and unsaturated fats, depending on their particular chemical composition. Saturated fatty acids have single bonds tying the carbon atoms together (e.g., -C-C-C-C-) whereas unsaturated fatty acids have one or more double bonds (e.g., -C=C-C=C-): The term polyunsaturated fatty acids is applied to those having more than one double bond. Although more than 100 fatty acids have been identified, only one has been determined to be dietary essential (linoleic acid). The two apparent functions of the essential fatty acids are: (a) precursors of prostaglandins and (b) structural components of cells.

Proteins

Proteins always contain carbon, hydrogen, oxygen, and nitrogen and sometimes iron, phosphorus or sulphur, or both. Protein is the only nutrient class that contains nitrogen. Proteins in feeds, on average, contain 16% nitrogen. This is why feeds are analyzed for the percent nitrogen in the feed, with the percent multiplied by 6.25 (100% ÷ 16% = 6.25) to convert it to percent protein. If, for example, a feed is 3% nitrogen, 100 g of the feed contains 3 g nitrogen. Multiplying 6.25 × 3 gives 18.75%, meaning that 100 g of this feed contains 18.75 g protein.

Proteins are composed of various combinations of some 25 amino acids. Amino acids are called the building blocks of the animal's body. The building blocks for growth (including growth of muscle, bone, and connective tissue), milk production, and cellular and tissue repair are amino acids that come from proteins in feed. The interstitial (between cells) fluid, blood, and lymph require amino acids to regulate body water and to transport oxygen and carbon dioxide. All enzymes are proteins, so amino acids are also required for enzyme production. Amino acids have an amino group (NH_2) in each of their chemical structures. There are many different combinations of amino acids that can be structured together. The chemical, or peptide bonding, of amino acids is illustrated using alanine and serine, which results in the formation of a dipeptide:

It can be seen that amino acids have both a basic portion, NH_2, and an acid portion,

$C{\stackrel{\nearrow}{\searrow}}{\scriptstyle O \atop OH}$, and it is because of these that they can combine into long chains to make pro-

teins. When digestion occurs, the action is at the peptide linkage to free amino acids from one another.

Some amino acids are provided in abundance in many proteins but others are quite limited. Thus, monogastric animals might have difficulty obtaining scarce amino acids that they cannot synthesize. Some of the 15 nonessential amino acids (alanine, aspartic acid, glutamic acid, hydroxyproline, proline, and serine) can be synthesized by monogastric animals if carbon, hydrogen, oxygen, and nitrogen are available. These amino acids are called nonessential amino acids. Other amino acids, however, either cannot be synthesized at all by monogastric animals (isoleucine, leucine, lysine, methionine, phenylalanine, threonine, tryptophan, and valine) or are produced so slowly that they are called semiessential and must be provided in the feed of growing animals (arginine, histidine—also glycine for the chick).

All amino acids are needed by all animals; the terms essential and nonessential merely refer to whether or not they must be supplied through the diet. The shortage of any particular amino acid can prevent an animal from using other amino acids for needed functions, and a protein deficiency results. Ruminants do not need a dietary supply of amino acids because the amino acids are synthesized in the ruminant stomach.

Minerals

Chemical elements, other than carbon, hydrogen, oxygen, and nitrogen are called minerals. They are inorganic because they contain no carbon; organic nutrients do contain carbon. Some minerals are referred to as macro (required in larger amounts) and others are micro or trace minerals (required in smaller amounts).

The macrominerals include calcium, chlorine, magnesium, phosphorus, potassium, sodium, and sulfur. Calcium and phosphorus are required in certain amounts and in a certain ratio to each other for bone growth and repair and for other body functions. The blood plasma contains sodium chloride; the red blood cells contain potassium chloride. The osmotic relations between the plasma and the red blood cells are maintained by proper concentrations of sodium chloride and potassium chloride. Sodium chloride may be depleted by excessive sweating that results from heavy physical work in hot weather. It is essential that salt and plenty of water be available under such conditions. The acid–base balance of the body is maintained at the proper level by minerals.

The essential microminerals for farm animals include cobalt, copper, fluorine, iodine, iron, manganese, molybdenum, selenium, and zinc. Microminerals may become a part of the molecule of a vitamin (for example, cobalt is a part of vitamin B_{12}) and they may become a part of a hormone (for example, thyroxine, a hormone made by the thyroid gland, requires iodine for its synthesis).

Hemoglobin of the red blood cells carries oxygen to tissues and carbon dioxide from tissues. Iron is required for production of hemoglobin because it is a part of the hemoglobin molecule. A small quantity of copper is also necessary for protection of hemoglobin (even though it does not normally become a part of the hemoglobin molecule in farm animals) because it apparently is necessary for normal iron absorption from the digestive tract and for release of iron to the blood plasma.

Certain important metabolic reactions in the body require the presence of minerals. Selenium and vitamin E both appear to work together to help prevent white muscle disease, which is a calcification of the striated muscles, the smooth muscles, and cardiac muscles. Both vitamin E and selenium are more effective if the other is present. Excesses of certain minerals may be quite harmful. For example, excess amounts of fluorine, molybdenum, and selenium are highly toxic.

Vitamins

Vitamins are organic nutrients needed in very small amounts to provide for specific body functions in the animal. There are 16 known vitamins that function in animal nutrition. Vitamins may be classed as either fat soluble or water soluble.

The fat-soluble vitamins are vitamins A, D, E, and K. Vitamin A helps maintain proper repair of internal and external body linings. Because the eyes have linings, lack of vitamin A adversely affects the eyes. Vitamin A is also a part of the visual pigments of the eyes. Vitamin D is required for proper use of calcium and phosphorus in bone growth and repair. A major function of vitamin D is to regulate the absorption of calcium and phosphorus from the intestine. Vitamin D is produced by the action of sunlight on steroids of the skin; therefore, animals that are exposed to sufficient sunlight make all the vitamin D they need. Vitamin K is important in blood clotting; hemorrhage might occur if the body is deficient in vitamin K.

The water-soluble vitamins are ascorbic acid (vitamin C), biotin, choline, cyanacobalamin (vitamin B_{12}), folic acid, niacin, pantothenic acid, pyridoxine (vitamin B_6), riboflavin (B_2) and thiamin (vitamin B_1). More diseases caused by inadequate nutrition have been described in the human than in any other animal, and among the best known are those caused by a lack of certain vitamins: beri-beri (lack of thiamin); pellagra (lack of niacin); pernicious anemia (lack of vitamin B_{12}); rickets (lack of vitamin D); and scurvy (lack of vitamin C).

In ruminant animals, all of the water-soluble vitamins are made by microorganisms in the rumen. Water-soluble vitamins also appear to be readily available to horses; perhaps some are made by fermentation in the cecum. Water-soluble vitamins cannot be synthesized by monogastric animals and must therefore be in their feed. Most fat-soluble vitamins are not synthesized by either ruminants or monogastrics and must be supplied in the diets of both groups (an exception is vitamin K, which is synthesized by rumen bacteria in ruminants). Many vitamins are supplied through feeds normally given to animals.

PROXIMATE ANALYSIS OF FEEDS

The nutrient composition of a feed cannot be determined accurately by visual inspection. A system has been devised by which the value of a feed can be approximated. Proximate analysis separates feed components into groups according to their feeding value. This analysis is based on a feed sample and analysis, and therefore is no more accurate than how representative the sample is of entire feed source.

The inorganic and organic components resulting from a proximate analysis are: water, crude protein, crude fat (sometimes referred to as ether extract), crude fiber, nitrogen-free extract, and ash (minerals). Figure 14.1 shows these components resulting from a

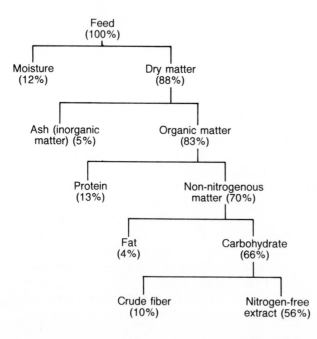

FIGURE 14.1.
Proximate analysis showing the inorganic and organic components of a feed (similar to wheat) on a natural or air-dry basis. Courtesy of CSU.

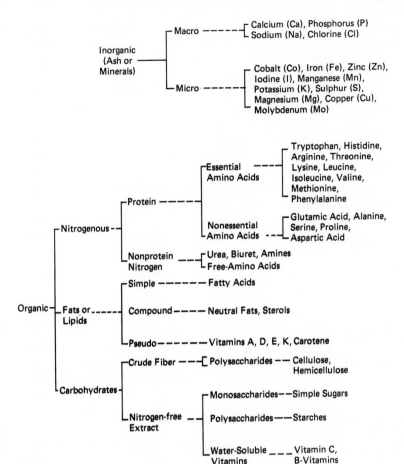

FIGURE 14.2.
Chemical analysis scheme of inorganic and organic nutrients. Courtesy of CSU.

feed that would have a laboratory analysis of 88% dry matter; 13% protein; 4% fat; 10% crude fiber; and 56% nitrogen-free extract (NFE) on a natural or air-dry basis. The analysis might be reported on a dry matter basis (no moisture) as: 14.8% protein; 4.5% fat; 11.4% crude fiber; and 63.6% nitrogen-free extract. Therefore, caution needs to be exercised in interpreting the proximate analysis results because different laboratories may report their analytical values on an air dry (or as-fed) basis or on a dry matter basis.

The proximate analysis for the six basic nutrients does not distinguish the various components of a nutrient. For example, ash content of a feed does not tell the amount of calcium, phosphorus, or other specific minerals. Figure 14.2 gives the chemical analysis for organic and inorganic nutrients. There are specific chemical analyses for each of these nutrients in a feed if such an analysis is needed.

DIGESTIBILITY OF FEEDS

Digestibility refers to the amount of various nutrients in a feed that are absorbed from the digestive tract. Different feeds and nutrients vary greatly in their digestibility. Many feeds have been subjected to digestion trials, in which feeds of known nutrient composition have been fed to livestock and poultry. Feces have been collected and the nutrients in the feces analyzed. The difference between nutrients fed and nutrients excreted in the feces is the apparent digestibility of the feed.

For example, the digestibility of protein is obtained by determining the digestibility of nitrogen in a feed. Digestibility is expressed as a percentage for nitrogen, for example, as follows:

$$\frac{\text{Nitrogen in feed} - \text{Nitrogen in feces}}{\text{Nitrogen in feed}} \times 100 = \text{Percentage digestibility}$$

As an example, if 100 g feed contains 3.2 g nitrogen and 100 g feces contains 0.8 g nitrogen, the percent digestibility of nitrogen is:

$$\frac{3.2 - 0.8}{3.2} \times 100 = 75\%$$

Note that the determination of 3.2 g nitrogen in 100 g feed enables one to estimate the percentage of protein in the feed as 20% ($3.20 \times 6.25 = 20$).

ENERGY EVALUATION OF FEED

Nutrients that contain carbon provide the energy for animals. Carbohydrates, fats, and proteins can all be used to provide energy; however, carbohydrates supply most of the energy, as they are a more economical energy source than proteins. Complete oxidation (burning or taking on oxygen) of carbon releases the energy. Energy is the force, or power, that is used to drive a variety of body systems.

Energy can be used to power movement of the animal, but most of it is used as chemical energy to drive reactions necessary to convert feed into animal products and to keep the body warm or cool.

Energy needs of animals generally account for the largest portion of feed consumed. Several systems have been devised to evaluate feedstuffs for their energy content. **Total digestible nutrients (TDN)** has been the most commonly used energy system, however it is being replaced by **metabolizable energy (ME)** and **net energy (NE).** TDN is typically expressed in pounds, kilograms, or percent after obtaining the proximate analysis and digestibility figures for a feed. The formula for calculating TDN is TDN = (digestible crude protein) + (digestible crude fiber) + (digestible nitrogen-free extract) + (digestible crude fat × 2.25). The factor of 2.25 is used to equate fat to a carbohydrate basis, since fat has 2.25 times as much energy as an equivalent amount of carbohydrate.

An example of calculating TDN in 100 grams of feed is shown in Table 14.1.

TDN is roughly comparable to digestible energy (DE) but it is expressed in different units. TDN and DE both tend to overvalue roughages.

Even though there are some apparent shortcomings in using TDN as an energy measurement of feeds, it works well in balancing rations for cows and growing cattle. There is more precision in the energy measurement of feeds by using the net energy (NE) system. This system usually measures energy values in megacalories per pound or kilogram of feed. The calorie basis, which measures the heat content of feed, is as follows:

Calorie (cal)—amount of energy or heat required to raise the temperature of one gram of water 1°C.

Kilocalorie (Kcal)—amount of energy or heat required to raise the temperature of one kilogram of water 1°C.

Megacalorie (Mcal)—equal to 1000 kilocalories or 1,000,000 calories.

TABLE 14.1. An Example of Calculating Total Digestible Nutrients (TDN)

Nutrient	Amount of Nutrient (g)	Digestibility (%)	Amount of Digestible Nutrient
Protein	20	75	15.00
Carbohydrates			
Soluble (NFE)	55	85	46.75
Insoluble (fiber)	10	20	2.00
Fat 5 ($\times$ 2.25)	11.15	85	9.50
Minerals	3.85	35	1.35
		TDN =	74.60

Figure 14.3 shows various ways that the energy of feeds is utilized by animals and the various energy measurements of feeds. Note that some energy in the feed is lost in the feces (not digested), the urine (digested but not used by the body cells), gases from microbial fermentation of the feed, and heat, resulting from digestion and metabolism of the feed.

Feeds provide energy which the animal uses to supply two basic functions: (a) maintenance energy and (b) production energy. Maintenance energy is used to keep the basal metabolism functioning, provide for the voluntary activity of the animal, generate heat to keep the body warm, and provide energy to cool the body.

Production energy becomes stored energy and energy needed for work. The stored energy exists in fetus development, semen production, growth, fat deposition, and milk, eggs, and wool production.

Gross energy (GE) is the quantity of heat (calories) released from the complete burning of the feed sample in an apparatus called a bomb calorimeter. GE has little practical value in evaluating feeds for animals because the animal does not metabolize feeds in the same manner as a bomb calorimeter. For example, oat straw has the same GE value as corn grain. Digestible energy (DE) is GE of feed minus fecal energy. Metabolizable energy (ME) is GE of feed minus energy in the feces, urine, and gaseous products of digestion. Net energy (NE) is the ME of feed minus the energy used in the consumption, digestion, and metabolism of the feed. This energy lost between ME and NE is called **heat increment.**

Another way to illustrate the several measures of feed energy and how they are utilized is shown in Fig. 14.4. In this example, 2000 kcal of GE (in approximately 1 lb of feed) is fed to a laying hen. The DE shows that 400 kcal was lost in the feces. The ME (1450 kcal) is used for heat increment, maintenance, and production (eggs and tissue). There are 300 kcal lost in the heat increment, which leaves 1150 kcal for maintenance and

Feed
Intake
Gross Energy
In Feces
Energy Lost
Digestible Energy
In Urine and Gases
Energy Lost
Metabolizable Energy
In Form of Heat
Energy Lost
Net Energy
Maintenance
Production

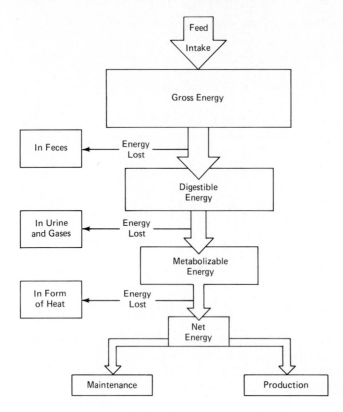

FIGURE 14.3.
Measures of energy and energy utilization.

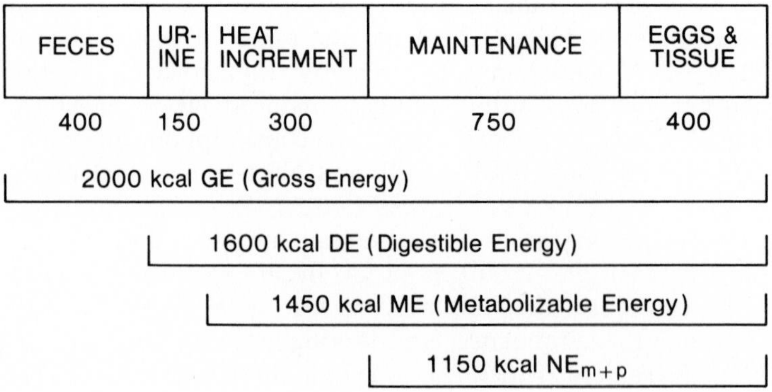

FECES
UR-INE
HEAT INCREMENT
MAINTENANCE
EGGS & TISSUE
400
150
300
750
400
2000 kcal GE (Gross Energy)
1600 kcal DE (Digestible Energy)
1450 kcal ME (Metabolizable Energy)
1150 kcal NEm+p

FIGURE 14.4.
Energy utilization by a laying hen. Data represents approximately one pound of feed containing 2000 kcal of gross energy. Courtesy of CSU Graphics.

TABLE 14.2. Comparison of TDN and NE Systems for a Yearling Steer with Varying Rates of Gain

Energy System	Rate of Gain (lb/day)		
	0	2	2.9
TDN (lb)	6.40	12.80	15.00
NEm (Mcal)	6.24	6.24	6.24
NEg (Mcal)		4.29	6.48
Total lb feed			
TDN basis	8.2	16.5	19.3
NE basis	7.2	15.0	19.0
Feed per lb gain (NE)		7.5	6.6
Feed per lb gain (TDN)		8.2	6.6

production. For maintenance, 750 kcal are needed, which leaves 400 kcal for egg production and tissue growth. Therefore only 20% ($\frac{400}{2000}$) of the GE is used for production.

Net energy for maintenance (NEm) and net energy for gain (NEg) are more commonly used for formulating diets for feedlot cattle than any other energy system. Net energy for lactation (NEl) is used in dairy cow ration formulation. NEm in animals is the amount of energy needed to maintain a constant body weight. Animals of known weight, fed for zero energy gain, have a constant level of heat production.

The NEg measures the increased energy content of the carcass after feeding a known quantity of feed energy. All feed fed above maintenance is not utilized at a constant level of efficiency. Higher rates of gain require more feed per unit of gain as composition of gain varies with rate of gain.

The TDN and NE systems are compared in Table 14.2. Information in the table is based on feeding a simple diet of ground ear corn (90%) and supplement (10%) to a yearling steer (772 lb) for different rates of gain.

FEEDS AND FEED COMPOSITION

Classification of Feeds

Feeds are naturally occurring ingredients in diets of farm animals that are used to sustain life. The terms "feeds" and "feedstuffs" are generally used interchangeably; however, feedstuffs is a more inclusive term. Feedstuffs can include certain nonnutritive products such as additives to promote growth and reduce stress, for flavor and palatability, to add bulk, or to preserve other feeds in the ration.

The National Research Council (NRC) classification of feedstuffs is as follows:

1. Dry roughages and forages
   Hay (legume and nonlegume)
   Straw
   Fodder
   Stover
   Other feeds with greater than 18% fiber (hulls and shells)

2. Range, pasture plants, and green forages
3. Silages (corn, legume, and grass)
4. Energy feeds (cereal grains, mill by-products, fruits, nuts, and roots)
5. Protein supplements (animal, marine, avian, and plant)
6. Mineral supplements
7. Vitamin supplements
8. Nonnutritive additives (antibiotics, coloring materials, flavors, hormones, preservatives, and medicants)

Roughages and forages are used interchangeably, although roughage usually implies a bulkier, coarser feed. In the dry state, roughages have more than 18% crude fiber. The crude fiber is primarily a component of cell walls that is not very digestible. Roughages are also relatively low in TDN, although there are exceptions, for example, corn silage, which has over 18% crude fiber and approximately 70% TDN.

Feedstuffs that contain 20% or more protein, such as soybean meal and cottonseed meal, are classified as protein supplements. Feedstuffs having less than 18% crude fiber and less than 20% protein are classified as energy feeds or concentrates. Cereal grains are typical energy feeds, which is reflected by their high TDN values.

Nutrient Composition of Feeds

Feeds are analyzed for their nutrient composition as discussed previously. The ultimate goal of nutrient analysis of feeds is to predict the productive response of animals when they are fed rations of a given composition.

The nutrient composition of some of the more common feeds is shown in Table 14.3 (ruminants) and Table 14.4 (monogastric animals). The information in Table 14.3 represents averages of numerous feed samples. Feeds are not constant in composition, and an actual analysis should be obtained whenever economically feasible. The actual analysis is not always feasible or possible because of lack of available laboratories and insufficient time to obtain the analysis. Therefore, feed analysis tables becomes the next best source of reliable information on nutrient composition of feeds. It is not uncommon to expect the following deviations of actual feed analysis from the table values for several feed constituents: crude protein ($\pm 15\%$), energy values ($\pm 10\%$), and minerals ($\pm 30\%$).

Digestible protein is included in some feed composition tables, but because of the large contribution of body protein to the apparent protein in the feces, digestible protein is more misleading than crude protein. For this reason, crude protein is more commonly found in feed composition tables and used in formulating diets for ruminants.

Digestible protein (DP) can be calculated from crude protein (CP) content by using the following equation (%DP and %CP are on a dry matter basis):

$$\%DP = 0.9\ (\%CP) - 3$$

Five measures of energy values (TDN, ME, NEm, NEg, and NE$_l$) are shown in Table 14.3 TDN is shown because there are more TDN values for feeds and because TDN has been a standard system of expressing the energy value of feeds. Some individuals desire to have ME (metabolizable energy) values for feed because the values are in calories rather than pounds. NEm and NEg values are used primarily to formulate feedlot diets and

TABLE 14.3. Nutrient Composition of Selected Feeds Commonly Used in Diets of Ruminants

Feed	Dry Matter (%)	TDN (%)	ME (Mcal/lb)	NEm (Mcal/lb)	NEg (Mcal/lb)	NE_l (Mcal/lb)	Crude Protein (%)	Vitamin A (Carotene) (mg/lb)	Calcium (%)	Phosphorus (%)
				On a Dry Matter Basis (moisture-free)						
Alfalfa hay (early bloom)	90.0	57.0	0.94	0.55	0.25	0.59	18.4	57.8	1.25	0.23
Barley	89.0	83.0	1.36	0.97	0.64	0.85	13.0	—	0.09	0.47
Bermudagrass (grazed)	31.0	66.0	—	0.67	0.38	0.61	14.6	150.2	0.49	0.27
Bluegrass (grazed)	30.6	58.0	0.95	0.56	0.27	0.73	17.0	86.4	0.39	0.39
Bone meal (steamed)	95.0	16.0	—	—	—	—	12.7	—	30.51	14.31
Brome (grazed early)	32.5	63.0	1.04	0.62	0.34	0.62	20.3	208.8	0.59	0.37
Buffalo grass	47.7	56.0	0.92	0.54	0.23	0.60	9.2	42.6	0.52	0.16
Corn (No. 2 dent)	89.0	91.0	1.50	1.04	0.67	0.96	10.0	0.91	0.02	0.35
Corn (ground ear)	87.0	90.0	1.48	1.01	0.63	0.85	9.3	—	0.05	0.31
Corn silage	27.9	70.0	1.15	0.71	0.45	0.80	8.4	—	0.28	0.21
Corn stover (no ears, husks)	87.2	59.0	0.97	0.55	0.25	0.52	5.9	—	0.49	0.09
Cottonseed meal	93.5	74.0	1.22	0.75	0.49	0.77	42.4	—	0.20	1.09
Dicalcium phosphate	96.0	—	—	—	—	—	—	—	23.10	18.65
Grama (early vegetation)	41.0	64.0	1.05	0.63	0.36	0.50	13.1	—	0.54	0.19
Limestone (ground)	100.0	—	—	—	—	—	—	—	33.84	0.02
Meadow hay (native)	92.9	51.0	0.84	0.50	0.14	0.56	9.1	—	0.57	0.17
Milk	12.0	130.0	2.14	2.09	0.91	1.52	25.8	—	0.93	0.75
Milo (sorghum)	89.0	80.0	1.31	0.84	0.56	0.81	12.4	—	0.04	0.33
Molasses (cane)	75.0	72.0	1.25	2.27	1.48	0.77	4.3	—	1.19	0.11
Oats (grain)	89.0	76.0	1.25	0.79	0.52	0.80	13.2	—	0.11	0.39
Prairie hay (midbloom)	91.0	51.0	0.84	0.50	0.14	—	8.1	9.1	0.34	0.21
Sorghum, Sudan grass (grazed)	22.7	63.0	1.04	0.62	0.35	0.72	8.7	—	0.43	0.35
Soybean meal (solvent)	89.0	81.0	1.33	0.88	0.59	0.84	51.5	—	0.36	0.75
Wheat (hard, red winter)	89.1	88.0	1.45	0.98	0.64	0.92	14.6	—	0.06	0.57
Wheat (grazed early)	21.5	73.0	1.20	0.75	0.49	0.79	28.6	236.4	0.42	0.40
Wheat (straw)	90.1	48.0	0.79	0.47	0.09	0.37	3.6	1.0	0.17	0.08
Wheatgrass, crested (early)	30.8	67.0	1.10	0.66	0.40	0.57	23.6	197.1	0.46	0.35

Source: Adapted from National Research Council, and Preston, 1984.

TABLE 14.4. Nutrient Composition of Selected Feeds Commonly Used in Rations of Monogastric Animals (air dry basis)

Feed	ME	Protein (%)	Calcium (%)	Phosphorus (%)	Iron (mg/lb)	Manganese (mg/lb)	Zinc (mg/lb)	A (IU/lb)	Niacin (mg/lb)	Pantothenic Acid (mg/lb)	Riboflavin (mg/lb)	Choline (mg/lb)	B_{12} (mg/lb)	Lysine (%)	Methionine (%)	Tryptophan (%)
			Minerals						Vitamins					Amino Acids		
Alfalfa meal (dehydrated)	1032	17.5	1.44	0.22	141	13	8	12,272	17	13	7	497	0.002	0.73	0.2	0.28
Barley	1304	11.6	0.05	0.36	23	4	8	—	29	4	1	450	—	0.40	0.2	0.14
Blood meal	876	85.0	0.30	0.25	1364	3	139	—	10	1	1	340	0.20	8.10	1.5	1.10
Bone meal	—	—	28.00	13.00	—	—	—	—	—	—	—	—	—	—	—	—
Corn	1511	8.8	0.02	0.28	16	2	4	454	15	3	0.5	241	—	0.24	0.2	0.05
Dicalcium phosphate	—	—	26.0	20.0	—	—	—	—	—	—	—	—	—	—	—	—
Feather meal	1032	86.4	0.20	0.80	—	10	—	—	12	4	1	405	0.27	1.10	0.4	0.50
Fish meal (menhaden)	1014	60.5	5.11	2.88	200	15	67	—	25	4	2	1389	0.07	4.83	1.8	0.68
Limestone	—	—	39.0	—	—	—	—	—	—	—	—	—	—	—	—	—
Meat and bone meal	1106	50.4	10.1	4.96	223	6	42	—	21	2	2	907	0.03	2.60	0.7	0.28
Milo (sorghum)	1468	8.9	0.28	0.32	18	6	6	—	19	5	0.5	308	—	0.22	0.1	0.10
Oats	1213	11.4	0.06	0.27	32	20	—	—	7	13	0.5	500	—	0.40	0.2	0.16
Skim milk (dried)	1527	33.5	1.28	1.02	23	1	18	—	5	15	10	568	—	2.40	0.9	0.44
Soybean meal (solvent)	1404	48.5	0.27	0.62	54	12	20	—	10	7	1	1295	—	3.18	0.7	0.67
Wheat, hard (red winter)	1464	14.1	0.05	0.37	23	28	6	—	25	6	2	495	—	0.40	0.2	0.18
Whey (dried)	1450	13.6	0.97	0.76	59	3	—	—	5	20	12	900	—	0.97	0.2	0.19

Source: Adapted from National Research Council.

diets for growing replacement heifers as these values offset the major problem associated with the TDN energy system. NE_l is used in formulating diets for dairy cows.

SELECTED REFERENCES

Church, D. C. 1986. *Livestock Feeds and Feeding*. Englewood Cliffs, NJ: Prentice-Hall.

Cullison, A. E. 1987. *Feeds and Feeding*. Reston, VA: Reston Publishing Co., Inc.

Ensminger, M.E. and Olentine, C. G., Jr. 1978. *Feeds and Nutrition*. Clovis, CA: The Ensminger Publishing Co.

Jurgens, M. H. 1982. *Animal Feeding and Nutrition*. Dubuque, IA: Kendall-Hunt Publishing Co.

Nutrient Requirements (of Beef Cattle, 1984), (of Dairy Cattle, 1978), (of Goats, 1981), (of Horses, 1978), (of Poultry, 1984), (of Sheep, 1985), and (of Swine, 1979). National Research Council (NRC). Washington, D.C.: National Academy Press.

CHAPTER 15

Digestion and Absorption of Feed

Animals obtain substances needed for all body functions from the feeds they eat and the liquids they drink. Before the body can absorb and use them, feeds must undergo a process called digestion. **Digestion** includes mechanical action, such as chewing and contractions of the intestinal tract; chemical action, such as the secretion of hydrochloric acid (HCl) in the stomach and bile in the small intestine; and action of enzymes such as maltase, lactase, and sucrase (which act on disaccharides), lipase (which acts on lipids), and peptidases (which act on proteins). Enzymes are produced either by the various parts of the digestive tract or by microorganisms. The effect of digestion is to reduce the size of molecules so that they may be absorbed into the blood.

CARNIVOROUS, OMNIVOROUS, AND HERBIVOROUS ANIMALS

Animals are classed as carnivores, omnivores, or herbivores according to the types of feed they normally eat. Carnivores, such as dogs and cats, normally consume animal tissues as their source of nutrients; herbivores, such as cattle, horses, sheep, and goats, primarily consume plant tissues. Humans and pigs are examples of omnivores, who eat both plant and animal substances.

Carnivores and omnivores are monogastric animals, meaning that the stomach is simple and has only one compartment. Some herbivores, such as horses and rabbits, are also monogastric. Other herbivores, such as cattle, sheep, and goats, are ruminant animals, meaning that the stomach is complex and contains four compartments. Animals classified as carnivores, omnivores, and herbivores can utilize certain feeds they do not normally consume. For example, animal products can be fed to herbivorous ruminants, and certain cereal products can be fed to carnivores.

The digestive tracts of pigs and humans are similar in anatomy and physiology; therefore, much of the information gained from studies on pig nutrition and digestive physiology can be applied to the human. Both the pig and the human are omnivores and both are monogastric animals. Neither can synthesize the B-complex vitamins or amino acids to a significant extent. Both pigs and humans tend to eat large quantities, which can result in obesity. Humans can control obesity by controlling food intake and exercising as a means of using, rather than storing, excess energy. Obesity in swine can be controlled by limiting the amount of feed available to them or through genetic selection of leaner animals. The latter has received the greater emphasis as pigs are typically fed **ad libitum.**

DIGESTIVE TRACT OF MONOGASTRIC ANIMALS

The anatomy of the digestive tract varies greatly from one species of animal to another. The basic parts of the digestive tract are mouth, stomach, small intestine and large intestine, or colon. The primary function of the parts preceding the intestine is to reduce the size of feed particles. The small intestine functions in splitting food molecules and absorption, and the large intestine absorbs water and forms indigestible wastes into solid form called feces. In a mammal having a simple stomach (such as the pig), the mouth has teeth and lips for grasping and holding feed that is masticated (chewed), and salivary glands that secrete saliva for moistening feed so it can be swallowed.

Feed passes from the mouth to the stomach through the esophagus. A sphincter (valve) is at the junction of the stomach and esophagus. It can prevent feed from coming up the esophagus when stomach contractions occur. The stomach empties its contents into that portion of the small intestine known as the duodenum. The pyloric sphincter, located at the junction of the stomach and the duodenum, can be closed to prevent feed from moving into or out of the stomach. Feed goes from the duodenum to the jejunum portion and then to the ileum portion of the small intestine. It then passes from the small intestine to the large intestine, or colon. The ileocecal valve, located at the junction of the small intestine and the colon, prevents material in the large intestine from moving back into the small intestine.

The small intestine actually empties into the side of the colon near, but not at, the anterior end of the colon. The blind anterior end of the colon is the cecum, or, in some animals, the vermiform appendix. The large intestine empties into the rectum. The anus has a sphincter, which is under voluntary control so that defecation can be prevented by the animal until it actively engages in the process. The structures of the digestive system of the pig are shown in Fig. 15.1.

Animals such as pigs, horses, and poultry are classed as monogastric animals, but they differ markedly in certain ways. For example, the horse (Fig. 15.2) has a large structure called the cecum in which much fermentation occurs. Because the cecum is posterior to the area where most feed is absorbed, horses do not obtain all of the nutrients made by microorganisms in the cecum.

The digestive tracts of most poultry species differ from the pig in several respects. Because they have no teeth, poultry break their feed into a size that can be swallowed by pecking with their beaks or by scratching with their feet. Feed goes from the mouth through the esophagus to the crop, which is an enlargement of the esophagus in which feed can be stored. Some fermentation may occur in the crop, but it does not act as a fermentation vat. Feed passes from the crop to the proventriculus, which is a glandular stomach in birds that secretes gastric juices and HCl but does not grind feed. Feed then goes to the gizzard, where it is ground into finer particles by strong muscular contrac-

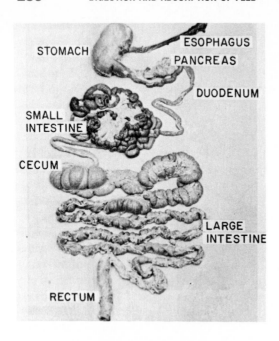

FIGURE 15.1.
Digestive tract of the pig as an example of the digestive tract of a monogastric animal. Reprinted with permission from Church and Pond, *Basic Animal Nutrition and Feeding,* Corvallis, Oregon: published by D. C. Church, copyright © 1974.

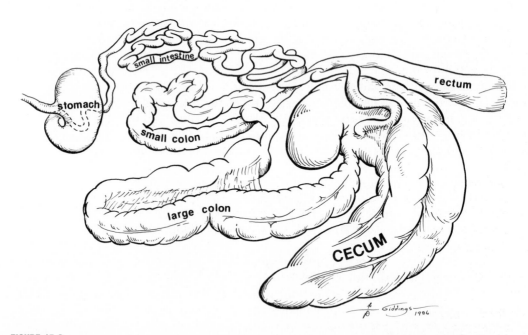

FIGURE 15.2.
Digestive system of the horse. The posterior view shows the colon or large intestine proportionally larger than the rest of the digestive tract. Note particularly the location of the cecum at the anterior end of the colon. Drawn by Dennis Giddings.

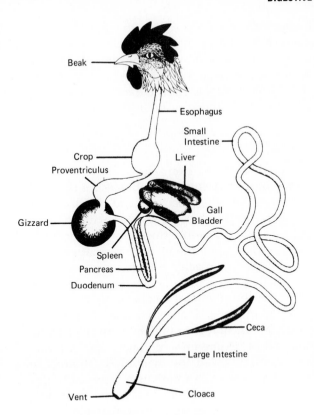

Beak

Esophagus

Small
Intestine

Crop

Liver

Proventriculus

Gall
Bladder

Gizzard

Spleen

Pancreas

Duodenum

Ceca

FIGURE 15.3.
Digestive tract of the chicken showing crop,
proventriculus, and gizzard, all of which are
characteristic of poultry. Courtesy of J. E.
Parker, Oregon State University.

Large Intestine

Vent

Cloaca

TABLE 15.1.   Digestive Tract Sizes and Capacities of Selected Monogastric Animals

Part of Digestive Tract	Species			
	Human	Pig	Horse	Chicken
Esophagus	—	—	4 feet	Total length of
Stomach	1 quart	2 gallons	4 gallons	digestive tract in
Small intestine	1 gallon	2 gallons	12 gallons	mature chickens is
			70 feet	approximately 7
Large intestine	1 quart	3 gallons		feet (beak to crop,
Cecum			8 gallons	7 in.; beak to
			4 feet	proventriculus, 14
Large colon			18 gallons	in.; duodenum, 8
			10 feet	in.; ileum and
Small colon			4 gallons	jejunum, 48 in.;
			12 feet	and cecum, 7 in.)

Source: Compiled from several sources.

tions. The gizzard apparently has no function other than to reduce the size of feed particles, because birds from which it is removed digest a finely ground ration. Feed moves from the gizzard into the small intestine. Material from the small intestine empties into the large intestine. At the junction of the small and large intestines are two ceca, which contribute little to digestion. Material passes from the large intestine into the cloaca, into which urine also empties. Material from the cloaca is voided through the vent (Fig. 15.3).

STOMACH COMPARTMENTS OF RUMINANT ANIMALS

In contrast to the single stomach of monogastric animals, stomachs of cattle, sheep and goats have four compartments—**rumen, reticulum, omasum,** and **abomasum** (Figs. 15.4 and 15.5). The rumen is a large fermentation vat in which bacteria and protozoa thrive and break down roughages to obtain nutrients for their use. It is lined with numerous papillae, which give it the appearance of being covered with a thick coat of short projections. The papillae increase the surface area of the rumen lining. The microorganisms in the rumen can digest cellulose and can synthesize amino acids from nonprotein nitrogen and can synthesize the B-complex vitamins. Later, these microorganisms are digested in the small intestine to provide these nutrients for the ruminant animal's use.

The reticulum has a lining with small compartments similar to a honeycomb, thus it is occasionally refered to as the honeycomb. Its function is to interact with the rumen in initiating the mixing activity of the rumen and providing an additional area for fermentation. The omasum has many folds, so it is often called the manyplies. The omasum may not have a major digestive function, although some individuals believe the folds produce a grinding action on the feed. The abomasum, or true stomach, corresponds to the stomach of monogastric animals and performs a similar digestive function.

The size and capacity of the ruminant stomach and intestinal tract are given in Table 15.2. The data in Table 15.2 are for mature ruminants, as the relative proportions of the stomach compartments are considerably different in the young lamb and calf. At birth, the abomasum comprises 60% of the total stomach capacity, whereas the rumen is only 25% of the total.

TABLE 15.2.   Digestive Tract Sizes and Capacities of Mature Ruminant Animals

Part of Digestive Tract	Species	
	Cow	Ewe
Stomach		
Rumen	40 gallons	5 gallons
Reticulum	2 gallons	2 quarts
Omasum	4 gallons	1 quart
Abomasum	4 gallons	3 quarts
Small intestine	15 gallons	2 gallons
	(130 feet)	(80 feet)
Large intestine	10 gallons	6 quarts

Source: Compiled from several sources.

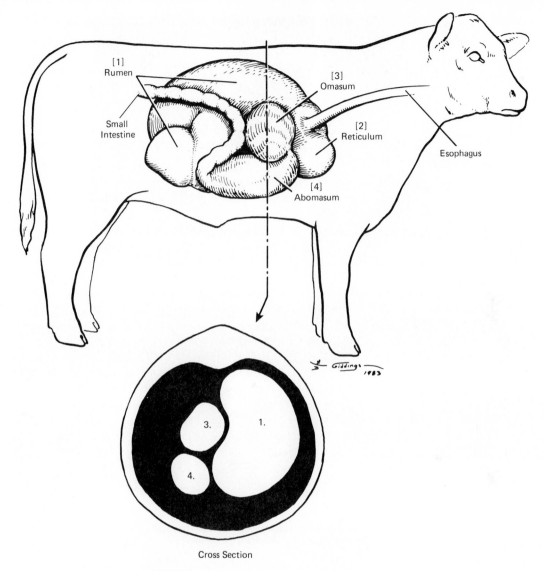

FIGURE 15.4.
Beef cattle digestive tract. Drawn by Dennis Giddings.

Animals that have the four-compartment stomach eat forage rapidly and later, while resting, regurgitate each bolus of feed known as the **cud.** The regurgitated feed is chewed more thoroughly, swallowed, then another bolus is regurgitated and chewed. This process continues until the feed is thoroughly masticated. The regurgitation and chewing of undigested feed is known as rumination. Animals that ruminate are known as ruminants. As feeds are fermented by microorganisms in the rumen, large amounts of gases (chiefly

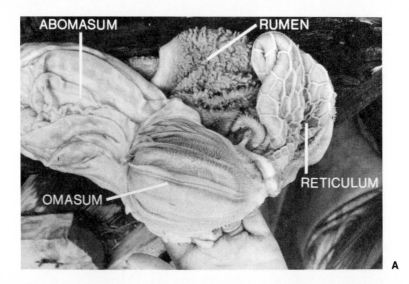

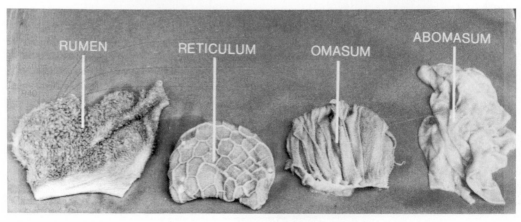

FIGURE 15.5.
Lining of the four compartments of the ruminant stomach (goat). (A) Compartments connected. (B) Compartments separated. Courtesy of George F.W. Haenlein, University of Delaware.

methane and carbon dioxide) are produced. The animal normally can eliminate the gases by controlled belching, also called **eructation.**

DIGESTION IN MONOGASTRIC ANIMALS

Feed that is ingested (taken into the mouth) stimulates the secretion of saliva. Chewing reduces the size of ingested particles and saliva moistens the feed. The enzyme amylase, which is present in saliva of some species including pigs and humans, acts on starch. Ruminants do not secrete salivary amylase. However, very little actual breakdown of starch into simpler compounds occurs in the mouth, primarily because feed is there for a short time.

TABLE 15.3. Important Enzymes in the Digestion of Feed

Enzyme	Substrate	Substances Resulting from Enzyme Action
Amylase	Starch	Disaccharides, dextrin
Chymotrypsin	Peptides	Amino acids and peptides
Lactase	Lactose	Glucose and galactose
Lipase	Lipids	Fatty acids and glycerides
Maltase	Maltose	Glucose and glucose
Pepsin	Protein	Polypeptides
Peptidases	Peptides	Amino acids
Sucrase	Sucrose	Glucose and fructose
Trypsin	Protein	Polypeptides

Source: Adapted from several sources.

An **enzyme** is an organic catalyst that speeds a chemical reaction without being altered by the reaction. Enzymes are rather specific; that is, each type of enzyme acts on only one or a few types of substances. Therefore, it is customary to name enzymes by giving the name of the substance on which it acts and adding the suffix -ase, which, by convention, indicates that the molecules so named are enzymes (Table 15.3). For example, lipase is an enzyme that acts on lipids (fats); maltase is an enzyme that acts on maltose to convert it into two molecules of glucose; lactase is an enzyme that acts on lactose to convert it into one molecule of glucose and one molecule of galactose; and sucrase is an enzyme that acts on sucrose to convert it into one molecule of glucose and one molecule of fructose. Some lipase is present in saliva but little hydrolysis of lipids into fatty acids and glycerides occurs in the mouth.

As soon as it is moistened by saliva and chewed, feed is swallowed and passes through the esophagus to the stomach. The stomach secretes HCl, mucus, and the digestive enzymes pepsin and gastrin. The strongly acidic environment in the stomach favors the action of pepsin. Pepsin breaks proteins down into polypeptides. The HCl also assists in coagulation, or curdling, of milk. Little hydrolysis of proteins into amino acids occurs in the stomach. Mucous secretions help to protect the stomach lining from the action of strong acids.

In the stomach, feed is mixed well and some digestion occurs; the mixture that results is called **chyme.** The chyme passes next into the duodenum, where it is mixed with secretions from the pancreas, bile, and enzymes from the intestine.

Secretion from the pancreas and discharge of bile from the gall bladder are stimulated by secretin, pancreozymin, and cholecystokinin, three hormones that are released from the duodenal cells. The enzymes from the pancreas are lipase, which hydrolyzes fats into fatty acids and glycerides; trypsin, which acts on proteins and polypeptides to reduce them to small peptides; chymotrypsin, which acts on peptides to produce amino acids; and amylase, which breaks starch down to disaccharides, after which the disaccharides are broken down to monosaccharides. The liver produces bile that helps emulsify fats; the bile is strongly alkaline and so helps to neutralize the acidic chyme coming from the stomach. Some minerals that are important in digestion also occur in bile.

By the time they reach the small intestine, amino acids, fatty acids, and monosacchar-

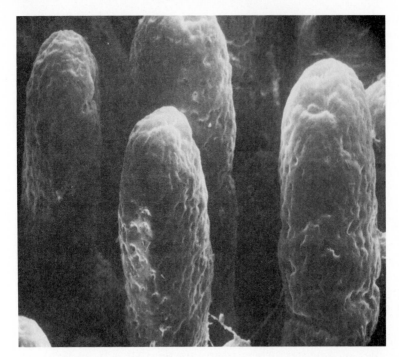

FIGURE 15.6.
Electron micrograph of the lining of the small intestine. These projections (villi) increase the surface area and are covered with cells that digest and absorb nutrients from the feed. Magnification is approximately ×200. Courtesy of Dr. G. L. Waxler (*Am. J. Vet. Res.* 33:1323).

ides (simple sugars or carbohydrates) are all available for absorption. Thus, the small intestine is the most important area for both digestion and absorption of feed. Absorption of feed molecules may be either passive or active. Passive passage results from diffusion, which is the movement of molecules from a region of high concentration of those molecules to a region of low concentration. Active transport of molecules across the intestinal wall may be accomplished through a process in which cells of the intestinal lining (**villi,** shown in Fig. 15.6) engulf the molecules and then actively transport these molecules to either the bloodstream or the lymph. Energy is expended in accomplishing the active transport of molecules across the gut wall.

When molecules of digested feed enter the capillaries of the blood system, they are carried directly to the liver. Molecules may enter the lymphatic system, after which they go to various parts of the body including the liver. The liver is an extremely important organ both for metabolizing useful substances and for detoxifying harmful ones.

In some monogastric animals, such as the horse, postgastric (cecal) fermentation of roughages occurs. In these animals, the feed that can be digested by a monogastric animal is digested and absorbed before the remainder reaches the cecum. These animals are perhaps more efficient than ruminants in their use of feeds such as concentrates. In the ruminant animals, the concentrates given along with roughages are used by the bacteria and protozoa. Because the microorganisms in ruminants use starches and sugars, little glucose is available to ruminants for absorption. The microorganisms do provide volatile fatty acids, which are absorbed by the ruminant and converted to glucose as an energy source. The postgastric fermentation that occurs in horses breaks down roughages, but this takes place posterior to the areas where nutrients are most actively absorbed; consequently, all nutrients in the feed are not obtained by the animal in postgastric fermentation.

DIGESTION IN RUMINANT ANIMALS

In mature ruminant animals (cattle, sheep and goats), predigestive fermentation of feed occurs in the rumen and reticulum. The bacteria and protozoa in the rumen and reticulum use roughages consumed by the animal as feed for their growth and multiplication; consequently, billions of these microorganisms develop. The rumen environment is ideal for microorganisms because moisture, a warm temperature, and a constant supply of nutrients are present. Excess microorganisms are continuously removed from the rumen and reticulum along with small feed particles that escape microbial fermentation and pass through the omasum into the abomasum. When feed passes into the abomasum, strong acids destroy the bacteria and protozoa. The ruminant animal then digests the microorganisms in the small intestine and uses them as a source of nutrients. The digested microbial cells provide the animal with most of its amino acid needs and some energy. Thus, ruminant animals and microorganisms mutually benefit each other. All digestive processes in ruminants are the same as those in monogastric animals after the feed reaches the abomasum, which corresponds to the stomach of monogastric animals.

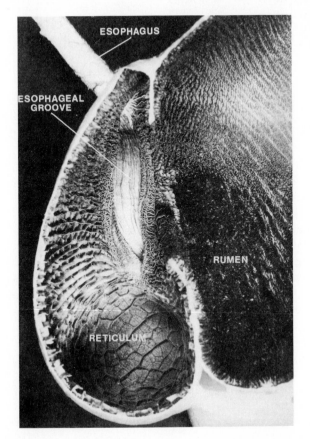

FIGURE 15.7.
The esophageal groove with its relative location to the esophagus, reticulum, and rumen. Courtesy of N. J. Benevenga et al. *J. Dairy Sci.* 52:1294.

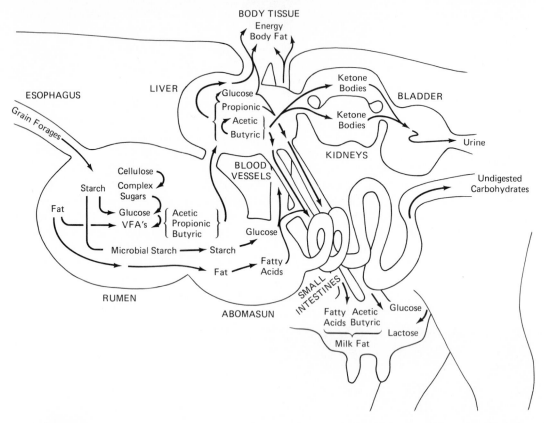

FIGURE 15.8.
Energy pathways in the ruminant. Courtesy of J. Bryant and B. R. Moss, Montana State University.

The rumen fermentation process also produces **volatile fatty acids** (acetic, propionic, and butyric acids), which are waste products of microbial fermentation of carbohydrates. The animal then uses these volatile fatty acids (VFA), which is its major source of energy. In the process of fermenting feeds, methane gas is also produced by the microorganisms. The animal releases the gas primarily through belching. Occasionally the gas-releasing mechanism does not function properly and gas accumulates in the rumen, causing **bloat** to occur. Death will occur owing to suffocation if gas pressure builds to a high level and interferes with adequate respiration.

A young, nursing ruminant consumes little or no roughage. Consequently, at this early stage of life, its digestive tract functions similar to a monogastric animal. Milk is directed immediately into the abomasum in young ruminants by the **esophageal groove** (Fig. 15.7). The sides of the esophageal groove extend upwards by a reflex action and form a tube through which milk passes directly from the esophagus to the abomasum. This allows milk to bypass fermentation in the rumen. Rumen fermentation is an inefficient use of energy and protein in a high-quality feed such as milk.

When roughage is consumed, it is directed into the rumen where bacteria and proto-

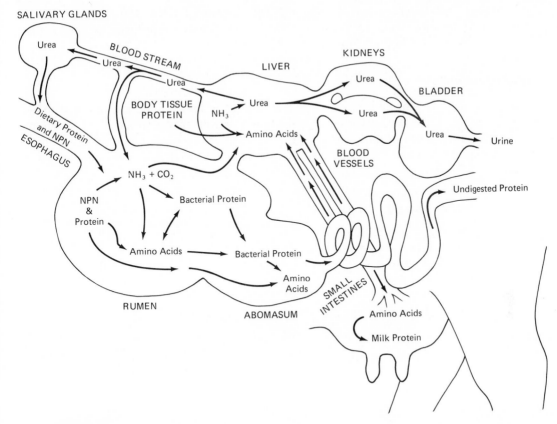

FIGURE 15.9.
Protein pathways in the ruminant. Courtesy of J. Bryant and B. R. Moss, Montana State University.

zoa break it down into simple forms for their use. The rumen starts to develop function-ally as soon as roughage enters it, but some time is required before it is completely func-tional. Complete development of the rumen, reticulum, and abomasum requires about 2 months in sheep and about 3–4 months in cattle. One can influence development of the rumen by the type of feed one gives to the animal. If only milk and concentrated feeds are given, the rumen shows little development. If very young ruminants are forced to live on forage, the rumen develops much more rapidly.

Energy Pathways

Figure 15.8 shows the digestion and utilization of carbohydrates and fats contained in the ingested forages and grains. The primary energy end products of glucose and fatty acids supply energy in the body tissues and become milk fat and lactose in the lactating rumi-nant. Excess energy is stored as body fat in the body tissues.

The primary organs and tissues in energy metabolism are shown in Fig. 15.8. These are the rumen, abomasum, small intestines, liver, blood vessels, mammary gland, and body tissues. Undigested carbohydrates (primarily complex carbohydrates such as lignin) are excreted through the large intestine. Other energy waste products such as ketone bodies are excreted through the kidneys in the urine.

Protein Pathways

The digestion, utilization, and excretion of dietary protein and **nonprotein nitrogen (NPN)** is shown in Fig. 15.9. The end products of protein and NPN are amino acids, ammonia (NH_3), and synthesized amino acids. Excess NH_3 can be formed into urea in the liver, then excreted through the urine, with some urea returning to the rumen as a component of saliva.

SELECTED REFERENCES

Church, D. C. 1986. *Livestock Feeds and Feeding.* Englewood Cliffs, NJ: Prentice-Hall.

Cullison, A. E. 1987. *Feeds and Feeding.* Reston, VA: Reston Publishing Co.

Ensminger, M. E. and Oletine, C. G., Jr. 1978. *Feeds and Nutrition.* Clovis, CA: The Ensminger Publishing Co.

Jurgens, M. H. 1982. *Animal Feeding and Nutrition.* Dubuque, IA: Kendall/Hunt.

CHAPTER 16

Providing Nutrients for Body Functions

Feeding animals is of fundamental importance to any farm production program, because animals must be healthy to function efficiently and yield maximum profits to the producer. The basic task of the producer, then, is to supply animals with feed that will satisfy their body functions for maintenance, growth, fattening, reproduction, lactation, egg laying, wool production, and work. Each of these functions has a unique set of nutrient requirements, and they are additive when more than one function is occurring.

Profits derived from any feeding program must be assessed against production costs. Knowledgeable producers can increase their profits by feeding their animals adequately but nevertheless economically.

BODY MAINTENANCE

The **maintenance** of the body requires that nutrients are supplied to keep the body functioning in a state of well-being. There is no gain or loss of weight or production. Maintenance functions that have a high priority for nutrients are: (a) body tissue repair, (b) control of body temperature, (c) energy to keep all vital organs (respiratory, digestive, etc.) functioning, and (d) maintain water balance.

Nutrients first meet maintenance needs before supplying any of the other body functions. Approximately half of all feed fed to livestock and poultry is used to fill the maintenance requirement. Feedlot animals on full feed may only use 30–40% of their nutrients for maintenance, while some mature breeding animals will need 90% of their feed for maintenance. Highly efficient dairy cows, producing over 100 lb of milk per day, will have a daily feed consumption four to five times their maintenance requirement.

Body Size and Maintenance

Maintenance needs are related to body size. A large animal obviously needs more feed than a small one, but maintenance requirements are not linearly related to body weight.

TABLE 16.1.   TDN Needed for Maintenance of Cattle in the Growing-Finishing Period

Weight of Cattle (lb)	TDN Needed Daily for Maintenance (lb)
400	5.7
600	7.7
800	9.7
1,000	11.4
1,100	12.3

Small animals require more feed per unit of body weight for maintenance than large ones. The approximate maintenance requirement in relation to weight is expressable as $Wt^{0.75}$ rather than $Wt^{1.00}$. Thus, if a 500-lb animal requires 15 lb of feed per day for maintenance, a 1,000-lb animal of the same type does not require twice as much feed even though the latter animal weighs twice as much as the first. The quantity of $1,000^{0.75}$ can be determined and applied to show that the 1,000-lb animal requires approximately 1.7 times as much feed for maintenance as the 500-lb animal. The 1,000-lb animal requires, therefore, approximately 25.5 lb daily (15 lb × 1.7 = 25.5 lb). Table 16.1 shows how the TDN requirement changes with increasing weight. Where corn is 91% TDN, on an air dry basis, it would take approximately 12.5 lb (12.5 lb × 0.91 = 11.4 lb TDN) of corn to fill only the maintenance requirement for a 1,000 lb animal.

GROWTH

Growth occurs when protein synthesis is in excess of its breakdown. Growth is accomplished by an increase in cell numbers, an increase in cell size, and a combination of these. Growth at the tissue level is accomplished primarily through the building of muscle, bone, and connective tissue.

There are several important nutrient requirements for growth, including protein, minerals, vitamins, and energy. The dry matter of muscle and connective tissue is composed largely of protein; therefore, young, growing animals, which need feed to sustain growth in addition to maintenance, have greater protein requirements. A young, growing animal is a muscle-building factory, and protein in the feed is the raw material for the manufacturing process. If provided with only a maintenance amount of feed for an extended period, a young animal may be permanently stunted.

Monogastric animals not only need a certain quantity of protein, but they must also have certain amino acids for proper growth. The protein needs of hogs, for example, are usually supplied by feeding them soybean meal as a supplemental source of amino acids. Young ruminant animals cannot consume enough roughage to make maximum growth. If young ruminant animals are being nursed by dams that produce adequate amounts of milk, the young will do well on good pastures, good quality hay, or both together.

The mineral needs of a young, growing animal include calcium and phosphorus for proper bone growth, salt, for a normal sodium level in the body, and any mineral that may be deficient in the area in which the animal lives. Calcium is usually plentiful in legume forages, and phosphorus is usually plentiful in grains, so a combination of hay and grain should provide all the calcium and phosphorus that young ruminant animals need. Animals fed on hay alone may need additional phosphorus, and those fed diets high in concentrates may need additional calcium. Some producers feed a mixture of steamed bone meal or dicalcium phosphate and salt at all times to assure that their animals have the necessary calcium and phosphorus.

Two minerals, iodine and selenium, require special consideration. Some areas may be deficient in one or both of these elements. An insufficient amount of iodine in the ration of pregnant females might cause an iodine deficiency in the fetus, which prevents thyroxine from being produced and thus causes a goiter in the newborn. Young with goiters die shortly after they are born. Iodized salt can be easily provided to the pregnant females to avoid the iodine deficiency.

A lack of selenium might cause the young to be born with white muscle disease; it can be prevented by giving the pregnant female an injection of selenium. The injectable selenium is distributed commercially and directions for proper dosages that are supplied by the distributor should be followed closely because an overdose in natural feeds or supplements can kill the animal.

Vitamins are needed by young, growing animals. Young ruminant animals are usually on pasture with their dams and are thus exposed to sunshine. The action of ultraviolet rays from the sun converts steroids in the skin into vitamin D, thus providing the animal with this vitamin. Vitamin D is needed for the proper use of calcium and phosphorus in bone growth. Because pigs, poultry, and rabbits are often raised inside, where sunshine is limited or lacking, they need some dietary source of vitamin D.

Most vitamins must be supplied to pigs and poultry through feeds. The only vitamin commonly fed to ruminant animals is vitamin A, and then only when they are on dry pasture, are they fed hay that is quite mature or hay that has been moistened in processing and has, therefore, been dried in the sun for several days. Vitamin A is easily lost in sunlight and during extended dry storage. The activity of vitamin A in silage is usually quite high because this vitamin is usually preserved by the acid fermentation that takes place when silage is made. Silage, however, is usually quite low in vitamin D because the plants used in making silage do not make this vitamin through the action of sunlight and they are not exposed to sunlight for long periods after they are cut.

Young animals need sufficient energy to sustain their growth, high metabolic rate, and activities. They obtain some energy from their mother's milk, and additional energy is supplied as carbohydrates (starch and sugar) and fats from grazed forage or supplemental feeds. Feed grains are high in carbohydrates and also contain some fats. Young ruminants on good pasture typically obtain sufficient energy from pasture and from milk.

The energy needs of young pigs and poultry are generally supplied by feeding them grains such as corn, barley, or wheat. Young horses can usually obtain their energy needs when they are on pasture with their dams since most mares produce much milk; however, for adequate growth after weaning, they need some grain in addition to good pasture or good quality hay.

FATTENING

Fattening is storing surplus feed energy as fat both within and around body tissues. Fattening is desirable to give meat some of its palatability characteristics and to provide energy reserves for postpartum reproductive performance.

Gain from growth is usually less costly than gain from fattening. It takes 2.25 times the energy to produce a pound of fat compared with a pound of protein tissue.

Fattening is the result of excess energy from carbohydrates, fats, or protein above what is required for maintenance and growth. Usually fattening animals are full-fed high energy rations during the last phase of the growing–finishing feeding program.

REPRODUCTION

The requirements for **reproduction** fall into two categories—requirements for gamete production and requirements for fetal growth in the uterus. In general, healthy males and females are capable of producing gametes. The energy needs for germ cell production are no greater than those needed to keep animals in a normal, healthy condition. For example, ruminant animals grazing on pastures of mixed grass and legumes are generally neither deficient in phosphorus nor lacking in fertility. A lack of phosphorus may cause irregular estrous cycles and impaired breeding in females.

Animals that are losing weight rapidly because of poor feed conditions and animals that are overly fat may be low in fertility. To attain optimum fertility from female animals, they should be in moderately low to moderate condition as breeding season approaches, but should, ideally, be increasing in condition (i.e., gaining weight) for 2–4 weeks before and during the breeding season.

The nutrients required by the growing **fetus** are much greater in the last trimester of pregnancy than earlier as little fetal growth occurs during the first two trimesters of pregnancy. Because the fetus is growing, its requirements are the same as those for growth of a young animal after it is born. Healthy females can withdraw nutrients from their bodies to support the growing fetus temporarily while the amount or quality of their feed is low, but reproductive performance will be lower if nutrition is inadequate for a lengthy time of 2–3 months in cattle and a few weeks in swine and sheep.

LACTATION

Among common farm animals, dairy cows and dairy goats produce the most milk; however, most all females are expected to produce milk for their young. Milk production requires considerable protein, minerals, vitamins, and energy. The need for protein is great because milk contains more than 3% protein. As an example, a cow that weighs 1,500 lb and produces 30 lb of milk per day needs at least 30 lb of feed per day, which contains 15% protein; this gives her 4.5 lb protein for her body and the milk she produces. If the protein she eats is 60% digestible, there is 2.7 lb of digestible protein, of which 1.5 lb is present in her milk. If a cow of this weight is to produce 100 lb of milk per day, she must now consume 50 lb of feed containing 15% protein to compensate for the 3 lb of protein in the 100 lb of milk that she gives. Generally during peak milk production, feed consumption cannot compensate for nutrient output and the cow does mobilize some body protein. Actually, more body energy is mobilized than body protein to meet the nutrient deficit.

Calcium and phosphorus are the two most important minerals needed for lactation. Milk is rich in these minerals, and their absence or imbalance may result in decreased lactation or even cause death. The dairy cow may develop milk fever shortly after calving if there is an exceptionally heavy drain of calcium from her system. Cows afflicted by milk fever might become comatose and die if not treated. An intravenous injection of calcium gluconate usually helps the cow recover in less than a day. Milk fever rarely occurs in species or animals that produce relatively small quantities of milk.

Dairy cows produce milk that contains considerable quantities of vitamin A and most B-complex vitamins. Because cows are ruminants, it is unnecessary to feed them B-complex vitamins, and they require vitamin D supplementation only if confined indoors.

Exceptions occur in situations in which beef cows give large amounts of milk. If they do not have adequate feed while they are nursing their calves, they may not conceive for the next calf crop. In those parts of the world where sheep and dairy goats are the principal dairy animals, the energy needs of these animals are quite similar to those of dairy cows that produce much milk.

The requirements for milk production in sows are usually provided by increasing the percentage of protein in the ration, by increasing the amount of feed allowed, and by providing a mineral mix (a combination of minerals that usually contains calcium, phosphorus, salt, and some trace minerals).

Energy is perhaps the most vital requirement for the production of much milk. The energy need is based on the amount of milk being produced. A lactating cow needs energy for body maintenance while she also produces milk and provides the energy stored in it. She cannot eat enough hay to obtain the quantity of energy needed so she must receive high-energy feeds such as concentrates; even then her production may be limited by the amount of feed she can eat. A high producing dairy cow may need three to four times the energy of a nonlactating cow of the same size. Even when fed large amounts of concentrates, a cow that is producing much milk often loses weight and body condition because she cannot consume enough feed to produce at her maximum level; therefore, she draws on her body reserves to supply part of her energy needs.

In the dairy cow, the roughage to concentrate ratio should be approximately 40:60, as a certain amount of roughage is needed to maintain the desired fat content in milk. Therefore, simply feeding more concentrates is not the only answer to increased milk production.

EGG LAYING

The nutrient requirements of poultry are dependent on the specific purposes of production—e.g., broilers primarily for growth with less emphasis on egg production; Leghorn-type (layers) with primary emphasis on eggs and less on growth. Nutrient requirements for growth was discussed earlier in this chapter.

Leghorn-type chickens are smaller in body size than broilers so their maintenance requirements are less. They are prolific in egg production so they are usually fed *ad libitum* during the growing and laying period. Because layers eat *ad libitum* to satisfy their energy needs rations need to have adequate concentrations of energy, protein (amino acids), vitamins, and minerals.

WOOL PRODUCTION

Nutrient requirements for wool production are in addition to nutrients needed for maintenance, growth, and reproduction. Insufficient energy, owing to amount or quality of feed, is usually the most limiting nutritional factor affecting wool production. As wool fibers are primarily protein in composition, the ration should be adequate in protein content.

Shearing removes the natural insulation and might cause an increase in energy requirements owing to heat loss. This is especially true when periods of cold weather occur shortly after shearing.

WORK

Animals used for work, either for pulling heavy loads or for being ridden, require large amounts of energy in addition to the needs for maintenance. Horses are the primary work animals in the United States, but elsewhere, donkeys, cattle, and water buffaloes are used.

Horses, mules, and donkeys rely partly on perspiration to remove nitrogenous wastes. If a horse is used for hard work for five days of the week and is not allowed to exercise the next two days, a strain is placed on the kidneys and illness may result.

The primary requirement is energy above that needed for maintenance and growth. If energy in the ration is not sufficient to meet the work needs, then body fat stores will provide the additional energy needs.

RATION FORMULATION

It is not intended to cover the details of ration formulation in this chapter. Attempting to superficially expose the reader to this area without providing the details can be very misleading. Books are written and courses taught on feeds and feeding or animal nutrition that provide an in-depth coverage of this topic. The reader who desires to pursue ration formulation should refer to one of the references at the end of this chapter.

Chapters 14, 15, and 16 of this book have given a brief background on nutrition which leads into ration formulation. The primary objective of ration formulation is economically matching the animal's nutrient requirements (Tables 16.2–16.8) with the available feeds, taking into consideration the nutrient content of the feeds. Additional considerations are the palatability of the ration, physical form of the feed, and other factors that affect feed consumption.

Additional material on feeding farm animals are contained in the following chapters on individual species: beef cattle (Chapter 22), dairy cattle (Chapter 24), swine (Chapter 26), sheep (Chapter 28), horses (Chapter 30), poultry (Chapter 32), and goats (Chapter 33).

NUTRIENT REQUIREMENTS OF RUMINANTS

The following are some major comparisons that demonstrate changes in nutrient requirements for maintenance growth, lactation, and reproduction:

1. For ewes (maintenance), note the increased requirement in dry matter, energy (TDN or ME), protein, calcium, phosphorus, vitamin A and vitamin D as body weight changes from 110 lb to 176 lb. More nutrients are needed to maintain a heavier body weight.
2. Compare requirements under ewes (last 6 weeks of gestation) with requirements of ewes (maintenance) and ewes (nonlactating and first 15 weeks of gestation). During the latter part of gestation, there are greater nutrient demands owing to rapid fetal growth. During the latter part of gestation compared with maintenance only, energy

TABLE 16.2. Daily Nutrient Requirements of Sheep (dry matter basis)

Weight (lb)	Gain (lb)	Dry Matter (lb)	TDN (lb)	ME (meat)	Total Protein (lb)	Ca (g)	P (g)	Vitamin A (IU)	Vitamin D (IU)
Ewes (maintenance)									
110	0.02	2.2	0.25	1.98	0.20	3.0	2.8	1,275	278
176	0.02	2.9	0.33	2.60	0.26	3.3	3.1	2,040	444
Ewes (nonlactating and first 15 weeks of gestation)									
110	0.07	2.4	0.27	2.16	0.22	3.0	2.8	1,275	278
176	0.07	3.3	0.37	2.96	0.30	3.3	3.1	2,040	444
Ewes (last 6 weeks of gestation)									
110	0.39	3.7	0.45	3.58	0.35	4.1	3.9	4,250	278
176	0.42	4.8	0.58	4.62	0.45	4.8	4.5	6,800	444
Ewes (first 8 weeks of lactation, suckling singles)									
110	−0.06	4.6	0.62	4.90	0.48	10.9	7.8	4,250	278
176	−0.06	5.7	0.77	6.10	0.59	12.6	9.0	6,800	444
Ewes (first 8 weeks of lactation, suckling twins)									
110	−0.13	5.3	0.71	5.63	0.61	12.5	8.9	4,250	278
176	−0.13	6.6	0.89	7.04	0.76	14.4	10.2	6,800	444
Lambs (finishing)									
66	0.44	2.9	0.38	2.99	0.32	4.8	3.0	765	166
88	0.55	3.5	0.51	4.04	0.39	5.0	3.1	1,020	222
110	0.48	4.0	0.57	4.54	0.44	5.0	3.1	1,275	278
121	0.44	4.2	0.60	4.80	0.46	5.0	3.1	1,402	305

Source: Nutrient Requirements of Sheep, NRC (1975).

TABLE 16.3. Daily Nutrient Requirements (NRC) for Breeding Heifers and Cows

Weight (lb)	Daily Gain (lb)	Minimum Dry Matter Consumption (lb)	Crude Protein (lb)	TDN (lb)	ME (Mcal)	Ca (g)	P (g)	Vitamin A (1000 IU)
Pregnant Heifers—Last 3–4 Months of Pregnancy								
800	0.9	16.8	1.4	9.2	15.2	21	15	21
950	0.9	19.0	1.5	10.3	16.9	23	17	24
Cows Nursing Calves—Average Milking Ability[a]—First 3–4 Months Postpartum								
900	0	18.8	1.9	10.8	17.7	24	19	33
1,100	0	21.6	2.0	12.1	19.9	27	22	38
1,400	0	25.6	2.3	14.0	23.0	31	26	46
Cows Nursing Calves—Superior Milking Ability[b]—First 3–4 Months Postpartum								
900	0	18.7	2.4	13.1	21.5	35	24	33
1,100	0	22.3	2.6	14.5	23.8	38	27	40
1,400	0	26.7	2.9	16.5	27.1	42	31	47
Dry Pregnant Mature Cows—Middle Third of Pregnancy								
900	0	16.7	1.2	8.2	13.4	14	14	21
1,100	0	19.5	1.4	9.5	15.6	17	17	25
1,400	0	23.3	1.6	11.4	18.7	21	21	30
Dry Pregnant Cows—Last Third of Pregnancy								
900	0.9	18.2	1.5	9.8	16.2	22	17	23
1,100	0.9	21.0	1.6	11.2	18.3	25	20	26
1,400	0.9	24.9	1.9	13.1	21.5	29	24	32

[a] Ten pounds of milk per day (equivalent of approximately 450 lb of calf at weaning if there is adequate forage).
[b] Twenty pounds of milk per day (equivalent of approximately 650 lb of calf at weaning if there is adequate forage).
Source: National Research Council, 1984. Nutrient Requirements of Beef Cattle.

TABLE 16.4.   Daily Nutrient Requirements for Growing-Finishing Heifers and Steers

Weight[a] (lb)	Daily Gain (lb)	Minimum Dry Matter Consumption (lb)	Protein (%)	Crude Protein (lb)	NEm (Mcal)	NEg (Mcal)	TDN (lb)	Ca (g)	I (g
Growing-Finishing Heifer Calves (medium-frame)									
500	1.0	11.8	9.4	1.10	4.84	3.81	10.1	21	1
500	2.0	11.8	11.4	1.35	4.84	5.37	11.9	27	2
800	1.0	16.7	8.1	1.36	6.24	2.52	11.2	15	1.
800	2.0	16.8	9.0	1.51	6.24	6.91	15.2	21	2•
1,000	2.0	19.8	8.1	1.61	7.52	6.71	16.3	19	1•
Growing-Finishing Steer Calves (medium-frame)									
500	1.0	12.3	9.5	1.16	4.84	2.53	8.8	18	1•
500	2.0	13.1	11.4	1.49	4.84	3.33	9.9	22	1
500	3.0	11.8	14.4	1.69	4.84	4.17	10.4	26	2
800	2.0	18.6	9.2	1.72	6.89	5.33	15.0	21	2•
800	3.0[b]	16.8	10.8	1.81	6.89	7.80	17.0	26	2.
1,000	2.0	22.0	8.4	1.85	8.14	7.73	18.1	21	2
1,000	3.0[b]	19.8	9.5	1.88	8.14	8.47	19.2	22	2.

[a] Average weight for a feeding period.
[b] Most steers of the weight indicated, and not exhibiting compensatory growth, will fail to sustain the energy intak
necessary to maintain this rate of gain for an extended period.
Source: National Research Council, 1984. Nutrient Requirements of Beef Cattle.

TABLE 16.5.   Daily Nutrient Requirements of Dairy Cows

Body Weight (lb)	NE$_l$ (Mcal)	TDN (lb)	Crude Protein (lb)	Ca (lb)	P (lb)	Vitamin A (IU)
Mature Lactating Cows (maintenance)						
882	7.16	6.95	0.82	0.033	0.029	30
1,102	8.46	8.21	0.96	0.040	0.033	38
1,323	9.70	9.42	1.08	0.047	0.038	46
1,544	10.89	10.56	1.20	0.053	0.042	53
Mature Dry Cows (last 2 months of gestation)						
882	9.30	9.04	1.55	0.058	0.040	30
1,102	11.00	10.67	1.81	0.069	0.049	38
1,323	12.61	12.24	2.06	0.082	0.058	46
1,544	14.15	13.74	2.28	0.093	0.066	53
Milk Production (Mcal or lb of nutrient per lb of milk for various fat percentages)						
Percentage fat						
2.5	0.268	0.260	0.072	0.0024	0.00165	
3.0	0.291	0.282	0.077	0.0025	0.00170	
3.5	0.313	0.304	0.082	0.0026	0.00175	
4.0	0.336	0.326	0.087	0.0027	0.00180	
4.5	0.354	0.344	0.092	0.0028	0.00185	
5.0	0.377	0.365	0.098	0.0029	0.00190	

Source: National Research Council, 1978. Nutrient Requirements of Dairy Cattle.

and protein requirements almost double with mineral and vitamin requirements show-ing significant increases.

3. During lactation, nutrient requirements are even higher than those during gestation. Even with larger amounts of dry matter being supplied the ewes are losing weight. Compare the requirements of ewes nursing single lambs versus twins and note the requirements are even higher.

4. The nutrient requirements for lambs being finished for slaughter show the gains approximately the same with increased nutrient requirements due to increased body weight. The increased nutrient requirements are due primarily to an increased maintenance requirement.

NUTRIENT REQUIREMENTS OF MONOGASTRIC ANIMALS

TABLE 16.6. Daily Nutrient Requirements (or percent of ration) for Swine

	Growing-Finishing (fed ad libitum)			Breeding Swine	
				Bred Gilts and Sows (4 lb daily air-dry feed intake)	Lactating Gilts and Sows (10.5 lb daily air-dry feed intake)
Live Weight (lb)	11 to 22	44 to 77	132 to 220		
Expected Daily Gain (lb)	0.66	1.32	1.76		
Expected FE (feed/gain)	1.67	2.50	3.75		
ME (kcal, daily)	1,700	4,740	9,480	5,760	15,180
Crude Protein, lb(%)[a]	0.22(20)	0.53(16)	0.86(13)	0.48	1.36
Indispensable amino acids					
Lysine g (%)	4.8	10.5	17.1	7.7	27.6
Arginine g (%)	1.3	3.0	4.8	0	19.0
Histidine g (%)	1.2	2.7	4.5	2.7	11.9
Isoleucine g (%)	3.2	7.5	12.3	6.7	18.5
Leucine g (%)	3.8	9.0	14.4	7.6	33.2
Methionine plus cystine g (%)	2.8	6.8	9.0	4.1	17.1
Phenylalanine plus tyrosine g (%)	4.4	10.5	17.1	9.4	40.4
Threonine g (%)	2.8	6.8	11.1	6.1	20.4
Tryptophan g (%)	0.8	1.8	3.0	1.6	5.7
Valine g (%)	3.2	7.5	12.3	8.3	26.1
Minerals (selected)					
Calcium g	4.0	9.0	15.0	13.5	35.6
Phosphorus g	3.0	7.5	12.0	10.8	232.8
Sodium g	0.5	1.5	3.0	2.7	9.5
Chlorine g	0.7	2.0	3.9	4.5	14.2
Magnesium g	0.2	0.6	1.2	0.7	1.9
Iron mg	70	90	120	144	380
Zinc mg	50	90	150	90	238
Manganese mg	2	3	6	18	48
Vitamins (selected)					
A IU	1,100	1,950	3,900	7,200	9,500
D IU	110	300	375	360	950
E IU	5.5	17	33	18	47.5
Riboflavin mg	1.5	3.9	7	5.4	14.2
Niacin mg	11	21	30	18	47.5
Panothenic acid mg	6.5	17	33	21.6	57
B$_{12}$ mg	11	17	33	27	71.2

[a] Pounds per day or percentage of the ration.
Source: Nutrient Requirements of Swine, NRC (1979).

TABLE 16.7. Daily Nutrient Requirements of Horses

	Weight (lb)	Daily Gain (lb)	Daily Feed[a] (lb)	Digested Energy (Mcal)	TDN (lb)	Crude Protein (lb)	Ca (g)	P (g)	Vitamin A (1000 IU)
Ponies (approximate mature weight, 440 lb)									
Maintenance	440	0	8.2	8.24	4.12	0.70	9	6	5.0
Mares, last 90 days of gestation	440	0.60	8.1	9.23	4.62	0.86	14	9	10.0
Lactating mares, first 3 months (18 lb of milk per day)	440	0	11.0	12.99	6.50	1.32	20	13	11.0
Yearling (12 months of age)	308	0.44	6.4	8.15	4.07	0.77	12	9	5.5
Horses (approximate mature weight, 880 lb)									
Maintenance	880	0	13.9	13.86	6.93	1.19	18	11	10.0
Mares, last 90 days of gestation	880	1.17	13.7	15.52	7.76	1.41	27	19	20.0
Lactating mares, first 3 months (26 lb of milk per day)	880	0	17.1	20.20	10.10	2.00	33	22	18.0
Yearling (12 months of age)	583	0.88	10.9	13.80	6.91	1.32	24	17	10.0
Horses (approximate mature weight, 1,320 lb)									
Maintenance	1,320	0	18.8	18.79	9.40	1.61	27	17	15.0
Mares, last 90 days of gestation	1,320	1.47	18.5	21.04	10.52	1.91	40	27	30.0
Lactating mares, first 3 months (40 lb of milk per day)	1,320	0	26.0	33.05	16.53	3.52	60	40	33.0
Yearling (12 months of age)	847	1.32	14.8	18.85	9.42	1.98	35	25	14.0

[a] Air dry basis.
Source: Adapted from Nutrient Requirement of Horses (NRC, 1978).

TABLE 16.8. Nutrient Requirements of Leghorn-Type Chickens and Broilers as Percentages or as Milligrams or Units per Kilogram (2.2 lb) of Diet

Energy Base kcal ME/kg Diet[a]	Leghorn-Type Chickens				Broilers	
	Growing		Laying			
	0–6 Weeks 2,900	14–20 Weeks 2,900	2,900	Daily Intake per Hen (mg)[b]	0–3 Weeks 3,200	6–8 Weeks 3,200
Protein (%)	18	12	14.5	16,000	23.0	18.0
Arginine (%)	1.00	0.67	0.68	750	1.44	1.00
Glycine and serine (%)	0.70	0.47	0.50	550	1.50	0.70
Histidine (%)	0.26	0.17	0.16	180	0.35	0.26
Isoleucine (%)	0.60	0.40	0.50	550	0.80	0.60
Leucine (%)	1.00	0.67	0.73	800	1.35	1.00
Lysine (%)	0.85	0.45	0.64	700	1.20	0.85
Methionine plus cystine (%)	0.60	0.40	0.55	600	0.93	0.60
Methionine (%)	0.30	0.20	0.32	350	0.50	0.32
Phenylalanine plus tyrosine (%)	1.00	0.67	0.80	880	1.34	1.00
Phenylalanine (%)	0.54	0.36	0.40	440	0.72	0.54
Threonine (%)	0.68	0.37	0.45	500	0.80	0.68
Tryptophan (%)	0.17	0.11	0.14	150	0.23	0.17
Valine (%)	0.62	0.41	0.55	600	0.82	0.62
Linoleic acid (%)	1.00	1.00	1.00	1,100	1.00	1.00
Calcium (%)	0.80	0.60	3.40	3,750	1.00	0.80
Phosphorus, available (%)	0.40	0.30	0.32	350	0.45	0.35
Potassium (%)	0.40	0.25	0.15	165	0.40	0.30
Sodium (%)	0.15	0.15	0.15	165	0.15	0.15
Chlorine (%)	0.15	0.12	0.15	165	0.15	0.15
Magnesium (mg)	600	400	500	55	600	600
Manganese (mg)	60	30	30	3.30	60.0	60.0
Zinc (mg)	40	35	50	5.50	40.0	40.0
Iron (mg)	80	60	50	5.50	80.0	80.0
Copper (mg)	8	6	6	0.88	8.0	8.0
Iodine (mg)	0.35	0.35	0.30	0.03	0.35	0.35
Selenium (mg)	0.15	0.10	0.10	0.01	0.15	0.15
Vitamin A (IU)	1,500	1,500	4,000	440	1,500	1,500
Vitamin D (ICU)	200	200	500	55	200	200
Vitamin E (IU)	10	5	5	0.55	10	10
Vitamin K (mg)	0.50	0.50	0.50	0.055	0.50	0.50
Riboflavin (mg)	3.60	1.80	2.20	0.242	3.60	3.60
Pantothenic acid (mg)	10.0	10.0	2.20	0.242	10.0	10.0
Niacin (mg)	27.0	11.0	10.0	1.10	27.0	11.0
Vitamin B_{12} (mg)	0.009	0.003	0.004	0.00044	0.009	0.003
Choline (mg)	1,300	500	?	?	1,300	500
Biotine (mg)	0.15	0.10	0.10	0.011	0.15	0.10
Folacin (mg)	0.55	0.25	0.25	0.0275	0.55	0.25
Thiamin (mg)	1.8	1.3	0.80	0.088	1.80	1.80
Pyridoxine (mg)	3.0	3.0	3.0	0.33	3.0	2.5

[a] These are typical dietary energy concentrations.
[b] Assumes an average daily intake of 110 g of feed/hen daily.
Source: Nutrient requirements of poultry, 1984. NRC.

SELECTED REFERENCES

Church, D. C. 1986. *Livestock Feeds and Feeding.* Englewood Cliffs, NJ: Prentice-Hall.

Cullison, A. E. 1987. *Feeds and Feeding,* 2nd edition. Reston, VA: Reston.

Ensminger, M. E. and Olentine, C. G., Jr. 1978. *Feeds and Nutrition.* Clovis, CA: The Ensminger Publishing Co.

Jurgens, M. H. 1982. *Animal Feeding and Nutrition.* Dubuque, IA: Kendall/Hunt.

Nutrient Requirements (of Beef Cattle, 1984), (of Dairy Cattle, 1978), (of Goats, 1981), (of Horses, 1978), (of Poultry, 1984), (of Sheep, 1985) and (of Swine, 1979). National Research Council. Washington, D. C.: National Academy Press.

United States-Canadian Tables of Feed Composition. 1982. National Research Council. Washington, D.C.: National Academy Press.

CHAPTER 17

Growth and Development

Profitable and efficient production of livestock and poultry involves understanding their growth and development. Manipulation of genetic and environmental factors can change growth patterns in farm animals.

Many aspects of growth and development are contained in other chapters of this book. The material on reproduction, genetics, nutrition, and products should be integrated with this chapter to more fully understand animal growth and development.

Generally speaking, growth is an increase in body weight until mature size is reached. This growth is an increase in cell size and cell numbers with protein deposition resulting. More specifically, growth is an increase in the mass of structural tissues (bone, muscle, and connective tissue) and organs accompanied by a change in form or composition of the animal's body.

Development is defined as the directive coordination of all diverse processes until maturity is reached. It involves growth, cellular differentiation, and changes in body shape and form. In this chapter, growth and development will be combined and discussed as one entity.

Prenatal (Livestock)

The three phases of prenatal life—sex cells, the embryo, and the fetus—are briefly discussed in Chapter 9. Embryological development is a fascinating process as a spherical mass of cells differentiates into specific cell types and eventually into recognizable organs (Figs. 17.1 and 17.2). The endoderm [Fig. 17.2(D)] differentiates into the digestive tract, lungs, and bladder; the mesoderm [Fig. 17.2(D)] into the skeleton, skeletal muscle, and connective tissue; while the ectoderm [Fig. 17.2(D)] differentiates into the skin, hair, brain, and spinal cord. The growth, development, and differentiation processes, involving primarily protein synthesis, are directed by DNA chains of chromosomes and the organizers in the developing embryo (see Chapter 11). Thus the nucleus is a center of activity for different types of cells, directing the growth and development process (Fig. 17.3).

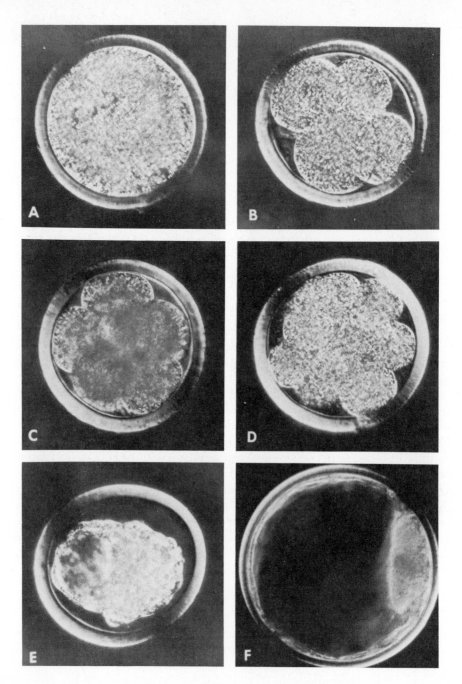

FIGURE 17.1.
Embryonic development during the first 8 days of pregnancy in cattle. (A) Unfertilized egg. (B) Four-cell embryo on day 2 of pregnancy (estrus = day 0). (C) Eight-cell embryo on day 3 of pregnancy. (D) An 8- to 16-cell embryo on day 4 of pregnancy. (E) Very early blastocyst stage (approximately 60 cells). (F) Expanded blastocyst (> 100 cells) recovered from the uterus on day 8 of pregnancy. Magnification approximately ×300. Courtesy of Seidel, G. E., Jr., 1981. Superovulation and embryo transfer in cattle. *Science* 211:351. Copyright 1981 by the American Association for the Advancement of Science.

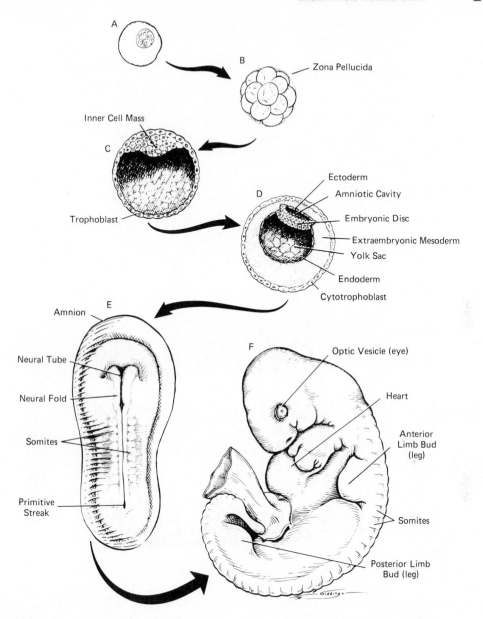

FIGURE 17.2.
The morphogenesis of a single egg cell (A) into a morula (B), then to a blastocyst (C). (D) The stage at which the two cavities have formed in the inner cell mass; an upper (amniotic) cavity and a lower cavity yolk sac. The embryonic disc containing the ectoderm and endoderm germ layers is located between cavities. (E) A cattle embryo showing the neural tube and somites. (F) The development of the 14-day cattle embryo. Drawn by Dennis Giddings.

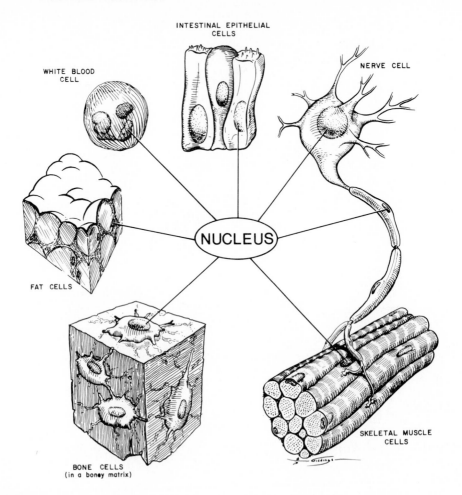

INTESTINAL EPITHELIAL
CELLS

NERVE CELL

WHITE BLOOD
CELL

NUCLEUS

FAT CELLS

SKELETAL MUSCLE
CELLS

BONE CELLS
(in a boney matrix)

FIGURE 17.3.
Cells of selected tissues showing similarity in cell structure, but not cell shape. The nucleus gives direction to differentiation of cells and their function. Courtesy of Dennis Giddings.

The fetus undergoes marked changes in shape and form during prenatal growth and development. Early in the prenatal period, the head is much larger than the body. Later, the body and limbs grow more rapidly than other parts. The order of tissue growth follows a sequential trend determined by physiological importance, starting with the central nervous system and progressing to bones, tendons, muscles, intermuscular fat, and subcutaneous fat.

During the first two-thirds of the prenatal period, most of the increase in muscle weight is due to hyperplasia (increase in size of fibers). During the last 3 months of pregnancy, hypertrophy (increase in number of fibers) represents most of the muscle growth. Individual muscles vary in their rate of growth, with larger muscles (those of legs and back) having the greatest rate of postnatal growth. Water content of fetal muscle declines with fetal age, and this decline in water content continues through postnatal growth as well.

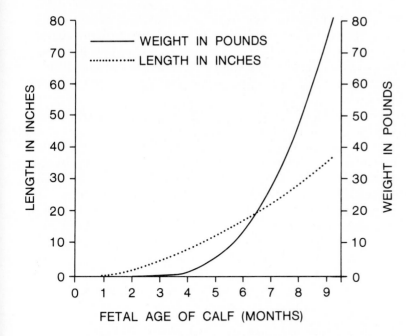

FIGURE 17.4.
Growth of the fetal calf.
Drawn by CSU Graphics.

The relative size of the fetus changes during gestation, with the largest increase in weight occurring during the last trimester of pregnancy (Fig. 17.4).

Birth (Livestock)

After birth, the number of muscle fibers does not appear to increase significantly, therefore postnatal muscle growth is primarily by hypertrophy. In red meat animals, all muscle fibers appear to be red at birth, but shortly thereafter some of them differentiate into white and intermediate muscle types.

At birth, the various body parts have considerably different proportions when compared with mature body size and shape. At birth, the head is relatively large; the legs are long; and the body is small, whereas in the mature animal, the head is relatively small; the legs are relatively short; and the body is relatively large. Birth weight represents approximately 5–7% of the mature weight, while leg length at birth is approximately 60%; height at withers is approximately 50% of those same measurements at maturity. Hip width and chest width at birth are approximately one-third of the same measurements at maturity. This shows that the distal parts (leg and shoulder height) are developed earlier than proximal parts (hips and chest).

POULTRY

Embryonic Development

The development of a chick differs from mammals because there is no connection with its mother. The chick develops in the egg, but entirely outside the hen's body. Embryonic development is much more rapid in chicks than the farm mammals.

Every egg whether fertile or nonfertile has a germ spot called the blastoderm (see Fig. 5.1). This is where the chick embryo will develop if the fertile egg is properly incubated.

A controlled environment must be maintained during incubation to produce live chicks from the fertile eggs. The major components of the controlled environment are: (a) temperature $99\frac{1}{2}$–100°F); (b) 60–75% relative humidity; (c) turning the egg every 1–8 hours; and (d) providing adequate oxygen.

Three hours after fertilization, the blastoderm divides to form two cells. Cell division occurs until maturity except during the holding period before incubating the eggs.

There are four membranes that are essential to the growth of the chick embryo (Figure 17.5). The **allantois** is the membrane that allows the embryo to breathe. It takes oxygen through the porous shell and oxygenates the blood of the embryo. The allantois removes the carbon dioxide, receives excretions from the kidneys, absorbs albumen used as food for the embryo, and absorbs calcium from the shell for use by the embryo. The **amnion** is a membrane filled with a colorless fluid that serves as a protection from mechanical shock. The **yolk sac** is a layer of tissue growing over the surface of the yolk. This tissue has special cells that digest and absorb the yolk material for the developing embryo. The **chorion** surrounds both the amnion and yolk sac.

Table 17.1 identifies some of the primary changes in the growth and development of

TABLE 17.1.   Major Changes in Weight, Form, and Function of the Chick Embryo (white Leghorn) during Incubation

Day	Weight (g)	Developmental Changes
1	00.0002	Head and backbone are formed; central nervous system begins
2	00.0030	Heart forms and starts beating; eyes begin formation
3	00.0200	Limb buds form
4	00.0500	Allantois starts functioning
5	00.1300	Formation of reproductive organs
6	00.2900	Main division of legs and wings; first movements noted
7	00.5700	
8	01.1500	Feather germs appear
9	01.5300	Beak begins to form; embryo beings to look bird-like
10	02.2600	Beak starts to harden; digits completely separated
11	03.6800	
12	05.0700	Toes fully formed
13	07.3700	Down appears on body; scales and nails appear
14	09.7400	Embryo turns its head toward blunt end of egg
15	12.0000	Small intestines taken into body
16	15.9800	Scales and nails on legs and feet are hard; albumen is near gone; yolk is main food
17	18.5900	Amniotic fluid decreases
18	21.8300	
19	25.6200	Yolk sac enters body through umbilicus
20	30.2100	Embryo becomes a chick; it breaks amnion, then breathes air in air cell
21	36.3000	Chick breaks shell and hatches

Source: Arizona Cooperative Extension Publication 8427.

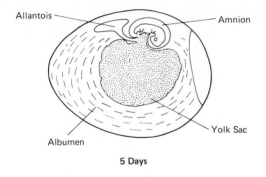

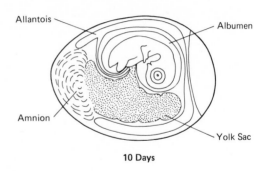

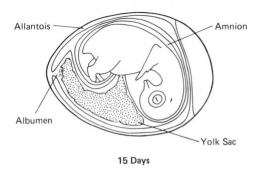

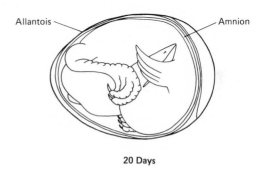

FIGURE 17.5.
Changes in the development of the chick embryo with associated changes in membranes and other contents of the egg.

the chick embryo. Many of these phenomenal changes occur rapidly, sometimes in only a few hours.

BASIC ANATOMY AND PHYSIOLOGY

In the developmental process, cells become grouped in appearance and function. Specialized groups of cells that function together are called **tissues.** The primary types of tissues are: (a) muscle, (b) nervous, and (c) connective and epithelial, with examples shown in Fig. 17.3.

Organs are groups of tissues which provide specific functions. For example, the uterus is an organ which functions in the reproductive process. A group of organs which function in concert to accomplish a larger, general function comprise a **system.** The reproductive system (Chapter 9), digestive system (Chapter 15), and mammary system (Chapter 18) are discussed in their respective chapters.

It is not the intent of this chapter to discuss all the different systems even though they are important in growth and development. A few additional systems are briefly surveyed that are considered important in understanding farm animals and their productivity. Fig-

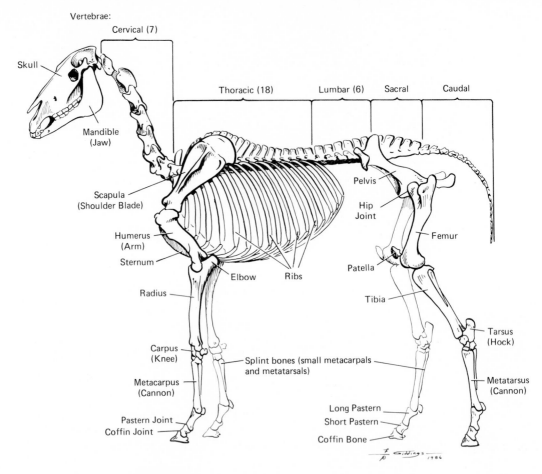

FIGURE 17.6.
Skeletal system of the horse. Courtesy of Dennis Giddings.

ures are provided showing selected systems for one or two species. The systems are similar and generally comparable among the various species of farm animals, although some large differences between poultry and farm mammals do exist.

Skeletal System

Figure 17.6 shows the skeletal system of the horse and Fig. 17.7 portrays the chicken's skeletal system. Even though only bones and some joints are shown in these figures, teeth and cartilage are also considered part of the skeletal system.

The skeleton protects other vital organs and gives a basic form and shape to the animal's body. Bones function as levers, store minerals, and the bone marrow is the site of blood cell formation.

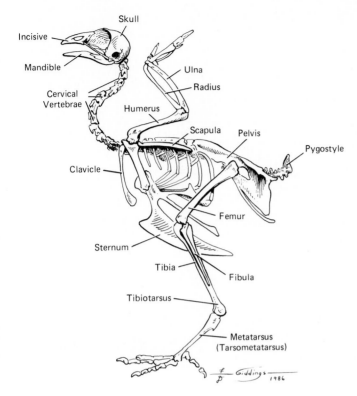

FIGURE 17.7.
Skeletal system of the chicken. Courtesy of Dennis Giddings.

Chicken bones are more pneumatic (bone cavities are filled with air spaces) harder, thinner, more brittle, and have a different ossification process than mammals.

Muscle System

There are three types of muscle tissue—skeletal, smooth, and cardiac. Skeletal muscle is the largest component of red meat animal products. Smooth muscle is located in the digestive, reproductive, and urinary organs. The heart is composed of cardiac muscle.

Figure 17.8 identifies some of the major muscles similar in name and location in the meat animal species and horses. Of special note is the longissimus dorsi which was previously mentioned in Chapter 2. The size of this muscle and marbling it contains are important factors in determining yield grades and quality grades of meat animals.

The primary muscles of the turkey are shown in Figure 17.9. The muscles of poultry are refered to as dark meat (legs and thighs) and white meat (breast and wings).

Circulatory System

Figure 17.10 shows the major components of the dairy cow's circulatory system. The circulatory systems of other farm animal species are similar to the dairy cow.

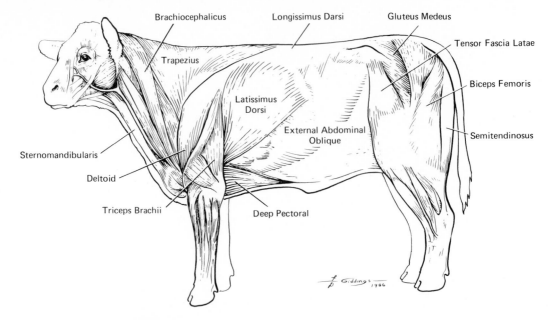

FIGURE 17.8.
Primary muscles of the beef steer. Courtesy of Dennis Giddings.

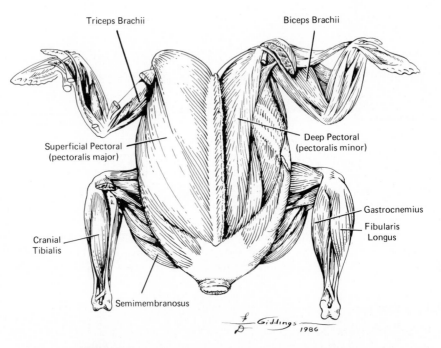

FIGURE 17.9.
Primary muscles of the turkey. Courtesy of Dennis Giddings.

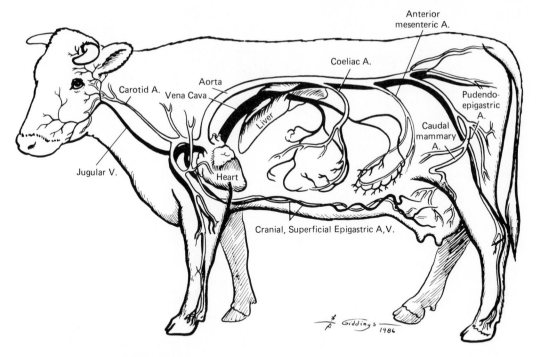

FIGURE 17.10.
Circulatory system of the dairy cow. Courtesy of Dennis Giddings.

The heart, acting as a pump, and the accompanying vessels comprise the circulatory system. Arteries are vessels transporting blood away from the heart, while the vessels carrying blood to the heart are called veins. The lymph vessels transport lymph (intercellular fluid) from tissues to the heart.

The circulatory system is very important in growth and development as the blood transports oxygen, nutrients, cellular waste products, and hormones.

Milk production in all species is dependent on the nutrients in milk arriving through the circulatory system. Dairy cows producing 20,000 to 40,000 lb milk per year (over 100 lb per day) must consume large amounts of feed, with these feed nutrients circulating in high concentration through the mammary blood supply. Approximately 400–500 lb of blood circulates through the dairy cow's mammary gland for each pound of milk produced.

Endocrine System

Growth and development is highly dependent on the endocrine system. This system consists of several endocrine (ductless) glands that secrete **hormones** into the circulatory system. Hormones are chemical substances that affect a gland (or organ) or, in some cases, all body tissues.

The major endocrine glands are shown in Fig. 17.11, and the hormones produced and

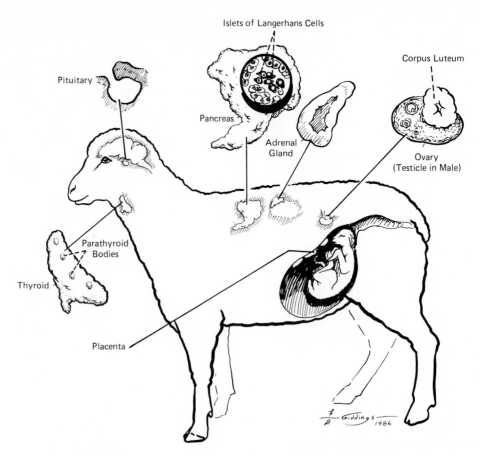

FIGURE 17.11.
Primary endocrine organs of the sheep. Courtesy of Dennis Giddings.

their major effects are identified in Table 17.2. The hormones affecting mammary gland growth, development, and function are also identified in Chapter 18.

GROWTH CURVES

Postweaning growth is a curved-line function regardless of how it is expressed mathematically. If growth is considered an increase in body weight, then most of the early growth follows a straight line or is linear in the age–weight relationship (Fig. 17.12). As the animal increases in age and approaches puberty, rate of growth usually declines, and true growth ceases when the animal reaches maturity.

After an animal reaches maturity, it may have large fluctuations in body weight simply by increasing or decreasing the amount of fat or water that is stored. This increase in weight owing to fattening is not true growth because no net increase in body protein

TABLE 17.2. Major Hormones Affecting Growth and Development in Farm Animals

Hormone	Source	Major Effect
Growth (Somatotrophic, STH)	Pituitary (anterior)	Body cell growth—especially muscle and bone cells
Adrenocorticotropic (ACTH)	Pituitary (anterior)	Stimulates adrenal cortex to produce adrenal cortical steroid hormones
Glucocorticoids	Adrenal (cortex)	Conversion of proteins to carbohydrates
Mineralocorticoids	Adrenal (cortex)	Regulates sodium and potassium balance; water balance
Thyroid-stimulating (TSH)	Pituitary (anterior)	Stimulates thyroid gland to produce thyroid hormones
Thyroid	Thyroid	Regulates metabolic rate
Testosterone	Testicles (interstitial cells)	Libido; accessory sex gland development; male secondary sex characteristics; spermatogenesis
Follicle-stimulating (FSH)	Pituitary (anterior)	Development of ovarian follicles; spermatogenesis
Luteinizing (LH)	Pituitary (anterior)	Ovum maturation; ovulation; formation of corpus luteum (CL); stimulates interstitial cells (testicle) to produce testosterone
Prolactin (luteotropic, LTH)	Pituitary (anterior)	Milk secretion (initiation and maintenance); CL maintenance during pregnancy
Estrogen	Ovary (follicle) Placenta	Female reproductive organ growth; female secondary sex characteristics; mammary gland duct growth
Progesterone	Corpus luteum; placenta	Uterine growth; maintenance of pregnancy; alveoli growth in mammary gland growth
Antidiuretic (ADH) (Vasopressin)	Pituitary (posterior)	Controls water loss in kidney
Oxytocin	Pituitary (posterior) (released with nursing reflex)	Causes uterine contractions during parturition; causes milk let-down in mammary gland
Relaxin	Ovary; placenta	Relaxes pelvic ligaments during parturition
Epinephrine	Adrenal (medulla)	Increases blood glucose concentration
Norepinephrine	Adrenal (medulla)	Maintains blood pressure
Insulin	Pancreas	Lowers blood sugar
Parathormone (PTH)	Parathyroid	Calcium and phosphorus metabolism
Glucagon	Pancreas	Raises blood sugar

Source: Compiled from several sources

occurs. In fact, animals tend to lose body protein as they grow older. The loss of body protein is one of the phenomena in the aging process.

Figures 17.13 and 17.14 show typical growth curves for most of the farm animals. The different curves for large and small breeds is primarily a function of differences in skeletal frame size. Maturity is reached at heavier weights in larger breeds, which have larger skeletal frame sizes than smaller framed animals.

The relatively straight-lined growth shown for turkeys and broilers (Figure 17.14) does

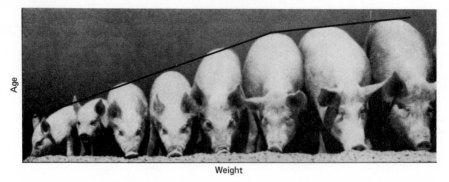

FIGURE 17.12.
Early growth in pigs is linear (straight line), whereas later in life the rate of growth decreases. Courtesy of Elanco Products Co.

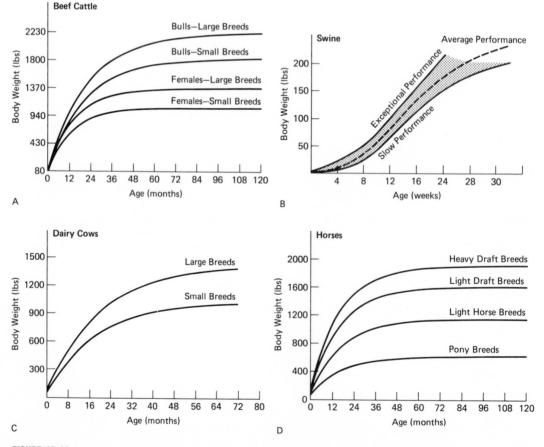

FIGURE 17.13.
Typical growth curves for beef cattle (A), swine (B), dairy cows (C), and horses (D). Courtesy of Battaglia, R. A. and Mayrose, V. B. 1981. *Handbook of Livestock Management Techniques.* © Macmillan Publishing Company.

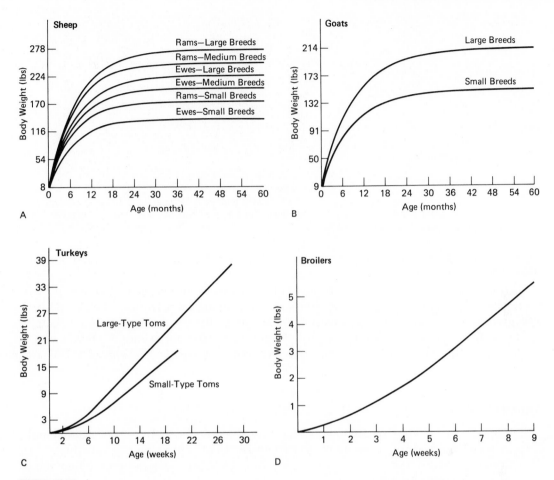

FIGURE 17.14.
Typical growth curves for sheep (A), goats (B), turkeys (C), and broilers (D). Courtesy of Battaglia, R. A. and Mayrose, V. B. 1981. *Handbook of Livestock Management Techniques.* © Burgess Publishing Company.

curve and level off as shown for the other species. This occurs when the birds become older which for broilers is approximately 11 weeks of age.

CARCASS COMPOSITION

Products from meat animals and poultry are composed primarily of fat, lean, and bone. A superior carcass is characterized by a low proportion of bone, a high proportion of muscle, and an optimum amount of fat. Understanding the growth and development of these animals is very important in knowing when animals should be slaughtered to produce the desirable combinations of fat, lean, and bone.

Figure 17.15 shows the expected changes in fat, muscle, and bone as animals increase in live weight during the linear growth phase. As the animal moves from the linear growth

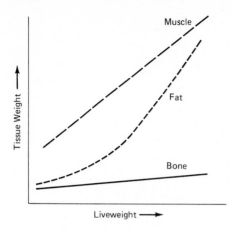

FIGURE 17.15.
Tissue growth relative to increased live weight. Courtesy of the American Society of Animal Science, Berg and Walters (1983).

phase into maturity, then the growth curves shown in Fig. 17.15 are extended similar to those shown in Fig. 17.13.

At light weights, fat deposition begins rather slowly, then increases geometrically when muscle growth slows as the animal approaches physiological maturity. Muscle comprises the greatest proportion and its growth is linear during the production of young slaughter animals. Bone has a smaller relative growth rate than either fat or muscle. Because of the relative growth rates, the ratio of muscle to bone increases as the animal increases in weight.

Effects of Frame Size

Different maturity types of animals (sometimes refered to as breed types or biological types) have a marked influence on carcass composition at similar live weights (Fig. 17.16). The earlier maturing types increase fat deposition at lighter weights than either the average or later maturing types.

Skeletal frame size, measured as hip height, is a more specific way of defining matu-

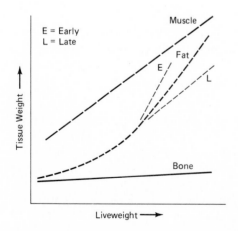

FIGURE 17.16.
Effects of different maturity types on carcass composition. Courtesy of the American Society of Animal Science, Berg and Walters (1983).

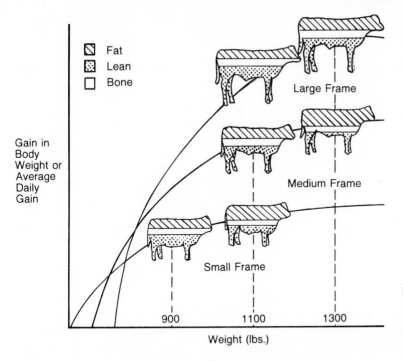

FIGURE 17.17.
The relationship of frame size and weight to carcass composition in beef steers. Drawn by CSU Graphics.

rity type. Figure 17.17 shows the difference in fat, lean, and bone composition for small, medium, and large frame beef steers at different live weights. Note that the three different frame sizes have a similar composition at 900 lb (small frame), 1100 lb (medium frame), and 1300 lb (large frame).

Effect of Sex

The effect of sex is primarily on the fat component, although there are differences among species. Heifers deposit fat earlier than steers or bulls, with bulls being leaner at the same slaughter weight [Fig. 17.18(A)]. Heifers are typically slaughtered at lighter weights (100–200 lb less) than steers to have a similar fat to lean composition.

Swine are different than cattle as barrows are fatter than gilts or boars at similar slaughter weights [Fig. 17.18(B)]. The reason for this species difference is not known.

At typical slaughter weights, the sex differences in sheep do not have a marked effect on fat and lean composition. Apparently sheep are slaughtered in an earlier stage of development (before puberty) than cattle and swine. Compositional differences in sheep are similar to cattle if sheep are fed longer than their present slaughter weights.

Effect of Muscling

Relatively large differences exist for muscling in the various species of farm animals. The proportion of muscle in the carcass varies indirectly with fat, and a higher proportion of fat is associated with a lower proportion of muscle. Muscle has a much faster relative

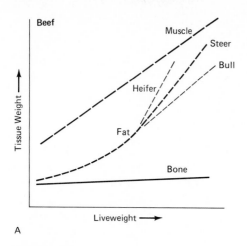
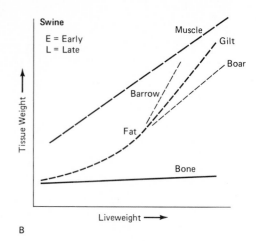

FIGURE 17.18.
Influence of sex on carcass composition in beef cattle (A) and swine (B). Adapted from the American Society of Animal Science, Berg and Walters (1983).

growth rate than bone. Muscle weight relative to liveweight or muscle to bone ratio can be used as a valuable measurement of muscle yield.

Most of the current cattle population in the U.S. do not have large differences in muscling. There are, however, muscling differences in 10% to 15% of the cattle which are best expressed by a muscle to bone ratio (pounds of muscle ÷ pounds of bone). Muscle to bone ratios can range from 3 lb of muscle to 1 lb of bone in thinly-muscled slaughter steers and heifers to 5 lb of muscle to 1 lb of bone in more thickly-muscled cattle. Muscle to bone ratios higher than 5:1 are found in double-muscled cattle. These differences in muscle to bone ratio are economically important, particularly if more boneless cuts are merchandised.

Some of the muscling differences are shown for cattle: lighter-muscled Holsteins [Fig. 17.19(A)], heavy-muscled [Fig. 17.19(B)], and double-muscled cattle [Fig. 17.19(C)]. Double-muscled cattle do not have a duplicate set of muscles but they have an enlargement (hypertrophy) of the existing muscles.

Extreme muscling, as observed in double-muscled or heavy-muscled types of animals, can be associated with problems in total productivity. Some heavy-muscled swine might die when subjected to stress conditions and reproductive efficiency is lower in some heavily-muscled animals. There is some latitude in increasing muscle to bone ratio in slaughter animals without encountering negative side effects.

Individual muscle dissection has shown a similarity of proportion of individual muscles to the total body muscle weight. Small differences in muscle distribution have been shown, but the differences do not seem to be economically significant for the industry. Thus it appears that increasing muscle weight in areas of high-priced cuts, at the expense of muscle weight in lower priced cuts, is not commercially feasible. Even though muscle distribution is not important, muscling as previously defined as muscle to bone ratio has high economic importance.

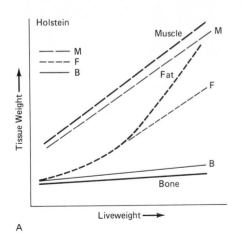

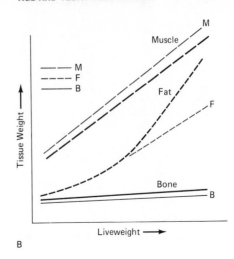

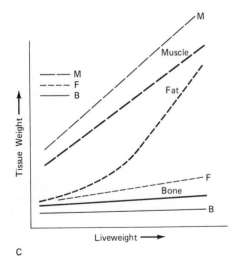

FIGURE 17.19.
Effect of muscling on fat, lean, and bone composition in cattle. A, light-muscled Holsteins; B, heavily muscled; C, double muscled. Courtesy of the American Society of Animal Science, Berg and Walters (1983).

AGE AND TEETH RELATIONSHIP

For many years, producers have observed teeth of farm animals to estimate their age or their ability to effectively graze with advancing age. Some animals are **mouthed** to classify them into appropriate age groups for showring classification. The latter has not always proven exact because of animal variation in teeth condition and the ability of some exhibitors to manipulate the condition of the teeth.

Table 17.3 gives some guidelines used in estimating the age of cattle by observing their teeth. The **incisors** (front cutting teeth) are the teeth used. There are no upper incisors, and the molars (back teeth) are not commonly used to determine age.

Figure 17.20 shows pictures of incisor teeth of cattle of different ages. If the grazing area is sandy and the grass is short, wearing of the teeth will progress faster than that

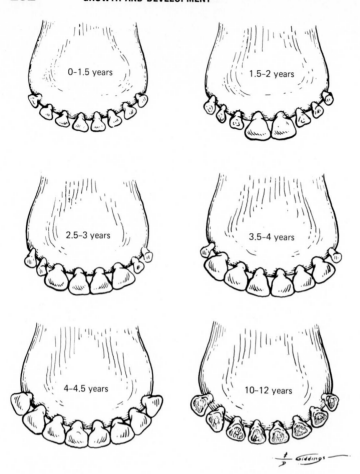

FIGURE 17.20.
Incisor teeth of cattle of different ages. Courtesy of Dennis Giddings.

TABLE 17.3.   **Description of Cattle's Teeth to Estimate Age**

Approximate Age	Description of Teeth
Birth	Usually only 1 pair of middle incisors
1 month	All eight, temporary incisors
1.5–2 yr	First pair of permanent (middle incisors)
2.5–3 yr	Second pair of permanent incisors
3.25–4 yr	Third pair of permanent incisors
4–4.5 yr	Fourth (corner) pair of permanent incisors
5–6 yr	Middle pair of incisors begins to level off from wear Corner teeth might also show some wear
7–8 yr	Both middle and second pair of incisors show wear
8–9 yr	Middle, second, and third pair of incisors show wear
10 yr and over	All eight incisors show wear

previously described. Also some cows may be **broken-mouth,** which means some of their permanent incisor teeth have been lost.

The age relationship to teeth condition for sheep and horses are given in Chapters 28 and 29, respectively.

SELECTED REFERENCES

Publications

Berg, R. T. and Butterfield, R. M. 1976. *New Concepts of Cattle Growth.* Sydney, Australia: Sydney University Press.

Berg, R. T. and Walters, L. E. 1983. The meat animal. Changes and challenges. *J. Anim. Sci.* 57:133(Suppl. 2).

Bone, J. F. 1982. *Animal Anatomy and Physiology.* Reston, VA: Reston.

Brown, C. J., Brown, A. H., Jr. and Johnson, Z. Studies in body dimensions of beef cattle. *Ark. Agric. Expt. Sta. Bull.* 863.

Frandson, R. D. 1981. *Anatomy and Physiology of Farm Animals.* Philadelphia: Lea and Febiger.

Hammond J., Jr., Robinson, T. and Bowman, J. 1983. *Hammond's Farm Animals.* Baltimore: Edward Arnold University Park Press.

Prior, R. L. and Lasater, D. B. 1979. Develoment of the bovine fetus. *J. Anim. Sci.* 48:1546.

Rollins, F. D. 1984. Development of the embryo. Arizona Coop. Ext. Serv. Public. 8427.

Swatland, H. J. 1984. *Structure and Development of Meat Animals.* Englewood Cliffs: Prentice-Hall.

Trenkle, A. H. and Marple, D. N. 1983. Growth and development of meat animals. *J. Anim. Sci.* 57:273 (Suppl. 2).

Visuals

Anatomy of the Fowl (sound filmstrip). Vocational Education Productions, Calif. Polytechnic State Univ., San Luis Obispo, CA 93407.

CHAPTER 18 ▮▮▮▮▮▮▮▮▮▮▮▮

Lactation

Lactation, the production of milk by the mammary gland, is a distinguishing characteristic of mammals whose young at first feed solely on milk from their mothers. Even after they start to eat other feeds, the young continue to nurse until they are weaned (separated from their mothers so they cannot nurse).

The mammary gland serves two functions: (1) provides nutrition to animal offspring and (2) is a source of passive immunity to the offspring. The importance of milk as a nutritional source to perpetuate each mammalian species has been known since the beginning of history. Only during the past few decades has the basic mechanisms of immunity through milk been determined.

Humans consume milk and recognize it as a palatable source of nutrients. Milk consumed in the U.S. comes primarily from dairy cows and, to a much lesser extent, from goats and sheep. In other countries, the human milk supply also comes from water buffalo, yak, reindeer, camel, donkey, and sow.

MAMMARY GLAND STRUCTURE

The mammary gland is an **exocrine gland** which produces the external secretion of milk transported through a series of ducts. The cow has four separate mammary glands that terminate into four teats; sheep and goats have two glands and two teats; and the mare has four mammary glands which terminate into two teats. The mammary glands of these species are located in the groin area.

Sows have 6–20 mammary glands located in two rows along the abdomen with each gland having a teat. Typically there are 10–14 of the sow's mammary glands that are functional (Fig. 18.1). There is evidence that teat number in swine is not related to litter size or litter weaning weight.

Figure 18.2 shows the basic structure of the cow's udder. The udder is supported horizontally and laterally by suspensory ligaments [Fig. 18.2(A)]. The internal structure

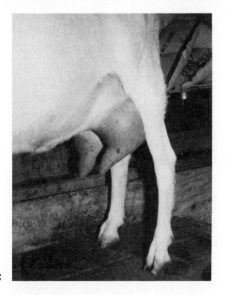
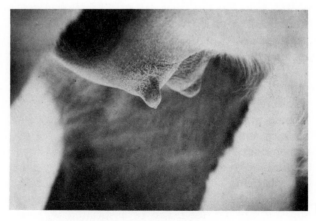

FIGURE 18.1.
Mammary glands of (A) cow, (B) sow, (C) goat, and (D) mare. Courtesy of (A) Hoards Dairyman, (B) University of Nebraska, (C) George F. W. Haenlein, and (D) Colorado State University.

of the udder is similar for all farm animal species, except for the number of glands and teats.

Secretory tissue of the mammary gland [Fig. 18.2(C)] is composed of millions of grape-like structures called **alveoli.** Each **alveolus** has its separate blood supply from which milk constituents are obtained by epithelial cells lining the alveolus. Milk collects in the alveolus lumen and, during milk let-down, travels through ducts to a larger collection area called the gland cistern. During milk let-down and the milking process, milk is forced into the teat cistern and through the streak canal to the outside of the teat.

MAMMARY GLAND DEVELOPMENT AND FUNCTION

Development

Mammary gland growth and development occurs rapidly as the female reaches puberty. The ovarian hormones (estrogen and progesterone) have a large effect on development. Estrogen is primarily responsible for duct and cistern growth, whereas progesterone stimulates growth of the alveoli.

Estrogen and progesterone are produced by the ovary under the stimulation of FSH and LH from the anterior pituitary (Fig. 18.3). The pituitary also has a direct influence on mammary growth through the production of growth hormone. Growth hormone stimulates general cell growth of the mammary gland.

Figure 18.3 also identifies the additional indirect effect (other than FSH and LH) on mammary growth and development. Thyroid hormones are produced under the influence of TSH, and the adrenal gland produces the corticosteriods when stimulated by ACTH from the pituitary. All of these hormones work in concert to produce mammary growth and function.

Milk Secretion

Growth hormone, adrenal corticoids, and prolactin are primarily responsible for the initiation of lactation. These hormones become effective as parturition nears and when estrogen and progesterone hormone levels decrease.

Through milking or nursing, the milk in the gland cistern is soon removed. There remains large amounts of milk in the alveoli, which is forced into the ducts by the contractions of the myoepithelial cells (Fig. 18.2). These cells contract under the influence of the hormone, oxytocin, which is secreted by the posterior pituitary (Fig. 18.3).

Oxytocin, which is released by the suckling reflex, can also be released by other stimuli. The milk let-down can be associated with feeding the cows or washing the udder. The latter is a typical management practice before milking cows. Oxytocin release can be inhibited by pain, loud noises, and other stressful stimuli.

Maintenance of Lactation

Lactation is maintained primarily through hormonal influence. Prolactin, thyroid hormones, adrenal hormones and growth hormone are all important in the maintenance of lactation.

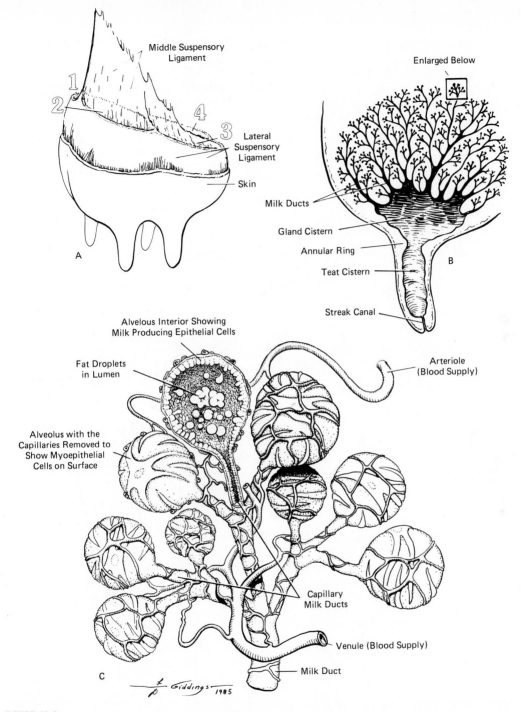

Middle Suspensory
Ligament

1
2
4
3

Lateral
Suspensory
Ligament

Skin

A

Enlarged Below

Milk Ducts

Gland Cistern

Annular Ring

Teat Cistern

Streak Canal

B

Alvelous Interior Showing
Milk Producing Epithelial Cells

Fat Droplets
in Lumen

Alveolus with the
Capillaries Removed to
Show Myoepithelial
Cells on Surface

Arteriole
(Blood Supply)

Capillary
Milk Ducts

Venule (Blood Supply)

Milk Duct

C

Giddings 1985

FIGURE 18.2.
(A) Basic structure of cow's udder showing suspensory ligaments and the location of the four separate quarters.
(B) Section through one of the quarters showing secretory tissue, ducts, and milk collecting cisterns. (C) An en-
larged lobe with several alveoli and their accompaying blood supply. Courtesy of Dennis Giddings.

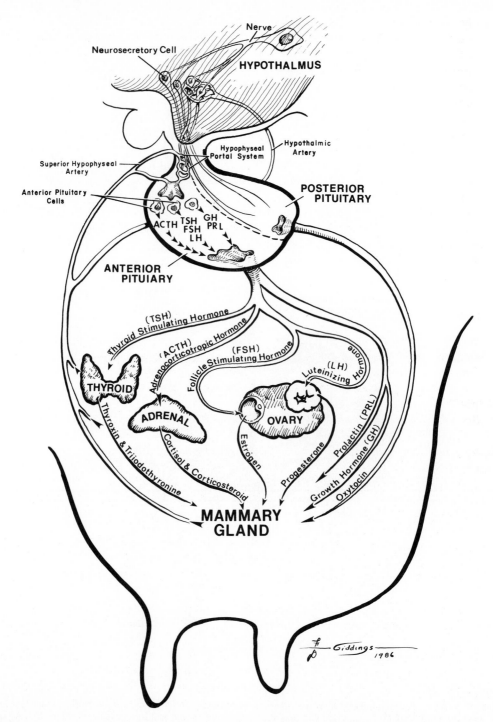

FIGURE 18.3.
Major hormones (and their source) influencing the anatomy and physiology of the mammary gland. Note Figure 17.11 for the location of the endocrine glands throughout the body. Courtesy of Dennis Giddings.

Daily milk production typically increases during the first few weeks of lactation, peaks at approximately 4–6 weeks, then decreases over the next several weeks of the lactation period. Persistency of lactation measures how milk production is maintained over time. For example in dairy cattle, persistency is determined by calculating the milk production this month as a percentage of last month's production.

Figure 18.4 shows lactation curves for the various species of farm animals. The curve for dairy cows [Fig. 18.4(A)] represents approximately 14,000 lb produced in 305 days.

Decreased milk production during the lactation period is due primarily to a decreased number of active alveoli and less secretory tissue (epithelial cells) in the alveoli. These and other changes are associated with hormonal changes.

When milking or suckling is stopped, the alveoli are distended and the capillaries are filled with blood. After a few days the secretory tissue becomes involuted (reduced in size and activity) and the lobes of the mammary gland consist primarily of ducts and connective tissue. The female then becomes **dry** as milk secretion is not occurring. After the dry period (approximately 2 months in the dairy cow) and when parturition approaches, hormones and other influences prepare the mammary gland to resume its secretion and production of milk.

FACTORS AFFECTING MILK PRODUCTION

Inheritance determines the potential for milk production, but feed and management determine whether or not this potential is attained. The best feed and care will not make a high-producing female out of one that is genetically a low producer. Likewise, a female with high genetic potential will not produce at a high level unless she receives proper feed and care. Production is also influenced by the health of the animal. For example, **mastitis** (an inflammation of the udder) in dairy cows can reduce production by 30% or more. Proper management of dairy cows, particularly adherence to routine milking and feeding schedules, will contribute to a high level of production.

In lactating farm animals, the level of milk production is important because milk from these females provides much of the required nutrients for optimal growth of the young. If beef cows, for example, are inherently heavy milkers, their milk-producing levels will be established by the ability of the young to consume milk. Normally, if one does not milk such a cow, she will adjust her production to the consumption level of the calf. However, if the cow is milked early in lactation, her milk flow may be far greater in amount than the calf can consume. This situation makes it necessary to either continue milking the cow until the calf becomes large enough to consume the milk, or to put another calf onto the cow.

Anything that causes a female to reduce her production of milk also usually causes some regression of the mammary gland, thereby preventing resumption of full production. If the young animal lacks vigor, it does not consume all the milk that its mother can supply, and her level of production is lowered accordingly. This is one reason why inbred calves grow more slowly and crossbred calves more rapidly during the nursing period. Crossbred calves are usually larger and more vigorous at birth and can stimulate a high level of milk production by their dams, whereas inbred calves are smaller and less vigorous and cause their dams to produce milk at a comparatively low rate.

Females with male offspring produce more milk than females with female offspring.

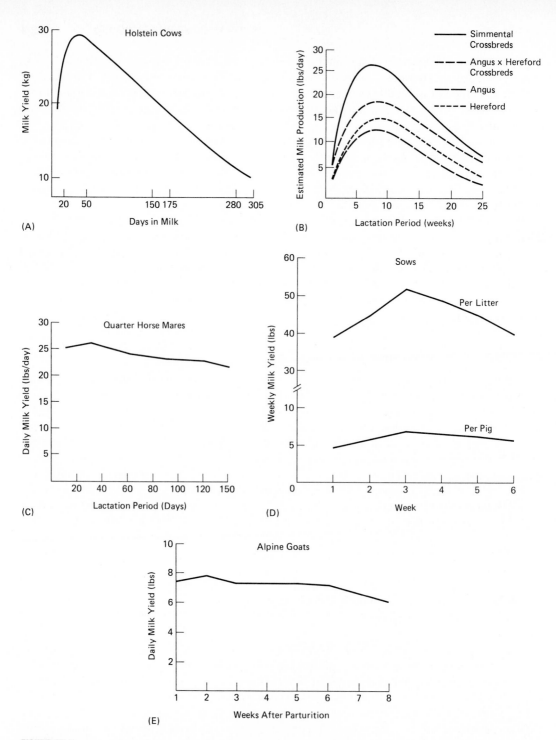

FIGURE 18.4.
Lactation curves for several species of farm animals. (A) Dairy cows. Courtesy of Colorado State University (CO, ID, and UT DHIA records). (B) Beef cows. Courtesy of Jenkins and Ferrell, 1984 Beef Cow Efficiency Forum. (C) Quarter horse mares. Courtesy of *J. Anim. Sci.* 54: 496. (D) Sow. Courtesy of *Missouri Agric. Research Bull.* 712. (E) Alpine goats. Courtesy of *J. Dairy Sci.* 63:1677.

Male offspring are usually heavier at birth and have a faster growth rate than females. This puts a greater demand on the milk-producing ability of the dam.

Females having multiple births usually produce more total milk than females with single births. Age of the female affects milk yield with younger and older females producing less milk compared to females who have had several lactations. For example, sows are usually at peak milk yields during third or fourth lactation, beef cows produce the most milk at 5–9 years of age, and dairy cows have the highest milk yields between 5 and 8 years of age.

Large amounts of nutrients are needed to supply the requirements of lactating females for high levels of milk production, because much energy is required for milk secretion and because milk contains large quantities of nutrients. The need for adequate nutrition is much greater for lactation than for gestation; for example, a cow producing 100 lb of milk daily could yield 4.0 lb of butterfat, 3.3 lb of protein, and almost 5.0 lb of lactose in that milk. If nutrition is inadequate in quality or quantity, a cow that has the inherent capacity to produce 100 lb of milk daily will draw nutrients from her own body as the body attempts to sustain a high level of milk production. Withdrawal of nutrients reduces the body stores. This withdrawal usually occurs to a limited extent in high producers even when they are well fed. Cows reduce their milk production in response to inadequate amounts or quality of feed, but usually do so only after the loss of nutrients is sufficiently severe to cause a loss in body weight. In gestation, a cow becomes relatively efficient in digesting her feed, so feed restrictions in gestation are less harmful than they might be at other times.

MILK COMPOSITION

Species Differences

Milk composition is markedly different for some of the mammalian species (Table 18.1). Milk from reindeer and aquatic mammals is exceptionally high in total solids, where mare's milk is low. Large differences in milkfat percentages exist in mammals; donkeys are low (1.4%) and fur seals are very high (53.3%).

TABLE 18.1. Average Milk Composition of Several Species

Species	Total Solids (%)	Fat (%)	Protein (%)	Lactose (%)	Minerals (%)
Human	13.3	4.5	1.6	7.0	0.2
Cow (Bos taurus)	12.7	3.9	3.3	4.8	0.7
Cow (Bos indicus)	13.5	4.7	3.4	4.7	0.7
Goat	12.4	3.7	3.3	4.7	0.8
Water buffalo	19.0	7.4	6.0	4.8	0.8
Ewe	18.4	6.5	6.3	4.8	0.9
Sow	19.0	6.8	6.3	5.0	0.9
Mare	10.5	1.2	2.3	5.9	0.4
Reindeer	33.7	18.7	11.1	2.7	1.2

Except in protein, milk from cows and goats is similar in composition to milk from humans. Milk from cows or goats is much higher in protein than milk from humans. This is not a problem in bottle feeding babies as the additional protein is not harmful.

In dairy cows, the amount or percentage of fat is easily changed through changing the diet fed. Carbohydrates (lactose) remain relatively constant even with dietary fluctuations. Varying the protein content of the ration has little change on the protein content of the milk.

COLOSTRUM

The fetus develops in a sterile environment. Its immune system has not yet been challenged to the microorganisms existing in the external environment. Before or shortly after birth, the fetal immune system must be made functional or death is imminent.

Immoglobulins (Ig) are involved in the passive immunity transfer maternally to the offspring. In some animals, the Ig are transferred *in utero* through the bloodstream, whereas in other animals immoglobulins are transferred through colostrum. Certain animals utilize both methods of Ig transfer; however, the colostrum method is most common for larger, domestic animals.

These Ig antibodies give the newborn protection from harmful microorganisms that invade the body and cause illness. The intestinal wall of the newborn is quite porous, permitting absorption of colostrum antibodies to enter the bloodstream. Within a few hours (no more than 24 h), the gut wall becomes less porous, allowing little absorption of the antibodies to occur. Thus passive immunity of the newborn is dependent on an adequate supply of antibodies in the colostrum and consuming the colostrum within a few hours after parturition.

SELECTED REFERENCES

Publications

Allen, A. D., Lasley, J. F. and Tribble, L. F. Milk production and related factors in sows. *Missouri Agric. Res. Bull.* 712.

Gibbs, P. G., Potter, G. D., Blake, R. W. and McMullan, W. C. 1982. Milk production of quarter horse mares during 150 days of lactation. *J. Anim. Sci.* 54:496.

Keown, J. F., Everett, R. W., Empet, N. B. and Wadell, L. H. 1986. Lactation curves. *J. Dairy Sci.* 69:769.

Larson, B. L. (editor). 1985. *Lactation.* Ames, IA: Iowa State University Press.

Offedal, O. T., Hintz, H. F. and Schryver, H. F. 1983. Lactation in the horse: Milk composition and intake by foals. *J. Nutr.* 113:2196.

Visuals

External Features of the Udder, Internal Features of the Udder, Removal of Milk and Proper Milking Practices, Abnormalities in Mammary Glands and *Milk Composition and Factors Affecting Milk Yield*. Videotapes. Agricultural Products and Services, 2001 Killebrew Drive, Suite 333, Bloomington, MN 55420.

CHAPTER 19

Adaptation to the Environment

Livestock throughout the world are expected to produce under an extremely wide range of environments. Variations in temperature, humidity, wind, light, altitude, feed, water, and exposure to parasites and disease organisms are some environmental conditions to which livestock are exposed (Fig. 19.1).

Under intensive management, such as integrated poultry enterprises, confinement swine operations, and large dairies, many environmental conditions are highly controlled by humans. Feed and water is plentiful, rations are carefully balanced, excellent health programs are carefully monitored, and, in many cases, temperature, humidity, and other weather influences are controlled.

Extensive management involves much less producer control over environmental conditions in which the animals are expected to produce. Many ruminant animals are extensively managed, being exposed to numerous climatic conditions as they graze the forage, which can be very sparse. Livestock that produce and survive under these conditions must have physiological mechanisms for adaptability.

Economics dictate to a large extent if animals will be managed intensively or extensively. Under intensive management, many environmental conditions are changed to fit the animals. Under extensive management, animals are changed (selected for adaptability) to fit the environment.

It is important to understand the major environmental effects on animal productivity. Then management decisions can be made to determine if changing the environment or changing the adaptability of animals is most economically feasible. In some situations, environmental changes and changes in animal adaptability might produce optimal results.

MAJOR ENVIRONMENTAL EFFECTS

Adjusting to Environmental Changes

As the seasons change, two major kinds of changes occur in the environment: changes in temperature and changes in length of daylight. As summer approaches, temperature

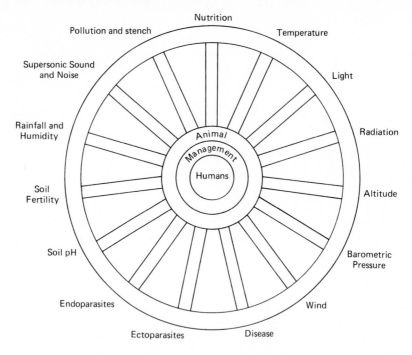

FIGURE 19.1.
The livestock ecology wheel. Human's management decisions with farm animals have a role in determining how the animals will interact with their environment. Courtesy of Bonsma, J. 1983. *Livestock Production.* Cody, WY: Agi Books.©

and length of daylight increase, thus increasing the amount of heat available to the animal. As autumn approaches, length of daylight and temperature decrease. Hormone changes help the animal respond physiologically to these seasonal changes.

The thyroid gland, the lobes of which are located on both sides of the trachea, regulates the metabolic activity of the animal. When the environmental temperature is cool, the air inhaled tends to cool the thyroid gland. The thyroid responds by increasing its production of the hormone thyroxine; the thyroxine in turn stimulates metabolic activity and heat is generated. The growing and shedding of insulating coats are slow changes that are also influenced by hormones. Sudden warm periods before shedding occurs in spring, and sudden cool periods before a warm coat has been produced in autumn can have severe effects on an animal. These sudden temperature changes are the greatest predisposers for the onset of respiratory diseases such as pneumonia, influenza, and shipping fever.

Temperature Zones of Comfort and Stress

The comfort zone, or **thermoneutral zone** (TNZ) identifies a range of temperatures where heat production and heat loss from the body are about the same. When the temperature is below the TNZ, animals increase their feed intake and reduce blood flow to the surface and to the extremities. Mammals generate heat by shivering. Fowl fluff their feathers to create a large space of dead air about themselves to prevent rapid heat transfer from their bodies. Some animals might hunch to expose less body surface. Cattle congregate in an area that provides protection from blizzards, and they huddle closely such that each animal receives heat from other animals. During a severe blizzard, some animals can be killed from trampling and suffocation.

To avoid excessive heat loss, animals maintain the temperature of their extremities at a level below that of the rest of the body through a unique mechanism called countercurrent blood flow action. Blood in arteries coming from the core of the body is relatively warm and blood in veins in the extremities is relatively cool. The blood in veins in the extremities cools the warmer arterial blood so that the extremities are kept at a cool temperature. With the extremities thus kept at a relatively cool temperature, loss of heat to the environment is reduced.

Between 65 and 80°F, dilation of blood vessels near the skin and in the extremities occurs so that the surface of the animal becomes warmer, water consumption increases, respiration increases, and, in animals that can sweat, perspiration increases. Above 80°F, animals that have the capacity of sweat keep their body surfaces wet with sweat so that evaporation can help cool them.

When the environmental temperature exceeds 90°F, farm animals may die, have lower daily gains and poor feed conversion, and have lower reproductive performance through higher embryonic deaths. During these high temperatures, animals become less active, reducing the amount of heat they generate, and lie down in the shade, reducing their exposure to the sun. Animals typically increase their consumption of water and excretion of urine. If the water consumed is cooler than the temperature of the animal, it can help considerably to cool the body.

CHANGING THE ENVIRONMENT

Adjusting Rations for Weather Changes

Nutrient requirements and predicted gains of farm animals are published from research studies in which animals are protected from environmental extremes. The most common environmental factor that alters both performance and nutrient requirements is temperature. Thus livestock producers should be aware of critical temperatures that affect performance of their animals and consider making changes in their feeding and management programs if economics so dictate.

The TNZ is the range in effective temperature where rate and efficiency of performance is optimum. Critical temperature is the lower limit of the TNZ, and is typified by the ambient temperature below which the performance of animals begins to decline as temperatures become colder. Figure 19.2 shows that the maintenance energy requirement increases more rapidly during cold weather than does rate of feed intake. This results in a reduction of gain and more feed required per pound of gain, which typically causes cost per pound of gain to be higher.

Effective ambient temperatures below the lower critical temperature (below the TNZ) constitute cold stress and those above the TNZ, heat stress. The term, "effective ambient temperature," is an index of the heating or cooling power of the environment in terms of

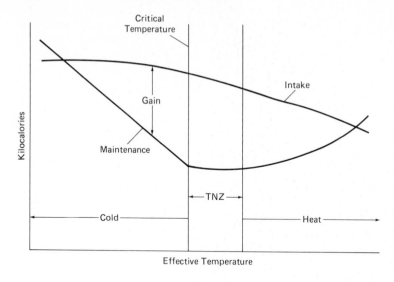

FIGURE 19.2.
Effect of temperature on rate of feed intake, maintenance energy requirement, and gain (Ames, 1980).

dry bulb temperature. It would include any environmental factor, such as radiation, wind, humidity, or precipitation that alters environmental heat demand.

The critical temperatures and optimal temperature ranges are shown for some animals in Fig. 19.3. Effective ambient temperatures have not been calculated for all farm animals, although the effects of some combined environmental factors are known. For example, the wind-chill index for cattle, is shown in Table 19.1. An example in Table 19.1 shows that with a temperature of 20°F and a wind speed of 30 mph, the effective ambient temperature is −16°F.

The lower critical temperature for cattle depends on how much insulation is provided by the hair coat, whether the animal is wet or dry, and how much feed the cow consumes. Table 19.2 shows some lower critical temperatures for beef cattle. For example, a cow being fed a maintenance ration may have a low critical temperature of 32°F if dry, but the lower critical temperature might be 60°F if the same cow is wet.

TABLE 19.1. Wind-Chill Factors for Cattle with Winter Coat

Wind Speed (mph)	Temperature (°F)												
	−10	−5	0	5	10	15	20	25	30	35	40	45	50
Calm	−10	−5	0	5	10	15	20	25	30	35	40	45	50
5	−16	−11	−6	−1	3	8	13	18	23	28	33	38	43
10	−21	−16	−11	−6	−1	3	8	13	18	23	28	33	38
15	−25	−20	−15	−10	−5	0	4	9	14	19	24	29	34
20	−30	−25	−20	−15	−10	−5	0	4	9	14	19	24	29
25	−37	−32	−27	−22	−17	−12	−7	−2	2	7	12	17	22
30	−40	−41	−36	−31	−26	−21	−16	−11	−6	−1	3	8	13
35	−60	−55	−50	−45	−40	−35	−30	−25	−20	−15	−10	−5	0
40	−78	−73	−68	−63	−58	−53	−48	−43	−38	−33	−28	−23	−18

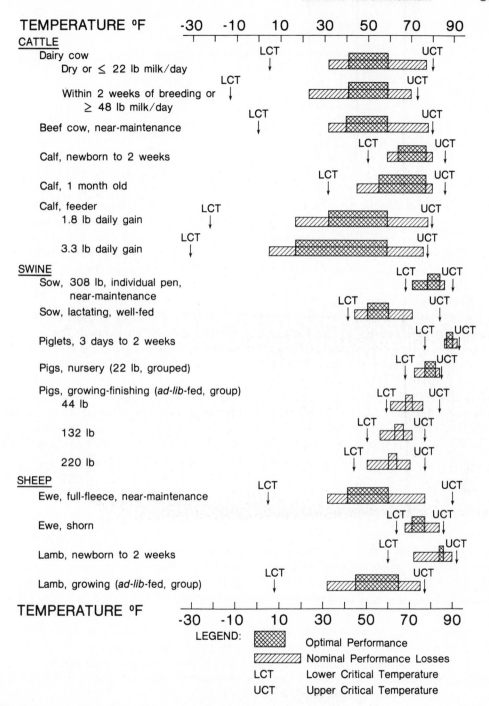

FIGURE 19.3.
Critical ambient temperatures and temperature zones for optimal performance in cattle *(Bos taurus),* swine, and sheep. Adapted from CRC Press, Inc., Boca Raton, FL.

TABLE 19.2.   Estimated Lower Critical Temperatures for Beef Cattle

Coat Description	Critical Temperature (°F)
Summer coat or wet	59
Fall coat	45
Winter coat	32
Heavy winter coat	18

Coldness of a specific environment is the value that must be considered when adjusting rations for cows. Coldness is simply the differences between effective temperature (wind-chill) and lower critical temperature. Using this definition for coldness instead of using temperature on an ordinary thermometer helps explain why wet, windy days in March might be colder for a cow than extremely cold but dry, calm days of January.

The major effect of cold on nutrient requirements of cows is increased need for energy, which usually means increasing the total amount of daily feed.

Feeding tables recommend that a 1200 lb cow receive 16.5 lb of good mixed hay to supply energy needs during the last one-third of pregnancy. How much feed should the cow receive if she is dry and has a winter hair coat but the temperature is 20°F with a 15 mph wind? The coldness is calculated by subtracting the wind-chill or effective temperature (4°F) from the cow's lower critical temperature (32°F). Thus the magnitude of coldness is 28°F. A rule of thumb (more detailed tables are available) is to increase the amount of feed 1% for each degree of coldness. A 28% increase of the 16.5 lb (original requirement) would mean that 21.1 lb of feed must be fed to compensate for the coldness.

This example is typical of many feeding situations; however, if the cow was wet, the same increase in feed would be required at 31°F wind-chill (28°F of coldness).

Inability of Animals to Cope with Heat

Animals are sometimes unable to control their body temperatures in conditions of extreme heat or cold. When the body temperature exceeds normal because the animal cannot dissipate its heat, a condition known as fever results. Fever is often caused by systemic infectious diseases and is often most severe when environmental temperatures are extremely high or low. Keeping the ill animal comfortable while medication is given will assist recovery.

When the temperature becomes excessively high, some animals (notably pigs) might lose normal control of their senses and do things that aggravate the situation. If nothing is done to prevent them from doing so, pigs that get too hot will often run up and down a fence line, squealing, until they collapse and die. Pigs seen doing this should be cooled with water and induced to lie down on damp soil.

Most dairy cows will have a reduction in milk production when the environmental temperature exceeds 80°F. This occurs primarily because appetite is depressed and feed consumption declines. Milk yields will improve when thermal stress is alleviated by providing shade, fans, or refrigerated air.

CHANGING THE ANIMALS

One of the most impressive examples of adaptability under extensive management are the small Black Bedouin goats that graze the Negev desert of Israel and the Eastern Sinai. Owing to the arid desert conditions they are watered only once in 2–4 days, which greatly increases their foraging range.

Ninety Bedouin goats (2970 lb total body mass) can be herded successfully 2 days away from water points, where only one fat-tail Awassi sheep (110 lb) or 1.5 Black Mediterranean goats (165 lb) would be able to live because of more frequent watering needs.

During water deprivation for 4 days, the Bedouin goats lose 25–30% of their body weight from reduction of total body water and blood plasma volume. Nevertheless, these goats have relatively high milk yields, as 35-lb goats produce over 4.5 lb of milk per day. The Bedouin goats also have relatively low daily caloric demands for maintenance per unit of metabolic weight. Milk production efficiencies have exceeded 33% of energy consumed. The Bedouin goats have developed this adaptability over several centuries of natural selection as they have survived and produced in these arid, desert environments.

Inability of Animals to Cope with Other Stresses

Some animals are unable to adjust to environmental stresses. This inability may be genetically controlled.

When cattle are grazed at high altitudes, 1–5% develop high mountain (brisket) disease. This disease usually occurs at altitudes of 7,000 feet or more. It is characterized as right heart failure (deterioration of the right side of the heart). It was considered that this disease was caused by chronic **hypoxia** (lack of oxygen), which leads to pulmonary **hypertension** (high blood pressure). As the animal becomes more affected, marked subcutaneous edema (filling of tissue with fluid) develops in the brisket area. Some recent studies at the U. S. Poisonous Plant Center at Logan, Utah, have shown that certain poisonous plants that grow only at high altitudes will cause brisket disease when fed to cattle maintained at low altitudes.

Some young cattle and sheep whose dams are fed plants in selenium-deficient areas develop **white muscle disease.** The disease is characterized by calcification of the heart and voluntary muscles, giving the muscles a white appearance. White muscle disease can be prevented by administering selenium to the mother during pregnancy and to the young at birth. In the absence of treatment, some animals are affected and some are not; therefore, resistance and susceptibility appear to be inherited.

SELECTED REFERENCES

Ames, D. R. Feeding beef cows in winter. *Angus J.*, Jan. 1981.
Ames, D. R. Livestock nutrition in cold weather. *Anim. Nutr. and Health*, Oct. 1980.
Bosma, J. 1983. *Livestock Production.* Cody, WY: Ag Books.

Effect of Environment on Nutrient Requirements of Domestic Animals. 1981. National Research Council. Washington, D. C.: National Academy Press.

Hahn, G. L. 1985. Weather and climate impacts on beef cattle. Beef Research Progress Report No. 2, MARC, USDA, ARS-42.

McDowell, L. R. (editor). 1985. *Nutrition of Grazing Ruminants in Warm Climates.* New York: Academic Press.

Yousef, M. K. 1985. *Stress Physiology in Livestock. Vol. II Ungulates.* Boca Raton, FL: CRC Press.

CHAPTER 20

Animal Health

Productive animals are typically healthy animals. Death loss **(mortality rate)** is the most dramatic sign of health problems, however lower production levels and higher costs of production, which are due to sickness **(morbidity),** are economically more serious.

Disease is defined as any deviation from normal health in which there are marked physiological, anatomical, or chemical changes in the animal's body. Noninfectious diseases and infectious diseases are the two major disease types. Noninfectious diseases result from injury, genetic abnormalities, ingestion of toxic materials, and poor nutrition. Microorganisms are not involved in noninfectious diseases. Examples of noninfectious diseases are plant poisoning, bloat, and mineral deficiencies.

Infectious diseases are caused by microorganisms such as bacteria, viruses, and protozoa. A **contagious disease** is an infectious disease that spreads rapidly from one animal to another. Brucellosis, ringworm, and transmissible gastroenteritis (TGE) are examples of infectious diseases.

PREVENTION

The old adage, "An ounce of prevention is worth a pound of cure," is an important part of herd health. Preventative herd health programs can eliminate or reduce most of the health-related livestock losses. Most major animal disease problems are associated with management.

The components of herd health programs include: (a) sanitation, (b) proper nutrition, (c) record analysis, (d) source of livestock, (e) physical facilities, (f) proper use of biologics and pharmaceuticals, (g) minimizing stress, and (h) personnel training. All these components require cooperative efforts between the producer and veterinarian.

Planning with Veterinarian

Veterinarians are professionally educated and trained in animal health. Veterinarians located near animal producers are familiar with disease and health problems that are primary concerns in the producer's area. Producers having the most successful herd health programs use veterinarians in planning such programs. Part of the plans include regularly scheduled visits by the veterinarian throughout the year to assess preventative programs, in addition to animal treatment. They also can train farm and ranch personnel in several simple herd health management practices and serve as a reference for new products.

In most livestock and poultry operations, it is more economically feasible to have veterinarians assist in planning and implementing preventative health programs than to use their services only in crisis situations.

Producers should keep cost-effective health records as recommended by the veterinarian. Records are needed to better assess problem areas and to develop a more effective herd health plan. Records might include dates and what was done in preventative and treatment programs, description of conditions during health problems, and deaths. Possible causes of death should be recorded with **necropsy** records obtained where economically feasible.

Sanitation

The severity of some diseases is dependent on the number and virulence of microorganisms entering the animal's body. As many microorganisms live and multiply outside the animal, the number of microorganisms can be reduced by implementing sanitation practices. These practices, in turn, will reduce the incidence of disease outbreaks.

Manure and other organic waste materials are ideal environments for proliferation of microorganisms. A good sanitation program includes the cleaning of organic materials from buildings, pens, and lots. This allows the effective destruction of microorganisms from high temperatures and drying. Buildings, pens, and pastures should be well drained, preventing prolonged wet areas or mud holes. These sanitation practices help both in disease prevention and controlling parasites.

Antiseptics and **disinfectants** should be carefully selected and effectively utilized in a good sanitation program. Antiseptics are substances, usually applied to animal tissue, that kill or prevent the growth of microorganisms. Disinfectants are products that destroy pathogenic microorganisms. They are usually agents used on inanimate objects. In the absence of disinfectants, sanitizing with clean water is helpful.

Other important sanitation measures include the prompt and proper disposal of dead animals, either to rendering plants or by burial.

Proper Nutrition

Well-nourished animals receive an adequate daily supply of essential nutrients. Undernourished animals usually have a weak immune system, thus making them more vulnerable to invading microorganisms.

Nutrition principles are presented in Chapters 14, 15, and 16. Feeding programs for the various species of farm animals are covered in Chapters 22, 24, 26, 28, 30, 32, and 33.

Record Analysis

Herd health needs to be approached from an economic standpoint. Is there a health problem or is it just a perception? Proper records permit health problems to be identified, determine what is causing them, then alternative methods of prevention and treatment can be assessed.

Physical Facilities

Physical facilities contribute to animal health problems by causing physical injury, stress, or allowing dissemination of pathogens through a group of animals. They can contribute to the spread of disease by not preventing its transmission, e.g., venereal disease transmission owing to poor or inadequate fences. Even proper facilities that are misused, for example at a feedlot, can provide an easier transmission of disease owing to crowding, stress, and poor sanitation.

Source of Livestock

Producers can reduce the spread of infectious diseases into their herds and flocks by: (a) purchasing animals (entering their herds) from other producers who have effective herd health programs; (b) controlling exposure of their animals to other people and vehicles; (c) providing clothing, boots, and disinfectant to other individuals who must be exposed to their animals and facilities; (d) isolating animals to be introduced into their herds so that disease symptoms can be observed for several weeks; (e) controlling insects, birds, rodents, and other animals that might carry disease organisms; and (f) keeping their animals out of drainage areas that run through their farm from other farms.

Use of Biologics and Pharmaceuticals

Drugs are classified as **biologicals** or **pharmaceuticals.** Biologicals are used primarily to prevent diseases whereas pharmaceuticals are used mainly to treat diseases. Both are needed in a successful herd health program.

Most biologicals are used to stimulate immunity against specific diseases. Vaccines are biological agents that stimulate active immunity in the animal.

Other biologicals stimulate the body's immune system to produce antibodies that fight diseases. This is similar to immunity through natural infection. An antibody is a protein molecule that circulates in the bloodstream and neutralizes disease-causing microorganisms. An animal is immunized when sufficient antibodies are produced to prevent the disease from developing. Specific antibodies must be produced for specific diseases, thus vaccination for several different diseases may be necessary. Also periodic revaccination might be needed to maintain antibodies in the blood at an adequate level.

Immunity from vaccination occurs only if the immune system properly responds. In some situations vaccination is not effective. Examples are in the newborn young in which, in most animal species, immune systems are not yet fully developed to respond. This is why their immunity is dependent on the maternal antibodies ingested from their mother's milk. Also, undernourished animals, stressed animals, or those animals already exposed to the disease will rarely give a positive response to vaccination.

The amount of vaccine, frequency, route of administration, and duration of immunity varies with the specific vaccine. The manufacturer's directions should be followed carefully.

Vaccines (biologicals) and drugs (pharmaceuticals) can be administered in the following ways:

FIGURE 20.1.
Using a balling gun to give a bolus or capsule to a cow. After the instrument is placed on the back of the tongue, the plunger is pushed to deposit the pill onto the back of the tongue. The animal then swallows the pill. Courtesy of Norden Laboratories.

1. Topically—applied to the skin.
2. Orally—through the mouth by feeding, drenching, or using balling guns. The latter is used to give pills such as capsules or bóluses (Fig. 20.1).
3. Injected directly into the animal's body by using a needle and syringe.

Types of injections include: (a) subcutaneous (under the skin but not into the muscle), (b) intramuscular (directly into the muscle), (c) intravenous (into a vein), (d) intramammary (through a cannula in the teat canal and injecting the drug into the milk cistern), (e) intraperitoneal (into the peritoneal cavity), and (f) intrauterine (using infusion pipette or tube through cervix). Some injection sites are shown in Figs. 20.2–20.4.

Pharmaceuticals are used to kill or reduce the growth of microorganisms in the treatment of diseases and infections. Pharmaceuticals come in a variety of forms: liquid, powder, boluses, drenches, and feed additives. They should be administered only after a specific diagnosis has been made. Examples of pharmaceuticals are antibiotics, steroids, sulfa compounds, hormones, and nitrofurans.

Other pharmaceuticals are used to control external and internal parasites. Losses owing to external and internal parasites are usually reduced weight gains, poor feed conversion, lower milk and egg production, reduced hide value, and excessive carcass trim, rather than high death losses.

Common external parasites are flies, lice, mites, and ticks, which live off the flesh and blood of animals. These parasites can also transmit infectious diseases from one animal to another. Insecticides are used to control external parasites. They can be applied as systemics, spraying, fogging, dipping, back rubbers, ear tags, or dust bags.

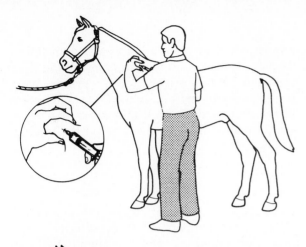

FIGURE 20.2.
Subcutaneous injection in the horse. Courtesy of Battaglia, R. A., and Mayrose, V. B. 1981. *Handbook of Livestock Management Techniques.* © Macmillan Publishing Company.

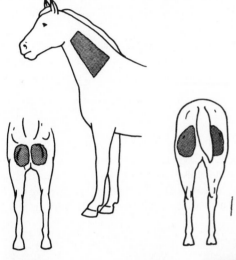

FIGURE 20.3.
Intramuscular injection sites in the horse. Courtesy of Battaglia, R. A. and Mayrose, V. B. 1981. *Handbook of Livestock Management Techniques.* © Macmillan Publishing Company.

FIGURE 20.4.
Intramuscular injection into the muscles of the hip. Courtesy of Norden Laboratories.

Roundworms and tapeworms are the most common internal parasites, with flukes causing problems in certain areas. Internal parasites are controlled by interrupting their life cycle by: (a) presence of unfavorable climatic conditions (wet, warm weather favors the proliferation of many internal parasites); (b) destroying intermediate hosts; (c) managing animals so they will not ingest the parasites or their eggs; and (d) giving therapeutic chemical treatment. **Anthelmintics** are drugs which are given to kill the internal parasites.

Stress

Stress is any environmental factor that might cause a significant change in the animal's physiological processes. Physical sources of stress are temperature, wind velocity, nutritional deficiency or oversupply, mud, snow, dust, fatigue, weaning, ammonia buildup, transportation, castration, dehorning, and abusive handling. Social or behavior-related stress can result from aggression or overcrowding.

Prolonged stress can impair the body's immune system, causing a reduced resistance to disease. Stressful conditions in animals should be minimized within economic constraints.

Personnel Training

Maintaining good animal health requires continuous training of personnel. This is one of the most difficult areas to accomplish, particularly in a livestock operation in which several employees are involved. The owner must understand what needs to be done, but this may differ from what the employees actually accomplish. Cooperative training programs and informative sessions with the veterinarian can assist in the implementation of effective herd health programs.

DETECTING UNHEALTHY ANIMALS

Visual Observation

Detecting sick animals and separating them from the healthy animals is an important key to a successful treatment program. Animals treated in the early stages of sickness will usually respond more favorably to treatment than animals whose illness has progressed to advanced stages.

Early signs of sickness are not easily detected; however, the following are some of the observable signs, several of which can be observed in Fig. 20.5:

1. loss of appetite
2. animal will appear listless and depressed
3. ears may be droopy or not held in an alert position
4. may have a hump in the back and the head held in a lower position
5. animal stays separated from the rest of the herd or flock
6. coughing, wheezing, or labored breathing.

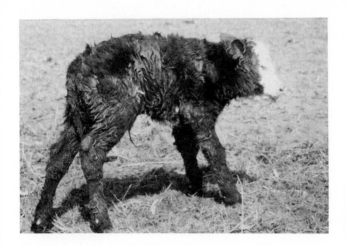

FIGURE 20.5.
This calf is sick with diarrhea (calf scours).
Note the visual signs of illness. Courtesy of
Norden Laboratories.

Vital Signs

Body temperature, respiration rate, and heart rate are the vital signs of the animal. Health problems are usually evident when one or more of the vital signs deviate from the normal range. The average and normal range for the vital functions are given in Table 20.1.

Body temperature is taken rectally. An elevated body temperature occurs in overheated animals or with most infectious diseases. A subnormal temperature indicates chilling or a critical condition of the animal.

TABLE 20.1. Vital Functions of Selected Livestock and Poultry Species

Animal	Rectal Temperature (°F)	Respiration Rate (per min)	Heart Rate (per min)
Cattle	101.5[a] (100.4–102.8)[b]	30[a]	50[a]
Swine	102.5 (101.6–103.6)	16	60
Sheep	102.3 (100.9–103.8)	19	75
Goat	102.3 (101.3–103.5)	15	90
Horse	100.0 (99.1–100.8)	12	45
Chicken	107.1 (105.0–109.4)	12–36	275
Turkey	105	28–49	165

[a] Average.
[b] Normal range.
Source: Battaglia, R. A. and Mayrose, V. B. 1981. *Handbook of Livestock Management Techniques.* New York: Macmillan Publishing Co., copyright.

Evaluating the vital signs, along with visual observation, can identify health problems in their early stages. This identifies the need for early treatment, which usually prevents serious losses.

MAJOR DISEASES OF FARM ANIMALS

Table 20.2 identifies some of the major diseases and health problems of beef cattle, dairy cattle, swine, horses, poultry, and sheep.

TABLE 20.2.   Selected Major Disease and Health Problems of Beef Cattle, Dairy Cattle, Swine, Horses, Poultry, and Sheep

Species/Disease/Cause	Signs	Prevention	Treatment
Beef Cattle			
Bovine viral diarrhea (BVD) Virus. Laboratory diagnosis imperative for accuracy	(Feeder Cattle) Ulcerations throughout digestive tract Diarrhea (often containing mucus or blood	Vaccination prior to exposure Avoid contact with infected animals	Symptomatic treatment; antibiotics; sulfonamids; "Hyper" serums; force feed
	(Breeding Cattle) Abortions	Vaccination prior to breeding	None Symptomatic treatment
Brucellosis bacteria	Abortions	Calfhood vaccination (in some states) between the ages of 2–12 months Test and slaughter reactors	None
Campylobacteriosis (vibriosis) bacteria	Repeat services Abortions (1–2%)	Vaccination of females and bulls prior to breeding Use of artificial insemination Untried bulls on virgin heifers Avoid sexual contact with infected animals	None
Leptospirosis Leptospira spp Several bacteria	(Breeding Cattle) Fever, off feed, abortions, icterus, discolored urine	Vaccination at least annually (in high risk areas, more frequently) Control rodents Avoid contact with wildlife and other infected animals	Streptomycin Penicillin Oxytetracycline (i.m.), 1 mg/lb. body weight/ 7 days Chlortetracycline, 1 mg/lb./body weight/7 days
	(Feeder Cattle)	Vaccination on arrival	Dihydrostreptomycin

Species/Disease/Cause	Signs	Prevention	Treatment
Infectious bovine rhinotracheitis (IBR) Virus. Laboratory diagnosis imperative for accuracy	Respiratory involvement Fever, vaginitis, abortion, eye and nose discharges Abortions or stillborn calves Pneumonia in calves	Vaccination prior to exposure	Chlortetracycline, 350 g/day Sulfamethazine, 350 mg to keep down secondary infections "Hyper" serums
Scours, calf colibacillosis "septicemia" E. coli (bacteria) plus stress factors	Diarrhea, weakness, and dehydration Rough hair and coat	Dry, clean calving areas Adequate colostrum	Fluid therapy Antibiotics to prevent secondary infections
Scours (viral) Virus (reovirus and corona virus)	Acute diarrhea, high mortality Affects calves shortly after birth (See Figs. 20.5 and 20.6).	Vaccination with specific viral vaccines	Fluid therapy not effective in severe cases
"Shipping fever" Para-influenza Virus plus bacteria plus stress May be accompanied by other viral diseases	Respiratory involvement	Chlortetracycline, 350 mg/head/day Sulfamethazone, 350 mg/head/day Oxytetracycline, 0.5–2 mg/head/day for 3 days	Sulfonamides; antibiotics; antiserums; serums; "hyper" serums; Tylan 200
Trichomoniasis protozoa	Differential diagnosis Repeat services Abortion	Use of artificial insemination Untried bulls on virgin heifers Avoid sexual contact with infected animals	

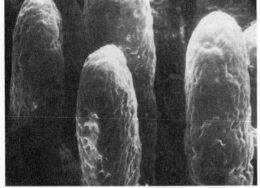 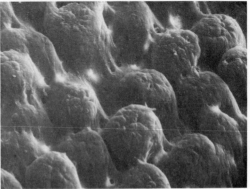

FIGURE 20.6.
Left: Normal villi that line the intestinal wall. **Right:** De-nuded villi after virus infection in pigs and calves. Severe diarrhea results as the animal has lost the ability to digest and absorb nutrients. If the animal survives, the villi will regain their natural form in 1–2 weeks. Courtesy of Dr. G. L. Waxler (*Am. J. Vet. Res.* 33:1323).

TABLE 20.2. *(Continued)*

Species/Disease/Cause	Signs	Prevention	Treatment
Dairy Cattle			
Mastitis Primarily infectious bacteria (*Streptococci, Staphylococci,* or *E. coli*)	Inflammation of the udder Decreased milk production	Use CMT (California mastitis test) or SCC (somatic cell counts)—checks for white blood cells in milk Avoid injury to udder Use proper milking techniques	Determine specific organism causing the problem, then select most effective antibiotic. For example, cloxacillin, penicillin, erythromycin, and neomycin are effective against several streptococcus and staphylococcus organisms
Milk fever (Parturient Daresis) Metabolic disorder (low blood calcium level associated with a deficiency of vitamin D and phosphorus)	Occurs at onset of lactation Muscular weakness; drowsiness; cow lies down in curled position	Proper nutrition and management of cows during dry period	Intravenous injection of calcium borogluconate

Respiratory diseases (See Infectious Bovine Rhinotracheitis, "Shipping Fever" parainfluenza under Beef Cattle)

Species/Disease/Cause	Signs	Prevention	Treatment
Uterine infections Usually result from bacterial contamination at the time of calving	Infections classified according to tissues involved Animal may be systemically or only locally affected **Endometritis:** inflammation of lining of uterus **Metritis:** infection of all tissues of uterus **Pyometra:** pus in the uterus.	Sanitation; cleanliness when giving calving assistance	Early detection; antibiotics, hormone therapy, prostaglandins

Venereal diseases (See BVD, IBR, leptospirosis, vibriosis and trichomoniasis under Beef Cattle)

Species/Disease/Cause	Signs	Prevention	Treatment
Swine			
Atrophic rhinitis Bordetella bronchiseptica (bacteria); Pasteurella (secondary invader)	Sneezing (most common); sniffling; snorting; coughing; nose shape (twisted to one side) Nasal infection (inflammation of	Monitor performance; monitor contact with animals from outside the herd; correct environmental deficiencies (sanitation, temperature,	Vaccinate against Bordetella and Pasteurella organisms; medicate sow feed with sulfamethazine or oxytetracycline; improve environment

Species/Disease/Cause	Signs	Prevention	Treatment
	membranes in nose) Diagnosis confirmed by observing turbinate bone atrophy (in nose) during postmortem examination (Fig. 20.7)	humidity, ventilation, dust, drafts, excessive ammonia and overcrowding)	
Colibacillosis E. coli (bacteria)	Diarrhea (pale, watery feces); weakness; depression Most serious in pigs under 7 days of age.	Good sanitation; good management practices (nutrition, pigs suckle soon after birth, prevent chilling); vaccinate sows to increase protective value of colostrum	Effective treatment is limited; identify strain of *E. coli*, then use specific antibacterial drug
Mycoplasmal infections (Pneumonia and arthritis) Mycoplasma bacteria	Pneumonia: death loss low; dry cough; reduced growth rate; lesions in lungs Arthritis: inflammation in lining of chest and abdominal cavity; lameness; swollen joints; sudden death	Pneumonia: reduce animal contact; good nutrition, warm and dust-free environment and parasite (ascarid and lung worm) control will minimize effects of the disease	Pneumonia: adequate treatment not available; sulfas and antibiotics prevent secondary penumonia infections Arthritis: no satisfactory treatment; can depopulate and restock with disease free animals
Pseudorabies Virus of herpesvirus group	Baby pigs (less than 3 weeks old): sudden death Pigs (3 weeks to 5 months): fever,	Prevent direct contact with infected swine; humans should wear clean clothes and disin-	Drugs and feed additives are not effective; quarantine; blood test and slaughter those in-

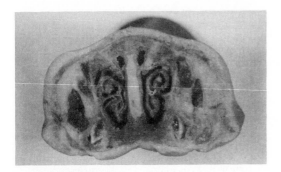 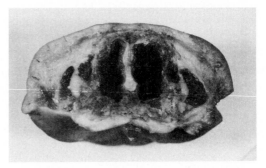

FIGURE 20.7.
Cross-sections through the hog's snout showing the effects of atrophic rhinitis. **Left:** Normal snout with turbinate bones. **Right:** Diseased snout showing severe atrophy or absence of turbinate bones. Courtesy of Elanco Products Company.

TABLE 20.2.   *(Continued)*

Species/Disease/Cause	Signs	Prevention	Treatment
	loss of appetite, labored breathing, tembling and incoordination of hindlegs Mature pigs: less severe fever, loss of appetite; abortion and reproductive failure.	fect boots when entering and leaving the premises Recovered swine are immune but should be considered carriers	fected; repopulate
Transmissable gastroenteritis (TGE) Virus of coronavirus group	Vomiting; diarrhea; weakness; dehydration; high death rate in pigs under 3 weeks of age (see Fig. 20.6)	Sanitation is most cost effective Prevent transfer of virus from infected animals through exposure to other pigs, birds, equipment, and people	No drugs are effective Fresh water and a draft-free environment will reduce losses Antibiotics may prevent some secondary infections
Horses			
Colic Noninfectious; a general term indicating abdominal pain	Looking at flank; kicking at abdomen; pawing; getting up and down; rolling; sweating	Parasite control, avoiding moldy or spoiled feeds; avoid overeating or drinking too much water when hot; have sharp points of teeth filed (floated) annually to assure proper chewing of feed	Quiet walking of horse; avoid undue stress; bran mash and aspirin; milk of magnesia; mild, soapy, warm water enema for impaction
Lameness Most common are stone bruises, puncture wounds, sprains, and navicular disease	Departure from normal stance or gait; limping; head bobbing; dropping of the hip; alternate resting on front feet; pointing (extends one front foot in front of the other); stiffness Navicular disease: soft tissue bursitis in the foot	Bruise: avoid running horse on graveled roads or rocky terrain Puncture wounds: avoid areas where nails and other sharp objects may be present Sprains: avoid stress and strains on feet and legs Navicular disease: exact cause not known; avoid continual concussion on hard surfaces; proper shoeing, and trimming	Varies with cause of lameness Bruises: soaking foot in bucket of ice water or standing horse in mud; aspirin Puncture wound: tetanus protection; antibiotic; aspirin; soaking foot in hot epsom salts Navicular disease: no known cure; drug therapy and corrective shoes for temporary relief; surgery

Species/Disease/Cause	Signs	Prevention	Treatment
Respiratory Three viral diseases: viral rhinopneumonitis; viral arteritis; influenza One bacterial disease: strangles	Nasal discharge; coughing; lung congestion; fever Rhinopneumonitis may cause abortion	Isolate infected animals; avoid undue stress; draft-free shelter; parasite control; avoid chilling Rhinopneumonitis, influenza and strangles vaccines are available	Follow prevention guidelines Strangles: drain abscess
Parasites 150 different internal parasites with strongyles (blood worms), ascarids (round worms), bots, pin worms, and stomach worms most common External parasites (most common are lice, ticks and flies)	Unthrifty appearance ("potbellied", rough hair coat); weakness; poor growth Lice, ticks, and flies can be visually observed on the horse	Good sanitation practices (clean feeding and watering facilities; regular removal of manure); periodic rest periods for pastures; avoid overcrowding of horses; avoid spreading fresh manure on pastures.	Internal parasites: worm horses twice a year (usually spring and fall); bot eggs can be shaved from the hairs of the legs External parasites: application of insecticides

Poultry

Species/Disease/Cause	Signs	Prevention	Treatment
Avian influenza Virus	Drop in egg production; sneezing; coughing; in the severe systemic form, deep drowsiness and high mortality are common	Vaccine (but has short immunity); select eggs and poults from clean flocks	No effective drug available
Coccidiosis Coccida	Depends on type of coccidiosis—there are nine or more types. Some signs, prevention, and treatment of three more common types are identified.		
	Weight loss; unthriftiness; pallor; blood in droppings; lesions in intestinal wall	Use coccidiostat (kills coccida organism)	Sulfa drugs in drinking water
Lymphoid leukosis Virus	Combs and wattles may be shriveled, pale and scaly; enlarged, infected liver Lesions common in liver and kidneys	Sanitation; development of resistant strains through breeding methods	None
Marek's disease Herpesvirus	Can cause high mortality in pullet flocks; paralysis; death might occur without any clinical signs	Vaccination	None

TABLE 20.2.   *(Continued)*

Species/Disease/Cause	Signs	Prevention	Treatment
Mycoplasma infections Mycoplasma organisms	There are several diseases caused by mycoplasma organisms. Most important are chronic respiratory disease (CRD) and infectious synovitis.		
	CRD—difficult breathing; nasal discharge; rattling in windpipe; death loss may be high in turkey poults; swelling of face in turkeys Synovitis—swollen joints, (Fig. 20.8) tendon sheaths and footpads; ruffled feathers	For both CRD and synovitis: use chicks or poults from disease free parent stock; sanitation (cleaning and disinfecting the premises)	Antibiotic in feed or drinking water
Newcastle Disease	Gasping, coughing, hoarse chirping; paralysis; mortality may be high (Fig. 20.9)	Vaccination	None

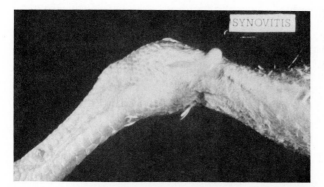

FIGURE 20.8.
Infectious synovitis. A swollen hock and lameness are characteristic. When opened, the hock usually contains a creamy exudate. Courtesy of Salisbury Laboratories.

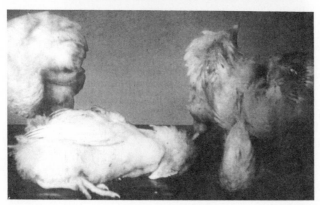

FIGURE 20.9.
Newcastle disease. Birds may have complete or partial paralysis of legs and wings. The photo shows common signs when nerves are affected. Courtesy of Salisbury Laboratories.

Species/Disease/Cause	Signs	Prevention	Treatment
Sheep			
C. pseudotuberculosis bacteria	Enlarged lymph nodes	Sanitation; reduce opportunity for wound infection at docking, castration and shearing	Antibiotics
Enterotoxemia (Overeating disease) bacteria	Often sudden death occurs without warning; sick lambs might show nervous symptoms, i.e., head drawn back; convulsions	Vaccination (most effective in young lambs under 6 weeks of age nursing heavy milking ewes and in weaned lambs on lush pasture or in feedlots)	None; antitoxins can be used but they are expensive and immunity is temporary (2–3 weeks)
Epididymitis bacteria	Most common in western U.S.; swelling of epididymis; poor semen quality; low conception rates	Rigid culling and vaccination	Rigid culling
Footrot bacteria	Lameness; interdigital skin is usually red and swollen	Remove from wet pastures or stubble pastures; vaccines in some cases	Disinfectants such as 5% formalin or 10% copper sulfate; antibiotics
Pneumonia Many types; caused by viruses, stress, bacteria, and parasites	Sudden death; nasal discharge; depression; high temperature	Reduce stress factors	Antibiotics; sulfonamides
Pregnancy toxemia Undernutrition in late pregnancy; stress associated with poor body condition	Listlessness; loss of appetite; unusual postures; progressive loss of reflexes; hypoglycemia; coma; death	Prevent obesity in early pregnancy; good nutrition during last 6 weeks of pregnancy; reduce environmental stresses	In early stages the administration of some glucogenic materials may reduce mortality; in advanced stages no treatment improves survival

SELECTED REFERENCES

Publications

Battaglia, R. A. and Mayrose, V. B. 1981. *Handbook of Livestock Management Techniques* (Chap. 10, Animal health management). New York: Macmillan.

Gaafar, S. M., Howard, W. E. and Marsh, R. E. (Editors). 1985. *Parasites, Pests and Predators*. New York: Elsevier Science.

Haynes, N. B. 1985. *Keeping Livestock Healthy*. Pownal, VT.: Storey.

Keeler, R. F., VanKampen, K. R. and James, L. F. (Editors). 1978. *Effects of Poisonous Plants on Livestock.* New York: Academic Press.

Kirkbride, C. A. 1986. *Control of Livestock Diseases.* Springfield, IL: Charles C Thomas.

Naviaux, J. L. 1985. *Horses in Health and Disease.* Philadelphia: Lea and Febiger.

Radostits, O. M. and Blood, D. C. 1985. *Herd Health.* Philadelphia: W. B. Saunders.

Sainsbury, D. 1983. *Animal Health.* Boulder, CO: Westview Press.

Visuals

Distinguishing Normal and Abnormal Cattle. Videotape. Agricultural Products and Services, 2001 Killebrew Dr., Suite 333, Bloomington, MN 55420.

CHAPTER 21

Beef Cattle Breeds and Breeding

A breed of cattle is defined as a race or variety, the members of which are related by descent and similar in certain distinguishable characteristics. More than 250 breeds of cattle are recognized throughout the world, and several hundred other varieties and types have not been identified with a breed name.

Some of the oldest recognized breeds in the United States were officially recognized as breeds during the middle to late 1800s. Most of these breeds originated from crossing and combining existing strains of cattle. When a breeder or group of breeders decided to establish a breed, distinguishing that breed from other breeds was of paramount importance; thus, major emphasis was placed on readily distinguishable visual characteristics, such as color, color pattern, polled or horned condition, and rather extreme differences in form and shape.

New cattle breeds, such as Brangus, Santa Gertrudis, and Beefmaster, have come into existence in the United States in the past 25 to 35 years. These breeds have been developed by attempting to combine the desirable characteristics of several existing breeds. Currently there are new breeds of cattle being developed. They are sometimes referred to as "composite breeds" or "synthetic breeds." In most cases, however, the same visual characteristics as previously mentioned are used to give the new breeds visual identity.

After some of these first breeds were developed, it was not long until the word **purebred** was attached to them. Herd books and registry associations were established to assure the "purity" of each breed and to promote and improve each breed. Purebred refers to purity of ancestry, established by the pedigree, which shows that only animals recorded in that particular breed have been mated to produce the animal in question. Purebreds, therefore, are cattle within the various breeds that have pedigrees recorded in their respective breed registry associations.

When viewing a herd of purebred Angus, Herefords, or other breed, one notes uniformity, particularly uniformity of color or color pattern. Because of this uniformity of one or two characteristics, the word purebred has come to imply genetic uniformity (homozygosity) of all characteristics. Cattle within the same breed are not highly homozygous because high levels of homozygosity occur only after many generations of close inbreeding. This close inbreeding has not occurred in the cattle breeds. If breeds were uniform genetically, they could not be improved or changed even if changes were desired.

WHAT IS A BREED?

The genetic basis of cattle breeds and their comparison is not well understood by most livestock people. Often the statement is made, "There is more variation within a breed than there is between breeds." The validity of this statement needs to be carefully examined. Considerable variation does exist within a breed for most of the economically important traits. This variation is depicted in Fig. 21.1, which shows the number of calves, of a particular breed, that fall within certain weight-range categories at 205 days of age. A bell-shaped curve is formed by connecting the high points of each bar. The breed average is represented by the solid line that separates the bell curve into equal halves. Most of the calves are near the breed average; however, at the outer edge of the bell curve are high- and low-weaning weight calves. Note that they are fewer in number at these extremes.

Figure 21.2 allows us to hypothetically compare three breeds of cattle in terms of weaning weight. The breed averages are different; however, the variation within each breed is comparable among all three. The statement, "There is more variation within a breed than between breed averages," is more correct than the statement, "There is more variation within a breed than there is between breeds."

Figures 21.1 and 21.2 are hypothetical examples; however, they are based on realistic samples of data obtained from the various breeds of cattle. Figure 21.3 shows percent milk fat differences in samples from the Holstein and Jersey breeds. Obviously, selecting certain Holsteins with high percent milk fat and low percent milk fat Jerseys could infer that Holsteins have a higher percent milk fat average than Jerseys. Random sample selec-

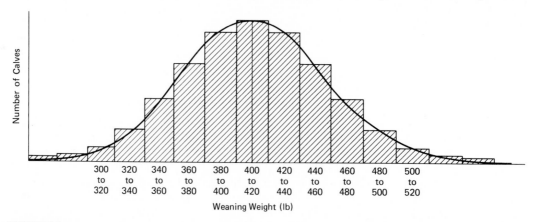

FIGURE 21.1.
Variation or differences in weaning weight in beef cattle. The variation shown by the bell-shaped curve could be representative of a breed or a large herd. The vertical line in the center is the average or the mean, which is 410 lb. Note that the number of calves is greater around the average and is less at the extremely light and extremely heavy weights. Courtesy of Colorado State University.

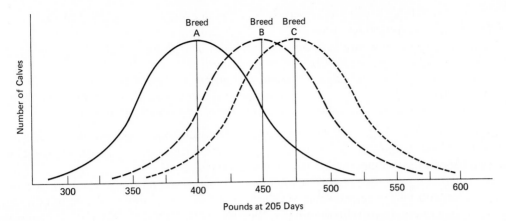

FIGURE 21.2.
Comparison of the breed averages and the variation within each breed for weaning weight in beef cattle. The vertical lines are the breed averages. Note that some individual animals in Breed B and Breed C can be lower in weaning weight than the average of Breed A. Courtesy of Colorado State University.

tions of cows from each breed, however, would show that Jerseys, on the average, do have a higher percent of fat in their milk than Holsteins. Occasionally a bull may seem to have a high level of productivity in a certain trait for which the information comes from selected data and not comparisons based on random samples. Only breed comparisons based on random samples are valid.

MAJOR BEEF BREEDS IN THE U. S.

Until the 1960s and 1970s, the number of cattle breeds in the United States had remained relatively stable at between 15 and 20. Today, more than 60 breeds of cattle are available to the United States beef producers. Why the large importation of the different breeds from several different countries? Following are several possible reasons:

1. Feeding more cattle larger amounts of grain, a practice started in the 1940s, resulted in many overfat cattle—cattle that had been previously selected to fatten on forage

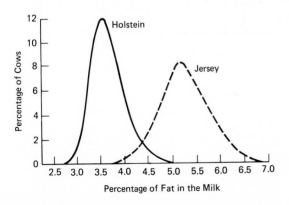

FIGURE 21.3.
Breed differences in percentage of fat in the milk of Holstein and Jersey cows. Courtesy of G. E. Shook and W. H. Freeman and Company, © 1974.

rations. Therefore, a need was established for grain-fed cattle that could produce a higher percentage of lean to fat at the desired slaughter weight.

2. Economic pressures to produce more weight in a shorter period of time demonstrated a need for cattle with more milk and more growth.

3. An opportunity was available for some promoters to capitalize on merchandising a certain breed as being the ultimate in all production traits. This opportunity could be realized easily because there was little comparative information on breeds.

Color plates I through L (following page 338) identify the most numerous breeds of beef cattle in the United States. Table 21.1 gives some distinguishing characteristics and brief background information for each of these breeds.

The relative importance of the various breeds' contributions to the total beef industry is best estimated by the registration numbers of the breeds (Table 21.2). Although registration numbers are for purebred animals, they reflect the commercial cow-calf producers' demand for the different breeds. Registration numbers show that Angus, Hereford, and Polled Hereford are the most important breeds of the beef cattle industry in the United States.

The numbers of cattle belonging to various breeds in this country have changed over the past years, as shown in Fig. 21.4. No doubt some breeds will become more numerous in future years, whereas other breeds will decrease significantly in numbers. These changes will be influenced by economic conditions and by genetic changes being made in the economically important traits in the breeds.

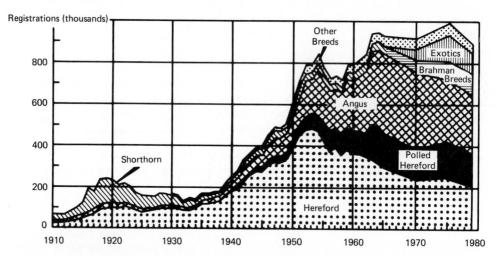

Prior to 1963: Some Polled Herefords were registered only in the American Hereford Association. Others were registered in the American Polled Hereford Association.

Prior to 1965: Brahman breeds were included in other breeds. Very few exotics registered.

After 1965: Shorthorn ends, included in other breeds.
Brahman breeds (Brahman, Brangus, Santa Gertrudis).
Exotics (Chianina, Limousin, Maine Anjou, Simmental).

FIGURE 21.4.
Registration numbers of beef cattle in the United States, 1910 through 1979. Courtesy of the USDA.

TABLE 21.1. Background and Distinguishing Characteristics of Major Beef Breeds in the United States

Breed	Distinguishing Characteristics	Brief Background
Angus	Black color and polled	Originated in Aberdeenshire and Angushire of Scotland Imported into the U.S. in 1873 to cross with Longhorn
Beefmaster	Has various colors	Developed in the U.S. from Brahman, Hereford, and Shorthorn breeds Selected for its ability to reproduce, produce milk, and grow under range conditions
Blonde d'Aquitaine	Fawn colored, sometimes with a reddish tinge, heavily muscled	Originated in France and live cattle imported in the U.S. in 1973 Semen imported prior to that time
Brahman	Various colors, with gray predominant. They are one of the Zebu breeds which have the hump over the top of the shoulder. Most Zebu breeds also have large, drooping ears and excess skin in the throat and dewlap	Major importations to the United States from India and Brazil Largest introductions in early 1900s These cattle are heat tolerant and well adapted to the harsh conditions of the Gulf Coast region
Brangus	Black and polled predominate, although there are Red Brangus	Breed developed around 1912 in the U.S. from three-eighths Brahman and five-eighths Angus
Charolais	White color with heavy muscling, horned or polled	One of the oldest breeds in France Brought into the U.S. soon after World War I, but its most rapid expansion occurred in the 1960s
Chianina	White color with black eyes and nose. Extremely tall cattle	An old breed originating in Italy Acknowledged to be the largest breed, with mature bulls weighing more than 3,000 pounds First used for breeding in U.S. in 1971 Early use was for draft
Devon	Dark red (North Devon) to light red or brown (South Devon)	North Devon is an old breed originating in England South Devon is more of a dual-purpose breed The Devon first came to America with the Pilgrims in 1623
Galloway	Polled with majority solid black in color, some dun-colored, and others white with black noses, ears, and feet. The Belted Galloway is black with distinctive white belt	An old breed from the Scottish province of Galloway Long, burly hair has made it adaptable to the harsh climates of the north The Belted Galloway has a separate breed association

TABLE 21.1. *(Continued)*

Breed	Distinguishing Characteristics	Brief Background
		It has the same origin as the Galloway but had an infusion of the belted cattle in the seventeenth or eighteenth century
Gelbvieh	Golden colored	Originated in Austria and West Germany
		Dual-purpose breed used for draft, milk, and meat
Hereford	Red body with white face and horned	Introduced in the U.S. in 1817 by Henry Clay. Followed the Longhorn in becoming the traditionally known range cattle
Limousin	Golden color with marked expression of muscling	Introduced into the U.S. in 1969, primarily from France
Longhorn	Multi-colored with characteristically long horns	Came to West Indies with Columbus. Brought to the U.S. through Mexico by the Spanish explorers
		Longhorns were the noted trail-drive cattle from Texas into the Plains States
Maine-Anjou	Red and white spotted	A large breed developed in France and introduced into Canada in late 1968
Murray Grey	Silver gray and polled	Developed in Australia from Shorthorns crossed on Angus
Pinzgauer	Reddish chestnut color with white markings on rump, back, and belly	A hardy breed developed in the Pinza Valley of Austria and areas of Germany and Italy. Introduced into North America in 1972
Polled Hereford	Red body with white face and polled	Bred in 1901 in Iowa by Warren Gammon who accumulated several naturally polled cattle from horned Hereford breeders
Red Angus	Red color and polled	Founded as a performance breed in 1954 by sorting the genetic recessives from Black Angus herds
Red Polled	Red color and polled	Introduced into the U.S. in the late 1800s. Originally a dual-purpose breed but now considered a beef breed
Santa Gertrudis	Red color and horned	First United States breed of cattle developed on the King Ranch in Texas
		Combination of five-eighths Shorthorn and three-eighths Brahman
Scotch Highland	Golden color with long, shaggy hair	Bred in the highlands of Scotland. Imported into U.S. in 1894

Breed	Distinguishing Characteristics	Brief Background
Shorthorn	Red, white, or roan in color. Both horned and polled	Introduced into the U.S. in 1783 under the name "Durham" Most prominent in the U.S. around 1920
Simmental	Yellow to red and white color pattern. Both polled and horned	A prominent breed in parts of Switzerland and France First bull arrived in Canada in 1967 Originally selected as a dual-purpose milk and meat breed
Tarentaise	Sold wheat-colored hair ranging from a light cherry to dark blond	Mountain cattle derived from an ancient Alpine strain in France Originally a dairy breed in which maternal traits have been emphasized

TABLE 21.2.   Major Beef Breeds Based on Yearly Registration Numbers

Breed	U.S. Breed Associations			Canadian Breed Associations
	1985[a]	1970	Date Association Formed	1985[b]
Angus	174,539	346,195	1883	19,817
Hereford	120,000	253,832	1881	72,409
Polled Hereford	96,456	168,021	1901	
Simmental	85,533		1969	
Limousin	44,484		1968	
Beefmaster	32,100	2,800	1961	
Brangus	31,031	8,238	1949	
Brahman	30,000	18,219	1924	
Charolais	30,000	158,132	1957	
Santa Gertrudis	28,000	16,837	1951	
Chianina	21,000	0	1972	
Shorthorn	20,000	32,647	1872	3,091
Polled Shorthorn	18,208	35,653		
Gelbvieh	16,086	0	1971	
Red Angus	12,427	5,290	1954	3,498

[a]Only breeds with more than 10,000 registrations are listed.
[b]Only breeds with more than 3,000 registrations are listed.
Source: 1985 North American Livestock Census.

IMPROVING BEEF CATTLE THROUGH BREEDING METHODS

Genetic improvement in beef cattle can occur by selection and also by using a particular mating system. Significant improvement by selection results when the selected animals are superior to the herd average, and the heritabilities of the traits are relatively high (40% and higher). It is important that the traits included in a breeding program be of economic importance and that the traits be measured as objectively as possible.

TRAITS AND THEIR MEASUREMENT

Most economically important traits of beef cattle can be classified under: (a) reproductive performance, (b) weaning weight, (c) postweaning growth, (d) feed efficiency, (e) carcass merit, (f) longevity (functional traits), (g) conformation, and (h) freedom from genetic defects.

Reproductive Performance

Reproductive performance has the highest economic importance when compared to the other traits. Most cow–calf producers have a goal for percent calf crop weaned (number of calves weaned compared to the number of cows in the breeding herd) of 85% or higher. Beef producers also desire each cow to calve every 365 days or less and have a calving season for the entire herd of less than 90 days. All of these are good objective measures of reproductive performance.

The heritability of fertility in beef cattle is quite low (less than 20%), so little genetic progress can be made through selection. Heritabilities for birth weight and scrotal circumference are higher, therefore selection for these traits will improve the percent calf crop weaned. The most effective way to improve reproductive performance is to improve the environment through, for example, adequate nutrition and good herd health practices. Selecting bulls based on a reproductive soundness examination, which includes semen testing, will also improve reproductive performance in the herd.

Reproductive performance can be improved through breeding methods by crossbreeding to obtain heterosis for percent calf crop weaned, using bulls with relatively light birth weights (heritability of birth weight is 40%), which decreases calving difficulty, and by selecting bulls that have a relatively large scrotal circumference. Scrotal circumference has a high heritability (40%), and bulls with a larger scrotal size (over 30 cm for yearling bulls) produce a larger volume of semen and have half-sister heifers that reach puberty at earlier ages than heifers related to bulls with a smaller scrotal size.

Weaning Weight

Weaning weight, as measured objectively by the scales, reflects the milking and mothering ability of the cow and the preweaning growth rate of the calf. Weaning weight is commonly expressed as the adjusted 205-day weight, where the weaning weight is adjusted for the age of the calf and age of the dam. This adjustment puts all weaning weight records on a comparable basis since older calves will weigh more than younger calves and mature cows (5–9 years of age) will milk heavier than younger cows (2–4 years of age) and older cows (10 years and older).

Weaning weights of calves are usually compared by dividing the calf's adjusted weight by the average weight of the other calves in the herd. This is expressed as a ratio. For example, a calf with a weaning weight of 440 lb, where the herd average is 400 lb, has a ratio of 110. This calf's weaning weight ratio is 10% above the herd average. Ratios should be used primarily for selecting cattle within the same herd in which they have had similar environmental opportunity. Comparing ratios between herds is misleading from a genetic standpoint because most differences between herds are caused by differences in the environment. Weaning weight will respond to selection because it has a heritability of 30%.

Postweaning Growth

Postweaning growth measures the growth from weaning to a weight that approaches slaughter weight. Postweaning growth might take place on a pasture or in a feedlot. Usually animals with relatively high postweaning gains make efficient gains at a relatively low cost to the producer.

Postweaning gain in cattle is usually measured in pounds gained per day after a calf has been on a feed test for 140–160 days. Weaning weight and postweaning gain are usually combined into a single trait, namely, adjusted 365-day weight (yearling weight). It is computed as follows:

Adjusted 365-day weight = (160 × average daily gain) + adjusted 205-day weight

Average daily gain for 140 days and adjusted 365-day weight both have high heritabilities (40%), therefore genetic improvement can be quite rapid when selection is practiced on postweaning growth or yearling weight.

Feed Efficiency

Feed efficiency is measured by the pounds of feed required per pound of liveweight gain. Specific records for feed efficiency can only be obtained by feeding each animal individually and keeping records on the amount of feed consumed. With the possible exception of some bull testing programs, determining feed efficiency on an individual animal basis is seldom economically feasible.

Interpretation of feed efficiency records can be rather confusing, depending on the endpoint to which the animals are fed. Feeding endpoint can be a certain number of days on feed (e.g., 140 days), to a specified slaughter weight (e.g., 1200 lb), or to a carcass compositional endpoint (e.g., low Choice quality grade). Most differences in feed efficiency, shown by individual animals, are related to the pounds of body weight maintained through feeding periods and the daily rate of gain or feed intake of each animal. Cattle fed from a similar initial feedlot weight (e.g., 600 lb) to a similar slaughter weight (e.g., 1100 lb) will demonstrate a high relationship between rate of gain and efficiency of gain. In this situation, cattle that gain faster will require fewer pounds of feed per pound of grain. Thus, a breeder can select for rate of gain and thereby make genetic improvement in feed efficiency. However, when cattle are fed to the same compositional endpoint (approximately the same carcass fat), there is little, if any, difference in the amount of feed required per pound of gain. This is true for different sizes and shapes of cattle, and even for various breeds that vary greatly in skeletal size and weight.

The heritability of feed efficiency, at similar beginning and final weights, is high (45%), so selection for more efficient cattle can be effective. It seems logical to use the genetic correlation between gain and efficiency where possible.

Carcass Merit

Carcass merit is presently measured by quality grades and yield grades, which were discussed in detail in Chapter 7. Many cattle breeding programs have goals to produce cattle that will grade Choice and have yield grade 2 carcasses. Objective measurements using backfat thickness probes on the live animal and hip height measurement can assist in predicting the yield grade at certain slaughter weights. Visual appraisal, which is more subjective, can be used to predict amount of fat or predisposition to fat and skeletal size. These visual estimates can be relatively accurate in identifying actual yield grades if cattle differ by as much as one yield grade.

Because quality grade cannot be evaluated accurately in the live animal, it is necessary to evaluate the carcass. Steer and heifer progeny of different bulls are slaughtered to best identify the genetic superiority or inferiority of bulls for both quality grade and yield grade. The heritabilities of most beef carcass traits are high (over 40%), so selection can result in marked genetic improvement for these traits.

Longevity

Longevity measures the length of productive life. It is an important trait, particularly for cows, when replacement heifer costs are high or a producer reaches an optimum level of herd performance and desires to stabilize it. Bulls are usually kept in a herd 3–5 years, or inbreeding might occur. Some highly productive cows remain in the herd at age 15 years or older, whereas other highly productive cows have been culled before reaching 8 or 9 years of age. These cows may have been culled because of such problems as skeletal unsoundness, poor udders, eye problems (i.e., cancer eye), and lost or worn teeth. Little selection opportunity for longevity exists because few cows remain highly productive past the age of 10 years.

Some producers need to improve their average herd performance as rapidly as possible rather than improve longevity. In this situation, a relatively rapid turnover of cows is needed. Some selection for longevity occurs because producers have the opportunity to keep more replacement heifers born to cows that are highly productive in the herd for a longer period of time than heifers born to cows that stay in the herd for a short time. Some beef producers attempt to identify bulls that have highly productive, relatively old dams. Certain conformation traits, such as skeletal soundness and udder soundness, may be evaluated to extend the longevity of production.

Conformation

Conformation is the form, shape, and visual appearance of an animal. How much emphasis to put on conformation in a beef cattle selection program has been, and continues to be, controversial. Some producers feel that putting a productive animal into an attractive package contributes to additional economic returns. It is more logical, however, to place more selection emphasis on traits that will produce additional numbers and pounds of lean growth for a given number of cows. Placing some emphasis on conformation traits such as skeletal, udder, eye, and teeth soundness is justified. Conformation differences such as fat accumulation or predisposition to fat can be used effectively to make meaningful genetic improvement in carcass composition.

Most conformation traits are medium to high (30–60%) in heritability, so selection for these traits will result in genetic improvement.

Genetic Defects

Genetic defects, other than those previously identified under longevity and conformation, need to be considered in breeding productive beef cattle. Cattle have numerous known hereditary defects; most of them, however, occur very infrequently and are of minor concern. Some defects increase in their frequency, and selection needs to be directed against them. Most of these defects are determined by a single pair of genes that are usually recessive. When one of these hereditary defects occurs, it is a logical practice to cull both the cow and the bull.

Some common occurring genetic defects in cattle today are double muscling, **syndactyly (mule foot), arthrogryposis (palate-pastern syndrome), osteopetrosis (marble bone disease), hydrocephalus,** and **dwarfism.** Double muscling is evidenced by an enlargement of the muscles with large grooves between the muscle systems of the hind leg. Double-muscled cattle usually grow slowly and their fat deposition in and on the carcass is much less than that of the normal beef animal. Syndactyly is a condition in which one or more of the hooves are solid in structure rather than cloven. Mortality rate is high in calves with syndactyly. Arthrogryposis is a defect in which the pastern tendons are contracted, and the upper part of the mouth has not properly fused together. Osteopetrosis is characterized by the marrow cavity of the long bones being filled with bone tissue. All calves having osteopetrosis have short lower jaws, protruding tongue, and impacted molar teeth. A bulging forehead where fluid has accumulated in the brain area is typical of the defect of hydrocephalus. Calves with arthrogryposis, hydrocephalus, or osteopetrosis usually die shortly after birth. The most common type of dwarfism is snorter dwarfism, in which the skeleton is quite small and the forehead has a slight bulge. Some snorter dwarfs exhibit a heavy, labored breathing sound. This defect was most common in the 1950s, and it has decreased significantly since that time.

BULL SELECTION

Bull selection must receive the greatest emphasis if optimum genetic improvement of a herd is to be achieved. Bull selection accounts for 80–90% of the herd improvement over a period of several years. This does not diminish the importance of good beef females, because genetically superior bulls have superior dams. However, most of the genetic superiority or inferiority of the cows will depend on the bulls previously used in the herd. Also the accuracy of records can be much higher in bulls than in cows and heifers.

All commercial and purebred producers should identify breeders who have honest, comparative records on their cattle. Most performance-minded seedstock producers will record birthweights, weaning weights and yearling weights of their bulls. Some breeders are also obtaining feedlot and carcass data on their own bulls, or they are using bulls from sires for whom these test data are known.

Purebred breeders should provide accurate performance data on their bulls, and commercial producers should request the information. Excellent performance records can be obtained and made available on the farm or ranch. The trait ratios are useful and comparative if the bulls have been fed and managed in similar environments.

Estimated Breeding Values

Breed associations have developed computer programs that utilize performance records on the individual calf and its relatives. This information is used to calculate estimated breeding values (EBVs). Breeding values are most frequently reported on birth weight, maternal weaning weight, weaning weight, and yearling weight. Using breeding values can improve the accuracy of identifying superior sires.

An EBV for an individual is determined by the heritability of the trait, performance records (individual and its relatives), weighted according to the genetic relationship between the individual and relatives. Figure 21.5 shows how breeding value relates to the phenotype of an animal. An EBV is an estimate of this breeding value.

Phenotype is the appearance or performance of an animal determined by the genotype (genetic makeup) and the environment in which it was raised. Genotype is determined by two factors: (a) breeding value (what genes are present) and (b) nonadditive (how genes are combined). Environmental factors influence the phenotype through known and unknown effects. Age of dam, resulting in different levels of milk production, is a known environmental effect adjusted for in an adjusted 205-day weight (phenotype). Unknown effects are such things as injury or health problems for which it is difficult to adjust the phenotypic record.

An example depicting breeding values is shown in Fig. 21.6. In the "brick wall concept," each brick represents a gene. Some genes have positive, additive effects whereas other genes have negative effects. The different sized bricks represent the magnitude of the gene's effect. In this example (Fig. 21.6), the 10 pairs of genes (20 bricks) are used to show a superior breeding value and a negative breeding value. The superior breeding value (higher brick wall) is where the sum of the additive gene effects (bricks) is above herd average.

Figure 21.7 shows how EBVs are reported on performance pedigrees by three different breed associations. The EBVs are expressed as ratios where 100 is average. Accuracy (ACC) figures show how closely the "estimated" breeding value predicts the "true" breeding value. An ACC above 0.90 is considered high, whereas an ACC of 0.75 and smaller is

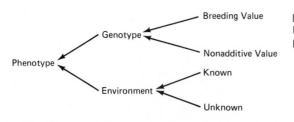

FIGURE 21.5.
Breeding value and its relationship to genotype and phenotype. Courtesy of BIF: FS2.

Angus

Beefmaster

Brahman

Brangus

Charolais

Chianina

PLATE I.

PLATES I-L.
Some of the breeds of cattle used for beef production in the United States. Courtesy of the American Angus Association, BBU, American Brahman Breeders Association, International Brangus Breeders Association, Charolais Journal, American Breeders Service, North American Gelbvieh Association, North American Limousin Foundation, American Polled Hereford Association, Dickinson Ranch, Santa Gertrudis Breeders International, Leachman Cattle Co., American Scotch Highland Breeders Association, American Shorthorn Association, and American Simmental Association.

Devon

Galloway

Gelbvieh

Hereford

Limousin

Longhorn

PLATE J.

Maine Anjou

Pinzgauer

Red Angus

PLATE K.

Santa Gertrudis

Shorthorn

Tarentaise

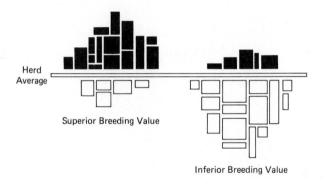

FIGURE 21.6.
Breeding value as depicted by the "brick wall concept." Courtesy of BIF: FS2.

low. The ACC indicates how much confidence can be placed on the EBV. As more progeny data become available the ACC figure will increase.

Sire Summaries

The development of sire summaries has been a big advancement in more effective sire selection. Most breed associations publish a sire summary that is updated annually.

American Angus Association

Estimated Breeding Values	Weaning		Maternal		Yearling	
	EBV	ACC	EBV	ACC	EBV	ACC
	108	.68	104	.63	114	.78

American Polled Hereford Association

Estimated Breeding Values		ACC
Birth	102	72%
Weaning	105	70%
Maternal	106	65%
Yearling	104	75%

American Simmental Association

FIGURE 21.7.
Example of how EBVs are shown on performance pedigrees of three different breeds. Courtesy of BIF: FS2.

TABLE 21.3.   Selected Data from the 1985 Angus Sire Evaluation Report

Sire	Birth Weight		Weaning Weight Direct		Weaning Weight Maternal				Yearling Weight	
					Milk			Combination		
	EPD	ACC	EPD	ACC	EPD	ACC	DTS[a]	Index	EPD	ACC
A	−1	0.95	+20	0.95	+1	0.93	331	+12	+39	0.94
B	+9	0.95	+47	0.95	−17	0.89	88	6	+74	0.90
C	−7	0.83	−12	0.85	−10	0.79	29	−16	−4	0.85
D	+1	0.95	+22	0.95	+3	0.92	230	+14	+54	0.94
E	1	0.73	+15	0.70	+12	0.42	0	+20	+45	0.66

[a]DTS is daughters.

Expected Progeny Difference (EPD) and **Accuracy (ACC)** are the important terms used in understanding sire summaries. An EPD combines into one figure a measurement of genetic potential based on the individual's performance and the performance of related animals such as the sire, dam, and other relatives. The most common EPDs reported for bulls, heifers, and cows are birthweight EPD, milk EPD, weaning growth EPD, maternal EPD (includes milk and preweaning growth) and yearling weight EPD. EPD is expressed as a plus or minus value, reflecting the genetic transmitting ability of a sire. EPD is equal to one-half of the EBV.

Accuracy (ACC) is a measure of expected change in the EPD as additional progeny data become available. EPDs with ACC of 0.95 and higher would be expected to change very little, whereas EPDs with ACC below 0.70 might change dramatically with additional progeny data.

Table 21.3 shows sire summary data for several Angus bulls.

If Bull B and Bull C were used in the same herd (each on an equal group of cows) the expected performance on their calves would be:

Bull B's calves to be 16 lb heavier at birth.
Bull B's calves to weigh 59 lb more at weaning.
Bull B's calves to weigh 78 lb more as yearlings.
Bull B's daughters to wean 7 lb more calf.

Bull A and Bull D have an optimum combination of all traits. Bull E is a promising young sire with an excellent combination of EPDs. However, the ACC are relatively low and could significantly change with more progeny and as daughters start producing.

Figure 21.8 shows how a breeder would select for yearling weight to produce a superior herd for that trait. Bulls which have high EPDs for yearling weight are selected in successive generations. This process is called "stacking pedigrees." Great Expectation has an EPD for yearling weight of 45 lb. This 45 lb resulted from =½ (65 lb EPD of Better Bull) +½ (25 lb EPD of Decent Cow).

Figure 21.9 shows EPDs for yearling weight where little, if any, selection emphasis has been placed on yearling weight. Selection may have been only for type or conformation characteristics.

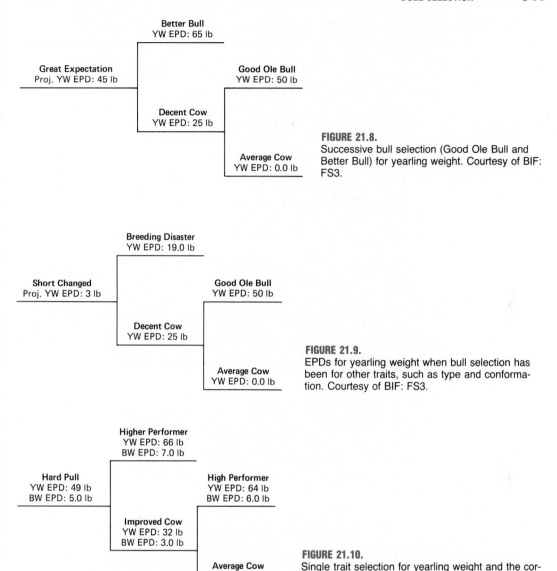

FIGURE 21.8.
Successive bull selection (Good Ole Bull and Better Bull) for yearling weight. Courtesy of BIF: FS3.

FIGURE 21.9.
EPDs for yearling weight when bull selection has been for other traits, such as type and conformation. Courtesy of BIF: FS3.

FIGURE 21.10.
Single trait selection for yearling weight and the correlated response in birth weight. Courtesy of BIF: FS3.

Bull selection becomes more complex in that desired genetic improvement is for a combination of several traits. "Stacking pedigrees" for only one trait may result in some problems in other traits. A good example is selecting for yearling weight alone, which results in increased birth weight because the two traits are genetically correlated. Birth weight is associated with calving difficulty so increased birth weight might be a problem. Figure 21.10 shows what can occur when selection emphasis is maximized for only one trait, yearling weights. Large increases in birth EPD can occur and calving difficulty might be serious.

The biggest challenge in bull selection is selecting bulls that will improve maternal traits. Frequently too much emphasis is placed on yearling growth, frame size, and large mature size. These traits can be antagonistic to maternal traits such as birthweight, early puberty, and maintaining a cow size consistent with an economical feed supply.

Table 21.4 identifies a bull selection program that would apply to a large number of commercial producers. First the breeding program goals and the maternal traits are identified. Second, the bull selection criteria are identified giving emphasis to the maternal traits. These traits would receive emphasis if a commercial producer was using only one breed, a rotational crossbreeding program as producing the cows used in a terminal cross-breeding program. Finally, Table 21.4 identifies the traits emphasized in selecting terminal cross bulls. In selecting terminal cross sires, little emphasis is placed on maternal traits as no replacement heifers from this cross are kept in the herd.

SELECTING REPLACEMENT HEIFERS

Heifers, as replacement breeding females, can be selected for several traits at different stages of their productive life. The objective is to identify heifers that will conceive early in the breeding season, calve easily, give a flow of milk consistent with the feed supply, wean a heavy calf, and make a desirable genetic contribution to the calf's postweaning growth and carcass merit.

Beef producers have found it challenging to determine which of the young heifers will make the most productive cows. Table 21.5 shows the selection process that producers are using increasingly to select the most productive replacement heifers. This selection process assumes that more heifers will be selected at each stage of production than the actual number of cows to be replaced in the herd. The number of replacement heifers that producers keep is based primarily on how many they can afford to raise. More heifers than the number needed should be kept through pregnancy-check time. Heifers are selected on the basis that they become pregnant early in life primarily for economic reasons rather than expected genetic improvement from selection.

COW SELECTION

Cows should be culled from the herd based on the productivity of their calves and additional evidence that they can be productive the following year, such as soundness of udders, eyes, skeleton, and teeth. Cow productivity is measured by pregnancy test, weaning and yearling weights (ratios) of their calves, and the EPD of the cow. Table 21.6 shows the weaning and yearling values of the high and low producing cows in a herd. Cow 1 is a high producing cow whereas cow 2 is the low producing cow. Cow 2 should be culled and replaced with a heifer of higher breeding potential.

CROSSBREEDING PROGRAMS FOR COMMERCIAL PRODUCERS

Most commercial beef producers use crossbreeding programs to take advantage of heterosis in addition to the genetic improvement from selection. A crossbreeding system should be determined according to which breeds are available and adapted to the commercial producers' feed supply, market demands, and other environmental conditions. A good

TABLE 21.4. Bull Selection Criteria for Commercial Beef Producers

Goals and Maternal Traits	Bull Selection Criteria for Maternal Traits	Selection Criteria for Terminal Cross Bulls
Breeding Program Goals	**Yearling frame size:** 4.0–6.0 (smaller frame size will adapt better to harsher environments—e.g., less feed, more severe weather, less care, etc.)	**Yearling frame size:** 6.0–8.0 (should be evaluated with frame size of cows so slaughter progeny will average 5.0 and 6.0)
1. Selecting for an **optimum** combination of **maternal traits** to **maximize profitability.** (Avoiding genetic antagonisms and environmental conflicts which come with maximum production or single trait selection.	**Mature weight:** Under 2000 lb in average condition (preference for future). Now we will likely have to consider bulls under 2500 lb)	**Mature weight:** No upper limit as long as birth weight and frame size are kept in desired range.
2. Provide genetic input so **cows** can be **matched with their environment**—cows that wean more lifetime pounds of calf without overtaxing the forage, labor or financial resources.	**Milk EPD:** 0 to +15 lb (prefer +10)	**Milk EPD:** Not considered
	Maternal EPD: +20 to +40 (prefer +25)	**Maternal EPD:** Not considered
3. Stack pedigrees for maternal traits.	**Body condition:** Backfat of 0.20–0.35 inches at yearling weights of 1100 to 1250 lb. Monitor reproduction of daughters	**Body condition:** Evaluate with body condition of cows so that progeny will be between 0.25 and 0.45 in at 1100–1300 lb slaughter weights.
Maternal Traits	**Scrotal circumference:** 34–40 cm at 365 days of age. Passed breeding soundness examination.	**Scrotal circumference:** 32–38 cm at 365 days of age. Passed breeding soundness exam.
Mature weight: Medium sized cows (1000 to 1250 lb under average feed)	**Birth weight EPD:** Preferably under +2.0 lb. Evaluate calving ease of daughters.	**Birth weight EPD:** Preferably no higher than +5.0; want calves birthweights in 85 to 95 lb range with few, if any, over 100 lb.
Milk production: Moderate (Wean 500 to 550 lb calves under average feed)	**Weaning weight EPD:** Approximately +20 lb	**Weaning weight EPD:** +40 lb or higher
Body condition ("fleshing ability"): "5" condition score at calving without high cost feeding.	**Yearling weight EPD:** Approximately +40 lb	**Yearling weight EPD:** +60 lb or higher
Early puberty/high conception: Calve by 24 months of age	**Accuracies:** All EPDs of 0.90 and higher (older, progeny tested sires).	**Accuracies:** All EPDs of 0.90 and higher (older, progeny tested sires)
Calving ease: Moderate birthweights (65 to 75 lbs hfrs; 75 to 90 lb cows). Calf shape (head, shoulders, hips) which relates to unassisted births.	**Functional traits:** Visual evaluation of bull and his daughters.	**Visual:**
Early growth and composition: Rapid gains—relatively heavy weaning and yearling weights within medium (4–6) frame size. Yield grade 2 (steers slaughtered at 1150 lb).	**Visual:** Functional traits Sufficiently attractive to sire calves which would not be economically discriminated against in the market place. Preference for "adequate middle" as medium frame size cattle need middle for feed capacity.	Functional traits: Disposition and structural soundness Sufficiently attractive to sire calves that would not be economically discriminated against in the market place.
Functional traits (longevity): Udders (shape, teats, pigment) Eye pigment Disposition Structural soundness	**Young bulls** without EPDs or with accuracies below 0.90—evaluate trait ratios and pedigree EPDs to predict mature bulls with above EPDs and accuracies. Select sons of bulls that meet EPDs listed above.	**Young bulls without EPDs** or with accuracies below 0.90—evaluate trait ratios and pedigree EPDs to predict mature bulls with above EPDs and accuracies. Select sons of bulls that meet EPDs listed above.

TABLE 21.5.   Replacement Heifer Selection Guidelines at the Different Productive Stages

Stage of Heifer's Productive Life	Emphasis on Productive Trait	
	Primary	Secondary
Weaning (7–10 months of age)	Cull only the heifers whose actual weight is too light to prevent them from showing estrus by 15 months of age. Also consider the economics of the weight gains needed to have puberty expressed.	Weaning weight ratio Weaning EPD Milk EPD Maternal EPD Predisposition to fatness Adequate skeletal frame Skeletal soundness
Yearling (12–15 months of age)	Cull heifers that have not reached the desired target breeding weight (e.g., minimum of 650–700 lb for small to medium-sized breeds or cross, minimum of 750 to 800 lb for large-sized breeds and crosses).	Milk, weaning, and maternal EPDs Yearling weight ratio Yearling EPD Predisposition to fatness Adequate skeletal frame Skeletal soundness
After breeding (19–21 months of age)	Cull heifers that are not pregnant and those that will calve in the latter one-third of the calving season	Milk weaning and maternal EPDs Yearling weight ratio Yearling EPD Predisposition to fatness Adequate skeletal frame Skeletal soundness
After weaning first calf (31–34 months of age)	Cull to the number of first-calf heifers actually needed in the cow herd based on the weaning weight performance of the first calf. Preferably all the calves from these heifers have been sired by the same bull. Heifers should also be pregnant.	

TABLE 21.6.   Performance Data on High- and Low-Producing Cows in the Same Herd

Cow Number	Number of Calves	Weaning Weight (lb)	Weaning Weight Ratio	Yearling Weight Ratio
1	9	583	112	108
2	9[a]	464	89	94

[a] One calf died before weaning (not computed in averages).

example of adaptability is the Brahman breed, which is more heat and insect resistant than most other breeds. Because of this higher resistance, the level of productivity (in the southern and Gulf regions of the United States) is much higher for the Brahman and the other breeds that include Brahman.

Most commercial producers travel 150 or fewer miles to purchase bulls used in natural mating. Therefore, a producer should assess the breeders with excellent breeding programs in a 150-mile radius, as well as available breeds. This assessment, in most cases, should be determined before planning which breeds to use and in which combination to use them.

Breeds should be chosen for a crossbreeding system based on how well the breeds complement each other. Table 21.7 gives some comparative ranking of the major beef breeds for productive characteristics. Although the information in Table 21.7 is useful, it should not be considered the final answer for decisions on breeds to use. First, a producer needs to recognize that this information reflects breed averages; therefore, there are individual animals and herds of the same breed that are much higher or lower than the ranking given. Producers need to use some of the previously described methods to identify superior animals within each breed. Second, it should be recognized that these average breed rankings can change with time, depending on the improvement programs used by the leading breeders within the same breed. Obviously, those traits that have high heritabilities would be expected to change most rapidly, assuming the same selection pressure for each trait. A careful analysis of the information in Table 21.7 shows that no one breed is superior for all important productive characteristics. This gives an advantage to commercial producers using a crossbreeding program if they select breeds whose superior traits complement each other. An excellent example of breed complementation is shown by the Angus and Charolais breeds, which complement each other for both quality grade and yield grade.

Most of the heterosis achieved in cattle as a result of crossbreeding is expressed by weaning time. The cumulative effect of heterosis on pounds of calf weaned per cow exposed is shown in Fig. 21.11, in which maximum heterosis is obtained when crossbred calves are obtained from crossbred cows. The traits that express an approximate heterosis of a 20% increase in pounds of calf weaned per cow bred are early puberty of crossbred heifers, high conception rates in the crossbred female, high survival rate of calves, increased milk production of crossbred cows, and a higher preweaning growth rate of crossbred calves.

Consistent high levels of heterosis can be maintained generation after generation if crossbreeding systems such as those shown in Figs. 21.12–21.14 are used. The crossbreeding system shown in Fig. 21.14 combines a two-breed rotation with a terminal cross. In this system, the two-breed rotation is used primarily to produce replacement females for the entire cow herd. In most cow herds, approximately 50% of the cows are bred to sires to produce replacement females with the remaining 50% being bred to terminal cross sires. All terminal cross calves are sold. This crossbreeding system maintains heterosis as high as the three-breed rotation system.

In the rotational crossing, breeds with maternal trait superiority (high conception, calving ease, and milking ability) would be selected. The terminal cross sire could come from a larger breed where growth rate and carcass cutability are emphasized. A primary advan-

TABLE 21.7. Breed Evaluation for the Productive Characteristics[a]

Breed	Age of Puberty	Weight at Puberty	Birth Weight	Wean Weight	Pounds of Calf Weaned per Cow Exposed	Post-wean Gain	Feed Efficiency Equal Age	Feed Efficiency Equal Weight	Feed Efficiency Equal Fat	Marbling Equal Age	Yield Grade Equal Age	Yield Grade Equal Ribeye Fat	Palatability Grain Fed and Equal Age
Angus	(1)[b]	2	2	4	3	4	3	3	2	(1)	4	4	(1)
Beefmaster	3	3	4	2	2	2	—	2	—	3	3	3	2
Blonde d'Aquitaine	4	4	4	3	2	2	2	2	—	4	2	3	1
Brahman	(5)	(5)	(4)	(1)	(1)	(3)	(3)	(3)	(3)	(5)	(3)	(3)	(3)
Brangus	(3)	(4)	(3)	(3)	(3)	(3)	—	(1)	—	(4)	(3)	—	2
Charolais	(4)	(5)	(5)	(1)	(4)	(1)	(1)	(1)	(3)	(5)	(1)	(1)	(1)
Chianina	(4)	(5)	(5)	(1)	(3)	(1)	(3)	(1)	(4)	(3)	(1)	(1)	(1)
Devon	(2)	(3)	(3)	(4)	(4)	(2)	(2)	(3)	(3)	(3)	(4)	(3)	(3)
Galloway	3	3	3	4	4	3	—	3	—	—	3	3	1
Gelbvieh	(1)	(2)	(4)	(1)	(2)	(2)	(2)	(2)	—	(4)	(3)	(3)	(1)
Hereford	(3)	(3)	(3)	(4)	(3)	(3)	(2)	(2)	(2)	(4)	(4)	(4)	(1)
Limousin	(4)	(4)	(4)	(3)	(4)	(3)	(1)	(2)	(3)	(5)	(1)	(1)	(1)
Longhorn	2	2	1	5	5	5	5	5	—	—	4	3	1
Maine-Anjou	(5)	(4)	(5)	(1)	(3)	(1)	(2)	(1)	—	(4)	(2)	(2)	(1)
Murray Grey	2	2	3	3	3	4	—	4	—	3	4	4	1
Pinzgauer	(1)	(2)	(4)	(2)	(2)	(2)	(2)	(2)	(3)	(4)	(3)	(3)	(1)
Polled Hereford	(3)	(3)	(3)	(4)	(3)	(3)	(2)	(2)	(2)	(4)	(4)	(4)	(1)
Red Angus	1	2	2	4	3	3	2	3	3	(1)	4	4	1
Red Poll	(1)	(1)	(2)	2	(3)	(3)	(4)	(3)	(3)	(4)	(4)	(4)	(1)
Santa Gertrudis	(3)	(5)	(4)	(2)	(4)	(3)	—	(3)	—	(4)	(4)	(3)	2
Scotch Highland	3	3	2	4	4	4	3	—	—	3	3	3	1
Shorthorn	3	3	3	4	3	3	3	3	2	4	5	5	1
Simmental	(2)	(4)	(5)	(1)	(3)	(1)	(2)	(2)	(2)	(4)	(2)	(1)	(1)
Tarantaise	(1)	(2)	(3)	(2)	(2)	(3)	(3)	(4)	(3)	(4)	(3)	(3)	(2)

[a] Ranking based on 1 (most desirable) through 5 (least desirable).

[b] Circled numbers are based primarily on extensive breed comparison data from the Meat Animal Research Center (MARC), Clay Center, Nebraska. Most of the MARC data are based on the various breeds of bulls being bred to Angus and Hereford cows. Rankings are not made where there are insufficient comparative data. Uncircled numbers are judgments based on less extensive, comparative data.

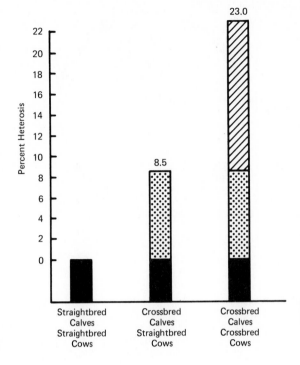

FIGURE 21.11.
Heterosis, resulting from crossbreeding, for pounds of calf weaned per cow exposed to breeding. Courtesy of the USDA.

tage of the rotational-terminal cross system is that smaller or medium-sized breeds can be used in rotational crossing and a larger breed could be used in terminal crossing.

Table 21.8 shows the advantage a commercial producer has over a purebred breeder in being able to use more of the breeding methods for genetic improvement. Commercial producers can use crossbreeding, whereas purebred breeders cannot use crossbreeding if they maintain breed purity.

Traits with a low heritability respond very little to genetic selection, but they show a marked improvement in a sound crossbreeding program. The commercial producer needs to select sires carefully to improve the traits with a high heritability.

TABLE 21.8. Heritability and Heterosis for the Major Beef Cattle Traits

Traits	Heritability	Heterosis
Reproduction	Low	High
Growth	Medium	Medium
Carcass	High	Low

FIGURE 21.12.
Two-breed rotation cross. Females sired by Breed A are mated to Breed B bulls, and heifers sired by Breed B are mated to Breed A bulls. This will increase the pounds of calf weaned per cow bred by approximately 15%. Courtesy of Colorado State University.

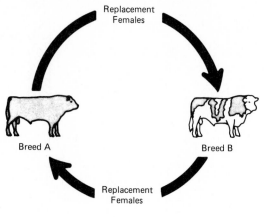

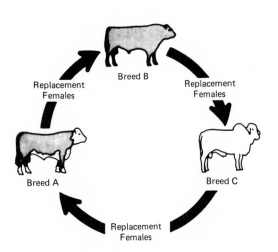

FIGURE 21.13.
Three-breed rotation cross. Females sired by a specific breed are bred to the breed of the next bull in rotation. This will increase the pounds of calf weaned per cow by approximately 20%. Courtesy of Colorado State University.

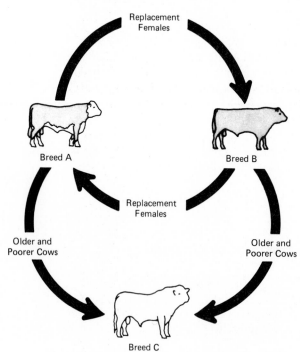

FIGURE 21.14.
Two-breed rotation and terminal sire cross-breeding system. Sires are used in the two-breed rotation primarily to produce replacement heifers. Terminal cross sires are mated to the less productive females. This system will increase the pounds of calf weaned per cow bred by more than 20%. Courtesy of Colorado State University.

SELECTED REFERENCES

Publications

Briggs, H. M. and Briggs, D. M. 1980. *Modern Breeds of Livestock.* New York: Macmillan.

Gibb, J., Wallace, R. and Wagner, W. 1985. Understanding performance pedigrees. *Beef Improvement Federation,* BIF:FS2.

Gregory, K. E. and Cundiff, L. V. 1980. Crossbreeding in beef cattle: evaluation of systems. *J. Anim. Sci.* 51:1224.

Henson, E. 1986. *North American Livestock Census.* Box 477, Pittsboro, NC 27312.

Lamm, D., Hixon, D. and Mankin, J. D. 1986. Culling the commercial cow herd. *Beef Improvement Federation,* BIF-FS7.

Lasley, J. F. 1987. *Genetics of Livestock Improvement.* Englewood Cliffs, NJ: Prentice-Hall.

Long, C. R. 1980. Crossbreeding for beef production: experimental results. *J. Anim. Sci.* 51:1197.

Mankin, J. D. 1985. Utilizing performance records in commercial beef herds. *Beef Improvement Federation,* BIF-FS4.

McGraw, R., Wallace, R. and Gosey, J. A. 1986. Modern commercial beef sire selection. *Beef Improvement Federation.* BIF-FS9.

Taylor, R. E. 1984. *Beef Production and The Beef Industry: A Commercial Producer's Perspective.* Minneapolis: Burgess.

Thomas, V. M. 1986. *Beef Cattle Production.* Philadelphia: Lea and Febiger.

Wagner, W., Gibb, J., Farmer, J. and Strohbehn, D. 1985. Understanding and using sire summaries. *Beef Improvement Federation,* BIF-FS3.

Visuals

Breeding Cattle Selection, Yearling Bull Test (sound filmstrips), Vocational Education Productions, California Polytechnic State University, San Luis Obispo, CA 93407.

Introduction to Beef Breed Selection, Breed Identification: British Breeds, Breed Identification: Continental Breeds, Breed Identification: Brahman and Brahman Crossbreeds, Selecting Beef Sires and *Designing a Beef Breeding System* (sound filmstrips), Prentice-Hall Media, 150 White Plains Rd., Tarrytown, NY 10591.

Cows that Fit Montana (1985 video tape; 12 min.). Department of Animal Science, Montana State University, Bozeman, MT 59715.

Basic Genetics in Beef Cattle Selection and *Selecting the Beef Heifer* (slide sets with audio tapes). Beef Improvement Federation, K. W. Ellis, Cooperative Extension Service, University of California, Davis, CA 95616.

CHAPTER 22

Feeding and Managing Beef Cattle

Commercial beef cattle production typically occurs in three phases, or operations: cow-calf, stocker-yearling, and feedlot. The cow-calf operator raises the young calf from birth to 7–9 months of age (400–500 lb), the stocker-yearling operator grows the calf to 600–800 lb primarily on roughage, and the feedlot operator typically uses high-energy rations to finish the cattle to a desirable slaughter weight of approximately 1000–1300 lb. Most steers and slaughter heifers are between 15–26 months of age when slaughtered. There are, however, alternatives to these three phases where several marketing transactions can occur. In an integrated operation, cattle may have only one owner from cow-calf through the feedlot, or ownership might change several times before the cattle are ready for slaughter. Alternative production and marketing pathways are shown in Fig. 22.1.

COW-CALF PRODUCTION

Cow-calf production centers around 33 million head of beef cows spread across the United States. Most of the cows are concentrated in areas in which forage is abundant. Figure 22.2 shows that the 11 states having over 1 million head of cows (50% of the United States total) are located primarily in the Plains and Corn Belt areas. There was a 22% decrease in calf numbers from 1975 to 1985 (Fig. 22.3). The decrease in numbers was due primarily to poor economic return. During this time there was a large supply of beef associated with a decreased consumer demand which depressed cattle prices. Drought conditions also decreased cattle numbers in certain areas of the U. S.

There are two basic kinds of cow-calf producers: (a) commercial cow-calf producers, who raise most of the potential slaughter steers and heifers, and (b) purebred breeders, specialized cow-calf producers who produce primarily breeding cattle and semen.

Cow-calf producers are interested in managing their operations as economical units. The profitability of a commercial cow-calf operation can be assessed easily by analyzing the following three criteria: (a) calf crop percent weaned (number of calves produced per 100 cows in the breeding herd), (b) average weight of calves at weaning (7–9 months of age), and (c) annual cow cost (number of dollars to keep a cow each year).

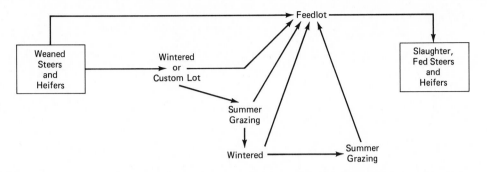

FIGURE 22.1.
Alternative feeding and marketing pathways for weaned calves. Some steers and heifers go to slaughter after the summer grazing period; however, the numbers are relatively small. Courtesy of Colorado State University.

An example of the economic assessment of a commercial cow-calf producer who has an 80% calf crop weaned, 420 lb weaning weights, and a $300 annual cow cost would be as follows:

Calf crop % (0.80) × weaning weight (420 lb) =
336 lb of calf weaned per cow in the breeding herd.
Annual cow cost ($300) ÷ lb of calf weaned (336 lb) = $89.28 per hundredweight.

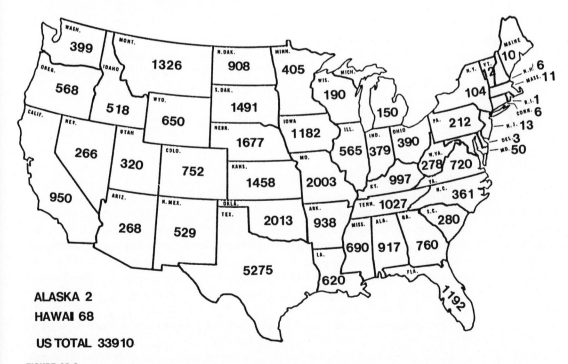

ALASKA 2

HAWAII 68

US TOTAL 33910

FIGURE 22.2.
Beef cows that have calved during 1986 (thousand head). For example, Texas is the leading state in beef cow numbers with 5,275,000 head. Courtesy of The Western Livestock Marketing Information Project.

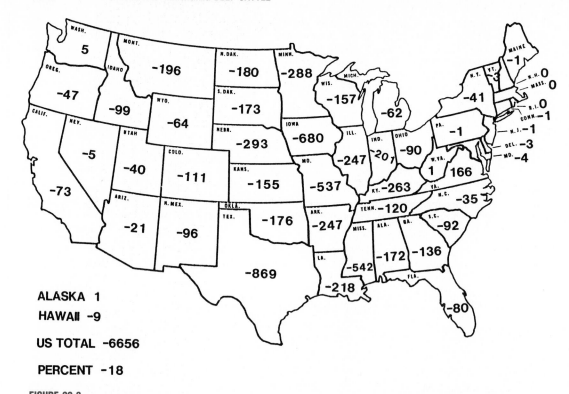

FIGURE 22.3.
Changes in beef calf crop, 1976–1986. For example, Texas had 869,000 fewer calves in 1986 compared to 1976, whereas Virginia had 166,000 more calves for the same time period. Courtesy of The Western Livestock Marketing Information Project.

A break-even price of $89.28 per hundredweight means that the producer would have to receive more than 89¢ per pound, or $89.28 per 100 lb of calf sold, to cover the yearly cost of each cow in the breeding herd. Table 22.1 shows the break-even price for several levels of calf crop percent, weaning weight, and annual cow cost. This information reflects different management levels in which the break-even price ranges from more than $1.00 per pound to less than $0.50 per pound. Profitability of the commercial cow-calf operation is determined by comparing the market price of the calves at the time they are sold with the break-even price.

Cow-calf operations are managed best by operators who know the factors that affect calf crop percent (Fig. 22.4), weaning weight, and annual cow cost. The primary management objective should be to improve pounds of calf weaned per cow and reduce or control the annual cow costs.

TABLE 22.1. Break-even Price (per Hundred Pounds) for Commercial Cow–Calf Operations with Varying Calf Crop Percentages, Annual Cow Costs, and Weaning Weights

Calf Crop Percent Weaned	Annual Cow Cost	Average Calf Weight at Weaning (lb)		
		350	450	550
95	$350	$105.26	$ 81.87	$66.99
	300	90.23	70.18	57.42
	250	75.19	58.48	47.85
85	350	117.65	91.50	74.87
	300	100.84	78.43	64.17
	250	84.03	65.36	53.48
75	350	133.33	103.70	84.85
	300	114.28	88.89	72.73
	250	95.24	74.07	60.61

MANAGEMENT FOR HIGH CALF CROP PERCENTAGES

The eleven primary management factors affecting calf crop percent are as follows:

1. Heifers need to be fed adequate levels of a balanced ration to reach puberty at 15 months of age if they are to calve at the desired age of 2 years. Heifers of English breeds and crosses (e.g., Angus and Hereford) should weigh 650–700 lb at 15 months of age. Cattle of the larger exotic breeds or crosses should weigh 100–150 lb more to assure a high percentage of heifers cycling at breeding.

2. Heifers should be bred to calve early in the calving season. Heifers calving early are more likely to be pregnant as 3-year-olds, whereas heifers calving late will likely not conceive during the next breeding season. Some producers save more heifers as potential replacements at weaning and as yearlings so selection for early pregnancy can be made at pregnancy test time.

3. Heifers typically have a longer postpartum interval than cows. This interval becomes shorter if first-calf heifers (heifers with their first calves) are separated from mature cows 60 days prior to calving and also after calving. This separation allows the heifers to obtain their share of the feed essential for a rapid return to estrus.

4. Feeding programs are designed to have cows and heifers in a moderate body condition (visually estimated by the fat over the back and ribs) at calving time. Table 22.2 shows a body condition scoring system (BCS) currently being used. Thin cows at calving usually have a longer postpartum interval (Fig. 22.5). Cows that are too fat reflect a higher feed cost than is necessary for high and efficient production.

5. Cows, particularly first-calf heifers, should be observed every few hours at calving time. Some females will have difficulty calving (**dystocia**) and will need assistance in delivery of the calf. Calving difficulties should be kept to a minimum to prevent potential death of calves and cows. Calving difficulty is also undesirable because cows given assistance will usually have longer postpartum intervals.

FIGURE 22.4.
Percent calf crop, as measured by a live calf born and raised per cow, is the most economically important trait for the cow-calf producer The bull, cow, calf, and producer each make a meaningful contribution to the level of productivity for this trait. Courtesy of Norden Laboratories, Inc.

6. Calving difficulty should be minimized but not necessarily eliminated. A balance should be maintained between number of calves born alive and the weight of the calves at weaning. Heavier calves at birth usually have heavier weaning weights. When calves are too heavy at birth, however, the death loss of the calves increases. Birth weight is the primary cause of calving difficulty; therefore, management decisions should be made to keep birth weights moderate. Bulls of the larger breeds or large frame sizes should not be bred to heifers, and large, extremely growthy bulls, even in the breeds known for calving ease, should not be bred to heifers. Birth weight within a herd is influenced by genetics. Genetic differences are more important than certain environmental differences such as amount of feed during gestation. Bulls, to be used artificially, should have extensive progeny test records for birth weight and calving ease in addition to an individual birth weight record.

7. The bulls' role in affecting pregnancy rate has a marked influence on calf crop percentage. Before breeding, bulls should be evaluated for breeding soundness by addressing physical conformation and skeletal soundness, palpating the genital organs, measuring scrotal circumference, and testing the semen for motility and morphology. **Libido** (sex drive) and mating capacity are additional important factors in how the bull affects pregnancy rate. These traits are not easily measured in individual bulls before breeding or even after breeding in multiple-sire herds. The typical cow-to-bull ratio is quoted by most cattle producers as 30 to 1. However, some bulls can settle more than 50 cows in a 60-day breeding season. In some large pastures with rough terrain, the cow-to-bull ratio might have to be only 10:1 to assure a high calf crop percent.

8. Crossbreeding affects percent calf crop in several ways. Crossbred heifers usually

TABLE 22.2.   System of Body Condition Scoring (BCS) for Beef Cattle

Group	BCS	Description
Thin condition	1	*Emaciated*—Cow is extremely emaciated with no palpable fat detectable over spinous processes, tranverse processes, hip bones, or ribs. Tail-head and ribs project quite prominently.
	2	*Poor*—Cow still appears somewhat emaciated but tail-head and ribs are less prominent. Individual spinous processes are still rather sharp to the touch but some tissue cover over dorsal portion of ribs.
	3	*Thin*—Ribs are still individually identifiable but not quite as sharp to the touch. There is obvious palpable fat along spine and over tail-head with some tissue cover over dorsal portion of ribs.
Borderline condition	4	*Borderline*—Individual ribs are no longer visually obvious. The spinous processes can be identified individually on palpation but feel rounded rather than sharp. Some fat cover over ribs, transverse processes, and hip bones.
Optimum moderate condition	5	*Moderate*—Cow has generally good overall appearance. On palpation, fat cover over ribs feel spongy and areas on either side of tail-head now have palpable fat cover.
	6	*High Moderate*—Firm pressure now needs to be applied to feel spinous processes. A high degree of fat is palpable over ribs and around tail-head.
	7	*Good*—Cow appears fleshy and obviously carries considerable fat. Very spongy fat cover over ribs and around tail-head. In fact "rounds" or "pones" beginning to be obvious. Some fat around vulva and in crotch.
Fat condition	8	*Fat*—Cow very fleshy and overconditioned. Spinous processes almost impossible to palpate. Cow has large fat deposits over ribs, around tail-head and below vulva. "Rounds" or "pones" are obvious.
	9	*Extremely Fat*—Cow obviously extremely wasty and patchy and looks blocky. Tail-head and hips buried in fatty tissue and "rounds" or "pones" of fat are protruding. Bone structure no longer visible and barely palpable. Animal's mobility might even be impaired by large fatty deposits.

Source: J. Anim. Sci. 62:300.

cycle earlier and have higher conception rates than their straightbred counterparts. Crossbred calves are more vigorous and have a higher survival rate. An effective crossbreeding program can increase the calf crop by 8–12%.

9. The primary nutritional factor influencing calf crop percent is energy expressed in terms of total pounds of feed. The amount of feed (TDN) is important in helping initiate puberty, maintaining proper body condition at calving, and keeping the postpartum interval relatively short. Other nutrients of major importance are protein, salt (sodium chloride), and phosphorous. Additional vitamins and minerals are important only in areas where the soil or feed are deficient.

10. Calf losses during gestation are usually low (2–3%) unless certain diseases are present in the herd. Serious reproductive diseases such as brucellosis, leptospirosis, vibriosis,

FIGURE 22.5.
The cow on the left is a body condition score 4, whereas the cow on the right is a score 6. Cows with body condition scores below 5 have a longer interval (days) between calving and pregnancy compared to cows having body condition scores of 5 and higher. Courtesy of Oklahoma State University.

and infectious bovine rhinotracheitis (IBR) can cause abortions, which may markedly reduce the calf crop percent. These diseases can be managed by blood testing animals entering the herd or vaccinating for the diseases. Herd health programs vary for different operations depending on the incidence of the diseases in the area. Details of these programs should be worked out with the local veterinarian.

11. Calf losses after 1–2 days following birth are usually small (2% to 3%) in most cow-calf operations. Severe weather problems, such as spring blizzards, can cause high calf losses where protection from the weather is limited. In certain areas and in certain years, health problems can also cause high death losses. Infectious calf scours and secondary pneumonia can occasionally reduce the potential calf crop 10–30%.

MANAGEMENT FOR HEAVY WEANING WEIGHTS

The seven primary management factors affecting calf weaning weights are as follows:

1. Calves born early in the calving season are heavier at weaning primarily because they are older. Calves are typically born over a several week period but are weaned together on one specified day. Every time the cow cycles during the breeding season and fails to become pregnant, the weaning weight of her calf is reduced by 30–40 lb. Most commercial producers have a breeding season of 90 days or less so the calves are heavier at weaning and can be managed in uniform groups (Fig. 22.6).

2. The amount of forage available to the cow and the calf has a marked influence on weaning weights. The cow needs feed to produce milk for the calf. The calf, after about 3 months of age, will consume forage directly in addition to the milk it receives.

3. Growth stimulants, commonly given to nursing calves, will increase the weaning weight by 5% to 15% (Fig. 22.7). Ralgro (Zeranol), Synovex C, and Compudose are commonly

FIGURE 22.6.
This calf is approximately 7 months old and soon will be weaned or separated from its dam. Commercial cow-calf producers manage their cows so that the calves will be heavy at weaning time. Courtesy of the Charolais Banner.

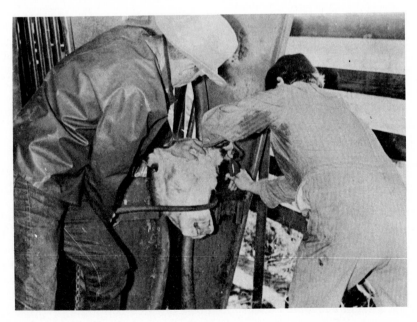

FIGURE 22.7.
The squeeze chute is an essential piece of equipment for restraining cattle. The head can be restrained for inserting ear tags, treating eyes, implanting growth stimulants, or administering medication orally. Note the numbered ear tag, which is the most common form of individual animal identification. Courtesy of Dr. A. T. Ralston, Oregon State University.

used growth stimulants which are implanted as pellets under the skin of the ear. The pellets dissolve over a period of several weeks and supply the growth-stimulating substance that is absorbed into the bloodstream. This implant, however, should not be used on bulls and heifers to be used for breeding purposes. The implant sometimes interferes with the proper development and functioning of the reproductive organs.

4. Providing supplemental feed to the calves where it is inaccessible to the cows will increase the weaning weight of the calves. This practice of **creep feeding** should be used with caution because it is not always economical. It can be used on calves that have the ability to grow and not fatten. It helps calves make the transition of the weaning process and is a feasible practice under drought or marginal feed supply conditions. Creep feeding of breeding heifer calves can impair the development of their milk secretory tissue and subsequently reduce milk production. This impairment is apparently caused by fat accumulating in the udder and crowding the secretory tissue.

5. Any diseases that affect the milk supply of the cow or growth of the calf will cause a reduction in the weaning weight of the calf.

6. Genetic selection for milk production and calf growth rate will increase calf weaning weight. Adjusted 205-day weight is the trait most commonly used to select for milk and growth. Effective bull selection will account for 80–90% of the genetic improvement in weaning weight, although weaning weight ratios and maternal breeding values can also be used in culling cows and selecting replacement heifers. It has been well demonstrated that effective selection can result in a 4-lb to 6-lb per year increase in weaning weight on a per calf basis.

7. Crossbreeding for the average cow-calf producer can result in a 20% increase in pounds of calf weaned per cow exposed in the breeding herd. Most of the increase occurs from improved reproductive performance; however, one-fourth to one-third of the 20% increase is due to the effect of heterosis on growth rate of the calf and increased milk production of the crossbred cow.

MANAGEMENT FOR LOW ANNUAL COW COSTS

In time of inflationary economic conditions, it may be difficult for a producer to lower annual cow costs. However, with careful attention, the increases can be moderated or, in some cases, kept at a similar level over several years. Adequate expense records must be maintained so that cost areas can be carefully analyzed (Fig. 22.8).

The greatest consideration should be given feed costs as they compose the largest part of annual cow costs, usually 60–75%. The period from the weaning of a calf until the last one-third of gestation in the next pregnancy is the time when cows can be maintained on comparatively small amounts of relatively cheap, low-quality feeds. Cow-calf operations having available crop aftermath feeds (e.g., corn stalks, grain stubble, and straw) will usually have the greatest opportunity to keep feed costs lower than other operations (Fig. 22.9). Nutrient costs of available feeds should be evaluated continually.

Labor costs will usually compose 15–20% of the annual cow costs. Labor costs are usually lower on a per cow basis as herd size increases and in areas where moderate weather conditions prevail. It typically takes 15–20 hours of labor per cow unit per year. Operations that use labor inefficiently require twice as much labor per cow.

FIGURE 22.8.
Cattle producers must keep accurate income and expense records to assess annual cow costs and the profitability of their operations. Courtesy of Duane Dailey, University of Missouri.

Interest charges on operating capital account for another 10–15% of the annual cow cost. Producers can reduce interest charges by carefully analyzing the costs of different credit sources.

Cows and heifers should be palpated at approximately 45 days after the end of the breeding season to determine if pregnancy has occurred. The producer should consider all marketing alternatives to maximize profits when selling open cows. Failing to check cows for pregnancy contributes to higher annual cow costs, lower calf crop percentages, and higher break-even costs of the calves produced.

FIGURE 22.9.
Corn stalk fields are available to cattle after the grain has been harvested. There are millions of acres of stalk fields and other crop aftermath feeds that can be grazed by cattle and other livestock. Courtesy of BEEF.

STOCKER-YEARLING PRODUCTION

Stocker-yearling producers manage cattle that are fed for growth prior to going into a feedlot for finishing. Replacement heifers that are intended to go into the breeding herd are typically included in the stocker-yearling category. Discussion here, however, will relate to steers and heifers being grown before going into the feedlot for finishing.

Several alternate stocker-yearling production programs are identified in Fig. 22.1. In some programs, one operator owns the calves from birth through the feedlot finishing phase, and the cattle are raised on the same farm or ranch. In some programs, one operator retains ownership, but the cattle are custom fed during the growth and finishing phases. In other programs, the cattle are bought and sold once, and still in other programs, they are bought and sold several times.

The primary basis of the stocker-yearling operation is to market available forage and high roughage feeds such as grass, crop residues (e.g., corn stalks, grain stubble, and beet tops), wheat pasture, and silage. Stocker-yearling operations also exist to use grazing areas that are usable for summer grazing only and are not adaptable for production of supplemental feed for winter.

The primary factors affecting the costs and returns of stocker-yearling operations are marketing (both purchasing and selling the cattle), the gaining ability of the cattle, amount and quality of available forage and roughage, and health of the cattle.

Stocker-yearling producers need to be aware of current market prices for both cattle they purchase or sell. They also need to understand the loss of weight of the cattle from the time of purchase until the cattle are delivered to their farm or ranch. This loss in weight is called **shrink** and can sometimes reflect the difference in the profit or loss of the stocker-yearling operation. Therefore it is common for calves and yearlings to shrink 3–12% from purchased weight to delivered weight. For example, yearlings purchased at 700 lb that shrink 8% will have a delivered weight of 644 lb. It typically takes 2–3 weeks to recover the weight loss.

The gaining ability of most stocker-yearling cattle is estimated visually. Cattle that are lightweight for their age, thin but healthy, with a relatively large skeletal frame size usually have a high gain potential. Cattle that are light for their age are typically most profitable for the stocker-yearling operator, whereas heavier cattle are most profitable for the cow-calf producer.

Stocker-yearling cattle that are purchased and sold several times encounter stress situations of fatigue, hunger, thirst, and exposure to many disease organisms. The more common diseases are shipping fever complex and other respiratory diseases. These stress conditions make it necessary for stocker-yearling producers to have effective health programs for newly purchased cattle. Producers who have poor herd health programs will typically experience higher costs of gain and higher death losses.

The primary objective of the stocker-yearling operation is to obtain the most pounds of cattle gain within economic reason, while having assurance that high-quality forage yields can be obtained consistently each year. Forage management to obtain efficient production and consumption of nutritious feed is another essential ingredient of a successful stocker-yearling operation. Time of grazing and intensity of grazing (number of animals per acre) are important considerations if maximum forage production and utilization are to be maintained.

FEEDLOT CATTLE PRODUCTION

Feedlot cattle are those cattle being fed for slaughter in small pens or areas where harvested feed is brought to them. Some cattle are finished to slaughter weights on pasture, but they represent only 10–15% of the slaughter steers and heifers. They are sometimes referred to as nonfed cattle as they are fed little, if any, grain or concentrate feeds.

The cattle feeding areas in the United States are shown in Fig. 22.10. These areas correspond to the primary feed-producing areas in which cultivated grains and roughages are grown. These locations are determined primarily by soil type, growing season, and amount of rainfall or irrigation water. Figure 22.11 shows where the approximately 13 million feedlot cattle are fed in the various states. Numbers are shown for only 13 states because very few cattle are fed in the other states. Figure 22.12 identifies states that have shown increases and decreases in fed cattle numbers over the past 10 years. These changes give evidence that the Plains states are becoming increasingly important as the major cattle feeding area of the United States.

TYPES OF CATTLE FEEDING OPERATIONS

Two basic types of cattle feeding operations are (a) commercial feeders and (b) farmer-feeders. These two types are generally distinguished by type of ownership and size of feedlot (Figs. 22.13 and 22.14). The farmer-feeder operation is usually owned and operated by an individual or family. The commercial feedlot might be owned by an individual or partnership, but quite often a corporation owns it, especially as feedlot size increases.

The two types of feeding operations are usually referred to as those over 1,000 head feedlot capacity (commercial) and those under 1,000 head (farmer-feeder). Approximately 70% of fed cattle are fed in feedlots with over 1,000 head capacity, whereas 30% of the cattle are fed in feedlots under 1,000 head capacity. A number of commercial feedlots have capacities of 40,000 head or higher, and a few have capacities of over 100,000 head. Some commercial feedlots custom-feed cattle where someone else owns the cattle and the commercial feedlot provides the feed and feeding service.

Each of the two types of feeding operations has different advantages and disadvan-

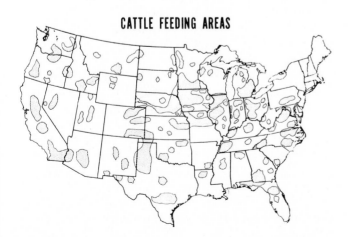

CATTLE FEEDING AREAS

FIGURE 22.10.
Cattle feeding areas in the United States. The areas represent location, but not volume of cattle fed. Courtesy of the USDA.

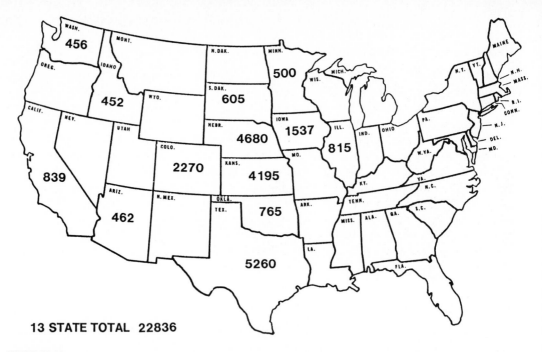

FIGURE 22.11.
Fed cattle marketing, 1986 (thousand head). For example, Nebraska marketed 4,680,000 cattle during 1986. Courtesy of The Western Livestock Marketing Information Project.

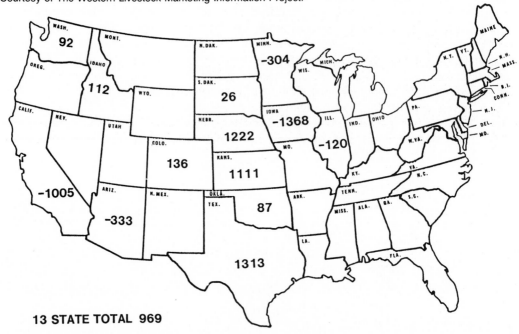

FIGURE 22.12.
Changes in fed cattle marketings, 1976 to 1986 (thousand head). For example, Kansas marketed 1,111,000 more cattle in 1986 than in 1976, whereas California marketed 1,005,000 fewer head of cattle during the same years. Courtesy of The Western Livestock Marketing Information Project.

FIGURE 22.13.
A large commercial feedlot where thousands of cattle can be easily fed in fenceline bunks using feed trucks. Courtesy of BEEF.

FIGURE 22.14.
A farmer feedlot. In this operation the feed is fed in troughs located within the pen. Courtesy of BEEF.

tages. Larger commercial feedlots usually have some economic advantages associated with size and have more professional expertise in nutrition, health, marketing, and financing. The farmer-feeder has the advantages of distributing labor over several enterprises using high-roughage feeds effectively, creating a market of home-grown feeds through cattle, and more easily closing down the feeding operation during times of unprofitable returns.

MANAGING A FEEDLOT OPERATION

The primary factors needed to analyze and properly manage a feedlot operation are the investment in facilities, cost of feeder cattle, feed cost per pound of gain, nonfeed costs per pound of gain, and marketing. A more detailed analysis of these factors is shown in Table 22.3.

TABLE 22.3. The Major Component Parts of a Feedlot Business Analysis

Major Component, with Primary Factors Influencing Them				
Investment in Facilities	Cost of Feeder Cattle	Feed Cost per Pound of Gain	Nonfeed Cost per Pound of Gain	Total Dollars Received
Land	Grade	Ration	Death loss	Market choice
Pens	Weight	Rate of Gain	Labor	Transportation
Equipment	Shrink	Feed efficiency	Taxes	Shrink
Feed mill	Transportation	Length of time	Insurance	Dressing
Office	Gain potential		Utilities	percentage
			Veterinary	Quality grade
			expenses	Yield grade
			Repairs	Manure value

Facilities Investment

The investment in facilities varies with type and location of feedlots. Larger commercial feedlots are quite similar regardless of where they are located in the United States. The general layout is an open lot of dirt pens with pen capacities varying from 100 to 500 head. The pens are sometimes mounded in the center to provide a dry resting area for cattle. The fences are pole, cable, or pipe. A feedmill, to process grains and other feeds, is usually a part of the feedlot. Special trucks distribute feed to fenceline feedbunks where cattle stand and eat inside the pens. Bunker trench silos hold corn silage and other roughages. Grains might be stored in these silos; however, they are more often stored in steel bins above the ground. The investment cost per head for this type of feedlot is approximately $150.

Feedlots for farmer-feeder operations vary from unpaved, wood-fenced pens to paved lots with windbreaks, sheds, or total confinement buildings. The latter might have manure collection pits located under the cattle, which stand on slotted floors. The feed might be stored in airtight structures. In most farmer-feeder operations, however, feeds are stored in upright silos and grain bins, particularly where rainfall is high. Feeds are typically processed on the farm and distributed with tractor-powered equipment to feedbunks located either inside or outside the pens. Investment costs for these feedlots will vary from $200 to $500 per head.

Cost of Feeder Cattle

Before buying feeder cattle, the feedlot operator first estimates anticipated feed costs and the price the fed slaughter cattle will bring. These figures are then used to project cost of feeder cattle or what the operator can afford to pay for them. Feeder cattle are priced according to weight, sex, **fill** (content of the digestive tract), skeletal size, thickness, and body condition. Most commercial feeders prefer to buy cattle with **compensatory gain.** These are cattle that are thin and relatively old for their weight. They have usually been grown out on a relatively low level of feed. When placed on feedlot rations, they gain very rapidly and compensate for their previous lack of feed.

The feeder cattle buyer typically projects a high gain potential in cattle that have a large skeletal frame and very little finish or body condition. However, not all cattle of this type will gain fast.

Heifers are usually priced a few cents a pound under steers of similar weight. The primary reason is that heifers gain more slowly, the cost per pound of gain is higher, and some feeder heifers are pregnant.

Feeder cattle of the same weight, sex, frame size, and body condition can vary several cents a pound in cost. This value difference is usually due to differences in fill. The differences in fill can amount to 10 to 40 lb in live weight of feeder cattle. Feed and water consumed before weighing, distance and time of shipping, temperature, and the manner in which cattle are loaded and transported are some major factors affecting the amount of shrink. Shrink results primarily from loss of fill, but weight losses can occur in other parts of the body as well.

Feed Costs

Feed costs per pound of gain form the major costs of putting additional weight on feeder cattle. Typically, feed costs will be 60–75% of the total costs of gain.

Feed costs per pound of gain are influenced by several factors, and the knowledge of these factors is important for proper management decisions. The choice of feed ingredients and how they are processed and fed are key decisions that affect feed costs.

Cattle that gain more rapidly and efficiently on the same feeding program will have lower feed costs. Some of these differences are genetic and can sometimes be identified with the specific producer of the cattle. Most feeder cattle receive feed additives (e.g., Rumensin®) and ear implants (e.g., Ralgro® or Synovex®) that improve gain and efficiency and eventually the cost of gain (Figs. 22.15 and 22.16).

Feed cost per pound of gain gets progressively higher as days on feed increase. Therefore, cattle feeders should avoid feeding cattle beyond their optimum combination of slaughter weight, quality grade, and yield grade.

Nonfeed Costs

Nonfeed costs per pound of gain is sometimes referred to as yardage cost. Yardage cost includes costs of gain other than feed. These costs can be expressed as either cost per pound of gain or cost per head per day. Obviously, cattle that gain faster will move in and out of feedlots sooner and accumulate fewer total dollar yardage costs.

Death loss and veterinary expenses caused by feeder cattle health problems can increase the nonfeed costs significantly. Most cattle feeders prefer to feed yearlings rather than calves, because the death loss and health problems in yearlings are significantly lower than in calves.

Gross Receipts

Total dollars received for slaughter cattle emphasizes the need for the cattle feeder to be aware of marketing alternatives and kinds of carcasses the cattle will produce. Most

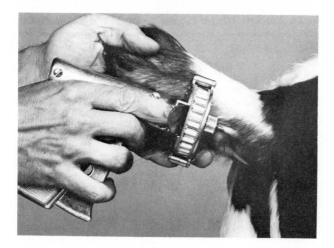

FIGURE 22.15.
Implanting a steer with the growth stimulant Ralgro®. Courtesy of International Minerals and Chemical Corporation.

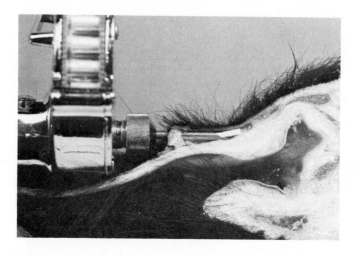

FIGURE 22.16.
Cut-away section of the ear showing the location of three Ralgro® pellets. Note that they are located approximately one inch from the base of the ear, just under the skin. Courtesy of International Minerals and Chemical Corporation.

slaughter cattle from feedlots are sold directly to the packer, through a terminal market when professionals make the marketing transaction, or through an auction where the cattle are sold to the highest bidder. Nearly 80% of the fed cattle are sold direct to a packer. This marketing alternative requires the cattle feeder to be aware of current market prices for the weight and grade of cattle being sold.

Many slaughter cattle at large commercial feedlots are sold on a standard shrink (pencil shrink) of 4%, with the cattle being weighed at the feedlot without being fed the morning of weigh day. Feeders who ship their cattle some distance before a sale weight is taken should manage their cattle to minimize the shrink.

Most slaughter steers and heifers are sold on a live weight basis with the buyer estimating the carcass weight, quality grade, and yield grade. Slaughter cattle of yield grades 4 and 5 usually have large price discounts. The price spread between the Good and Choice quality grades can vary considerably over time. Some marketing alternatives will not show a price differential between Good and Choice if the cattle have been well-fed for a minimum number of days (120 days for yearling feeder cattle).

Cattle feeders who manage their cattle consistently for profitable returns know how to purchase high-performing cattle at a reasonable purchase price. These cattle feeders will formulate rations that will optimize cattle performance with feed costs. Also their cattle have minimum health problems with a low death loss. The feeder feeds cattle the minimum number of days required to assure carcass acceptability and palatability and develops a marketing plan that will yield the maximum financial returns.

Some cattle are sold on a grade and yield basis (carcass weight and carcass grade) where total value is determined after the cattle have been slaughtered. This method of marketing is most useful to producers who know the carcass characteristics of their cattle.

SELECTED REFERENCES

Publications

Lasley, J. F. 1981. *Beef Cattle Production.* Englewood Cliffs, NJ: Prentice-Hall, Inc.

Nutrient Requirements of Beef Cattle. 1984. National Research Council. Washington, D. C.: National Academy Press.

Price, D. P. 1981. *Beef Production.* Dalhart, Texas: Southwest Scientific.

Ritchie, H. D. 1985. *Calving Difficulty in Beef Cattle.* Beef Improvement Federation, BIF-FS6.

Taylor, R. E. 1984. *Beef Production and the Beef Industry: A Commercial Producers Perspective.* New York: Macmillan Publishing Co.

Thomas, V. M. 1986. *Beef Cattle Production.* Philadelphia: Lea and Febiger.

Thompson, G. B. and O'Mary, C. C. 1983. *The Feedlot.* 3rd ed. Philadelphia: Lea and Febiger.

Visuals

Beef Production Systems (sound filmstrips) covering: *Purebred Operations; Cow/Calf Production* and *Feedlot Production;* Vocational Education Productions, California Polytechnic State University, San Luis Obispo, CA 93407.

Beef Management Practices (sound filmstrips) covering: *Basic Beef Cattle Nutrition, Preventative Health Care, Handling Equipment and Facilities, Beef Cattle Castration, Dehorning Beef Cattle, Beef Cattle Identification* and *Calving Management.* Vocational Education Productions, California Polytechnic State University, San Luis Obispo, CA 93407.

Beef Selection Kit (sound filmstrips) covering: *Breeding Cattle Selection, Market and Feeder Cattle Selection* and *Fitting and Showing Beef Cattle.* Vocational Education Productions, California Polytechnic State University, San Luis Obispo, CA 93407.

CHAPTER 23

Dairy Cattle Breeds and Breeding

The dairy cow could be considered a foster mother because many human babies have started early life by consuming cows' milk from a bottle. Milk is an important food in nutrition, particularly for infants, young children and the elderly.

There are several ways of indicating the importance of dairying to the states of the U. S. One way is number of cows in states, and this is illustrated in Fig. 23.1. The five leading states in thousands of dairy cows are: Wisconsin (1,814), New York (901), Minnesota (830), California (825), and Pennsylvania (685). The production per cow for the five leading states in 1985 is Washington (16,892 lb), California (16,667 lb), New Mexico (16,060 lb), Arizona (15,674 lb), and Oregon (14,380 lb). The total pounds of milk produced for the five leading states (given in million pounds) is Wisconsin (23.2), California (14.5), New York (11.1), Minnesota (10.3), and Pennsylvania (9.3). Production of milk per cow has been increased markedly in the past 50 years by improvements in breeding, feeding, sanitation and management. The application of genetic selection, coupled with the extensive use of artificial insemination, has contributed to successful dairy herd improvement program.

CHARACTERISTICS OF BREEDS

Six major breeds of dairy cattle (Holstein, Ayrshire, Brown Swiss, Guernsey, Jersey, and Shorthorn) are used for milk production in the United States. These breeds are shown in Fig. 23.2 and Plate M (following p. 370). Table 23.1 gives production characteristics and other information about the breeds.

Holstein cows produce extremely large amounts of milk—up to 150 lb/day at peak lactation. Thus, in high milk producing cows, great stress is placed on udder ligaments, which can break down and no longer support the udder. If an udder breaks down, it is more susceptible to injury and disease, often necessitating culling the cow.

Ayrshires and Brown Swiss produce milk over a greater number of years than Holsteins, but their production levels are lower. Efficiency of milk production per 100 lb body weight is about the same for the major breeds of dairy cows.

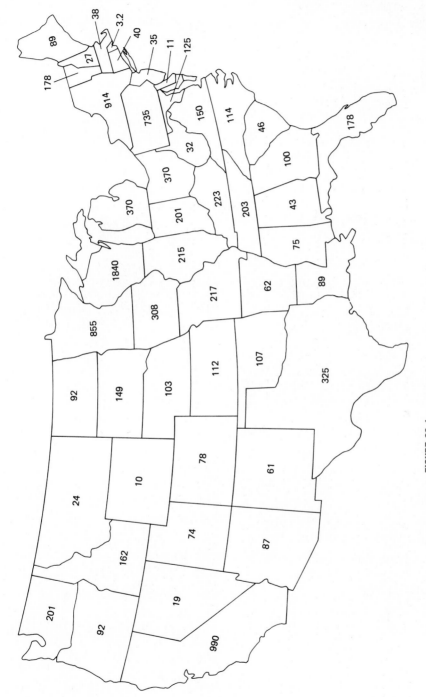

FIGURE 23.1.
The number of milking cows by states ($\times$ 1,000).

Jersey

Holstein

Guernsey

Ayrshire

Brown Swiss

Milking Shorthorn

FIGURE 23.2.
Major breeds of dairy cows used for milk production in the United States. Courtesy of Hoard's Dairyman and Agri-Graphic Services.

Ayrshire

Brown Swiss

Guernsey

Holstein

Jersey

Milking Shorthorn

PLATE M.
Major breeds of dairy cows used for milk production in the United States. Courtesy of Agri-Graphics.

Appaloosa

Arabian

Morgan

Paint

PLATE N.
Major breeds of light horses. Courtesy of Appaloosa Horse Club, Inc., International Arabian Horse Association, American Morgan Horse Association, Inc., and American Paint Horse Association.

Quarter Horse

Standardbred

Thoroughbred

Tennessee Walking Horse

PLATE O.
Major breeds of light horses (continued). Courtesy of The American Quarter Horse Association, the U.S. Trotting Association, Gainesway Farm, and Tennessee Walking Horse Breeders and Exhibitors Association.

DRAFT HORSES

PONIES

Belgian

Miniature

Clydesdale

Shetland

Percheron

Welsh

PLATE P.
Major breeds of draft horses and major breeds of ponies. Courtesy of Belgian Draft Horse Corporation, Betty Groves, Percheron Horse Association of America, Thomas Nebbia ©1985 National Geographic Association, Amercian Miniature Horse Registry, and Welsh Pony Society of America.

TABLE 23.1. Characteristics of Dairy Cattle Breeds

	Aryshire	Brown Swiss	Guernsey	Holstein	Jersey	Shorthorn
Origin	Scotland	Switzerland	Guernsey Island	Holland	Jersey Island	England
Weight						
Male	1,850 lb	2,000 lb	1,600 lb	2,200 lb	1,500 lb	2,000 lb
Female	1,200 lb	1,400 lb	1,100 lb	1,500 lb	1,000 lb	1,400 lb
Color	Mahogany and white spotted, may have pigmented legs	Solid blackish; hairs dark with light tips	Light red and white, yellow skin	Black and white	Blackish hairs have white tips to give gray color, or red tips to give fawn color; also can be solid black or white spotted	Red, roan, or white, or red and white or roan and white
Yearly milk yield (1985) for cows with official milk records	12,909 lb	13,588 lb	11,852 lb	16,648 lb	11,268 lb	11,758 lb
Percent fat	3.95%	4.08%	4.64%	3.64%	4.82%	3.65%
Udders	Large, strong	Large, strong	Medium size, strong	Very large	Small, strong	Large, strong

Guernsey and Jersey cows produce milk having high percentages of milk fat and solids-not-fat, but the total amount of milk produced is relatively low. The efficiency of energy production of different breeds varies less among the breeds than do either total quantity of milk produced or percentage of fat in the milk.

DAIRY TYPE

Some descriptive terms describing ideal dairy type are stature, angularity, level rump, long and lean neck, milk veins, and strong feet and legs. In the past, dairy producers have placed great emphasis on dairy type, but research studies indicate that some components of dairy type may have little or no value in improving milk production and might even be deleterious if overstressed in selection. However, a properly attached udder and strong feet and legs are good indicators that a cow will remain a high producer for a long time.

Figure 23.3 shows the parts of a dairy cow and also gives a description of the component parts (general appearance, dairy character, body capacity and udder) of preferred dairy type. The points on the score card show greatest emphasis for general appearance (especially feet and legs) and udder, where each receive 35 points out of a possible 100 points.

Some dairy breed associations have a classification program to evaluate the type traits. For example, the Holstein Association Linear Classification Program has twenty-nine primary and secondary linear descriptive traits that measure functional conformation or type. A classifier, approved by the Association, scores each cow or bull for the several traits and gives a final score if these scores are to be official records recognized by the Association.

In the Holstein Association program a final score is calculated from rating the four major categories of classification traits: general appearance, dairy character, body capacity, and mammary system. The emphasis for each category for cows and bulls is shown in Table 23.2.

The final score represents the degree of physical perfection of any given animal. It is expressed in the following numbers and words:

Excellent (EX) 90–100	Good (G) 75–79
Very Good (VG) 85–89	Fair (F) 65–74
Good Plus (G+) 80–84	Poor (P) 50–64

TABLE 23.2. Classification Traits Used for Final Scores in Holstein Cattle

	Emphasis	
Trait	Cows	Bulls
General appearance	30%	45%
Dairy character	20%	30%
Body capacity	20%	25%
Mammary system	30%	—

Source: Holstein Association Linear Classification Program.

DAIRY COW UNIFIED SCORE CARD

Copyrighted by The Purebred Dairy Cattle Association, 1943. Revised, and Copyrighted 1957, 1971, and 1982.

Breed characteristics should be considered in the application of this score card	Perfect Score

Order of observation

1. GENERAL APPEARANCE — 35
(Attractive individuality with feminity, vigor, stretch, scale and harmonious blending of all parts with impressive style and carriage.)

BREED CHARACTERISTICS — (see reverse side) — 5
STATURE — height including moderate length in the leg bones with a long bone pattern throughout the body structure. — 5
FRONT END — adequate constitution with strength and dairy refinement. **SHOULDER BLADES** and elbow set firmly and smoothly against the chest wall and withers to form a smooth union with the neck and body. **CHEST** deep and full with ample width between front legs. — 5
BACK — straight and strong; **LOIN** — broad, strong and nearly level **AND RUMP** — long, wide and nearly level with pin bones slightly lower than hip bones. **THURLS** high and wide apart; **TAIL HEAD** set nearly level with topline and with tail head and tail free from coarseness. — 5
LEGS AND FEET — bone flat and strong. **FRONT LEGS** straight, wide apart and squarely placed; **HIND LEGS**, nearly perpendicular from hock to pastern from a side view and straight from the rear view; **HOCKS** cleanly molded free from coarseness and puffiness; **PASTERNS** short and strong with some flexibility and **FEET** short, well rounded with deep heel and level sole. — 15

2. DAIRY CHARACTER — 20
(Angularity and general openness without weakness, freedom from coarseness, and evidence of milking ability with udder quality giving due regard to stage of lactation)

NECK — long, lean and blending smoothly into shoulders; clean cut throat, dewlap, and brisket; **WITHERS** — sharp with chine prominent; **RIBS** — wide apart, rib bones wide, flat and long; **THIGHS** — incurving to flat and wide apart from the rear view, providing ample room for the udder and its rear attachment, and **SKIN** — thin, loose and pliable.

3. BODY CAPACITY — 10
(Relatively large in proportion to size, age and period of gestation of animal, providing ample capacity, strength and vigor)

CHEST — large, deep and wide floor with well sprung fore ribs blending into the shoulders; crops full. **BODY** — strongly supported, long, deep and wide; depth and spring of rib tending to increase toward the rear; **FLANKS** — deep and refined.

4. UDDER — 35
(Strongly attached, well-balanced with adequate capacity possessing quality indicating heavy milk production for long period of usefulness)

FORE UDDER — strongly and smoothly attached, moderate length and uniform width from front to rear. — 6
REAR UDDER — strongly attached, high, wide with uniform width from top to bottom and slightly rounded to udder floor. — 8
UDDER SUPPORT — udder carried snugly above the hocks showing a strong suspensory ligament with clearly defined halving. — 11
TEATS — uniform size of medium length and diameter, cylindrical, squarely placed under each quarter, plumb, and well spaced from side and rear views. — 5
BALANCE, SYMMETRY AND QUALITY — symmetrical with moderate length, width and depth, no quartering on sides and level floor as viewed from the side; soft, pliable and well collapsed after milking; quarters evenly balanced. — 5

Because of the natural undeveloped udder in heifer calves and yearlings, less emphasis is placed on udder and more on general appearance, dairy character and body capacity. A slight to serious discrimination applies to overdeveloped, fatty udders in heifer calves and yearlings.

TOTAL — **100**

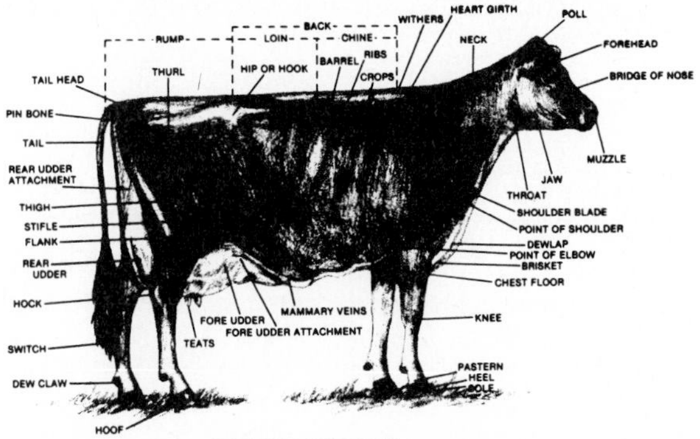

PARTS OF A DAIRY COW

FIGURE 23.3.
Parts of the dairy cow and a description of the preferred dairy type. Courtesy of the Purebred Dairy Cattle Association.

The final score is used in computing PDT (predicted difference type). The PDT identifies genetic differences in sires that can be considered in selection programs to improve dairy cattle type.

Type has value from a sales standpoint. Although some components of type score are negatively related to milk production, type is important as a measure of the likelihood that a cow will sustain a high level of production over several years.

IMPROVING MILK PRODUCTION

Great strides have been made over the past 50 years in improving milk production through improved management and breeding. For example, the amount of milk produced in the United States in 1979 was about the same as that produced in 1945, yet the number of dairy cows in 1979 was less than half the number in 1945. Thus, average production per cow in 1979 was more than double that in 1945. From 1975 to 1985, milk production per cow has increased 20% (Fig. 23.4). During this same time period, cow numbers have remained approximately the same, thus total milk production has had a dramatic increase. In the past 5 years alone, annual milk production by the major breeds of dairy cattle in the United States has increased by a total of 54 lb per cow per year.

SELECTION OF DAIRY COWS

The average productive life of a dairy cow is short (approximately 3–4 years), because many cows are culled primarily because of reproductive failure, low milk yield, udder breakdown, feet and leg weaknesses, and mastitis. Dairy cows to be culled because of low milk production should be culled during or following the first lactation.

When used to replace cows that are **culled** for low production, heifers whose ancestral records indicate they will be high producers will raise the level of production in the herd significantly. Such heifers should also be used to replace cows that leave the herd because of infertility, mastitis, or death, although the improvement gained thereby is generally modest.

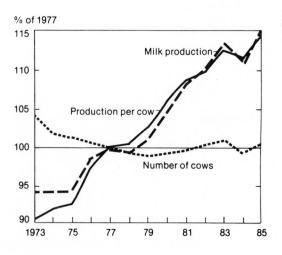

FIGURE 23.4.
Milk production, number of cows, and milk per cow. Courtesy of the USDA.

A basis for evaluating a dairy cow is the quantity of milk (lb) and quality (total solids) that she produces. For a dairy operator to know which cows are good producers and which cows should be culled, a record of milk production is essential.

The National Cooperative Dairy Herd Improvement Program (NCDHIP) is a national industry-wide dairy production testing and recordkeeping program. It is often referred to as the Dairy Herd Improvement (DHI) program. In this program USDA and Extension Service personnel work with dairy producers to help them improve milk production and dairy management practices. Records obtained are analyzed so dairy operators know how each cow compares with all cows in the herd and how the herd compares with other herds in the area. Producers can use several testing plans; some are official and others are unofficial, but any of them may be useful to a dairy operator interested in improving production in a herd. Six important types of testing plans follow.

Dairy Herd Improvement Association (DHIA) is the most common official testing plan in the NCDHIP. All cows must be properly identified and all Dairy Herd Improvement (DHI) rules enforced. The results of official records might be published and are then used by the USDA to evaluate sires.

Dairy Herd Improvement Registry (DHIR) is another official testing program in NCDHIP. The same rules and procedures apply as in official DHIA records. The main difference is that records are sent to the offices of the breed associations and additional "surprise tests" may be conducted. DHIA and DHIR regulations are the same, and both are official testing programs. The DHIR program is for registered cattle. The following rules are to be met for both programs.

1. All cows in the herd must be entered into the official testing program.
2. All animals in the herd must be permanently identified.
3. Copies of pedigrees of all registered cows must be made available for DHIR.
4. Testing is done each month with not less than 15 days nor more than 45 days allowed between test periods.
5. Testing is conducted over a 24-hour period.
6. An independent supervisor must be present for supervising the weighing of the milk and the sampling of the milk for milk fat and other determinations.
7. Milk or milk fat records that are above values established by the breed association require retesting of the cow to assure that an error was not made. An owner may request retesting if he or she feels the test does not properly reflect the production of the cows.
8. Surprise tests may be made if the supervisor suspects that they are needed to verify previous tests.
9. Any practice that is intended to or does create an inaccurate record of production is considered a fraudulent act and is not allowed.

Figure 23.5 shows an individual cow's record used in the DHIR program. Especially note the milk and milkfat recorded under the heading, "305 ME Season Production." Records are standardized to a lactation length of 305 days, two milkings per day (2x), and to a mature age of cow (ME = mature equivalent). Extensive tables are available for ME adjustments which consider the cow's breed, age, month of calving, and region of the country. Adjustments are made to records so they can be more accurately compared on a standardized basis.

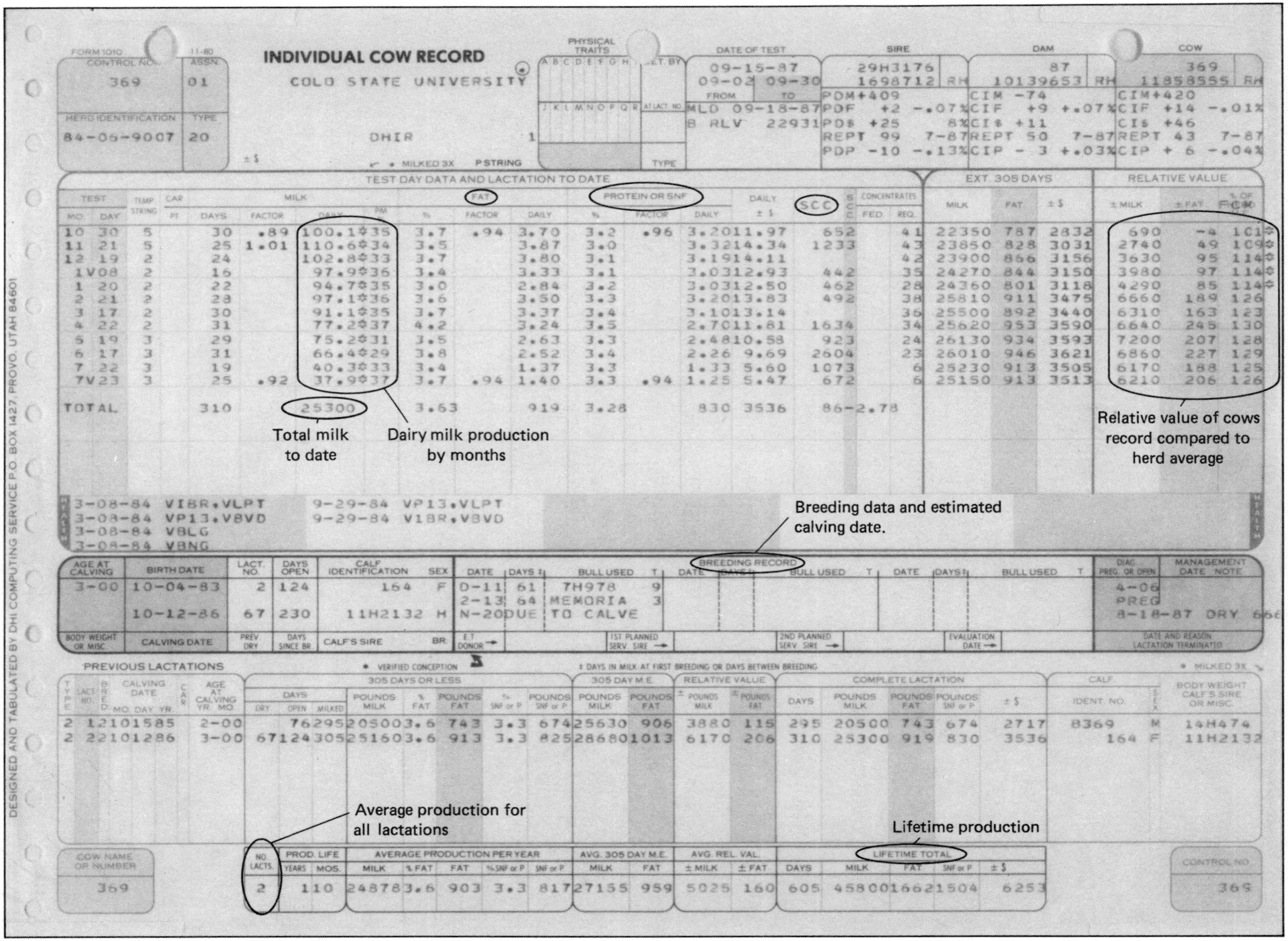

FIGURE 23.5.
An individual cow record used in the DHIA program (some selected information is highlighted). Courtesy of Colorado State University.

Owner-sampler records form a program in which the herd owner, rather than the DHIA supervisor, records the milk weights and takes the samples. The information recorded is the same as in official tests, but the records are for private use and are not published.

Tester-sampler records are similar to official records in that the DHIA supervisor samples and weighs the milk. The records are unofficial, however, and the enforcement of cow identification and other DHIA rules is less rigid than in official tests. These records are not for publication.

"A.M.–P.M." records can be official if a milking-time monitor is installed in the milking facilities. In this plan the herd is tested each month throughout the year, but the supervisor takes only the morning (A.M.) milking for 1 month and the afternoon (P.M.) milking for the 2nd month. Each daily milk weight is doubled to determine the daily milk weight for calculation by the test interval method. The daily milk weights printed on the herd report are the average daily milk weights for the last two consecutive tests (one A.M. and one P.M.).

Milk Only records are unofficial records in which only milk weights are used; no tests are made. Each cow's record and the herd summary are based only on amount of milk produced. Unofficial records may also be combined in that owner-sampler records can be A.M.–P.M. or Milk Only. It is also possible to combine Milk Only and A.M.–P.M.

A registry association for purebred animals exists for each dairy cattle breed in the United States. In addition, certain selective registries honor cows with outstanding production performance and bulls with daughters that are outstanding in production. The Holstein selective registry, for example, has the "gold medal dam" and the "gold medal sire" awards. A gold medal cow must produce 100,000 lb of milk or an average of 12,500 lb of milk and 500 lb of milk fat per year during her productive life; she must also have a minimum type score of 80 points as a 2- or 3-year-old, 81 minimum as a 4- or 5-year-old, and 82 minimum at 6 years of age or older on the basis of 100 points for perfection, when points are accumulated on a score card for scores assigned to the various parts of the animal. The gold medal cow should also have three daughters that meet the requirements of milk production mentioned and three male or female offspring that meet the classification (score) standards. A gold medal sire must have 10 or more daughters that produce an average of at least 13,000 lb of milk and 585 lb of fat per year during their productive lives. Each of these daughters must score 82 points or higher in classification. Other breed associations have selective registries as a means of encouraging improvement in production.

BREEDING DAIRY CATTLE

The heritability of traits is indicative of the progress that can be made by selection. The susceptibilities to cystic ovaries, ketosis, mastitis, and milk fever are all lowly heritable (5–10%). The percentages of fat, protein, and solids-not-fat are all high (50%). Yearly milk (M.E.), protein, solids-not-fat, and fat yields are medium in heritability (25–30%). With the exception of teat locations and spacings, which are moderately heritable (25–30%), all udder characteristics are lowly heritable (10–20%).

Strength of head and upstandingness are highly heritable (45 and 50%). Body weight, type score, levelness of rump, height of tail setting, depth of body, tightness of shoulders, and dairy characters are all moderately heritable (25–35%). Strength of pasterns, arch of back, heel depth, and straightness of hocks are lowly heritable (10–20%).

The genetic correlation between two traits is indicative of the amount of change in trait A that might be expected from a certain amount of selection pressure applied to trait B. The most important genetic correlations are those that might be associated with milk yields during the first lactation. Fat, solids-not-fat, protein yield, lifetime milk yields, and length of productive life have high genetic correlations with first lactation milk yield (0.70–0.90). Overall type score, levelness of rump, udder texture, strength of fore and rear udder are all negatively correlated with first lactation milk yield (−0.20 to −0.40). Dairy character and udder depth are positively correlated (0.35–0.40) with first lactation yield. Inbreeding tends to increase mortality rate and to reduce all production traits except fat percentage of milk and mature body weight.

There are several inherited abnormalities known in dairy cattle. This does not mean that dairy cattle have a higher number of inherited abnormalities than other farm animals, but that more is known about dairy cattle than about most other farm animals because dairy cattle are observed more closely. A list of inherited abnormalities that have been observed is presented in Table 23.3 with the mode of inheritance and a short description of each.

No one breed of dairy cattle carries the inheritance for all known inherited abnormalities. Also, no breed of dairy cattle is free from all inherited abnormalities.

Usually the occurrence of inherited abnormalities is rather infrequent. However, occasionally an outstanding sire might carry an abnormality or a specific dairy herd may have several genetically abnormal calves. As most inherited abnormalities result from recessive genes, both the sire and dam of genetically abnormal calves carry the undesirable gene. Breeding stock should not be kept from either parent.

One reason such rapid progress could be made in improving milk production is that genetically superior sires have been identified and used widely in artificial insemination programs. Genetically superior bulls today might become the sires of 100,000 calves or more each in their productive lives by use of artificial insemination. Generally, the better sires will sire more calves than ordinary sires because dairymen know the value of semen from outstanding bulls. Thus, selection is enhanced markedly for greater milk production by use of artificial insemination (Fig. 23.6).

Milk production, milk composition, efficiency of production, and characteristics that indicate that a cow will likely remain productive for several years are all highly important in selecting dairy cows. These traits are usually emphasized according to their relative heritability and economic importance. Milk and milk fat production are 20–30% heritable, and udder attachment is 30% heritable. Fertility is extremely important but low in heritability, so marked improvement in this trait is more likely to be accomplished environmentally through good nutrition and management than through selection.

Most dairy cattle in the U. S. are straightbred, because crossing breeds has failed to make a significant improvement in milk production. No combination of breeds, for example, equals the straightbred Holstein in total milk production. Because milk production is controlled by many genes and the effect of each of the genes is unknown, it is impossible to manipulate genes that control milk production by the same method that one can use in case of simple inheritance. Also, milk production is a sex-limited trait expressed only in the female. Furthermore, milk production is highly influenced by the environ-

TABLE 23.3. Inherited Abnormalities in Dairy Cattle

Abnormality	Mode of Inheritance	Breeds Having Trait	Brief Description
Achondroplasia	Recessive	Guernsey and Jersey	Lethal, short bones
Amputated	Recessive	Brown Swiss	One or more legs or parts missing
Albinism, complete	Recessive	Guernsey	No pigment in skin, hair, or eyes
Atresia ani	Recessive	Holstein	No anal opening
Blindness	Recessive	Jersey	Cataract development in fetus
Cerebral hernia	Recessive	Holstein	Brain protrudes through opening in skull
Congenital dropsy	Recessive	Ayrshire	Excessive fluid retention in fetus. Lethal
Congenital spasms	Recessive	Jersey	Muscle spasms of newborn. Lethal
Congenital porphyria (pink tooth)	Recessive	Ayrshire Holstein Shorthorn	Pink teeth, body sunburns easily
Dwarfism	Recessive	Jersey	Latent lethal
Flexed pasterns	Recessive	Jersey	Feet turned back
Fused teats	Recessive	Guernsey	Teats on same side of udder fused
Hairless	Recessive	Guernsey Holstein	Almost no hair on calf. Lethal. Can't control temperature
Hydrocephalus	Recessive	Ayrshire Holstein Jersey	Water on brain, bulging forehead
Hypoplasia of cerebellum	Recessive	Holstein	Calves lack sense of balance, lethal
Impacted molars	Recessive	Jersey Milking Shorthorn	Impacted molar teeth
Imperfect skin	Recessive	Ayrshire Holstein Jersey	Defective skin on lower legs. Infections lead to death
Muscle contracture	Recessive	Holstein	Birth difficult—rigid muscles. Delivery by cutting calf into pieces taken out one at a time
Prolonged gestation	Recessive	Guernsey	Calves carried 1–3 months overdue. Lethal
Splayed feet	Recessive	Jersey	Toes on front feet spread wide apart
Syndactylism (mule foot)	Recessive	Holstein	Only 1 toe on each foot. Usually calf becomes lame
Wry face	Recessive	Jersey	Twisted face. Asymmetry of face

FIGURE 23.6.
Cows being inseminated with semen from a genetically superior bull. A well-managed AI program can make a significant contribution to improving milk production in dairy herds. Courtesy of Colorado State University.

ment. To improve milk production by genetic selection, the environment must be standardized among all animals present to ensure, insofar as possible, that differences between animals are due to inheritance rather than environment.

In addition to a high level of milk production, characteristics of longevity, regularity of breeding, ease of milking, and quiet disposition are important in dairy cows. Although it is difficult to predict whether or not a young cow will have longevity of production, such longevity can possibly be enhanced by replacing culled cows with heifers born to dams that are highly productive when mature. Milk production, milk composition, efficiency of production, and characteristics that indicate that a cow will likely remain productive for several years are all highly important in selecting dairy cows.

Bulls are evaluated for their ability to transmit the characteristic of high-level milk production both by considering the production level of their ancestors (pedigree) and by considering the production level of their daughters (progeny testing). An index is used as a predictive evaluation of the bull's ability to transmit the characteristic of high-level milk production.

The **Sire Index,** computed by the USDA, is based on comparing daughters of a given sire with their contemporary herd mates. Each sire is assigned a **predicted difference** (PD) between himself and other sires based on the superiority or inferiority of his daughters to their herd mates. Many sires have daughters in 50–100 different herds, so the predicted difference (PD) for milk, fat, and type is generally highly reliable and provides a sound basis for the selection of semen. The USDA publishers predicted differences among sires semi-annually.

SIRE SELECTION

The dairy industry (breed associations, National Artificial Breeders Association, and the USDA) now use the ''Best Linear Unbiased Prediction'' **(BLUP)** method for estimating predicted differences among sires. This method was developed largely at Cornell University by Dr. C. R. Henderson. The BLUP procedure accounts for genetic competition among bulls within a herd, genetic progress of the breed over generations, pedigree information available on young bulls, and differing numbers of herdmate's sires, and partially accounts for the differential culling of daughters among sires. In the BLUP method, direct comparisons are made among bulls that have daughters in the same herd. Indirect comparisons are made by use of bulls that have daughters in two or more herds. An example presented by the Holstein Association (1980) depicts the use of three different sires in two herds:

Herd 1	**Herd 2**
Daughters of Sire A	Daughters of Sire B
Daughters of Sire C	Daughters of Sire C

Direct comparisons can be made between sires B and C in herd 2. Indirect comparisons between sires A and B can be made by using the common sire C as a basis for comparison.

Pedigree information is used in the BLUP procedure to group bulls based on a pedigree index of one-half of the sire-predicted difference plus one-fourth of the maternal-grandsire predicted difference. The pedigree information is especially valuable for evaluating bulls that have few progeny, and whose evaluations are therefore of low repeatability.

The BLUP method seems to correct the problem of the herdmate comparison method, which overestimates bulls with low predicted differences and underestimates bulls with high predicted differences. This improvement in accuracy of prediction results largely from the BLUP procedure of accounting for genetic progress of a herd over generations. Figure 23.7 depicts a Holstein bull evaluated as genetically superior for milk production and other traits on an extensive evaluation of records.

In addition to BLUP procedures for estimating predicted differences in milk production, fat percentage, and type, the Holstein Association identifies sires that are carriers of seven undesirable genes: **bulldog** (a lethal trait characterized by large bulging head, and thick shoulders); **dwarfism,** hairless, imperfect skin, **mule-foot** (having one instead of two toes), **pink tooth** (congenital porphyria, pink-gray teeth, susceptibility to sunburn), and prolonged gestation.

The Total Performance Index **(TPI)** combines the predicted difference for milk production, the predicted difference for fat percentage, and the predicted difference for type

FIGURE 23.7.
S-W-D Valiant is an outstanding Holstein bull. He ranks in the top 5-15% of all Holstein bulls in his Predicted Difference in transmitting production and type to his offspring. Courtesy of American Breeders Service.

FIGURE 23.8.
Beecher Arlinda Ellen is an outstanding Holstein cow. She has a 305 day milk record of 55,543 pounds with a type score of Excellent-91. Courtesy of Agri-Graphics.

into a single value. Since milk production and fat percent are negatively related, it is important that fat percentage be included. The Holstein Association combines these three traits by use of a ratio of 3:1:1 of predicted difference for milk, predicted difference for fat percentage, and predicted difference for type (that is, 3 times the predicted difference for milk/1 predicted difference for fat percentage/1 predicted difference for type). Using this ratio, fat percentage is maintained whereas milk production and type are improved. Figure 23.8 shows the Holstein cow that ranked first out of 700,000 cows for the best combination of production and type.

SELECTED REFERENCES

Publications

Bath, D. L. Dickinson, F. N., Tucker, H. A. and Appleman, R. D. 1985. *Dairy Cattle: Principles, Practices, Problems, Profits.* Philadelphia: Lea and Febiger.

Campbell, J. R. and Marshall, R. T. 1975. *The Science of Providing Milk for Man.* New York: McGraw-Hill.

Holstein Friesian Association of America. 1980. New Developments in Holstein Type Evaluations. BLUP. Best Linear Unbiased Prediction, TPI-Summaries. Brattleboro, Vermont.

Lasley, J. F. 1987. *Genetics of Livestock Improvement.* Englewood Cliffs, NJ: Prentice-Hall.

Trimberger, G. W., Etgen, W. E., and Galton, D. M. 1987. *Dairy Cattle Judging Techniques.* Englewood Cliffs, NJ: Prentice-Hall.

Visuals

Fitting and Showing Dairy Heifers (Sound Filmstrip), *The Dairy Judging Kit* (slides, manual and cassette), Vocational Education Production, California Polytechnic State Univ., San Luis Obispo, CA 93407.

Dairy Breed Selection, Selecting Dairy Females and *Dairy Breeding Systems* (sound filmstrips). Prentice-Hall Media, 150 White Plains Rd., Tarrytown, NY 10591.

Animal Acquisition/Reproduction (*Introduction to Animal Acquisition - Breed Identification/Advantages and Disadvantages, Heat Detection in Dairy Cows, Artificial Insemination* and *The Calving Process*). Videotapes. Agricultural Products and Services, 2001 Killebrew Drive, Suite 333, Bloomington, MN 55420.

CHAPTER 24

Feeding and Managing Dairy Cattle

The U. S. dairy industry has changed greatly from the days of the family milk cow to become a highly specialized industry that includes production, processing, and distribution of milk. A large investment in cows, machinery, barn, and milking parlor is necessary. Dairy operators who produce their own feed need additional money for land on which to grow the feed. They also need machinery to produce, harvest, and process the crops.

Although size of operations varies from 30 milking cows or less to more than 5000 milking cows, the average dairy has approximately 100 milking cows, 30 dry cows, 30 heifers, and 25 calves. These average dairy producers farm 200–300 acres of land, raise much of the forage and market the milk through cooperatives, of which they are members. These producers will sell about 3 tons of milk daily, or about 2.2 million lb annually, worth about $230,000. Their average total capital investment may exceed a half million dollars. The average producer likely has a partnership with a brother, wife, son, or outside person, which makes management of time and resources easier.

The dairy operator must provide feed and other management inputs to keep animals healthy and at a high level of efficient milk production. Feed and other inputs must be provided at a relatively low cost compared with the price of milk if the dairy operation is profitable.

NUTRITION OF LACTATING COWS

The average milk production per cow in the U. S. for a lactation period of 305 days is approximately 13,000 lb. Some herd averages exceed 20,000 lb, whereas some top-producing cows yield more than 30,000 lb of milk per year. Thus, some lactating cows may produce more than 150 lb of milk, more than 5 lb of milk fat, and more than 4.5 lb of protein per day. A great need for energy and total amount of feed is created by lactation. For example, a cow weighing 1,400 lb that produces 40 lb milk daily needs 1.25 times as much energy for lactation as for maintenance. If she is producing 80 lb of milk daily, she will need 2.5 times the energy for milk production than is needed for maintenance.

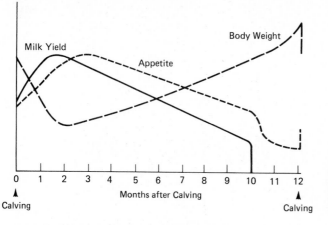

FIGURE 24.1.
Milk production, appetite (dry matter intake), and body weight during the lactational cycle.

Providing adequate nutrition to lactating dairy cows is challenging and complex. Some of the reasons for the complexity is shown in Fig. 24.1. It is difficult during the first 2–3 months after calving to provide adequate nutrition because milk yield is very high and intake is limited. As the nutrient intake is less than the nutrient demand for milk, the cow uses her body fat and protein reserves to make up the difference. Thus the cow is in a negative energy balance and usually loses body weight during the months of heavy milk production. This same period of time can also be challenging to reproduction as conception rates are usually lower when the cows are losing weight.

Body condition scores (1 = thin, 3 = average, 5 = fat) can be used to monitor nutrition, reproduction, and health programs for dairy herds. Condition scores of low 3s and high 2s are acceptable during the first few weeks after calving, whereas body condition scores in the low 4s are necessary as the cows move into the dry period. Scores lower or higher than these signal potential management problems. Cows can be grouped by condition score and stage of lactation to provide adequate nutrition at effective costs.

At 2–3 months into the lactation period, daily milk production peaks and then starts to decline; feed intake is adequate or higher than milk production demands. This contributes to bodyweight gain. The energy content of the ration should be reduced after approximately 5 months of lactation to prevent the cow from becoming too fat.

Different types and amounts of feeds can be used to provide the needed energy, protein, vitamins, and minerals to lactating cows. Availability, palatability and relative costs of different feeds are primary factors influencing ration composition. Basic nutrition principles, presented in Chapters 14, 15, and 16, provide the foundation for developing dairy cattle feeding programs.

Most dairy cattle rations are based on roughages (hays and silages). Roughages are usually the cheapest source of nutrients; usually they are produced by most dairy farms, but increasing herd size, greater managerial demands, higher land-tax costs, and cheaper feeds are changing this traditional role. In some cases, roughages are fed free-choice and separate from the concentrates, which are fed in restricted amounts. In other cases, silages and concentrates are mixed together before being fed to the cows; this is known as a total mixed ration, which is becoming the prefered way of feeding dairy cows. Chopped hay and silage can be delivered to managers on each side of an open alleyway (Fig. 24.2).

Lactating dairy cows cannot obtain an adequate supply of nutrients from an all-roughage

FIGURE 24.2.
Arrangement of feeding chopped hay and silage to dairy cows. The feed is delivered to the feeding area from a truck as the truck proceeds along the open alleyway. Courtesy of Dr. Lloyd Swanson, Animal Science Department, Oregon State University.

ration. Concentrates are supplied in amounts consistent with the level of milk production, body weight of the cow, amount of nutrients in the roughage, and the nutrient content of the concentrates. Concentrates are usually provided to cows while they are being milked. However, high-producing cows requiring large amounts of concentrates do not have sufficient time in the milking parlor to consume all the necessary grain; therefore additional concentrates are usually fed at another location. In large dairies, several groups of cows must be moved through the milking facilities two or three times a day.

The computer is involved with dairy feeding programs today. Many dairy farms have feeding stations that use computers to determine how much concentrate a cow will be fed based on need for milk production and to control the rate (lb/hour) the concentrate is made available to the cow. These devices have two advantages: (a) to prevent cows from overconsuming concentrates and (b) prevent cows from eating concentrates too rapidly, which prevents metabolic upsets.

There are economic advantages in feeding concentrates in relation to quantity of milk produced by each cow. However, this method tends to feed a cow that is declining in production too generously and to feed a cow that is increasing in production inadequately. It is a poor economic practice to allow cows in their last 2 months of lactation to have all the concentrates that they can consume. Heavy feeding at this stage of lactation does not result in increased milk production.

Young (2-year-old) cows that are genetically capable of high production should be fed large amounts of concentrates to provide the nutrition they need to grow as well as produce milk. Without adequate nutrition, their subsequent breeding and lactation may be hindered.

Forages vary immensely in nutrient concentrations. Legumes (alfalfa) have high cal-

TABLE 24.1.   Nutrient Content of Common Feeds Compared to Nutrient Requirements for Dairy Cows

Feed	Net Energy (Mcal/lb)	Percentage of Dry Matter				
		Protein	Ca	P	K	Mg
Legumes (early cut)	0.60	20	1.30	0.30	2.00	0.25
Grasses (early cut)	0.55	15	0.60	0.25	1.50	0.20
Corn silage	0.70	9	0.30	0.20	1.00	0.20
Corn	0.85	9	0.02	0.35	0.60	0.25
Wheat	0.80	11	0.05	0.41	0.60	0.15
Soybean meal	0.75	48	0.35	0.75	2.20	0.30
Dairy cow diet (required)	0.70	15	0.60	0.40	1.50	0.25

cium, protein, and potassium, relative to animal nutrient requirements whereas other forages such as grasses and corn silage are considerably lower. Concentrates (grains) usually are low in calcium and higher in phosphorus (Table 24.1). The amount of protein supplement (soybean meal) and mineral supplement that must be added to a concentrate formulation obviously depends upon type and amount of forage being fed. Trace mineralized salt and vitamins A, D, and E are usually added to concentrates.

DRY COWS

How dry cows are fed and managed may influence their milk production level and health in the next lactation.

A common practice of drying off lactating cows is to abruptly stop milking the cow; with high producers, however, this may be traumatic and dry-off may need to be more gradual (intermittent milking). The buildup of pressure in the mammary gland causes the secretory tissues to stop producing milk. At the last milking, the cow should be infused with a treatment for preventing mastitis.

Dairy producers plan for a 50–60 day dry period. Short dry periods usually reduce future milk yield because the cow has not adequately improved body condition and the mammary tissue has not properly regenerated. Long dry periods can lower milk yield owing to the cow becoming overly fat, and profitability may be less because feed costs are increased.

Dry cows should be separated from the lactating cows so they can be fed and managed consistent with their needs. Dry cows need less concentrates than lactating cows. If dry cows overeat, they will likely become fat. This excessive body fat may cause health problems and lower future milk yield.

These health problems include fatty liver, ketosis, mastitis, retained placenta, metritis, milk fever, and even death.

REPLACEMENT HEIFERS

Young heifers 5 months of age and older can meet most of their nutritional needs from good quality legume or grass-legume pasture in addition to 2–3 lb of grain dairy. If the pasture quality is poor, they will need good-quality legume hay and 3–5 lb of grain. In winter, good-quality legume hay or silage and 2–3 lb of grain are needed. If the forage is grass (hay or silage), they will need 3–5 lb of grain to keep them growing well.

Ideally, heifers should be large enough to breed at about 15 months of age and calve as 2-year-olds. Guernsey and Jersey heifers should weigh 600 lb when they are first bred; Holstein and Brown Swiss heifers, 800 lb. If their weights are considerably less, first breeding should be delayed until they are at the desired breeding weights.

BULLS

Young bulls should be fed sufficiently to grow without becoming fat. A good pasture provides most nutritional needs, but it may be necessary to feed limited amounts of concentrates when the pasture is less than ideal. Bulls being used heavily for breeding should be fed limited concentrates, legume hay, or silage.

BREEDING DAIRY COWS

Proper management of dairy cows at the time of breeding, at calving, during milking operations, and in herd health is essential for a profitable dairy operation.

The time of breeding is an important phase in dairy cattle management. Because they are milked each day, dairy cows are more closely observed than beef cows, and therefore easily detected when in heat. When in estrus, dairy cows may show restlessness, enlarged vulvas, and a temporary decline in milk production. Also, when cows are in standing heat they will permit other cows to mount them.

Technicians are available to artificially inseminate cattle, but dairy producers or employees who have mastered the insemination procedure may breed the cows. The use of semen from genetically proven sires is highly desirable, even though this semen may cost more than semen from an average bull. Considering the additional milk production that can be expected from heifers sired by a good bull, the extra investment can return high dividends. Semen costing $50–100 per unit from genetically superior bulls may be a better investment than semen from less desirable bulls at $10–20 per unit. Bulls should not run with the milking cows, because bulls are often dangerous.

If cows are to be bred naturally, proper facilities for handling the bull must be provided. The bull should be kept in a well-fenced pen. The bull should have a ring in his nose so that two ropes can be snapped into the ring. This allows two people to lead the bull to breed the cow as the two people can stay apart so if the bull tries to charge one of them, the other can prevent him from doing this by pulling on his rope. A rod with a snap on the end is good for reaching through the fence to snap onto the ring in the bull's nose. One should never trust a bull and one should never take chances; caution should be used at all times. Bulls that provide semen for artificial insemination should be handled with caution also.

CALVING OPERATIONS

Dairy cows that are close to calving should be separated from other cows, and each should be placed in a maternity stall that has been thoroughly cleaned and bedded with clean bedding. The cow in a maternity stall can be fed and watered there. A cow that delivers her calf without difficulty should not be disturbed, however, assistance may be necessary if the cow has not calved by 4–6 hours from the start of labor. Extreme difficulties in delivering a calf may require the services of a veterinarian who should be called as early as possible.

As soon as a calf arrives, it should be wiped dry. Any membranes covering its mouth or nostrils should be removed, and its **navel** should be dipped in a tincture of iodine solution to deter infection. Producers should be sure newborn calves have an adequate amount of colostrum produced by the dam, because colostrum contains antibodies to help the calf resist any invading microorganisms that might cause illness. The cow should be milked to stimulate her to produce milk.

Many commercial dairy operators dispose of bull calves shortly after the calves are born. Some bull calves are kept for veal, while others are castrated and fed for beef. Heifer calves are grown, bred, and milked for at least one lactation to evaluate their milk producing ability.

Dairy calves rarely nurse their dams. Usually they are removed from their dam and fed colostrum for several days. Calves that are separated from one another usually have fewer health problems and thus a higher survival rate (Fig. 24.3).

Milk that is not salable or milk substitutes are fed to the young calves. Grain and leafy hay are provided to young calves to encourage them to start eating dry feeds and to stimulate rumen development. As soon as calves are able to consume dry feeds (45–60 days of age), milk or milk replacer is removed from the diet because of its high cost.

MILKING AND HOUSING FACILITIES FOR DAIRY COWS

Dairy farming is very labor intensive, and labor-saving machines are commonly found as part of the facilities. Dairy cows usually are managed in groups of 20–50 head; they have

FIGURE 24.3.
An example of calf separation and isolation. These individual fiberglass calf huts are used for dairy calves from birth to weaning. The huts are 3½ ft. × 7½ ft. with an attached wire pen in front. Courtesy of Colorado State University.

FIGURE 24.4.
A water flush-free stall barn with individual lock-in stanchions. These facilities provide for inside feeding of totally mixed rations. Courtesy of Colorado State University.

access to stalls but are not tied in free-stall or loose housing systems. These facilities reduce cleaning, feeding and handling time, and labor. In tie-stall barns, cows are tied in a stanchion and remain there much of the year, feeding and milking are done individually in the stanchion. In this case, a pipeline milking system is used. Cows are milked in place and milk is carried by a stainless-steel or glass pipe to the bulk tank, where it is cooled and stored. In loose housing systems, such as loaf shed or free stall (Fig. 24.4), cows are brought to a specialized milking facility, called a parlor, for milking.

All housing systems require daily or twice daily removal of manure and waste feed; this may be done mechanically (tractor-blade) or by flushing with water. Manure, urine, and other wastes usually are applied directly to land and/or stored in a lagoon, pit, or storage tank for breakdown before being sprayed or pumped (irrigated) onto land.

MILKING OPERATIONS

The modern milking parlor basically is a concrete platform raised about 30 inches above the floor (pit) of the parlor (Fig. 24.5). It is designed to speed the milking operation, reduce labor, and make milking easier for the operator. Cows enter the parlor in groups or individually, depending on the facility. The udder and teats are washed clean, dried, and massaged for about 20–30 seconds. Milking begins in about 1 more minute and continues for about 6–8 min. After milking is completed, the milking unit is removed manually or by computer, and the teats are dipped in a weak iodine solution.

FIGURE 24.5.
A trigon milking parlor equipped with computer units and automatic milker take offs. Two sides of the milking parlor will accommodate six cows each, while the third side handles four cows. Courtesy of Colorado State University.

Modern milking machines (Figs. 24.6. and 24.7) are designed to milk cows gently, quickly, and comfortably. Some have a take-off device to remove the milking cup once the rate of flow drops below a designated level. Removal of the milking unit as soon as milking is complete is important as prolonged exposure to the machine can cause teat and udder damage and lead to disease problems (mastitis).

The milking machine operates on a two vacuum system: one vacuum inside the rubber liner and the other vacuum outside the rubber liner of the teat cup (Figure 24.8). There is a constant vacuum on the teat to remove the milk and keep the unit on the teat. The intermittent vacuum in the pulsation chamber causes the rubber liner to collapse around the teat. This assists blood and lymph to flow out of the distal teat into the upper part of the teat and udder.

The milking parlor and its equipment must be kept sanitary with the pipes being cleaned and sanitized between milkings.

Regular milking times are established with equal intervals between milkings. Most cows are milked twice per day, however high producing cows produce more milk if they are milked three times a day. The latter is labor intensive and only economically feasible in herds with very high levels of productivity.

CONTROLLING DISEASES

Dairy and beef cattle are afflicted by the same diseases but some diseases are more serious in dairy cattle than in beef cattle. Certain disease organisms can be transmitted through milk and can be a major problem for people who consume unpasteurized milk.

Several diseases, such as tuberculosis and brucellosis, can affect health of dairy cattle and humans consuming their milk. Requirements for the production of grade A milk are

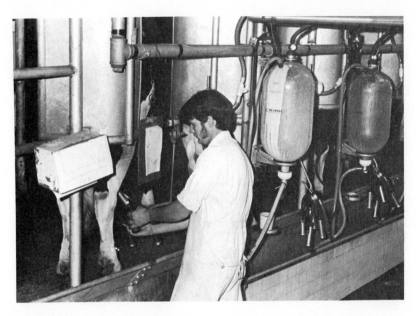

FIGURE 24.6.
The suction cups are applied to the teats of the cow so that she can be milked by machine after she has come to her place in the milking parlor. Courtesy of Dr. Lloyd Swanson, Animal Science Department, Oregon State University.

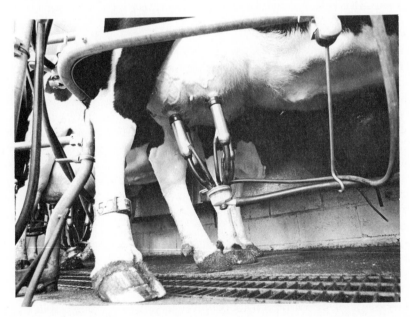

FIGURE 24.7.
Modern milker on cow. The personnel in charge of milking observe closely to be certain that the suction cups are removed as soon as the cow has been milked unless the machine used automatically reduces its suction at that time. Injury to the udder can result if suction continues for some time after milking has been completed. Courtesy of De Laval Agricultural Division, Alfa-Laval, Inc., Poughkeepsie, NY.

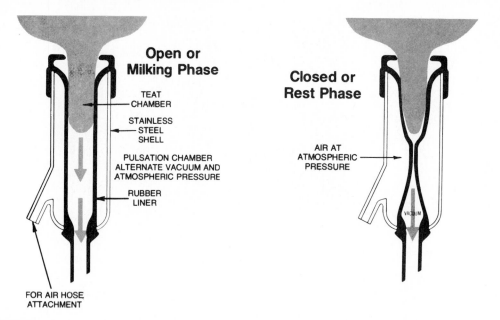

Open or Milking Phase

TEAT CHAMBER

STAINLESS STEEL SHELL

PULSATION CHAMBER ALTERNATE VACUUM AND ATMOSPHERIC PRESSURE

RUBBER LINER

Closed or Rest Phase

AIR AT ATMOSPHERIC PRESSURE

VACUUM

FOR AIR HOSE ATTACHMENT

FIGURE 24.8.
Cross section of the teat cup of a milking machine. (A) During the open or milking phase, the vacuum level is the same on both sides of the liner, permitting it to be in an open position. (B) During the rest phase, atmospheric air is admitted into the pulsayion chamber by the pulsator. The pressure differential between the inside of the liner and the outside causes it to collapse. Courtesy of De Laval Agricultural Division, Alfa-Laval, Inc.

that the herd must be checked regularly with the ring test for tuberculosis and must be bled and tested for brucellosis. Milking and housing facilities must be clean and meet certain specifications. Grade A milk must have a low bacteria count and somatic cell count (low mastitis). Grade A milk sells at a higher price than lower grades of milk; therefore, most dairy producers strive to produce grade A milk.

Bang's disease (Brucellosis) is caused by Brucella abortus organism. It can markedly reduce fertility in cows and bulls and it can be contracted by humans as a disease known as undulant fever. Heifers can be calfhood vaccinated, however some dairymen prefer to use a testing program. Since most dairy cows are bred by artificial insemination, there is little danger of a clean herd becoming contaminated from breeding. Persons drinking unpasteurized milk are taking a serious risk unless they know that the milk comes from a clean herd.

Perhaps the most troublesome disease in dairy cattle is mastitis—inflammation and infection of the mammary gland. The disease destroys tissue, impedes milk production, and lowers milk quality. Mastitis costs the U. S. dairy industry more than $1.5 billion each year or approximately $200 per cow annually.

In its early stage of development, subclinical mastitis is undetectable by the human eye. This type of mastitis can be detected only with laboratory equipment (Somatic Cell Counter at the DHI testing laboratory) or with a cow-side test called the California Mastitis Test (CMT). In the latter test, a reagent is added with a paddle to small wells, then

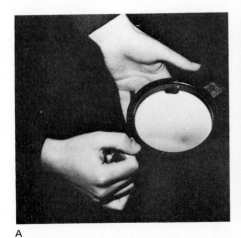 

A   B

FIGURE 24.9.
(A) Normal milk. (B) Ropey milk in strip cups. Courtesy of Dr. Lloyd Swanson, Oregon State University.

milk is squirted from each quarter into a specific well. If a reaction occurs, then the relative degree of subclinical mastitis infection can be determined (Fig. 24.9). As mastitis progresses, small white clots appear in the milk. At this stage it is called clinical mastitis and it is easily observed. One can also collect milk using a strip cup, which has a fine screen on a dark background such that white flakes, strings, or blood (from infected areas) can be seen. As mastitis progresses, the milk shows more clots, watery milk, and acute infection. At this stage of mastitis, the signs of mastitis in the cow are obvious; the udder is swollen, red, and hot and gives pain to the cow when touched. In the final stages of mastitis, the gel formation becomes dark and is present in a watery fluid. At this stage, the udder has been so badly damaged that it no longer functions properly.

Susceptibility to mastitis may be genetically related to a certain degree, but many environmental factors such as bruises, improper milking, and unsanitary conditions are more prevalent causes of mastitis. High-producing animals may be more prone to stress and mastitis than low-producing animals. Also, cows with **pendulous** udders or odd-shaped teats are more likely to develop mastitis (Fig. 24.10). Housing facilities and milking procedures are other important factors.

The best approach to controlling mastitis is using good management techniques to prevent its outbreak. These include using routine mastitis tests (DHI or CMT), treating infected animals, preventing infection from spreading, dipping teats after milking, using dry cow therapy, and practicing good husbandry for cleanliness. Chronically infected cows may have to be sold.

The Wisconsin mastitis test is more quantitative than the California test in that it measures the agglutination that occurs in the milk when the test is applied, but it cannot be used as a barn test. Perhaps the somatic cell count, in which cells in the milk are counted, is the more effective test. Milk always contains some cells, but the number of **leucocytes** (white blood cells) increases when the udder is infected.

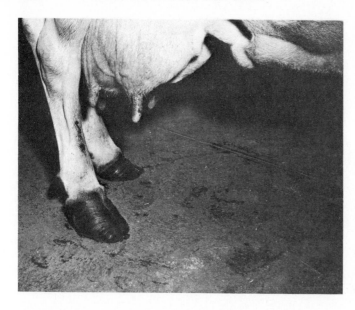

FIGURE 24.10.
A pendulous udder afflicted with mastitis. An udder is pendulous when the udder floor is below the hock and the fore and rear attachments are broken down. Courtesy of Dr. H. P. Adams, Oregon State University.

When mastitis is diagnosed, treatment must be started immediately if the udder is to be saved. The udder can be treated by inserting a small tube into the teat canal and infusing the antibiotic. Depending on the antibiotics used, milk from treated cows should not be sold for human use for 3–5 days after the last treatment.

Other diseases and health practices are discussed in Chapter 20.

SELECTED REFERENCES

Publications

Adams, R. S. 1980. *Stillage for Dairy Cattle*. Pennsylvania State Extension Paper DSE-80-5.

Babson Brothers Dairy Research Service. 1980. *The Way Cows Will Be Milked on Your Dairy Tomorrow*. 8th Edition. Oak Brook, IL: Babson Bros. Dairy Research Service.

Bath, D. L., Dickinson, F. N., Tucker, H. A. and Appleman, R. D. 1985. *Dairy Cattle: Principles, Practices, Problems, Profits*. Philadelphia: Lea and Febiger.

Berger, Larry L. 1981. *Nutritional Value of Distillers Feeds for Livestock*. Beltsville, MD: Agricultural Biomass Energy Workshop.

Campbell, J. R. and Marshall, R. T. 1975. *The Science of Providing Milk for Man*. New York: McGraw-Hill.

Miller, W. J. 1979. *Dairy Cattle Feeding and Nutrition*. New York: Academic Press.

Nutrient Requirements of Dairy Cattle. 1978. National Research Council. Washington, D. C.: National Academy Press.

The Changing Dairy Farm. 1986. *New York's Food and Life Sciences Quarterly*, vol. 16.

Whitlow, L. W. 1984. Dry cow feeding and management. North Carolina Agricultural Extension Publication D-22 (Dairy Handbook).

Wilcox, C. J., Van Horn, H. H., Harris, Jr. B., Head, H. H., Marshall, S. P., Thatcher, W. W., Webb, D. W. and Wing, J. M. 1979. *Large Dairy Herd Management*. Gainesville: University Presses of Florida.

Visuals

Nutrition *(Common Feeds for Dairy Cattle, Nutrition Related Problems, Storage and Handling of Feed)*, Herd Health/Husbandry *(Distinguishing Normal and Abnormal Cattle, Normal Movements and Body Positions of Dairy Cattle, Animal Handling* and *Young Stock Management)* and Waste Management/ Buildings and Equipment ("Waste Storage" and "Buildings"). Videotapes. Agricultural products and Services, 2001 Killebrew Drive, Suite 333, Bloomington, MN 55420.

CHAPTER 25

Swine Breeds and Breeding

Breeds of swine are genetic resources available to swine producers. Most market pigs in the United States today are crossbreds, but these crossbreds have their base and continued perpetuation from seedstock herds that maintain pure breeds of swine. A knowledge of the breeds, their productive characteristics, and their crossing abilities is an important part of profitable swine production.

CHARACTERISTICS OF SWINE BREEDS

In the early formation of swine breeds, breeders distinguished one breed from another primarily through visual characteristics. These visual, distinguishing characteristics for swine were, and continue to be, hair color and erect or drooping ears. Figure 25.1 shows the major breeds of swine and their primary distinguishing characteristics.

Relative importance of swine breeds is shown by registration numbers recorded by the different breed associations (Table 25.1). Although the registration numbers are for registered purebreds, the numbers do reflect the demand for breeding stock by commercial producers who produce market swine primarily by crossbreeding two or more breeds. Note how the popularity of breeds has changed with time. This change has resulted, in part, from using more objective measurements of productivity and the breeds' ability to respond to the demand and selection for higher productivity and profitability.

TRAITS AND THEIR MEASUREMENTS

Sow Productivity

Sow productivity, which is of extremely high economic importance, is measured by litter size, number weaned per litter, 21-day litter weight, and number of litters per sow per year. A combination of pig number (born alive) and the litter weight at 21 days best reflects sow productivity (Fig. 25.2). Litter weight at 21 days is used to measure milk production because the young pigs have consumed very little supplemental feed before

Landrace
(white, large
drooped ears)

Chester White
(white, drooped
ears)

Yorkshire
(white, erect
ears)

Hampshire
(black with
white belt;
erect ears)

Poland China
(black with
white on legs,
snout, and tail;
drooped ears)

Berkshire
(black with
white on legs,
snout, and tail;
erect ears)

Duroc
(red, drooped
ears)

Spotted
(black and
white spots,
drooped ears)

Boar Power (Hybrid boar)
(hybrids are not considered to be purebred
breeds but are used similarly to purebred
breeds. Foundation breeding of hybrids
comes from several breeds and hybrids are
bred as pure lines; no consistent color or
ear set.)

FIGURE 25.1.
The major breeds of swine in the United States. Courtesy of Poland China Record Association, Knoxville, Illinois
(Poland China); National Spotted Swine Record, Bainbridge, Indiana (Spotted); American Landrace, Lebanon,
Indiana (Landrace); USDA (Hampshire and Chester White); American Yorkshire Club, West Lafayette, Indiana
(Yorkshire); American Berkshire Association, Springfield, Illinois (Berkshire); United Duroc Swine Registry, Peoria,
Illinois (Duroc); and Boar Power (boar/gilt breeding system by Monsanto), Des Moines, Iowa (typical hybrid boar).

TABLE 25.1. Registration Numbers for Major Breeds of Swine

	U.S. Breed Associations			Canadian Breed Association (1985)[b]
Breed	1985[a]	1970	Date Association Formed	
Yorkshire	215,995	42,235	1935	11,087
Duroc	206,215	62,830	1957	1,867
Hampshire	109,878	74,101	1893	1,492
Spotted	87,512	13,974	1914	191
Chester White	41,145	16,456	1930	—
Landrace	38,619	6,123	1950	7,393
Poland China	22,000	11,191	1876	—
Berkshire	20,000	7,210	1875	—

[a]Only breeds having 20,000 or more registrations are listed.
[b]An association which registers all breeds shows 14,676 registrations.
Source: Henson (1986), North American Livestock Census

that time. Sow productivity traits are lowly heritable, which means these traits can be best improved through environmental changes rather than through selective breeding. Eliminating sows low in productivity while keeping highly productive sows is still a logical practice for economic reasons but not to make major genetic change. It is extremely important for commercial producers, using effective crossbreeding programs, to use breeds that rank high in maternal or sow productivity traits.

Growth

Growth rate is economically important to most swine enterprises and has a heritability of sufficient magnitude (35%) to be included in a selection program. Seedstock producers need a scale to measure weight in relation to age (Fig. 25.3). Growth rate can be expressed in several ways and typically is adjusted to some constant basis such as days required for swine to reach 230 lb.

Feed Efficiency

Feed efficiency measures the pounds of feed required per pound of gain. This trait is economically important because feed costs account for 60–70% of the total production costs for commercial producers. Obtaining feed efficiency records requires keeping individual or group feed records. However, feed efficiency can be improved without keeping feed records by testing and selecting individual pigs for gain and backfat. This improvement occurs because fast-gaining pigs and lean pigs tend to be more efficient in their feed utilization. Obtaining feed efficiency records is done in some central boar testing stations; however, costs and other factors would need to be considered if similar records are to be taken on the farm.

Carcass

Carcass traits are typically measured by ham-loin percentages, lean cut percentages, or both. Fortunately, these measures of carcass composition can be predicted reliably by

FIGURE 25.2.
This crossbred sow farrowed 17 pigs and raised 15 of them. The number of pigs raised per sow and the weight of the litter at 21 days of age effectively measures sow productivity. Courtesy of D. C. England, Oregon State University.

FIGURE 25.3.
Using the scales to measure growth rate of individual boars is an essential part of a sound breeding system. Courtesy of Iowa State University.

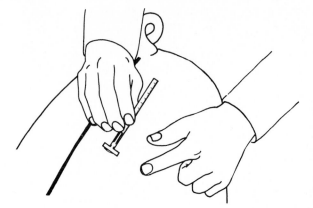

FIGURE 25.4.
A metal ruler is being used to probe this pig. The amount of backfat, measured in inches, will allow the swine breeder to predict the carcass desirability of a boar's offspring. From Battaglia and Mayrose, *Handbook of Livestock Management Techniques,* Macmillan Publishing Company, 1981, p. 264.

backfat measurements taken on live pigs by either a metal ruler or ultrasonics (Fig. 25.4). The metal probe is considerably less expensive and as accurate as ultrasonic machines. Experienced individuals can visually predict significant differences in lean-to-fat composition in live pigs with a fairly high degree of accuracy.

Soundness

Structural soundness, the capacity of breeding and slaughter animals to withstand the rigors of confinement rearing and breeding, is a vital need of the swine industry today. Breeders commonly consider unsoundness as one of the results of confinement, but in reality, confinement rearing only makes unsoundness more noticeable. Some seedstock producers raise their breeding stock in confinement (similar to the manner in which most commercial producers raise their offspring) and then cull the unsound ones.

Recent studies report unsoundness to be medium in heritability; therefore, improvement can be made through selection. Soundness can be improved through visual selection if breeders decide to cull restricted, too straight legged, unsound boars lacking the proper flex at the hock, set of the shoulder, even size of toes or proper curvature and cushion to the forearm and pastern.

Inherited defects and abnormalities generally occur with a low degree of frequency across the swine population. It is important to know they do exist and that certain herds may experience a relatively high incidence. Certain structural unsoundness characteristics might be categorized as inherited defects. **Cryptorchidism** (retention of one or both testicles in the abdomen), umbilical and scrotal **hernias** (rupture), and inverted **nipples** (nipples or teats don't protrude but they are inverted into the mammary gland), and **PSE** carcasses (pale, soft and exudative) are considered the most common occurring genetic defects and abnormalities. The PSE condition and the porcine stress syndrome (PSS) have occurred in recent years from selecting for extreme muscling. Most important genetic abnormalities are inherited as a simple recessive pair of genes.

SELECTING REPLACEMENT FEMALES

Sow productivity is the foundation of commercial pork production. The sow herd also contributes half of the genetic composition to the growing-finishing pigs. These two fac-

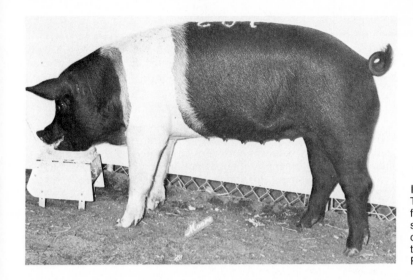

FIGURE 25.5.
This gilt has outstanding performance records for genetic superiority in the economically important traits. Courtesy of the Hampshire Swine Registry.

tors show the importance of replacement gilt selection so that highly productive gilts can be retained in the swine herd. Fast growing, sound, moderately lean gilts with body capacity from large litters should be selected for sow herd replacements (Fig. 25.5). Among sows that have farrowed and will rebreed, those that have physical problems, bad dispositions, extremely small litters (two pigs below herd average), and poor mothering records should be culled.

A balance between sow culling and gilt selection needs to be established. Replacement gilts should be available in sufficient numbers to replace culled sows. Gilts, replacing sows, represent a major opportunity for genetic change in the sow herd. As sows generally produce larger litters of heavier pigs than gilts, replacing large numbers of sows with gilts may reduce production levels. This production differential and the low relationship between the performance of successive litters argue for low rates of culling based on sow performance to maintain high levels of production.

The higher productive rate of sows over gilts must be balanced against the genetic change made possible by bringing gilts into production. A total gilt replacement level of 20–25% is suggested for each farrowing. Pork producers may find economic advantages in timing the culling of sows to take advantage of high prices for sows.

The gilt selection and sow culling scheme suggested assumes that there are no major genetic antagonisms between maternal traits (litter size and milking ability) and rate of gain and low-backfat thickness. Some evidence indicates that the so-called "meaty gilt" does not make a good sow. There is, however, no documented evidence that selecting fast-growing, low-backfat gilts that are not too extreme in muscling, will adversely affect sow performance. Guidelines for gilt selection by age and weight are given in Table. 25.2.

BOAR SELECTION

Boar selection is extremely important in making genetic change in a swine herd. Over several generations, selected boars will contribute 80–90% of the genetic composition of

TABLE 25.2.   Gilt Selection Calendar

When	What
Birth	Identify gilts born in large litters. Hernias, cryptorchids, and other abnormalities should disqualify all gilts in a litter from which replacement gilts are to be selected. Record birth dates, litter size, identification. Equalize litter size by moving boar pigs from large litters to sows with small litters. Pigs should nurse before moving. Keep notes on sow behavior at time of farrowing and check on: (a) disposition, (b) length of farrowing, (c) any drugs such as oxytocin administered, (d) condition of udder, (e) extended fever.
3–5 weeks	Take 21-day weight of litter. Wean litters. Feed balanced, well-fortified diets for excellent growth and development. Screen gilts identified at birth by examining underlines and reject those with fewer than 12 well-spaced teats. If possible, at this time select and identify about two to three times the number of gilts needed for replacement.
180–200 lb	Weigh and backfat-probe potential replacement gilts. Evaluate for soundness. Select for replacements the fastest growing, leanest gilts that are sound and from large litters. Save 25–30% more than needed for breeding. Remove selected gilts from market hogs. Place on restricted feed. Allow gilts to have exposure to a boar along the fenceline that separates them. Observe gilts for sexual maturity. If records are kept, give advantage to those gilts that have had several heat cycles prior to final selection.
Breeding time	Make final cull when the breeding season begins and keep sufficient extra gilts to offset the percentage of nonconception in your herd. Make sure all sows and gilts are ear-marked, ear-tagged or otherwise identified.

Source: Pork Industry Handbook, PIH-27, Guidelines for Choosing Replacement Females

the herd. This contribution does not diminish the importance of female selection. Boars should have dams that are highly productive sows. Productivity of replacement gilts is highly dependent on the level of sow productivity passed on by the boars.

The review of performance records is a must in selecting boars effectively (Fig. 25.6). The heritability and economic importance of some of their available records is shown in Table 25.3. Based on this comparison of Boar A with Boar B given in Table 25.3, if Boar A sired 700 pigs, he would return to the producer an extra $903 ($1.29 × 700). The superiority of replacement gilts retained in the herd would increase Boar A's value even more. This comparison shows the value of genetically superior boars and the justification for paying more for them.

Boars should be selected from the top 50% of the test group regardless of whether the selected boars come from a central test station or from the farm. Boars meeting the standards shown in Table 25.4 should be considered as potential herd sires.

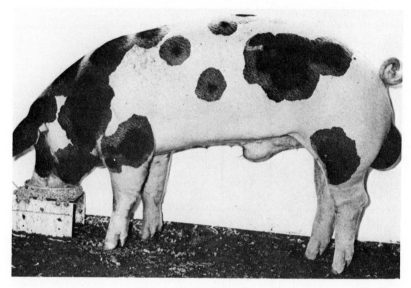

FIGURE 25.6.
This boar has outstanding records for growth, feed efficiency, and backfat thickness. He is out of a sow that has produced several large litters of heavy pigs at weaning time. Courtesy of the National Spotted Swine Record, Inc.

TABLE 25.3.   Economic Value of Production Records of Boars

	Daily Gain (lb/day)	Feed Efficiency (lb F/G)[a]	Adjusted Backfat (in.)
Heritability	0.30	0.40	0.50
Economic value/unit change	$4.00/lb/day	$12.00/lb F/G	$3.50/in.
Records			
Boar A	2.26	2.53	0.89
Group average	2.06	2.71	1.00
Boar B	1.86	2.89	1.11
Superiority of A over B	+0.40	−0.36	−0.22

Value of Boar A over Boar B

Trait	(Superiority) ×	(Heritability) ×	(Genetic Influence) ×	(Economic Value) =	Added Value/ 230-lb Hog
Daily gain	0.40	0.30	0.50	$ 4.00	$0.24
Feed efficiency	0.36	0.40	0.50	12.00	0.86
Backfat	0.22	0.50	0.50	3.50	0.19
					$1.29

[a]Pound feed per lb of gain
Source: Pork Industry Handbook, PIH-9, Boar Selection Guidelines for Commercial Producers.

TABLE 25.4.   Suggested Selection Standards for Replacement Boars

Trait	Standard
Litter size	10 or more pigs farrowed; 8 or more pigs weaned.
Underline	12 or more fully developed, well-spaced teats.
Feet and legs	Wide stance both front and rear, free in movement, good cushion to both front and rear feet, equal-sized toes.
Age at 230 lb	155 days or less.
Feed/gain, boar basis (60–230 lb)	275 lb of feed/hundred pounds (cwt) of gain.
Daily gain (60–230 lb)	2.00 lb/day or more.
Backfat probe (adj. 230 lb)	1.0 in. or less.

Source: Pork Industry Handbook, PIH-9, Boar Selection for Commercial Production.

EFFECTIVE USE OF PERFORMANCE RECORDS

Records are valuable when used in breeding programs, whereas those records obtained only for promotion are eventually self defeating. The primary reason for obtaining records is to improve accuracy of selection. Records of different pigs should be obtained under a similar environment for the records to be comparable and useful. Breeders who give a small group of pigs preferential treatment are only deceiving themselves. If this procedure is used, breeders can no longer compare accurately even the individuals in their own herds, and records thus obtained will yield false conclusions.

The ultimate objective in an improvement program is to predict an animal's breeding value. Breeding value is a measure of the animal's ability to transmit desired genetic traits to the resulting offspring. Proper records on the individual and on his or her relatives can help a great deal in predicting breeding values.

Heritability estimates are medium (20% to 39%) for growth traits and high (40% and higher) for carcass traits. Genetic principles indicate that the most rapid rate of improvement for these traits is through objectively measuring the traits and selecting pigs based on their superior individual performance. Thus, the performance value of an animal for most traits can be predicted at the young age of 5–6 months and will generally result in more rapid genetic progress than a selection scheme based on sib or progeny tests or on pedigree information. Regardless of the types of facilities available, individual performance tests can be conducted on the farm if all animals are treated similarly. On-the-farm testing is essential to a program of rapid genetic improvement because it permits testing of a larger sample of the potential breeding population than is possible in central testing stations.

The principle of using on-the-farm records to identify genetically superior animals is quite simple. The environment should be standardized and then traits are measured which estimate each animal's potential breeding value. If the animal is superior because of environment, this animal will breed poorer than its record would suggest. By contrast, an animal that has received a poorer than average environment will probably breed better than its record would indicate. Therefore, it is important to determine if an animal is superior because of environment or because of genetics. In comparing animals in different herds, this determination is difficult to make.

To compare animals within a group more objectively, indexes have been developed for use in testing stations and farm programs. The index is a single numerical value that is determined from a formula that combines the values for several traits. One such index, currently in use, combines the performance of daily gain, feed efficiency, and backfat into one numerical value for individual boars. In the formula which follows, $\overline{ADG}$ (average daily gain), $\overline{F/G}$ (feed per pound of gain), and $\overline{BF}$ (backfat) represent the averages for all the pigs in the group that are tested together. The index is as follows:

$$I \text{ (index)} = 100 + 60 \text{ (averaging daily gain of individual} - \overline{ADG})$$
$$- 75 \text{ (feed efficiency of individual or group of litter mates} - \overline{FG})$$
$$- 70 \text{ (backfat probe or sonoray of individual} - \overline{BF})$$

An example of this index would be the index of boar A in Table 25.3. Computing this index shows that boar A is 33 points superior for a combination of daily gain, feed efficiency, and backfat when compared with the average of other boars in the test group. The index of the average boar is 100. About 20% of the boars are expected to exceed 120 index points, and 20% are expected to fall below the recommended minimum culling level of 80.

CROSSBREEDING FOR COMMERCIAL SWINE PRODUCERS

The primary function of purebred or seedstock breeders is to provide breeding stock for commercial producers. Breeding programs for seedstock producers are designed to genetically improve the economically important traits. Within a specific breed, rate of improvement will be dependent on heritability of the traits, how much selection is practiced, and how quickly generation turnover occurs. The most rapid genetic changes that have occurred over the past few decades are growth rate and leanness.

Commercial producers and seedstock producers select for the same economically important traits. Commercial producers have an added advantage over seedstock producers, in making genetic change, because crossing two or more swine breeds will increase productivity. This genetic phenomenon of heterosis or hybrid vigor is shown in Table 25.5 The 41% improvement in total litter market weight of crossbreds is a cumulative effect of

TABLE 25.5. Effect of Crossbreeding on Swine Productivity (Expressed as Percent of Purebred Breeds)

Trait	Purebred	First Cross (two breeds)	Multiple Crosses (crossbred sow bred to purebred boar)
Number of pigs born alive	100	101	108
Litter size at weaning	100	110	124
21-day litter weight	100	110	127
Days to 230 lb	100	107	107
Total litter market weight	100	122	141

larger litters, high pig survival and increased growth rate of individual pigs. This marked increased in productivity tells why 90% of the market hogs in commercial production are crossbreds. Crossbreeding allows genetic improvement in some traits with low heritabilities such as sow productivity. There is little hybrid vigor for traits with high heritabilities, for example, carcass traits.

An effective crossbreeding program takes advantage of using hybrid vigor and selecting genetically superior breeding animals within breeds that best complement one another. A producer needs to know the relative strengths and weaknesses of the available breeds. This information is shown in Tables 25.6 and 25.7. These breed comparisons can change over time as a result of genetic change made in the individual breeds. Thus, breed comparisons need to be made on a continuing basis.

The crossbred female is an integral and important part of an effective crossbreeding program. Breed-cross rankings for sow productivity are shown in Table 25.8. Even though crossbreeding provides an opportunity to reap the benefits of many genetic sources, an unplanned crossing program will not yield success or profit for the pork producer. A crossbreeding system must be selected that will capitalize on heterosis, take advantage of breed strengths, and fit the individual producer's management program.

Two basic crossbreeding systems may be considered, namely, the rotational cross and the terminal cross. The rotational cross system combines two or more breeds where a different breed of boar is mated to the replacement crossbred females produced the previous generation (Figs. 25.7 and 25.8). In a terminal cross system, a two-breed single or rotational cross female is mated to a boar of a third breed. Thus, all the pigs produced are sired by the same breed of boar with all offspring marketed. The female stock is usually purchased through a system primarily emphasizing reproductive performance, while growth and carcass traits are emphasized in the sire line. This latter system will maximize heterosis; however, purchasing replacement female increases the risk of introducing disease into the herd.

TABLE 25.6   Relative Reproductive Performance of Female Breeds[a]

	Breed					
Traits	Berkshire	Chester White	Duroc	Hampshire	Landrace	Yorkshire
Number of pigs born alive	87	100	89	84	94	100
Litter size weaned	83	100	86	84	89	100
21-day litter weight	74	100	84	86	96	100
21-day litter weight/female exposed	84	100	78	94	89	94

[a]Best breed performance is given 100 and compared to each of the other breeds.
Source: Adapted from Pork Industry Handbook, PIH-39, and the National Hog Farmer, July 15, 1979.

TABLE 25.7. Sire Breed Influence on Various Reproduction, Production, and Carcass Traits[a]

	Breed							
Trait	Berk-shire	Chester White	Duroc	Hamp-shire	Landrace	Poland China	Spotted	York-shire
Number of pigs born alive	—	100	96	98	—	—	—	100
Litter size weaned	—	100	96	92	—	—	—	100
21-day litter weight	—	100	90	90	—	—	—	100
21-day litter weight per female exposed	—	—	91	81	—	—	—	100
Daily gain	96	92	100	98	93	95	98	98
Feed efficiency	94	96	100	98	90	95	96	99
Backfat thickness	87	88	88	100	78	95	86	87
Carcass length	96	95	97	98	99	94	97	100

[a]Best breed performance given 100 and compared to each breed where adequate data are available.
Source: Adapted from Pork Industry Handbook, PIH-39, and the National Hog Farmer, July 15, 1979.

TABLE 25.8. Relative Reproductive Performance of Crossbred Females[a]

Female Breed Cross	Litter Size Born Alive	Litter Size Weaned	Litter Size 21-day Weight	Litter Size 21-day Weight/or per Female Exposed
Chester × Duroc	83	79	86	—
Chester × Hampshire	92	81	77	—
Chester × Yorkshire	100	100	100	100
Chester × Landrace	100	100	100	100
Hampshire × Landrace	100	95	95	88
Hampshire × Yorkshire	91	87	87	81
Berkshire × Yorkshire	90	85	83	93
Berkshire × Landrace	92	90	87	91
Berkshire × Hampshire	81	77	76	74
Duroc × Yorkshire	93	85	82	85
Duroc × Landrace	92	93	86	79
Duroc × Hampshire	86	82	79	76
Duroc × Berkshire	93	82	79	77

[a]Best breed performance is given 100 and compared to each of the other breeds.
Source: Adapted from Pork Industry Handbook, PIH-39, and the National Hog Farmer, July 15, 1979.

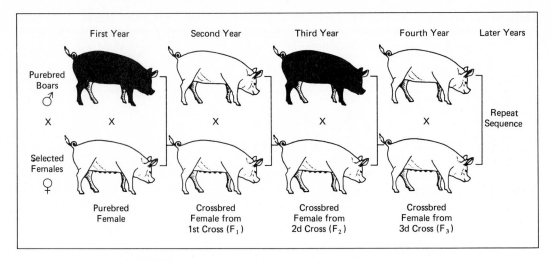

FIGURE 25.7.
Two-breed rotational system. Courtesy of Pork Industry Handbook.

A two- or three-breed rotational crossbreeding system may better fit a producer's management program even though heterosis is less than a terminal cross system. A two-breed rotation results in 67% of the heterosis of a terminal cross, whereas a three-breed rotation gives 86% of the heterosis of a terminal crossing system.

The three-breed rotational cross is probably the most popular crossbreeding system. It combines the strong traits of a third breed not available in the other two breeds. Sires from three breeds are systematically rotated each generation and replacement crossbred

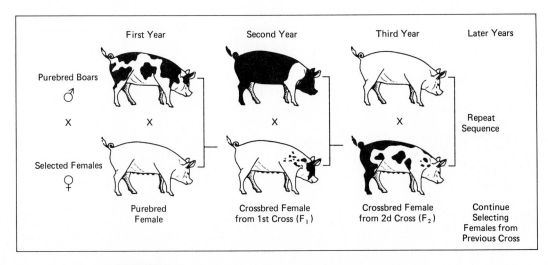

FIGURE 25.8.
Three-breed rotational system. Courtesy of Pork Industry Handbook.

females are selected each generation. These females are mated to the sire breed to which they are least related. Because reproductive performance is to be stressed in the initial two-breed cross, growth, feed efficiency, and superior carcass composition may be lower than desired. These traits can be emphasized in the individual boars selected from the third breed, although the reproduction and maternal traits should also receive attention.

Although limited research information is available on the use of crossbred boars, crossbred boars are apparently more aggressive breeders, have fewer problems in leg soundness, and improve overall breeding efficiency in comparison to straightbred boars. A crossbred boar could combine those traits that may not be available in one straightbred breed.

Hybrid boars sold by some commercial companies should not be confused with crossbred boars sold by private breeder concerns. Hybrid boars are developed from specific line crosses. These lines have been selected and developed for specific traits. When specific crosses are made, the hybrid boar must be used on specific cross females to maximize heterosis in their offspring.

Commercial pork producers have many selection tools, crossing systems, and genetic breeding stock sources for their use. Producers must capitalize on heterosis and breed strengths and must require complete performance records on all selected breeding stock. Although there is no one best system, breed, or source of breeding stock, producers must evaluate their total pork production program and integrate the most profitable combination of factors associated with a crossbreeding program.

SELECTED REFERENCES

Publications

Battaglia, R. A. and Mayrose, V. B. 1981. *Handbook of Livestock Management Techniques*. New York: Macmillan Publishing Co.

Briggs, H. M. and Briggs, D. M. 1980. *Modern Breeds of Livestock*. New York: Macmillan.

Guidelines for Uniform Swine Improvement Programs. 1981. USDA Program Aid 1157.

Johnson, R. K. 1981. Crossbreeding in swine: experimental result. *J. Anim. Sci.* 52:906.

Krider, J. L., Conrad, J. H. and Carroll, W. E. *Swine Production*. Hightstown, NJ: McGraw-Hill Book Co.

Lasley, J. F. 1987. *Genetics of Livestock Improvement*. Englewood Cliffs, New Jersey: Prentice-Hall.

National Hog Farmer, 1999 Shepard Road, St. Paul, Minnesota 55116.

Pond, W. G. and Maner, J. H. 1984. *Swine Production and Nutrition*. Westport, CT: AVI Publishing Co., Inc.

Pork Industry Handbook, 1980. Cooperative Extension Service. Oklahoma State University, Stillwater, Oklahoma.

Warwick, E. J. and Legates, J. E. 1979. *Breeding and Improvement of Farm Animals*. New York: McGraw-Hill.

Visuals

Breeding Herd Selection and Management (sound filmstrip). Vocational Education Productions, California Polytechnic State University, San Luis Obispo, CA 93407.

Swine Breed I.D. Slide Set. Vocational Education Productions, California Polytechnic State University, San Luis Obispo, CA 93407.

Swine Breed Identification, Female Selection in Swine, Boar Selection and *Swine Breeding Systems.* (sound filmstrips). Prentice-Hall Media, 150 White Plains Rd., Tarrytown, NY 10591.

The Swine Industry in the U. S., Breeds of Swine, Types of Breeding Programs, Selection of Breeding Stock, Low Cost Swine Breeding Units, How to Handle Newly Purchased Breeding Stock and Breeding Management. Videotapes. Agricultural Products and Services, 2001 Killbrew Drive, Suite 333, Bloomington, MN 55420.

CHAPTER 26

Feeding and Managing Swine

Swine production in the United States is concentrated heavily in the nation's midsection known as the Corn Belt. This area produces most of the nation's corn, which is the principal feed used for swine. The Corn Belt states of Iowa, Missouri, Illinois, Indiana, Ohio, northwest Kentucky, southern Wisconsin, southern Minnesota, eastern South Dakota, and eastern Nebraska produce nearly 75% of the nation's swine, with Iowa alone accounting for almost 25% (Fig. 26.1).

Swine operations have been getting larger over the past 25 years as the number of farms selling hogs has dropped from 2.1 million to 450,000. Farms selling more than 1,000 hogs annually have increased more than 300% in the last 25 years, and these large operations now account for 25–30% of all hogs sold. Most swine producers farm 200–500 acres of land and have other livestock enterprises.

TYPES OF SWINE OPERATIONS

There are four primary types of swine operations: (a) feeder pig production in which the producer maintains a breeding herd and produces feeder pigs for sale at an average weight of approximately 40 lb, (b) feeder pig finishing in which feeder pigs are purchased and then fed to slaughter weight, (c) farrow-to-finish in which a breeding herd is maintained where pigs are produced and fed to slaughter weight on the same farm, and (d) purebred or seedstock operations that are similar to farrow-to-finish except their salable product is primarily breeding boars and gilts.

Figure 26.2 shows the comparative economic returns of farrow-to-finish operations vs. feeder-to-finish operations. One system is not superior to the other and both have had profitable and unprofitable periods during 1985 and 1986.

FARROW-TO-FINISH OPERATIONS

Since farrow-to-finish is the major type of swine operation, primary emphasis will be given to it in this chapter. The information presented here is also pertinent to other operations, as most of them represent a certain phase of a farrow-to-finish operation.

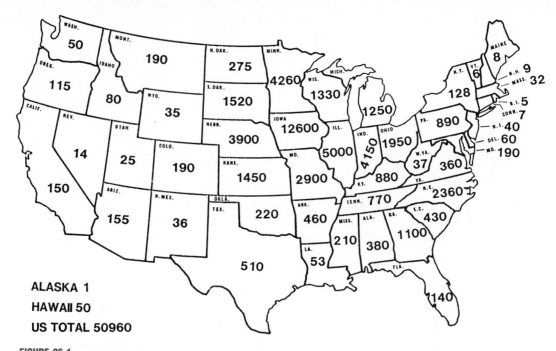

FIGURE 26.1.
Total number of hogs and pigs, 1986 (1,000 head). For example, Iowa is the leading state with 12.6 mil head. Courtesy of Western Livestock Marketing Information Project.

Selection and breeding practices to produce highly productive swine are presented in Chapter 25. The primary objective in the farrow-to-finish production phase is to feed and manage sows, gilts, and boars economically to assure a large litter of healthy, vigorous pigs from each breeding female.

Boar Management

Boars should be purchased at least 60 days before the breeding season so they can adapt to their new environment. A boar should be purchased from a herd that has an excellent herd health program and then isolated from the new owner's swine for at least 30 days. During this time, the boar can be treated for internal and external parasites if the past owner has not recently done this. The boar should be revaccinated for erysipelas and leptospirosis. Boars not purchased from a validated brucellosis-free herd should have passed a negative brucellosis test within 30 days of the purchase date. An additional 30-day period of fenceline exposure to the sow herd is recommended to develop immunities before the boar's use.

Younger boars (8–12 months of age) should be fed 5–6 lb of a balanced ration containing 14% protein, whereas older boars should receive 4–5 lb/day. The amount of feed should be adjusted to keep the boar in a body condition that is neither fat nor thin.

Young, untried boars should be test-mated to a gilt or two before the breeding season, as records show that approximately 1 of 12 young boars are infertile or sterile. The breed-

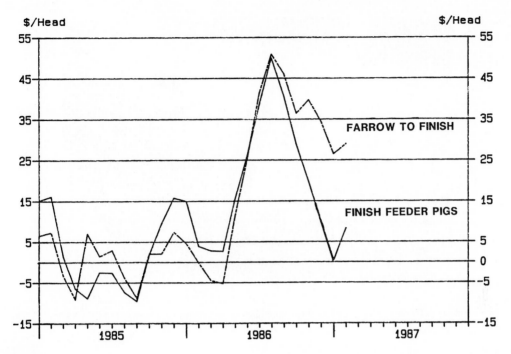

FIGURE 26.2.
Economic returns comparing two types of swine operations. Courtesy of Western Livestock Marketing Information Project.

ing aggressiveness, mounting procedure, and ability to get the gilts pregnant should be evaluated.

Successful breeding of a group of females requires an assessment of the number of boars required and the breeding ability of each boar. Generally, a young boar can pen-breed 8–10 gilts during a 4-week period. A mature boar can breed 10–12 females. Hand-mating (producers individually mate the boar to each female) takes more labor but extends the breeding capacity of the boar. Young boars can be handmated once daily, and mature boars can be used twice daily or approximately 10–12 times a week. Handmating can be used to mate heavier, mature boars with the use of a breeding crate. The breeding crate takes most of the boar's weight off the female. Breeding by handmating should correspond with time of ovulation (approximately 40 hours from the beginning of estrus) or litter size will be reduced.

Boars are to be considered dangerous at all times and handled with care. It is unsafe for a boar to have tusks as he may inflict injury on the handler or on other pigs. All tusks should be removed prior to the breeding season and every six months thereafter. Tusks can be removed with a bolt cutter.

Management of Breeding Females

Proper management of females in the breeding herd is necessary to achieve maximum reproductive efficiency. Gilts may start cycling as early as 5 months of age. It is recom-

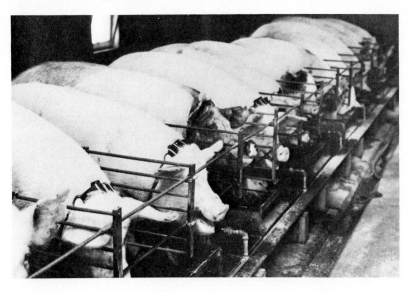

FIGURE 26.3.
Sows in individual feeding units to control the feed intake for each sow. Courtesy of the University of Illinois.

mended that breeding be avoided during their first heat cycle as the number of eggs ovulated is usually less than in subsequent heat cycles. Mixing pens of confinement-reared gilts and regrouping them with direct boar contact initiates early puberty.

Hot weather is detrimental to a high reproductive rate in the breeding females. High temperatures (above 85°F) will delay or prevent the occurrence of heat, reduce ovulation rate and increase early embryonic deaths. The animals will suffer heat stress when they are sick and have an elevated body temperature as well as when the environmental temperature is high.

Besides reducing litter size, diseases such as leptospirosis, pseudorabies, and those associated with stillbirths, mummified fetuses, and embryonic deaths may also increase the problem of getting the females pregnant. Attention to a good herd health program will assure a higher reproductive performance over the years.

It is important that sows and gilts obtain adequate nutrients during gestation to assure optimum reproduction. Feeding in excess is not only wasteful and costly but will likely increase embryonic mortality. A limited feeding system using balanced, fortified diets is recommended (Fig. 26.3). It assures that each sow gets her daily requirements of nutrients without consuming excess energy. As a rule of thumb, 4 lb of a balanced ration will provide adequate protein and energy; however, during cold weather an additional pound of feed may prove beneficial, especially for bred gilts. With limited feeding, it is important that each sow gets her portion of feed and no more. The individual feeding stall for each sow is best because it prevents the "boss sows" from taking feed from slower-eating or timid sows. During gestation, gilts should be fed so they will gain about 70–80 lb, and sows should gain approximately 30–40 lb.

Management of the Sow During Farrowing and Lactation

Farrowing is the process of the sow or gilt giving birth to her pigs. On the average, there are approximately 8.5 pigs farrowed per litter with approximately 7.2 pigs weaned per litter. Well-managed swine herds will have three to four more pigs per litter than these average figures. Extremely large litters are usually undesirable because the pigs are small and weak at birth and death losses can be high.

Proper care of the sow during farrowing and lactation is necessary to assure large litters of pigs at birth and at weaning. Sows should be dewormed about 2 weeks before being moved to the farrowing facility. The sow should be treated for external parasites, at least twice, a few days prior to farrowing time. The farrowing facility should be cleaned completely of organic matter, disinfected, and left unused for 5–7 days before sows are placed in the unit.

Before the sow is put in the farrowing pen or stall, her belly and teats should be washed with a mild soap and warm water. This helps eliminate bacteria that can cause diarrhea in nursing pigs.

Sows should be in the farrowing facility at the right time. Breeding records and projected farrowing dates should be used to determine the proper time. If farrowing is to occur in a crate or pen, the sow should be placed there no later than the 110th day of gestation. This gives some assurance the sow will be in the facility at farrowing time as the normal gestation is 111–115 days in length. Farrowing time can be estimated by observing an enlarged abdominal area, swollen vulva, and filled teats. The presence of milk in the teats indicates that farrowing will usually occur within 24 hours.

Before farrowing, while the sows are in the farrowing facilities, they can be fed a ration similar to the one fed during gestation. A laxative ration can be helpful at this time to prevent constipation. The ration can be made bland or bulky by the addition of 10 lb/ton of epsom salts, part of the protein as linseed meal, or the addition of oats, wheat bran, alfalfa meal, or beet pulp.

Sows need not be fed for 12–24 hours after farrowing, but water should be continuously available. Two or three pounds of a laxative feed may be fed at the first post-farrow feeding; the amount fed should be gradually increased until the optimum feed level is reached by 10 days after farrowing. Full feeding from the day of farrowing can be done successfully. Sows that are thin at farrowing may benefit from generous feeding early after farrowing. Sows nursing fewer than eight pigs may be fed a basic maintenance amount (6 lb/day) with an added amount, such as 0.5 lb for each pig being nursed. It is unnecessary to reduce feed intake before weaning. Regardless of level of feed intake, milk secretion in the udder will cease when pressure reaches a certain threshold level.

Baby Pig Management from Birth to Weaning

Most sows and gilts are farrowed in farrowing stalls or pens to protect baby pigs from the female lying on them (Fig. 26.4). Producers who give attention at farrowing will decrease the number of pigs that die during birth or within the first few hours after birth. Pigs can be freed from membranes, weak pigs can be revived, and other care can be given that will reduce baby pig death loss (Fig. 26.5). Manual assistance of the birth process

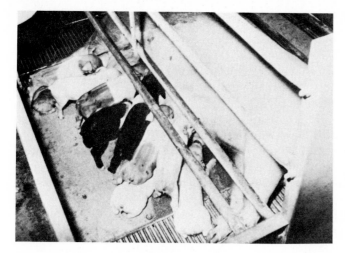

FIGURE 23.4.
Sow and litter of baby pigs in a farrowing crate. The farrowing crate is a pen (approximately 5 ft × 7 ft) that confines the sow to a relatively small area. The panel, a few inches off the floor, separates an area for the baby pigs where the sow cannot lay on them. There is ample space for the pigs to nurse the sow. Courtesy of the University of Illinois.

should not be given unless obviously needed. Duration of labor ranges from 30 min to more than 5 hours, with pigs being born either head or feet first. The average interval between births of individual pigs is approximately 15 min but can vary from almost simultaneously to several hours in individual cases. An injection of oxytocin can be given to speed the rate of delivery after one or two pigs have been born.

Manual assistance, using a well-lubricated gloved hand, can be used to assist difficult

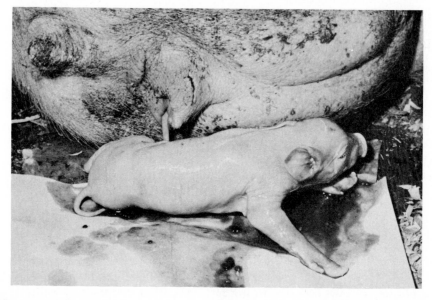

FIGURE 26.5.
A baby pig just farrowed with its umbilical cord still extending into the vulva. Proper attention by the swine producer at farrowing time can assure a higher number of pigs weaned per litter. Courtesy of R. W. Henderson, Oregon State University.

deliveries. The hand and arm should be inserted into the reproductive tract as far as is needed to locate the pig; the pig should be grasped and gently but firmly pulled to assist delivery. Difficult births often enhance the occurrence of the symptoms of "**MMA**"—**Mastitis** (inflammation of the udder), **Metritis** (inflammation of the uterus), and **Agalactia** (inadequate milk supply). Antibacterial solutions, such as nitrofurazone, are infused into the reproductive tract after farrowing, and sometimes intramuscular injections of antibiotics are used to decrease or prevent some of these infections.

It is very important that the newborn pig receive colostrum to give immediate and temporary protection against common bacterial infections. Pigs from extremely large litters can be transferred to sows having smaller litters to equalize the size of all litters. Care should be given to ensure that baby pigs receive adequate colostrum before the transfer takes place.

Air temperature in the farrowing facility should be 70–75°F with the creep area for baby pigs equipped with zone heat (heat lamps, gas brooders, or floor heat) having a temperature of 85–95°F. Otherwise the pigs will chill and die or become susceptible to serious health problems. Moisture should be controlled or removed from the farrowing facility without creating draft on the pigs.

Management of environmental temperature from birth to 2 weeks of age is critical. At 2 weeks of age, the baby pig has developed the ability to regulate its own body temperature.

Soon after birth, the navel cord should be cut to 3–4 in. from the body and then treated with iodine. This disinfects an area which could easily allow bacteria to enter the body and cause joints to swell and abscesses to occur.

The producer should also clip the 8 sharp needle teeth of the baby pig. This prevents the sow's udder from being injured and facial lacerations from occurring when the baby pigs fight one another. Approximately one-half of each tooth can be removed by using side-cutting pliers or toenail clippers.

Good records are important to the producer interested in obtaining optimum production efficiency. The basis of good production records is pig identification. Ear notching pigs at 1–3 days of age with a system shown in Fig. 26.6 provides positive identification for the rest of the pig's life.

Baby pig management from 3 days to 3 weeks of age includes anemia and scour control, castration, and tail docking. Sow's milk is deficient in iron so iron dextran shots are given intramuscularly at 3–4 days of age and again at 2 weeks of age (Fig. 26.7).

Baby pig scours are ongoing problems for the swine producer. Colostrum and a warm, dry, draft-free, and sanitary environment are the best preventative measures against scours. Orally-administered drugs determined to be effective against specific bacterial strains have been found to be the best control measures for scours. Serious diarrhea problems resulting from diseases such as **transmissible gastroenteritis (TGE)** and swine **dysentery** should be treated under the direction of a local veterinarian.

Castration is usually done before the pigs are 2 weeks old. Pigs castrated at this age are easier to handle, heal faster, and are not subjected to as much stress as those castrated later. The use of a clean, sharp instrument, low incisions to promote good drainage, and using antiseptic procedures make castration a simple operation.

Tail docking has become a common management practice to prevent tail biting during the confinement raising of pigs. Tails are removed $\frac{1}{4}$–$\frac{1}{2}$ in. from the body, and the stump should be disinfected along with the instrument after each pig is docked.

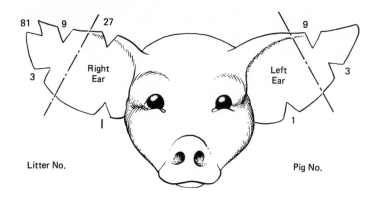

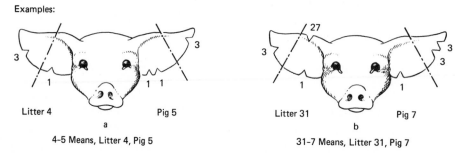

Examples:

Litter 4   Pig 5
a
4-5 Means, Litter 4, Pig 5

Litter 31   Pig 7
b
31-7 Means, Litter 31, Pig 7

FIGURE 26.6.
An ear notching identification system for swine. Individual pig identification is necessary for meaningful production and management records. Courtesy of Dennis Giddings.

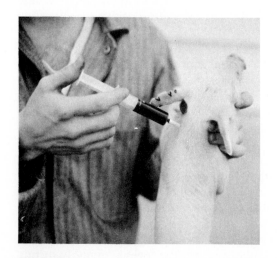

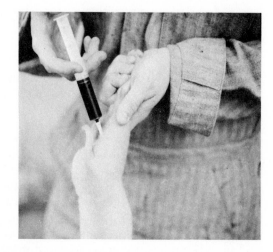

FIGURE 26.7.
Iron shots being given to baby pigs. These injections can be given in either the ham or neck muscles. Courtesy of E. R. Miller, Michigan State University.

FIGURE 26.8.
Baby pigs in a modern nursery unit. The pigs were weaned from the sow and placed into these pens when they weighed 15 to 20 lb. The pigs will be moved into pens for growing and finishing when they weigh approximately 40 lb. Courtesy of the University of Illinois.

Pigs should be offered feed at 1–2 weeks of age in the creep area. These starter feeds can be placed on the floor or in a shallow pan. By 3–4 weeks after farrowing, the sow's milk production has likely peaked; however, the baby pigs should be eating the supplemental feed and growing rapidly (Fig. 26.8).

Pigs can be weaned from their dams at a time that is consistent with the available facilities and the management of the producer. Pigs weaned at young ages should receive a higher level of management. General weaning guidelines are: wean only pigs over 12 lb, wean pigs over a 2- to 3-day period, wean the heavier pigs in the litter first, group pigs according to size in pens of 30 pigs or less, provide one feeder hole for 4–5 pigs and one waterer for each 20 to 25 pigs, and limit feed for 48 hours and use medicated water if scours develop.

Feeding and Management from Weaning to Market

A dependable and economical source of feed is the backbone of a profitable swine operation. Since approximately 55–75% of the total cost of pork production is feed, the swine producer should be keenly aware of all aspects of swine nutrition.

The pig is an efficient converter of feed to meat. With today's nutritional knowledge, modern meat-type hogs can be produced with a feed efficiency of 3.3 lb of feed per pound of gain from 40 lb to market. To obtain maximum feed use, it is necessary to feed well-balanced rations designed for specific purposes.

Swine rations for growing-finishing market pigs, as well as rations for breeding stock, are formulated around cereal grains that are the largest component of swine rations. The most common cereal grains that provide the basic energy source are corn, milo, barley, wheat, and their byproducts. Because of its abundance and readily available energy, corn is used as the base cereal for comparing the nutritive value of other cereal grains. Milo (grain sorghum) is very similar in nutritional content to corn and can completely replace corn in swine rations. Its energy value is about 95% the value of corn. Barley contains more protein and fiber than corn, but its relative feeding value is 85–95% of corn and it is less palatable than corn. Wheat is equal to corn in feeding value and is very palatable. However, for best feeding results, wheat should be mixed half and half with some other grain. Which grain or combination of grains to use can best be determined by availability and the relative costs.

Grinding the grains improves all grains for feeding, especially those high in fiber, such as barley and oats. Pelleting the ration may increase gains and efficiency of gains from 5–10%. However, advantages of pelleting are usually offset by the higher cost of pelleting the ration.

Cereal grains usually contain lesser quantities of proteins, minerals, and vitamins than swine require; therefore, rations must be supplemented with other feeds to increase these nutrients to recommended levels. Soybean meal has been proven to be the best single source of protein for pigs when palatability, uniformity, and economics are considered. Other feeds rich in protein (meat and bone meal or tankage) can compose as much as 25% of the supplemental protein, if economics so dictate. Cottonseed meal is not recommended as a protein source unless the gossypol has been removed. **Gossypol,** in relatively high levels, is toxic to pigs. Table 26.1 shows recommended protein levels in corn-soybean oil meal diets for different weights of pigs being grown and finished for slaughter.

If economics dictate, a producer can supplement specific amino acids in place of some of the protein supplement. Typically, lysine may be supplemented separately because lysine is the most limiting amino acid in swine rations.

Calcium and phosphorus are the minerals most likely to be deficient in swine rations, so care should be exercised to assure adequate levels in the ration. The standard ingredients for supplying calcium and phosphorus in the swine diet are limestone and either dicalcium phosphate or defluorinated rock phosphate. If an imbalance of calcium and zinc exists, a skin disease called **parakeratosis** is likely to occur.

Swine producers may formulate their swine rations several ways: (a) using grain and a complete swine supplement, (b) using grain and soybean meal plus a complete base mix carrying all necessary vitamins, minerals, and antibiotics, or (c) using grain, soybean meal, vitamin premix, trace mineral premix, calcium and phosphorus source, and antibiotic and salt mixed together. Which ration is used depends on the producer's expertise,

TABLE 26.1. Percent Protein Requirement for Varying Weights of Pigs from Weaning to Slaughter

Period and Weight	Percent Protein in Ration
Starter	
10 to 40 lb	18–20
Grower	
40 to 100 lb	16
Finisher I	
100 to 160 lb	14
Finisher II	
160 to 240 lb	13

costs, and availability of ingredients. The trace mineral premix typically includes vitamin A, vitamin D, vitamin E, vitamin B_{12} and other B vitamins (niacin, pantothenic acid, and choline).

Feed additives are used by most swine producers because the additives have the demonstrated ability to increase growth rate, improve feed efficiency, and reduce death and sickness from infectious organisms. Feed additives available to swine producers fall into three classifications: (a) **antibiotics** (compound coming from bacteria and molds that kill other microorganisms, (b) **chemotherapeutics** (compounds that are similar to antibiotics but are produced chemically rather than microbiologically, and (c) **anthelmintics** (dewormers).

Feed additive selection and the level to be used will vary with the existing farm environment, management conditions, and stage of the production cycle. Disease and parasite levels will vary from farm to farm. Also, the first few weeks of a pig's life are far more critical in terms of health protection. Immunity acquired from colostrum diminishes by 3 weeks of age, and the pig does not begin producing sufficient amounts of antibodies until 5–6 weeks of age. Also during this time, the pig is subjected to stress conditions of castration, vaccinations, and weaning. Therefore, at this age the additive level of the feed is much higher than when the pig is more than 75 lb and growing rapidly towards slaughter weight. Producers should be cautious because some additives have withdrawal times which means the additive should no longer be included in the ration. This is necessary to prevent additives from occurring as residues in meat.

MANAGEMENT OF PURCHASED FEEDER PIGS

Feeder pigs that are marketed one producer to another producer are subjected to more stress than pigs produced in a farrow-to-finish operation. These stresses are fatigue, hunger, thirst, temperature changes, ration changes, different surroundings, and social problems. Almost every group has a shipping fever reaction. Care of newly arrived pigs must be directed to relieving stresses and treating shipping fever correctly and promptly. Management priorities for newly purchased feeder pigs should be: (a) a dry, draft-free, well-bedded barn or shed with no more than 50 pigs per pen, (b) a specially formulated starter ration, having 12–14% protein, more fiber, and a higher level of vitamins and antibiotics than a typical starter ration, (c) an adequate intake of medicated water containing sulpha products and electrolytes or water-soluble antibiotics and electrolytes, and (d) prompt and correct treatment of any sick pigs.

MARKET MANAGEMENT DECISIONS

Slaughter weights of market hogs and which market is selected can have a marked effect on income and profitability of the swine enterprise. For many years the recommended weight to sell barrows and gilts was, in most instances, 200–220 lb. The primary reasons for selling these animals at a maximum weight of 220 lb were increased cost of gain and increased fat accumulations at heavier weights. Market discounts were common on pigs weighing over 220 lb because of the increased amount of fat. Today swine producers can maximize income by marketing pigs weighing up to 250 lb. These heavier pigs are leaner, and they are not subject to the same market discounts of fatter pigs marketed several

FIGURE 26.9.
A group of pigs in a growing and finishing production unit. The building is totally enclosed and the floor is totally slatted. The manure from the pigs drops through the slats into a 36-inch deep pit. The waste is removed from the pit through a water method. Courtesy of the University of Illinois.

years ago (Fig. 26.9). Decisions to market pigs at heavier weights should be based on feed costs and the gaining ability of the pigs.

Prices for slaughter pigs can vary among markets. Also marketing costs, such as selling charges, transportation, and shrink (loss of live weight) can also be different. If more than one market is available, producers should occasionally patronize different markets as a check against their usual marketing program. No single market is continually and consistently the best market.

SELECTED REFERENCES

Publications

Krider, J. L., Conrad, J. H. and Carroll, W. E. 1971. *Swine Production.* 5th edition. Hightstown, NJ: McGraw-Hill.

National Hog Farmer. 1976. *Questions and Answers on Managing Swine: Hog Information Please,* Vol. III, St. Paul: National Hog Farmer.

Nutrient Requirements of Swine. 1979. National Research Council. Washington, D. C.: National Academy Press.

Pond, W. G. and Maner, J. H. 1984. *Swine Production and Nutrition.* Westport, CT: AVI Publishing Co., Inc.

Pork Industry Handbook. 1980. Cooperative Extension Service, Oklahoma State University, Stillwater, OK.

Van Arsdall, R. N. 1978. *Structural Characteristics of the United States Hog Production Industry.* Agricultural Economic Report No. 415. USDA.

Visuals

The Swine Management Series (Separate Sound Filmstrips) covering: *Introduction to Swine Management; Brood Sow and Litter; Swine Health Care; Swine Nutrition; Fitting and Showing Swine; Feeder Pig*

Selection and *Swine Nutrition*. Vocational Education Productions, California Polytechnic State University, San Luis Obispo, CA 93407.

Feeding Equipment Choices, Swine Manure Handling Systems, Environmental Control in Swine Buildings, Swine Buildings and Equipment: Material Alternatives, Swine Building Systems: Labor and Cost Comparisons, Gestation Management and Health, Sow Management During Farrowing, Baby Pig Management, Farrowing House Equipment, Farrowing Systems, Swine Nursery Facilities, Purchasing Feeding Pigs/Transporting Feeder Pigs, Facilities for a Feeder-to-Finish Swine Enterprise/Nutrition, Health, and Management of a Feeder-to-Finish Swine Enterprise, Personal Capabilities of Feeder Pig Producers. Videotapes. Agricultural Products and Services, 2001 Killebrew Drive, Suite 333, Bloomington, MN 55420.

CHAPTER 27

Sheep Breeds and Breeding

THE SHEEP INDUSTRY

World

Sheep and goats are important ruminants in temperate and tropical agriculture. They provide wool, milk, hides, and meat for people in many parts of the world. Sheep and goats are better adapted to arid tropics than cattle; this is likely due to their superior water and nitrogen economy. Sheep and goats also produce useful fibers as well as meat and milk, which gives them an advantage on small farms in developing countries. Cattle, sheep, and goats often are grazed together because they utilize different plants. Goats graze browse and some forbs, cattle graze tall grasses and some forbs, and sheep graze short grasses and some forbs. Most of the forbs in many grazing areas are broad-leaf weeds.

The population of sheep in the world is 1,140 mil head (Table 27.1) which is 2.5 times the number of goats (460 mil). Australia and the U.S.S.R. each have approximately 140–145 mil sheep. The major sheep producing areas are within the 35–55°N latitudes in Europe and the Near East and between the latitudes of 30–45°S in South America, Australia, and New Zealand. There is a lesser area of sheep concentration between 5 and 35°N in India, the Near East, and the highlands of Eastern Africa.

More than 60% of all sheep are in temperate zones and less than 40% are in tropical zones. Goats, on the other hand, are mostly (80%) in the tropical or subtropical zones (0–40°N). Temperature and type of vegetation appear to be primary factors that encourage sheep production in temperate zones and goat production in tropical zones.

Sheep apparently evolved in the dry mountainous areas of southwest and central Asia. Present-day domesticated sheep were derived from wild animals that existed in Asia some 8,000–10,000 years ago. It is believed that domestication occurred during the early civilizations of southwest Asia. The Romans introduced a white, wooled sheep into western Europe, and these sheep contributed heavily to the British breeds. Some 100–200 years

TABLE 27.1.   World Sheep Numbers

Area or Country	Mil Head	Area or Country	Mil Head
Africa	188.6	Europe	145.6
South Africa	31.3	England	34.8
Ethiopia	23.4	Romania	18.4
North and Central America	20.1	Oceania	209.6
U.S.	11.4	Australia	139.2
Mexico	6.4	New Zealand	70.3
South America	107.4	U.S.S.R.	145.3
Uruguay	23.3	Developed Countries	544.2
Brazil	17.5	Developing Countries	595.3
Asia	322.9	World Total	1,139.5
China	98.9		
Turkey	48.7		

Source: 1984 FAO Production Yearbook.

ago, Europeans took the Merino and British breeds of sheep to South America, Australia, South Africa, and New Zealand.

As sheep originated in a dry, alternately hot-and-cold climate of southwest Asia, adaptive changes for sheep to succeed in tropical areas were ability to lose heat, to resist diseases, and to survive under an adverse nutritional environment. For sheep to lose heat easily, they need a large body surface-to-mass ratio. Such a sheep is a small, long-legged animal. Also, a hairy coat allows ventilation and also protects the skin from the hot sun and from abrasions. In temperate areas, sheep with large, compact bodies and a heavy fleece covering and storage of subcutaneous fat had an advantage. Sheep can also constrict or relax blood vessels to the face, legs, and ears for control of heat loss.

The productivity of sheep in temperate areas is much greater than that of sheep raised in tropical environments. This difference results from a more favorable environment (temperature and feed supply) and also a greater selection emphasis for growth rate, milk production, lambing percentage, and fleece weight. Sheep from temperate environments do not adapt well to tropical environments; therefore, it may be more effective to select sheep in their production environment rather than introduce sheep from other different environments.

Sheep contribute mutton and lamb to the total meat supply in Asia (1.4 mil tons), Europe (1.03 mil tons), Oceania (1.61 mil tons), and the U.S.S.R. (0.96 mil tons). Annual per capita meat consumption in Oceania is 110 lb, with most of the meat supplied by sheep.

Sheep make significant contributions of milk to the people of Europe (3.5 mil tons), Asia (3.1 mil tons), and Africa (0.6 mil tons). Milk contributed by sheep and goats combined is 23 lb per person per year for Europe. Sheep contribute essentially no milk for human use in the U.S.

Sheep skins are an important contribution of sheep in Asia, Europe, Oceania, Africa, and U.S.S.R. but not in the U.S. Wool production is very important in New Zealand and

Australia (Oceania), which produce 1.0 mil tons annually. Other important areas in wool production are the U.S.S.R., South America, Asia, Europe, and Africa.

United States

The number of sheep in the U.S. reached a maximum of 56 million in 1942, after which ensued a steady decline to a level of 10 million in 1985 (Fig. 27.1). Although the price received for lambs increased from 15¢ to 70¢ per pound between 1961 and 1985, this increase has been largely offset by increased production costs, scarcity of good sheep herders, and heavy losses from disease, predators, and adverse weather.

The numbers of ewes on farms (1985) are presented in Fig. 27.2. The leading states based on ewe numbers are Texas (1.26 mil), California (0.72 mil), Wyoming (0.57 mil), South Dakota (0.41 mil), and Utah (0.40 mil). The midwest and west coast states are important in farm flock sheep production, whereas the other western states are range flock areas. The midwestern states generally raise more lambs per ewe than the western states, although Washington, Oregon, and Idaho resemble the midwestern states in this regard.

Most of the decline in U. S. sheep numbers during the past 50 years has occurred in the West. However, this region, with its extensive public and private range lands, still produces 80% of the sheep raised in this country today. Most U. S. sheep growers have small flocks (50 or fewer sheep) and most raise sheep as a secondary enterprise. About

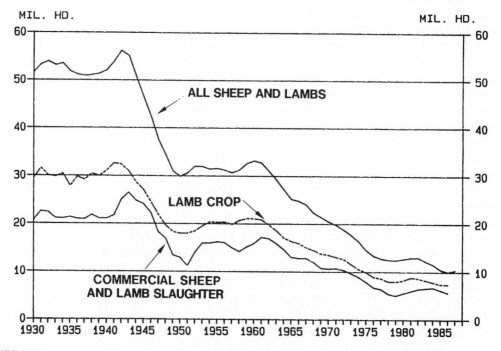

FIGURE 27.1.
Sheep and lamb numbers in the United States, 1930–1985. Courtesy of the Western Livestock Marketing Information Project.

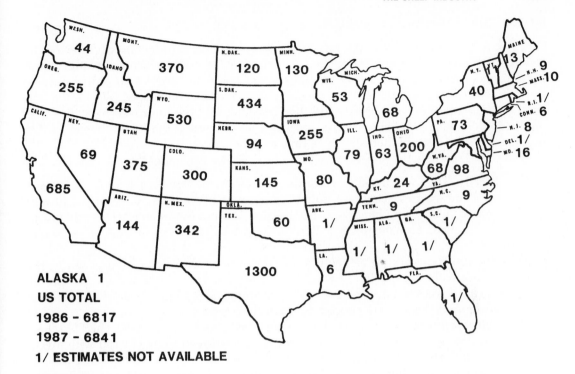

FIGURE 27.2.
Ewe numbers, by state, January 1, 1987. Courtesy of the Western Livestock Marketing Information Project.

40% of the West's sheep producers maintain flocks of more than 50 sheep and these flocks contain about 93% of the sheep in that region. Only about one-third of the operators in the West who have flocks of 50 or more sheep specialize in sheep, whereas the other two-thirds have diversified livestock operations.

Sheep are produced on farms in which pastures are generally good and where meadow aftermaths, cereal grains, and grass grown for seed can provide nutritional needs at certain times. Sheep are also produced under range conditions where vegetation is sparse and much of the feed is **forbs** (broad leaf plants) and **browse.** Farm-flock operations and range operations each have their own unique problems and methods.

About 23% of all sheep born in the western United States and 13% of those born in the north-central region of this country are lost before they are marketed. In the West, predators, especially coyotes, and weather are the most important causes of death. In the north-central region, the most important causes are weather and disease before **docking,** and disease and attacks by dogs after docking.

The range operator produces some slaughter lambs if good mountain range is available, but most range-produced lambs are feeders (lambs that must have additional feed before they are slaughtered). Feeder lambs are either fed by the producers, sold to feedlot operators, or fed on contract by feedlot operators. Colorado feeds more lambs for finishing than any other state. Some feedlots have a capacity for 10,000 or more lambs on feed at one time, and many feedlots that have a capacity of 1,000 are in operation.

MAJOR SHEEP BREEDS IN THE U.S.

Sheep have been bred in the past for three major purposes: production of fine wool for making high-quality clothing; production of long wool for making heavy clothing, upholstering, and rugs; and production of mutton and lamb. In recent years, dual-purpose sheep breeds (which produce both wool and meat) have been developed by crossing fine-wool breeds with long-wool breeds and then selecting for both improved meat and wool production. Some breeds serve specific purposes; an example is the Karakul breed, which supplies pelts for clothing items such as caps and Persian coats.

Scott, 1982, divides the sheep breeds into three major categories:

1. **Ewe Breeds.** They are usually white faced, fine or medium wool, or long wool or crosses of these types. They are noted for reproductive efficiency, wool production, size, milking ability, and longevity. The Delaine Merino, Rambouillet, Debouillet, Corriedale, Targhee, **Finnsheep,** and Border Leicester are breeds in the "Ewe Breed" category.
2. **Ram Breeds.** These are meat-type breeds. They are raised primarily to produce rams for crossing with ewes of the "Ewe Breed" category. They are noted for growth rate and carcass characteristics. The Suffolk, Hampshire, Shropshire, Oxford, Southdown, Montadale, and Cheviot are classed as "Ram-Breeds."
3. **Dual Purpose Breeds** are used either as ewe breeds or as ram breeds. The Dorset is a good example of a breed that is often used as a ewe breed to be crossed with a good "ram breed." The Columbia, Dorset, Lincoln, and Romney are often used as "ram breed" to cross on other breeds of ewes to improve milking ability and fertility of the ewe flock.

Characteristics

Breeds of sheep and their characteristics are listed in Table 27.2. The sheep breeds commonly used in the United States are shown in Figs. 27.3 and 27.4. The Merino and Rambouillet (fine wool) breeds and the dual-purpose breeds that possess fine-wool characteristics have a herding instinct, so one sheepherder with dogs can handle a band of 1,000 ewes and their lambs on the summer range. Winter bands of 2,500–3,000 ewes are common. Ram breeds, by contrast, tend to scatter over the grazing area. They are highly adaptable to grazing fenced pastures where feed is abundant. Rams of the ram breeds (meat type) are often used for breeding ewes of fine-wool breeding on the range for the production of market lambs. Crossbreeding in sheep has been used for more than a half century.

The Columbia and Targhee breeds were developed by the U. S. Sheep Experiment Station, Dubois, Idaho. Both breeds have proven useful on western ranges because they have the herding instinct and when bred to a meat breed of ram they raise better market lambs than do fine-wool ewes. The U. S. Sheep Experiment Station has made great strides in improving the Rambouillet, Columbia, and Targhee breeds. These three breeds were faulted by having skin folds, or wrinkles, that made shearing extremely difficult, and by **wool blindness** (wool covers the face preventing the sheep from seeing). Wool blindness prevented the sheep from observing dangers and interfered with their ability to locate

TABLE 27.2.   Characteristics of Breeds of Sheep Within Each Type.

Breed	Size[a]	Carcass Conformation	Wool Fineness	Wool Length[b]	Fleece Weight[c]	Color	Horns or Polled	Other
Fine-wool breeds								
Merino	Small to medium	Poor	Very fine	Medium	Heavy	White	Rams horned, ewes polled	Good herding instinct, skin folds
Rambouillet	Large	Medium	Fine	Medium	Heavy	White	Rams horned, ewes polled	Good herding instinct
Debouillet	Medium	Good	Fine	Medium	Heavy	White	Rams horned, ewes polled	Good herding instinct
Long, coarse-wool breeds								
Romney	Medium large	Medium	Coarse	Long	Heavy	White	Polled	Lambs do not fatten at small size
Lincoln	Large	Good	Coarse	Long	Heavy	White	Polled	Lambs do not fatten at small size
Leicester	Large	Good	Coarse	Long	Heavy	White	Polled	Lambs do not fatten at small size
Cotswold	Large	Good	Coarse	Long	Heavy	White	Polled	Lambs do not fatten at small size
Ram (meat) breeds								
Suffolk	Large	Excellent	Medium	Short	Light	White with black bare face and legs	Polled	Bare bellies; black fibers
Hampshire	Large	Excellent	Medium	Medium	Medium	White with black face	Polled	Wool blindness, good milkers, black fibers
Shropshire	Medium to large	Good	Medium	Medium	Medium	White, dark face and legs	Polled	Wool blindness, excellent milkers
Southdown	Very small	Excellent	Medium	Short	Very light	White with brown face and legs	Polled	Used in hot house lamb production

429

TABLE 27.2. Characteristics of Breeds of Sheep Within Each Type (Continued)

Breed	Size[a]	Carcass Conformation	Wool Fineness	Wool Length[b]	Fleece Weight[c]	Color	Horns or Polled	Other
Dorset	Medium	Good	Medium	Medium	Medium	White	Polled or horned	Highly fertile; good milkers
Cheviot	Small	Excellent	Medium	Medium	Medium	White	Polled	Very rugged
Oxford	Large	Excellent	Medium	Medium	Medium to heavy	White with brown face and legs	Polled	
Tunis	Medium	Medium	Medium	Medium	Medium	Red or tan face	Polled or horned	
Finnsheep	Small to medium	Poor	Medium	Medium	Medium	White	Polled	Very fertile
Dual-purpose breeds								
Corriedale	Medium	Good	Medium	Medium long	Heavy	White	Polled	Herding instinct
Columbia	Large	Good	Medium	Medium long	Heavy	White	Polled	Rugged, herding instinct
Targhee	Medium to large	Medium to good	Medium to fine	Medium	Heavy	White	Polled	Herding instinct
Panama	Large	Good	Medium	Medium long	Heavy	White	Polled	
Romeldale	Medium to large	Good	Medium	Medium	Heavy	White	Polled	
Montadale	Medium to large	Good	Medium	Medium	Medium	White	Polled	
Miscellaneous breeds								
Navajo	Medium	Poor	Coarse	Long	Medium	Variable	Polled or horned	Wool for making Navajo rugs
Karakul	Large	Poor	Coarse	Long	Medium	Black or brown	Rams horned, ewes polled	Used for pelts

[a]Small: ewes weigh 120–160 lb and rams weigh 160–200 lb; medium: ewes 140–180 lb and rams 180–250 lb; large: ewes 150–200 lb and rams 225–350 lb.

[b]Short: 2.0–3.5 in; medium: 2.5–4.5 in.; long: 4–6 in. or longer.

[c]Light: 5–8 lb; medium: 6–10 lb; heavy: 10–18 lb.

Dorset

Montadale

Polled Dorset

Cheviot

Hampshire

Suffolk

FIGURE 27.3.
Some breeds of sheep commonly used either as straightbreds or for crossbreeding to produce market lambs. Courtesy of Continental Dorset Club, Hudson, IA (Dorset and Polled Dorset); Montadale Sheep Breeders' Association, Indianapolis, IN (Montadale); American Cheviot Sheep Society, Lebanon, VA (Cheviot); American Hampshire Sheep Association, Columbia, MO (Hampshire); and National Suffolk Sheep Association, Logan, UT (Suffolk).

Rambouillet

Cotswold

Lincoln

Romney

Targhee

Finnsheep

FIGURE 27.4.
Some breeds of sheep commonly used either as straightbreds or for crossbreeding to produce market lambs. Courtesy of American Rambouillet Sheep Breeders Association, San Angelo, TX (Rambouillet); American Cotswold Record Association, Rochester, NH (Cotswold); National Lincoln Sheep Breeders' Association, West Milton, OH (Lincoln); American Romney Breeders Association, Corvallis, OR (Romney); USDA, ARS, Western Region, U.S. Sheep Experiment Station, Dubois, ID (Targhee); and Animal Science Department, University of Minnesota (Finnsheep).

forage. Thus, wool-blind ewes wean lambs that weigh less than those produced by open-faced ewes.

Selection against wool blindness and skin folds has been accomplished by using subjective scores because there was no way to measure these traits more objectively. After several years of selection, improvement has been noted in sheep having fewer skin folds and the sheep being more **open faced** (less wool around the eyes).

The most important trait in meat breeds of sheep is weight at 90 days of age, with some emphasis also given to conformation and finish of the lambs. Weight at 90 days of age can be calculated for lambs that are weaned at younger or older ages than 90 days as follows:

$$\text{90-day weight} = \frac{\text{weaning weight} - \text{birth weight}}{\text{age at weaning}} \times 90 + \text{birth weight}$$

If a lamb weighs 10 lb at birth and 100 lb at 100 days, the 90-day weight is:

$$\frac{100 \text{ lb} - 10 \text{ lb}}{100 \text{ days}} \times 90 \text{ days} + 10 \text{ lb} = 91 \text{ lb}$$

If birth weight was not recorded, an assumed birth weight of 8 lb can be used.

Approximately 85–90% of the total income from sheep of the meat breeds is derived from the sale of lambs, with only 10–15% coming from the sale of wool. By contrast, the sale of wool accounts for 30–35% of the income derived from fine-wool and long-wool breeds. Even with wool breeds, the income from lambs produced constitutes the greater portion of total income.

The U. S. Sheep Experiment Station, Dubois, Idaho, has developed a new synthetic breed of sheep, called the **Polypay,** that is superior in lamb production and in carcass quality. Four breeds (Dorset, Targhee, Rambouillet, and Finnsheep) provided the basic genetic material for the Polypay. Targhees were crossed with Dorsets and Rambouillets were crossed with Finnsheep, after which the two cross-bred groups were crossed. The offspring produced by crossing the two crossbred groups were intermated and rigid selection program was practiced. This population was closed to outside breeding. The Rambouillet and Targhee breeds contributed hardiness, herding instinct, size, a long breeding season, and wool of high quality. The Dorset contributed good carcasses, high milking quality, and a long breeding season. The Finnsheep contributed early puberty, early postpartum fertility, and high lambing rate.

This breed has shown outstanding performance in conventional once-a-year lambing under range conditions and superior performance in twice-a-year lambing when compared with other breeds or breed crosses.

There are several U. S. breeds of sheep that are not listed in Table 27.1. Some of these breeds not listed possess certain genes that may be useful for improving the most popular U. S. breeds either through systematic crossbreeding or in the establishment of new breeds.

Popularity of Breeds

Based on registration numbers, Suffolk and Hampshire are the most numerous breeds of sheep (Table 27.3). These two breeds provide rams that are used extensively in crossbreeding programs to produce market lambs both in farm flocks and on the range. The

TABLE 27.3.   Major Sheep Breeds Based on Annual Registration Numbers

	U.S. Breed Associations			Canadian Breed Associations
Breed	1985[a]	1970	Date Association Formed	1985[b]
Suffolk	80,553	38,920	1929	5,232
Hampshire	16,892	22,556	1889	813
Dorset	13,932	9,336	1898	2,262
Rambouillet	11,026	4,500	1889	
Columbia	8,007	6,442	1942	
Southdown	4,776	5,677	1882	
Shropshire	3,699	4,689	1884	
Polypay	3,249	0	1979	
Montadale	3,006	3,000	1945	
Cheviot	2,682	2,921	1891	757

[a]Only breeds with more than 2,500 annual registrations are listed.
[b]Only breeds with more than 500 annual registrations are listed.
Source: 1985 North American Livestock Census.

most popular ewe breeds on the range are Rambouillet, Columbia, and Corriedale. In a farm flock lamb production system, the Dorset, Shropshire, Finnsheep, and Polypay are important ewe breeds.

BREEDING SHEEP

Reproduction

Sheep differ from many farm animals in having a breeding season that occurs mainly in the Fall of the year. The length of the breeding season varies with breeds. Ewes of breeds with a long breeding season show heat cycles from mid to late Summer until mid-Winter. The breeds in this category are Rambouillet, Merino, and Dorset. Breeds with an intermediate breeding season (Suffolk, Hampshire, Columbia, and Corriedale) start cycling in late August or early September and continue to cycle until early winter. Ewes of breeds having long or intermediate breeding seasons are more likely the ones that will fit into a program of three lamb crops in 2 years.

Ewes of breeds with short breeding season (Southdown, Cheviot, and Shropshire) do not start cycling until early fall and discontinue cycling at the end of the fall period.

Puberty is reached at 5–12 months of age and is influenced by breed, nutrition, and date of birth. The average length of **estrous** cycles is slightly over 16 days, and the length of estrus (when the ewe is receptive to the ram) is about 30 hours. The length of gestation (time from breeding until lamb is born) averages 147 days but varies; medium-wooled and meat breeds have shorter a gestation period whereas fine-wooled breeds have longer gestation periods.

Several factors affect **fertility** in sheep. Lambing rates vary both within and between breed, with some ewes consistently producing only one lamb per year and others producing three or four lambs each year. To improve lambing rate, one needs to keep replacements that are produced by ewes that consistently produce two to four lambs each year.

Fertility of rams used in a breeding program should be checked by examining the semen. Semen can be collected by use of an artificial vagina and examined with a microscope. Ram fertility is evaluated by: (a) checking for abnormal sperm (tailless, bent tails, no heads, etc.); (b) observing the percent live sperm (determined by a staining technique of fresh semen); and (c) by checking the sperm motility and concentration in freshly collected semen. Two semen collections 3–4 days apart should be examined. When rams have not ejaculated for some time, there may be dead sperm in the **ejaculate.**

Other Factors Affecting Reproduction

Selection and Crossbreeding.   Crossbred ewe lambs, when adequately fed, are usually bred to lamb at 1 year of age. Generally crossbred lambs are more likely to conceive as lambs than purebred lambs.

Age.   Mature (3–7 year old) ewes are more fertile and raise a higher percentage of lambs born than younger or older ewes.

Light.   Light, temperature, and relative humidity affect reproduction in sheep. Sheep respond to decreased day lengths both by showing greater proportion of ewes in estrus and by higher **conception** rates.

Temperature.   Temperature has a marked effect on both ewes and rams. High temperatures cause heat sterility in rams because the testicles must be at a temperature below normal body temperature for viable sperm production. Embryo survival in the ewe is also influenced by temperature. When ambient temperature exceeds 90°F, embryo survival decreases during the first 8 days after breeding.

Health.   Environmental factors affecting the health and well-being of sheep, such as disease, **parasites,** lack of feed, or imbalance of the ration will reduce the number of lambs produced. The producer can increase his income and profit from a sheep operation by controlling diseases and parasites and seeing that the sheep have an adequate supply of good feed. Sheep in moderate body condition are usually more productive than fat sheep. Ewes that are in moderate condition and are gaining weight before and during the breeding season will have more and stronger lambs than ewes that are either overfat or **emaciated** at breeding.

Estrus Synchronization and AI.   In some intensively managed sheep operations, hormones can be used to synchronize estrus. Hormones can also be used to bring ewes into estrus at times other than during their normal breeding season.

Estrus can be synchronized if progesterone is given for a 12–14 day period either in the feed, as an implant, a pessary (intravaginal device), or by daily injections followed by injections of pregnant mare serum (PMS) the day progesterone is discontinued. Conception rates at this estrus are low; therefore a second injection of pregnant mare serum 15–17 days later will bring the ewes into estrus and fertile matings will occur. There must be sufficient ram power to breed a large number of ewes if they are synchronized or artificial insemination can be used.

This system of synchronizing estrus can also be used to obtain pregnancies out of the normal breeding season and to obtain a normal lamb crop from breeding ewe lambs. Also one can use estrus synchronization for accelerated lambing if it is desired to raise two lamb crops per year.

If the ewes are to be artificially inseminated, high quality fresh semen should be diluted with egg yolk-citrate diluter shortly before insemination. Ram semen has been frozen but conception rates have been low.

Estrogen in Feeds.   Sometimes a flock of sheep may show very low fertility because of high estrogen content of the legume pasture or hay that the sheep are eating. Producers are advised to have the legume hay or legumes in the pasture checked for estrogen if they are experiencing fertility problems in their sheep.

The Breeding Season

Sheep may be hand mated or pasture bred. If they are **hand mated,** some **teaser rams** are needed to locate ewes that are in heat. Either a **vasectomized** ram (each vas deferens has been severed so sperm are prevented from moving from the testicles to become a part of the semen) can be used, or an apron can be put on the ram such that a strong cloth prevents copulation.

Prior to the breeding season, rams to be used for breeding should be shorn to prevent sterility caused by high summer temperatures. Fertility of rams should be determined by examining the semen from two ejaculates collected 2 or 3 days apart.

Ewes should be checked twice daily for **heat.** Ewes normally stay in heat for 30 hours and ovulate near the end of heat. It is desirable to breed ewes the morning after ewes are found in heat in the afternoon. Ewes that are found in heat in the morning should be bred in the late afternoon.

If ewes are to be handmated, the ram should breed them once. Recently bred ewes should be separated from other ewes for 1 or 2 days so that the teaser ram does not spend his energies breeding the same ewe repeatedly.

If ewes are to be pasture mated, they are sorted into groups according to the rams to which they are to be mated. It is best to have an empty pasture between breeding pastures so not to entice rams to be with ewes in another breeding group.

Sheep can be identified by using numbered branding irons to apply scourable paint. The brisket of the breeding ram can be painted with scourable paint so that he marks the ewe when he mounts her. A light-colored paint should be used initially, so that a dark color can be used about 14 days later to detect ewes that return in heat. An orange paint can be used initially followed, successively, with green, red, and black. The paint color should be changed each 14 days. Because the ewes are numbered, they can be observed daily and a record of the breeding date of each can be entered in the record book. Also, a ram that is not settling his ewes can be detected and replaced. Occasionally a ram may be low in fertility even though his semen was given a satisfactory evaluation before the breeding season.

A breeding season of 40 days results in a lamb crop of uniform age and identifies ewes for culling that have an inherent tendency for late lambing.

The Purebred Breeder

The goal of purebred breeders is to make genetic change in the economically important traits. These breeders use selection as their method for genetic improvement as crossbreeding is limited primarily to commercial producers. If the purebred breeder has a large operation, it may be desirable for such a breeder to close his flock and select ewe and ram replacements from within this closed flock.

Normally, purebred breeders have selection programs to produce superior rams for commercial producers. At the same time, commercial producers prefer rams that will contribute outstanding performance in a commercial crossbreeding program.

In general, great progress in sheep improvement can be made by selection within a breed for traits that are highly heritable and economically important. Crossbreeding can give significant genetic improvement in traits of low heritability such as fertility. Traits that are only moderately heritable can be improved by selecting genetically superior breeding animals and also by using crossbreeding.

Highly heritable traits (40% or more) include mature body size, yearling type score, face cover, **skin folds,** clean fleece yield, yearling staple length, gestation length, loin eye area, fat weight, and retail cut weight. Lowly heritable traits (below 20%) include weaning type score, weaning condition score, multiple births, number of lambs weaned, fat thickness over loin, carcass weight per day of age, carcass grades, and dressing percent. Moderately heritable traits (20–39%) include birth weight, 90-day weight, rate of gain, neck folds, grease fleece weight, fleece grade, lambing date, milk production, carcass length, and bone weight.

Principles applicable to breeding and improving sheep were presented earlier in the book. They can be reviewed in Chapter 11 (Genetics), Chapter 12 (Selection) and Chapter 13 (Systems of Mating).

Commercial Sheep Production

Lambs born in December or January and sold as small, finished lambs for Easter or early spring markets can be produced by breeding ewes of a highly fertile breed that produces large quantities of milk (such as Dorset) to a ram of a small breed that has excellent meat conformation (such as the Southdown). The lambs produced by such a mating receive ample milk to make rapid growth. Smaller size and excellent conformation contributed by the Southdown cause lambs to have desirable conformation and to finish at a relatively small size. They should be sufficiently finished for slaughter at 70–80 lb and make excellent small carcasses.

Lambs produced for slaughter directly from pasture at weaning time are usually produced in areas where good pastures are available. The producer may be raising lambs of only one breed as straightbred animals; but more likely a producer who has good pastures and is interested in marketing finished lambs at 90–100 days will use some type of crossbreeding program. By crossbreeding, the producer can use ewes of a breed or breeds that are strong in traits such as high fertility, wool that is fairly long and fine, high milking ability, and long lived, such as Dorset, Rambouillet, Columbia, and Finnsheep. The producer can use rams of breeds that are rapidly growing, have desirable carcass conformation, and that finish well at 90–115 lb live weight, such as Suffolk or Hampshire.

Most lambs produced in farm-flock operations are marketed at weaning (when they are approximately 100 days of age and weigh from 90 to 120 lb). They are sufficiently finished by then to grade USDA Choice or Prime if they have good carcass conformation and have been adequately fed. The best method for producing such lambs is to use a three-breed rotational crossbreeding system. One can use breeds such as Dorset, Suffolk, and Columbia in areas where relatively good grazing conditions exist. Under more rigorous environmental conditions, the North Country Cheviot is preferable to the Columbia because of its ruggedness and adaptability. On the average, larger lambs are produced if Columbia sheep are used in rotation, rather than Cheviots, but more slaughter lambs result from using Cheviot when a smaller sheep is needed.

Some range sheep operations have summer grazing areas that are sufficiently productive to supply nutrients needed for lambs to grow and fatten on the range. Other range areas produce only enough forage for sheep to raise lambs that must be finished after weaning by placing them in feedlots and feeding grain and some hay for a period of 50–80 days. The use of good meat-type rams for breeding to ewes under this less-desirable condition as well as for breeding to ewes under highly desirable range conditions is well worthwhile. Under the more favorable summer range conditions, finished lambs at weaning can be produced. Under less desirable summer range conditions, **feeder lambs** can be produced when desirable meat-type rams are used. A range-sheep operator could use Rambouillet, Columbia, Panama, or Targhee ewes and breed them to Suffolk or Hampshire rams. All lambs would be marketed either as fat or feeder lambs. Replacement ewes could be purchased or raised by breeding some of the best range ewes to good rams of the same breed as that of the ewes.

Most market lambs are crossbreds (produced either from crossing breeds or from mating crossbred ewes with purebred rams), but the key to the success of **crossbreeding** is the improvement in production traits made by the purebred breeders in their selection programs. Crossbreeding can be used to combine meat conformation of the sire breeds with lambing ability and wool characteristics of the ewe breeds and also to obtain the advantage of fast growth rate of lambs that results from heterosis. The additive as well as the heterotic effects are evident in crossbred lambs. Maximum **heterosis** is obtained in three- or four-breed crosses, either rotational or terminal.

Crossing meat-type sheep with **Finnsheep** offers a means for rapidly increasing efficiency of lamb meat production. The Finnsheep breed was introduced into the United States in 1968, and its numbers are increasing rapidly in this country. Finnsheep (Fig. 27.5) are extremely prolific (two to six lambs per lambing), and studies show that crossbred ewes of 25–50% Finnsheep breeding produce more lambs than straightbred or crossbred ewes of the meat breeds (Dickerson, 1978).

There are some research studies dealing with raising two lamb crops per year or if feed and other conditions do not support such intensive production to raise three lamb crops every 2 years. To achieve intense production of this type, one needs a breed of

FIGURE 27.5.
One of the first Finnsheep ewes imported into the United States with her first lambs, illustrating the high productivity of Finnsheep. Courtesy of Dwight and Mae Holaway.

ewes that will breed most any time of the year. At present, the Finnsheep seem to fit into such a program fairly well. Also Rambouillet and Dorset have a tendency to breed out of season.

To raise three lamb crops in 2 years, ewes should be bred in late summer, probably for a 30-day period (August), for lambs to arrive in January (lamb crop 1). The ewes and lambs would need to be well fed so the lambs would be sufficiently finished to go to market at 90 days of age (in April). The ewes would be bred in April (30 day breeding period) to lamb in August (lamb crop 2) and the ewes and lambs would need good feed so the lambs would be well finished and sufficiently large to be marketed at 90 days of age (November). The ewes would be bred in November to lamb in April (3rd lamb crop) and the lambs would be marketed at 90 days of age in July.

Three-Breed Terminal Crossbreeding

In some production schemes, crossbred ewes are bred to rams of a third breed as a terminal cross (all offspring are marketed). Some producers cross Columbias with Dorsets to develop large ewes that have long, fairly fine fleece, are hardy and produce much milk. These crossbred ewes are bred to good Suffolk or Hampshire rams. A high percentage of lambs from these crosses are finished at weaning and weigh 90–100 lb at about 90–100 days of age. The entire lamb crop from this terminal mating is marketed. All the two-breed crossbred ewes that are productive are kept until they reach an age at which their production of lambs declines. Replacement ewes are produced by crossing

Columbias and Dorsets. A producer carrying out such a program keeps a small flock of either straightbred Dorsets or Columbias to produce the two-breed crossbred ewes. Pure-bred Suffolk or Hampshire rams are purchased as needed and they are used for breeding as long as they are sound and inbreeding is not a problem.

INHERITED ABNORMALITIES

Certain genetic abnormalities are important to guard against when managing sheep. Although exceptions occur, sheep showing obvious genetic defects are usually culled. Some of the inherited abnormalities follow.

Cryptorchidism is inherited as a simple recessive; therefore a ram with only one testis descended into the scrotum should never be used for breeding. In addition, one should cull rams that sire lambs having cryptorchidism, as well as the ewes that produced the lambs.

Dwarfism is inherited as a recessive and is lethal; therefore, ewes and rams producing dwarf offspring should be culled.

The mode of inheritance of **entropion** (turned-in eyelids) has not been determined, but it is known to be under genetic control. A record should be made of any lamb having entropion so that the lamb can be marketed.

Sheep have lower front teeth but lack upper front teeth. They graze by closing the lower teeth against the dental pad of the upper jaw. If the lower jaw is either too short **(overshot** or **Parrot mouth)** or too long **(undershot),** the teeth cannot close against the dental pad and grazing is difficult. The mode of inheritance of these jaw abnormalities is unknown, but these conditions are under genetic control. Sheep having abnormal jaws should be culled.

Rectal prolapse is more common in black-faced sheep. It is a serious defect but both inheritance and the environment are influential in its occurrence. Lambs on heavy feeding or lush pastures are more likely to show rectal prolapse. If surgery is used to correct this condition, the animal should not be used as a breeding animal.

Skin folds, open-faced, closed-faced, and **wool blindness** are inherited and selection against these traits should be practiced.

Many wool defects are known such as black fibers, black tipped fibers, hairiness, fuzziness, and ''high belly wool'' in which wool that is typical of wool on the belly is present on the sides of the sheep. Selection against all of these fleece defects should be practiced as a means of reducing the frequence of occurrence of these defects.

SELECTED REFERENCES

Publications

Dickerson, G. E. 1978. *Crossbreeding Evaluation of Finnsheep and Some U. S. Breeds for Market Lamb Production.* North Central Regional Publication #246. ARS, USDA and University of Nebraska.

Ensminger, M. E. and Parker, R. O. 1986. *Sheep and Goat Science.* Danville, IL: Interstate Printers and Publishers, Inc.

Henson, E. North American Livestock Census. Box 477, Pittsboro, NC, March, 1986.

Land, R. B. and Robinson, D. W. 1985. *Genetics of Reproduction in Sheep.* Boston: Butterworths.

Lasley, J. F. 1987. *Genetics of Livestock Improvement.* Englewood Cliffs, NJ: Prentice-Hall.

Neimann-Sorenson, A. and Tube, D. E., Editors in Chief. *World Animal Science. C. Production–System Approach*. 1982. I. E. Coop, editor. *Sheep and Goat Production*. New York: Elsevier Scientific Publishing Co.

Scott, George E. 1983. *The Sheepman's Production Handbook*. Sheep Industry Development Program, Inc. Denver, Colorado.

Tomes, G. L., Robertson, D. E. and Lightfoot, R. J. (editors). 1979. *Sheep Breeding*. Boston: Butterworth & Co. Ltd.

Visuals

Sheep Breed Identification Slide Kit (15 breeds of sheep). Vocational Education Productions, California Polytechnic State University, San Luis Obispo, CA 93407.

An Introduction to Sheep Breed Identification (sound filmstrip). Prentice-Hall Media, 150 White Plains Rd., Tarrytown, NY 10591.

Farm Flock Sheep Production *(Farm Flock Production Systems* and *Personal Requirements/Personal Rewards)*. Videotapes. Agricultural Products and Services, 2001 Killebrew Dr., Suite 333, Bloomington, MN 55420.

NSIP—Range Flock (10 min. video). National Sheep Improvement Program, Dept. of Animal Science, Iowa State University, Ames, IA 50010.

NSIP—Farm Flock (10 min. video). National Sheep Improvement Program, Dept. of Animal Science, Iowa State University, Ames, IA 50010.

CHAPTER 28 ▮

Feeding and Managing Sheep

Sheep feeding and management are essential for success of an operation. These areas must be integrated with a knowledge of how breeding and environmental factors affect sheep productivity and profitability.

PRODUCTION REQUIREMENTS FOR FARM FLOCKS

The objective of farm-flock operators is to efficiently produce slaughter lambs ready for market in May, at which time prices are usually high (Fig. 28.1). Several areas of feeding and management affecting efficient production are discussed.

Pastures

Good pastures are essential to the typical farm-flock operator. Grass-legume mixtures such as rye grass and subterranean clover or orchard grass and lespedeza are ideal for sheep. Sheep can also graze on temporary pastures of such plants as Sudan grass or rape, which are often used to provide forage in the dry part of summer when permanent pastures may show no new growth. Crops of grain and grass seed may also provide pasture for sheep in the autumn and early spring. Sheep do not trample wet soil as severely as do cattle; pasturing sheep on grass-seed and small-grain crops in winter and early spring when the soil is wet does not cause serious damage to the plants or soil.

Fencing

A woven-wire or electric fence is necessary to contain sheep in a pasture. Forage can be best utilized by "rotating" (moving) sheep from one pasture to another. This also assists in the control of internal parasites. Some sheep operators use temporary fencing such as electric fencing to divide pastures and for predator control. It is necessary to use two electrified wires, one located low enough to prevent sheep from going underneath, and the other located high enough to prevent them from jumping over.

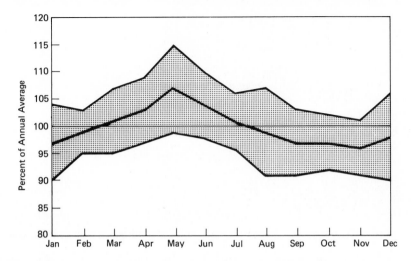

FIGURE 28.1.
Seasonal lamb price index (1970–1984). Dark line is the average monthly price compared with the average for all years (100) Shaded area shows the range in prices by years. Courtesy of Western Livestock Marketing Information Project.

Corrals and Chutes

It is occasionally necessary to put sheep in a small enclosure to sort them into different groups or to treat ailing individuals. Proper equipment is helpful, because sheep are often difficult to drive, particularly when ewes are being separated from their lambs. A **cutting chute** can be constructed to direct sheep into various small lots (Fig. 28.2). A well-designed cutting chute is sufficiently narrow to keep sheep from turning around and can be blocked so that sheep can be packed together closely for such purposes as treating diseases and reading ear tags. The chute can be constructed of lumber that is nailed to wooden posts set into the ground and properly spaced so that the proper width (14 to 16

FIGURE 28.2.
Sheep being directed into a special pen by use of a cutting chute.

FIGURE 28.3.
Sheep being unloaded directly onto pasture by use of a portable loading chute.

in.) is provided when the boards are nailed on the inside. Pens used to enclose small groups of sheep can be constructed of a woven-wire fencing. A loading chute is used to place sheep onto a truck for hauling. A portable loading chute (Fig. 28.3) is ideal because it can be moved to different locations where loading and unloading sheep is necessary.

Shelters

Sheep do not normally suffer from cold because they have a heavy wool covering; therefore, open sheds are excellent for housing and feeding wintering ewe lambs, pregnant ewes, and rams. Although the ewes can be lambed in these sheds, the newborn need an enclosed and heated room when the weather is cold.

Lambing Equipment

Small pens, about 4×4 ft. can be constructed for holding ewes and their lambs until they are strong enough to be put with other ewes and lambs. Four-foot panels can be constructed from 1×4-in. lumber, and the two panels can be hinged together. The lambing pens (called **lambing jugs**) can be made along a wall by wiring these hinged panels. Heat lamps are extremely valuable for keeping newborn lambs warm. A heat lamp above each lambing jug can be located at a height that provides a temperature of 90°F at the lamb's level (Fig. 28.4).

It is advisable to identify each lamb and to record which lambs belong to which ewes. Often, a ewe and her lambs are branded with a scourable paint to identify them. Numbered ear tags can be applied to the lamb at birth. If newborn lambs are to be weighed, a dairy scale and a large bucket are needed. The lamb is placed in the bucket, which hangs from the scale, for weighing. It is advisable to immerse navel cords of newborn lambs in a tincture of iodine.

Feeding Equipment

All sheep are given hay in the winter feeding period. Feeding mangers (Fig. 28.5) should be provided, because hay is wasted if it is fed on the ground. Some sheep may require

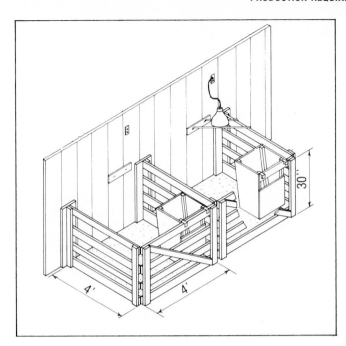

FIGURE 28.4.
A lambing jug with heat lamp for ewes with newborn lambs. Courtesy of Mr. John H. Pedersen, Midwest Plan Service, Agricultural Engineering, Iowa State University.

FIGURE 28.5.
Sheep eating from hay bunk.

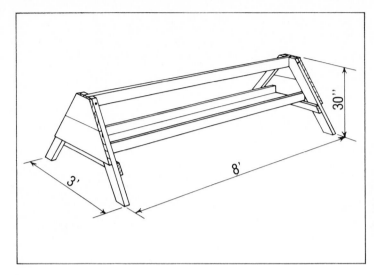

FIGURE 28.6.
A feed trough for providing concentrates for sheep. Note the top board that prevents sheep from getting into the trough. Courtesy of Mr. John H. Pedersen, Midwest Plan Service, Agricultural Engineering, Iowa State University.

limited amounts of concentrates. Concentrates can be fed in the same bunk as hay if the hay bunk is properly constructed, or separate feed troughs (Fig. 28.6) may be preferable.

Additional concentrates can be provided for lambs by placing the concentrate in a **creep** (Fig. 28.7) that is constructed with openings large enough to allow the lambs to enter but small enough to keep out the ewes. In addition to concentrates, it is advisable to provide good-quality legume hay, a mineral mix containing calcium, phosphorus, and salt. A heat lamp, in the creep area, may also be included if the weather is cold.

Water is essential for sheep at all times. It can be provided by tubs or automatic waterers. Tubs should be cleaned once each week. Large buckets are usually used for watering ewes in lambing jugs.

A separate pen equipped with a milk feeder may be needed for orphan lambs. Milk or a milk replacer can be provided free-choice if it is kept cold. Heat lamps kept some distance from the milk feeder can be provided.

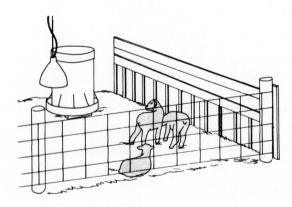

FIGURE 28.7.
Creep with heat lamp for lambs. From Battaglia and Mayrose, *Handbook of Livestock Management Techniques,* Macmillan Publishing Company, 1981, p. 381.

Feed Storage

Areas should be provided for storage of hay and concentrates so that feeds can be purchased in quantity or home-grown feeds can be stored where they will remain dry. An open shed is satisfactory for hay storage. Some operators prefer a feed bin for concentrates that is designed so that the feed can be put in at the top and removed from the bottom.

TYPES OF FARM-FLOCK PRODUCERS

Some people raise purebred sheep and sell **rams** for breeding. Commercial producers whose pastures are productive can produce slaughter lambs on pasture, whereas those whose pastures are less desirable produce feeder lambs. Lambs that are neither sufficiently fat nor large for slaughter at weaning time usually go to feedlot operators where they are fed to slaughter weight and condition; however, most feedlot lambs are obtained from producers of range sheep.

Purebred Breeder

Purebred sheep are usually given more feed than commerical sheep because the purebred breeder is interested in growing sheep that express their growth potential. Records are essential to indicate which ram is bred to each of the ewes and which ewe is the mother of each lamb. All ram lambs are usually kept together to identify those that are desirable for sale. The ram lambs are usually retained by the purebred breeder until they are a year of age, at which time they are offered for sale. Buyers usually want to obtain rams from purebred breeders in early summer.

Purebred breeders have major responsibilities to the sheep industry, because they determine the genetic productivity of commercial sheep. The purebred breeder should rigidly select animals that are kept for breeding and should offer only those rams for sale that will contribute improved productivity for the commercial producer.

Commercial Slaughter Lamb Producers

The producer of slaughter lambs, whose feed and pasture conditions are favorable, strives to raise lambs that finish at 90–120 days of age weighing approximately 110 lb each. Any lamb that is small or lacks finish requires additional feeding. Usually, the price decline that occurs from June through September offsets the improvement that is made in value of lambs by feeding, so that feeding the cull farm-flock lambs is generally unprofitable. Creep feeding of concentrates early in the life of lambs helps, but, because lambs may be disoriented when their creep is moved, they may not come to the creep after being put onto a new pasture. Therefore, creep feeding on pasture on which pasture rotation is practiced may not be beneficial. Keeping sheep healthy while providing adequate water on lush pasture will result in early market lambs.

Commercial producers can give some assurance to raising heavy, well-finished market lambs at weaning by using good ram selection and crossbreeding programs. Rams that are heavy at 90 days of age are more likely to sire lambs that are heavy at this age than are rams that are light in weight at 90 days of age. The breeds to use in a three-breed rotation should include one that is noted for milk-producing ability, one that is noted for rapid growth, and one that is noted for ruggedness and adaptability. The Dorset could be considered for milk production, either the Hampshire or Suffolk for rapid growth, and the Cheviot for its ruggedness. All of these breeds make desirable carcasses.

Commercial producers castrate all ram lambs. Young ewes that are selected to replace old ewes should be large and growthy and should be daughters of ewes that produce relatively many lambs. Ewes that have started to decline in production and poorly productive ewes should be culled.

Commercial Feeder Lamb Producers

Some commercial sheep are produced where pasture conditions are insufficient for growing the quantity or quality of feed needed for heavy slaughter lambs at weaning. Lambs produced under such conditions may either be fed out in the summer or carried through the summer on pasture with their dams and finished in the fall. Sudan grass or rape can be seeded so that good pasture is available in the summer, and lambs can be finished on pasture by giving them some concentrates. If good summer pasture cannot be made available, it is advisable to wait until autumn, at which time the lambs are put on full feed in a feedlot.

Commercial Feedlot Operator

Lambs that come to the feedlot for finishing are treated for internal **parasites** and for certain diseases, particularly **overeating disease.** They are provided water and hay initially, after which concentrate feeding is allowed and increased until they receive all the concentrates they will consume. Death losses may be high in unthrifty lambs.

The feedlot operator hopes to profit by increasing the value of the lambs per unit of weight and by increasing their weight. Lambs on feed gain about 0.5–0.8 lb per head per day when they are gaining satisfactorily. Feeder buyers prefer feeder lambs weighing 70–80 lb over larger lambs because of the need to put 20–30 lb of additional weight on the lambs to finish them. A feeding period of 40–60 days should be sufficient for finishing thrifty lambs.

Feeding lambs is riskier than producing them because of death losses, price fluctuations of feed and sheep, and the necessity of making large investments.

FEEDS AND FEEDING

Mature, pregnant ewes probably need nothing more than good-quality legume hay during the first half of pregnancy, after which some concentrates should be fed. Any grains such as corn, barley, oats, milo, and wheat are satisfactory feeds. Sheep usually chew these grains sufficiently so that grinding or rolling is not essential. Rolled or cracked grains may be digested slightly more efficiently and may be more palatable, but finely ground grains are undesirable for sheep unless the grains are pelleted.

Sheep need energy, salt, iodine, phosphorus, and vitamins A, D, and E. In some areas selenium is deficient and must be supplied either in the feed or by injection. Mature, ruminating sheep have little need for quality protein or B vitamins, but immature sheep have variable requirements. Rumen microorganisms can synthesize protein from nonprotein nitrogenous substances in the ration.

Sheep will perform poorly or die quickly when the water supply is inadequate, so the importance of supplying water cannot be overlooked. Water that is not clean will not be well accepted by sheep. Water intake is influenced by amount of food eaten, the protein intake, environmental temperature, mineral intake, water temperature, pregnancy, water content of feed eaten including rain and dew on pastures, and the odor and taste of the water.

Energy is perhaps the most common limiting nutrient for ewes. Underfeeding is the primary cause for a deficiency of energy because most feed materials are high in energy. Dry range grasses and mature forages such as grain straws may be high in gross energy but so low in digestibility that sheep cannot obtain all their energy needs from them.

The amount of protein in the ration for sheep is of greater importance than the quality of protein because sheep can make the essential amino acids by action of microorganisms in the rumen. The oil meals (soybean, linseed, peanut, and cottonseed) are all high in protein. Soybean meal is the most palatable and its use in a ration encourages a high feed intake. Nonprotein sources of nitrogen such as urea and **biuret** can be used to supply a portion but not all of the nitrogen needs of sheep. Not more than one-third of the nitrogen need should be supplied by urea or biuret, and these materials are not recommended for young lambs with developing rumens or for range sheep on low-energy diets. If protein intake is limited by mixing with salt, adequate water must be provided.

The mineral needs for sheep are calcium, phosphorus, sulfur, potassium, sodium, chlorine, magnesium, iron, zinc, copper, manganese, cobalt, iodine, molybdenum, and selenium. All of these minerals are found in varying amounts in different tissues of the body. For example, 99% of the calcium, 80–95% of the phosphorus, and 70% of the magnesium occurs in the skeleton and more than 80% of the iodine is found in the thyroid gland.

The requirement for minerals is influenced by breed; age, sex and growth rate of young animals, reproduction and lactation of the ewe; level and chemical form injected or fed; climate; and balance and adequacy of the ration.

Salt is important in sheep nutrition and it can become badly needed when sheep are grazing on lush pastures. Also sheep may consume too much salt when they are forced to drink brackish water.

In some areas, iodine and selenium are deficient. The use of iodized salt in an area where iodine is deficient may supply the iodine needs. Selenium can be added to the concentrate mixture that is being fed or it can be given by injection.

The two most critical periods of nutritional needs for the ewe are at breeding and just before, during, and shortly after lambing. A lesser but very important period for the ewe is during lactation. After the lamb is weaned, the ewe normally can perform satisfactorily on pasture or range without any additional feed. If accelerated lamb production is practiced, the ewes should be well-fed after the lamb is weaned so the ewe can be bred again.

Ewes are normally run on pasture or range where they obtain all their nutritional needs from grasses, browse, and forbs, except during the winter or dry periods when these plants are not growing. During these two periods when there is no plant growth, the sheep must be given supplemental feed.

Occasionally lambs are raised on milk replacer or on cow's milk if the ewe dies during lambing or her udder becomes nonfunctional. During these two periods, frozen colostrum should be available for the lambs. If milk or milk replacer is used, it can be bottle-fed twice daily or it can be self-fed if it is kept cold. It is important to feed cold milk when it is self fed to prevent the lamb from overconsumption. There should be heat lamps not far from where the lambs consume the cold milk so they can go to the heated area to become warm and sleep.

Rams should be fed to keep them healthy but not fat. A small amount of grain along with good-quality hay satisfies their nutritional needs in winter. Bred, young ewes should be fed some grain along with all the legume hay they will consume because they grow and also have some reserve for the subsequent drain of lactation. Ewe lambs should be grown out but not fattened in the winter. Limited grain feeding along with legume hay satisfies their nutritional needs.

At lambing time, the grain allowance of ewes needs to be increased to assist them in producing a heavy flow of milk. Also, the lambs need to be fed a high-energy ration in the creep. Lambs obtain sufficient protein in the milk given by their mothers, but they need more energy. Rolled grains provided free-choice in the creep are palatable and provide the energy needed.

Lambs on full feed are allowed some hay of good quality and all the concentrates they will consume. Some feedlot operators pellet the hay and concentrates, whereas others feed loose hay and grains. After the grasses and legumes start to grow in the spring, all sheep generally obtain their nutritional needs from the pasture.

CARE AND MANAGEMENT OF FARM FLOCKS

The handling of sheep is extremely important. A sheep should never be caught by its wool as the skin is pulled away from the flesh, causing a bruise. When a group of sheep is crowded into a small enclosure, the sheep will face away from the person who enters the enclosure. When the sheep's rear flank is grasped with one hand, the sheep starts walking backward; this allows one to reach out with the other hand and grasp the skin under the sheep's chin. When a sheep is being held, grasping the skin under the chin with one hand and grasping the top of the head with the other hand enables the holder to pull the sheep forward so its brisket is against the holder's knee. If a sheep is to be moved forward, the skin under the chin can be grasped with one hand and the **dock** (the place where the tail was removed) can be grasped with the other hand. Putting pressure on the dock makes the sheep move forward while holding its chin with the other hand prevents it from escaping.

Lambing Operations

Before the time the ewes are due to lamb, wool should be clipped from the dock, udder, and vulva regions. This process is called **crutching.** If weather conditions permit, the ewes may be completely shorn. Also all **dung tags** (small pieces of dung that stick to the wool) should be clipped from the rear and flank of the pregnant ewes. Young lambs will try to locate the teat of the ewe and may try to nurse a dung tag if it is present.

Ewes should be checked periodically to locate those that have lambed. The ewe and newborn lamb should be placed in a lambing jug. The ewe that is about ready to lamb can be watched carefully but with no interference if delivery is proceeding normally. In a normal presentation, the head and front feet of the lamb emerge first. If the rear legs emerge first **(breech presentation),** assistance may be needed if delivery is slow, because the lamb can suffocate if deprived of oxygen too long. If the front feet are presented but not the head, the lamb should be pushed back enough to bring the head forward for presentation.

As soon as the lamb is born, membranes or mucus that may interfere with its breathing should be removed. When the weather is cold, it may be necessary to dry the lamb by rubbing it with a dry cloth. If a lamb has become chilled, it can be immersed in warm water from the neck down to restore body temperature, after which it should be wiped dry. The lamb should be encouraged to nurse as soon as possible. A lamb that has nursed and is dry should survive without difficulty if a heat lamp is provided. The lamb can be identified by an ear tag, a tattoo, or both.

Some ewes may not want to claim their lambs. It may be necessary to tie a ewe with a rope halter so she cannot butt or trample her lamb.

Castrating and Docking

Because birth is a period of stress for lambs, it is best to wait for 3 or 4 days to castrate and **dock** lambs. To use the elastrator method of castration and docking, a tight rubber band is placed around the scrotum above the testicles (for castration) and around the tail about an inch from the buttocks (for docking). Some death losses can occur when tetanus-causing bacteria invades the tissue where the elastrator was applied. Another castration and docking practice is to surgically remove the testicles and tail. The emasculator is also useful for docking; the skin of the tail is pulled toward the lamb, the emasculator is applied about an inch from the lamb's buttocks and the tail is cut loose next to the emasculator. A fly repellent should be applied around any wound to lessen the possibility of **fly strike** (fly eggs are deposited during warm weather).

Occasionally, ewes develop a vaginal or uterine **prolapse** (the reproductive tract protrudes to the outside through the vulva). This condition is extremely serious and leads to death if corrective measures are not taken by trained personnel. The tissue should be pushed back in place, even it it is necessary to hoist the ewe up by her hind legs as a means of reducing pressure that the ewe is applying to push the tract out. After the tract is in place, the ewe can be harnessed so that external pressure is applied on both sides of the vulva. In some cases, it may be necessary to suture the tract to make certain that it stays in place. Once a ewe has prolapsed her reproductive tract, the tract shows a weakness that is likely to recur; therefore, a record should be kept so that the ewe can be culled after she weans her lamb.

Shearing

Sheep are usually shorn in the spring before hot weather. Shearing is usually done by professional personnel, but shearing classes are available to teach the operator. The usual method of shearing is to clip the fleece from the animal with power-driven shears, which leaves sufficient wool covering to protect the sheep's skin. In the shearing operation, the sheep is set on its dock and cradled between the shearer's legs, which are used to maneuver the sheep into the positions needed to make shearing easy.

The wool that is removed usually hangs together. It is spread with the clipped side out, rolled with the edges inside, and tied with paper twine. It is usually best to remove dung tags and coarse material that is clipped from the legs and put these items in a separate container. The tied fleece is put into a huge sack. When buyers examine the wool, they can obtain core samples from the sack. The core sample is taken by inserting a hollow tube that is sharp at the end into the sack of wool. The sample obtained is examined to evaluate the wool in the sack rather than having to remove the fleeces to examine them. If undesirable material is obtained in the core sample, the price offered will be much lower than if only good wool is found.

If the wool is kept for some time before it is sold, it should be stored in a dry place and on a wooden or concrete floor to avoid damage from moisture.

FACILITIES FOR PRODUCTION OF RANGE SHEEP

Sheep differ from cattle in that they graze weedy plants and brush as well as grasses and legumes. Because of the different grazing patterns, cattle and sheep can be effectively grazed together in some range areas. Total pounds of liveweight produced can be higher compared to grazing the separate species on the same range.

Range sheep are produced in large flocks primarily in arid and semiarid regions. Sheep of fine-wool breeding tend to stay together as they graze, which makes herding possible in large range areas. Range sheep are moved about either in trucks or by trailing so they can consume available forage at various elevations. Requirements for the production of range sheep are usually different from those for farm-flock operations. One type of range sheep operation is described herein, but it must be noted that variations exist.

Range sheep are usually bred to lamb later than sheep in farm flocks; therefore, they can be lambed on the range. Few provisions are needed for lambing when sheep are lambed on the range, but under some conditions a tent or a lambing shed may be used to give range sheep protection from severe weather at lambing time. Temporary corrals can be constructed using snow fences and steel posts when it is necessary to contain the sheep at the lambing camp or at lambing sheds.

Sheep on range are usually wintered at relatively low elevations where little precipitation occurs. Wintering sheep are provided **feed bunks** if hay is to be fed and windbreaks to give protection from cold winds. Some producers of range sheep provide pelleted feed to supplement the winter forage. Pellets are usually placed on the ground but are sometimes dispersed in grain troughs.

A sheep camp or sheep wagon is the mobile housing for the sheep herder. The camp is moved by truck or horses as the sheep need to be moved over large grazing areas. A sheep herder usually has a horse and dogs to assist in herding the sheep. The sheep are brought together to a night bedding area each evening.

MANAGING RANGE SHEEP

Range sheep are grazed in three general areas: the winter headquarters, which is in relatively low and dry areas, which sometimes provides forage for winter grazing; the spring-fall range, which is somewhat higher and which receives more precipitation; and the summer grazing area, which is at high elevations in the mountains and which receives considerable precipitation that results in lush feeds.

The Winter Headquarters

The forage on the winter range, where there are usually fewer than 10 in. of precipitation annually, is composed of sagebrush and grasses. The grasses are cured on the ground from the growth of the previous summer; consequently, winter forage is of lower quality than green forage because the plants in the winter forage are mature and because they have lost nutrients. The soil in these areas is often alkaline and the water is sometimes alkaline.

Because forage in the wintering area is of poor quality, supplemental feeding that provides needed protein, carotene, and minerals such as copper, cobalt, iodine, and selenium is usually necessary. A pelleted mixture made by mixing sun-cured alfalfa leaf meal, grain, solvent-extracted soybean or cottonseed meal, beet pulp, molasses, bone meal or dicalcium phosphate, and trace-mineralized salt is fed at the rate of 0.25–2.0 lb per head per day, depending on condition of the sheep. The feed may be mixed with salt to regulate intake so that feed can be before the sheep at all times. If the intake of feed is to be regulated through the use of salt, trace-mineralized salt should be avoided. Adequate water must also be provided at all times, because heavy salt intake is quite harmful if sheep do not have water for long periods of time.

The ewes are brought to the winter headquarters about the first of November. Rams are turned in with the ewes for breeding in November if lambing is to take place in sheds or if a spring range that is not likely to experience severe weather conditions is available for lambing. Otherwise, the rams are put with the ewes in December for breeding.

January, February, and March are critical months because the sheep are then in the process of exhausting their body stores and because severe snow-storms can occur. If sheep become snowbound, they should each be given 2 lb of alfalfa hay plus 1 lb of pellets containing at least 12% protein. Adequate feeding of ewes while they are being bred and afterward results in at least a 30% increase in lambs produced and perhaps a 10% increase in wool produced. In addition, death losses are markedly reduced. Sheep that are stressed by inadequate nutrition either as a result of insufficient feed or a ration that is improperly balanced are highly susceptible to pneumonia. Heavy death losses can result.

The Spring–Fall Range

Pregnant ewes are shorn at the winter headquarters (usually in April). They are then moved to the spring–fall range, where they are lambed. If they are lambed on the range, a protected area is necessary. An area having scrub oak or big sagebrush on the south slopes of foothills and ample feed and water is ideal for range lambing. Portable tents can be used if the weather is severe.

Ewes that have lambed are kept in the same area for about 3 days until the lambs become strong enough to travel. The ewes that have lambs are usually fed a pelleted ration that is high in protein and fortified with trace-mineralized salt, and either bone meal or dicalcium phosphate. Feeding at this time can help prevent sheep from eating poisonous plants.

Ewes with lambs are kept separate from those yet to lamb until all ewes have borne their lambs. In addition, ewes that are almost ready to lamb are separated from those that will not lamb for some time yet. Thus, after lambing gets underway, three separate groups of ewes are usually present until lambing is completed.

If the ewes are bred to lamb earlier than is usual for range lambing and a crested wheatgrass pasture is available, ewes may be lambed in open sheds. If good pasture is unavailable, the ewes may be confined in yards around the lambing sheds, starting a month before lambing. In this event, the ewes must be fed alfalfa or other legume hay and 0.50–0.75 lb of grain per head per day. The ewes should have access to a mixture of trace-mineralized salt and bone meal or dicalcium phosphate.

Although shed lambing is more expensive than range lambing, higher prices for lambs marketed earlier have made shed lambing advantageous. Fewer lambs are lost in shed lambing, and lambing can take place earlier in the year. The heavy market lambs that result produce enough income to more than offset the costs of shed lambing.

The Summer Range

Sheep are moved to summer range shortly after lambing is completed if weather conditions have been such that snow has melted and lush plant growth is occurring. The sheep are put into bands of about 1,000 to 1,200 ewes and their lambs. In some large operations, the general practice is to put ewes with single lambs in one band and ewes with twins in another. The ewes with twin lambs are given the best range area so the lambs will have added growth from the better forage supply. Sheep are herded on the summer range to assist them in finding the best available forage.

In October, prior to the winter storms, the lambs are weaned. Lambs that carry sufficient finish are sent to slaughter and other lambs are sold to lamb feeders. It is the general practice among producers of Rambouillet, Columbia, or Targhee sheep to breed some of the most productive and best-wooled ewes to rams of the breed being used to raise replacement ewe lambs. Most of these ewe lambs are kept and grown out, and only the less desirable ones are culled. The remainder of the ewes are bred to meat-type rams, such as Suffolk or Hampshire, and all their lambs are marketed for slaughter, as feeders, or as stockers (animals used in the flock for breeding).

The Fall Range

As soon as the lambs are weaned, the ewes are moved to the spring-fall range. Later they go to the winter headquarters for wintering.

The number of ewes that can be bred per ram during the breeding period of about 2 months is 15 for ram lambs, 30 for yearling rams, and 35 for mature, but not aged, rams. These numbers are general and depend greatly on the type of conditions existing on the range.

CONTROLLING DISEASES AND PARASITES

Sheep raised by most producers are confronted with a few serious diseases, several serious internal parasites and some external parasites. Some common diseases of sheep are the following:

Enterotoxemia ("overeating disease") is most often serious when sheep are in a high nutritional state (for example, lambs in the feedlot), but it can affect sheep that are on lush pastures. The disease can be prevented by administering Type D toxoid. Usually, three treatments are given; two about 4 weeks apart and a booster treatment 6 months later. Losses from this disease among young lambs can be prevented by vaccinating pregnant ewes.

E. coli complex - Clostridium E. At least three organisms are involved: (1) *E. coli,* (2) *Clostridium perfringens,* and (3) a virus called *rotavirus.* This disease affects lambs from birth to a few days of age. This disease can be prevented by vaccinating all pregnant ewes twice in later pregnancy with type C and D *Clostridium perfringens* toxoid. The lambing pen should be thoroughly cleaned. Broad spectrum **antibiotics** given to afflicted lambs help. The ewe flock should be vaccinated with *Clostridium perfringens* type C and D toxoid. Enterotoxemia may be caused by *Clostridium perfringens* type D or C. The *Clostridium perfringens* type CD toxoid given to the ewe will give the lamb protection from both types of enterotoxemias.

Lamb dysentery *C. perfringens* type B occurs very early in life, more likely during wet weather. It can be prevented by vaccinating ewes with Type "BCD" vaccine.

Footrot is one of the most serious diseases of the sheep industry because of its common occurrence. The disease can be treated with systemic medication, or it can be cured by severe trimming so that all affected parts are exposed, treating the diseased area with a solution of one part formalin solution to nine parts of water, and then turning the sheep into a clean pasture so reinfection does not occur. A tilting squeeze is useful for restraining sheep that need treatment (Fig. 28.8). Formaldehyde must be used with caution because the fumes are damaging to the respiratory system of both the sheep and the person applying the formaldehyde.

Once all sheep in the flock are free of footrot, it can best be prevented by making sure

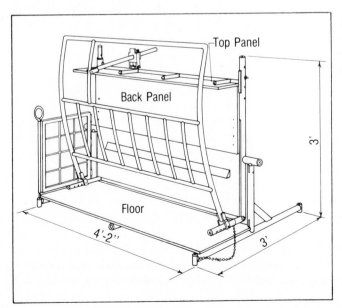

FIGURE 28.8.
A tilting squeeze chute for restraining sheep. Courtesy of Mr. John H. Pedersen, Midwest Plan Service, Agricultural Engineering, Iowa State University.

that it is not introduced into the flock again. Rams introduced for breeding should be isolated for 30–60 days for observation. If footrot develops, rams should continue in isolation until free of the disease. A vaccine is available for footrot.

Pneumonia usually occurs when animals have been stressed by other diseases, parasites, improper nutrition, or exposure to severe weather conditions. When pneumonia is recognized early, sulfonamides and antibiotics are usually effective against it. The afflicted animal should be given special care and kept warm and dry.

Ram epididymitis is an infection of the epididymis that reduces fertility. No effective treatment is available.

Sore mouth usually affects lambs rather than adult sheep. It is caused by a virus and can be contracted by humans. It can be controlled by vaccination, and sheep should be vaccinated as a routine practice.

Sheep are subject to a nutritional disease known as **white muscle disease.** To prevent it, pregnant ewes should be given an injection of selenium during the last one-third of pregnancy and the lambs should be given an injection of selenium at birth.

Selenium can be added to the feed of pregnant ewes to prevent white muscle disease of their lambs. If trace mineralized salt is provided the sheep, it should contain sufficient selenium to supply the needs of sheep.

Shipping fever is a highly infectious and contagious disease complex and usually affects lambs after stress of transportation. The most common type of shipping fever is Pasteurellosis. Bacterial agents associated with this disease are *Pasteurella hemolytica, Pasteurella multocida, Cornybacterium pyogenes* and to a lesser extent micrococci, streptococci, and pseudomonas. Antibiotics and **sulfonamides** are usually effective as treatment. Care in transporting lambs to prevent stress will help to prevent this disease.

Caseous lymphadenitis is a disease that occurs in greater frequency as sheep increase in age from lambs to old animals. It is due to a bacterium, *Corynebacterium pyogenes* var. *pseudotuberculosis,* which grow in lymph glands and cause a large development of caseous (a thick, cheeselike accumulation) material to form. It is a serious disease and one that is difficult to control. The abscesses can be opened and flushed with a solution of equal parts of 0.2% nitrofurazone solution and 3% hydrogen peroxide. One should not open an abscess and let the thick pus go onto the floor or soil where well sheep will be travelling because this might be a way of spreading the disease. All infected animals should be culled and they should be isolated from noninfected sheep immediately on appearance of being infected.

Salmonellosis causes dysentery in lambs and abortion by ewes. The organism is found in feces of animals and may contaminate water in stagnant pools. It usually occurs after lambs or other sheep have been stressed. There is no cure for this disease. If lambs show symptoms of the disease, they should be isolated from pregnant ewes and from well lambs.

Polyarthritis (stiff lamb disease, stiffness is usually associated with overeating) affects lambs. It responds to treatment with the oxytetracycline, chlortetracycline, tylosin, and penicillin.

Milk fever is due to hypocalcemia. Afflicted animals can be treated with injection of calcium salts. A 20% solution of calcium borogluconate given **intravenously** at the rate of 100 ml per sheep should have an afflicted animal up and in good condition within 2 hours.

Abortion may occur in late pregnancy due to vibriosis, enzootic abortion, or leptspirosis. Most of these are controlled by vaccination.

Sheath rot may occur in rams. There is no treatment that gives satisfactory results.

Urinary calculi (kidney stones) occurs when the salts in the body that are normally excreted in the urine are precipitated and form stones that may lodge in the kidneys' ureters, bladder, or urethra. Providing a constant supply of clean water helps immensely in preventing calculi formation. The ideal ratio of calcium to phorphorus is 1.6 calcium to 1.0 phosphorus. Ammonium chloride can be included in the ration to help prevent urinary calculi.

Pregnancy disease (ketosis) is a metabolic disease that affects ewes in late pregnancy, particularly if they are carrying twins or triplets. The problem is that ewes carrying twins or triplets must break down body fat to provide their energy needs in later pregnancy. It is possible that use of fat breakdown will not provide the glucose needs and hypoglycemia will result. Pregnancy disease may be prevented by feeding some "high-energy" grain such as corn, barley, or milo, but it is aggravated by one or more large lambs reducing rumen space.

Grass tetany (or grass staggers) is due to a deficiency of magnesium at a particular time and is most frequently seen in the spring when lactating ewes are put onto lush pasture where there is insufficient magnesium available to the sheep. An injection of 50–100 ml of 20% calcium borogluconate or injection of magnesium sulfate should give rapid recovery.

Johnes disease and **progressive ovine pneumonia** are two debilitating diseases that affect sheep. The lamb may be infected through colostrum of the ewe that has Johnes disease but no ill effects of the disease is exhibited until later when the animal becomes unthrifty and its condition deteriorates. Afflicted animals that are in thin condition do not sell well, but they must be culled because they can help to spread the disease to other sheep and they are unlikely to produce lambs when they are emaciated.

Sheep have internal and external parasites with internal parasites being the more serious of the two. The most important internal parasites are coccidiosis, stomach worms, nodular worms, liver flukes, lungworms, round worms, and tapeworms.

The treatments that are effective for controlling internal parasites of sheep are summarized in Table 28.1.

Common external parasites of sheep include blowfly maggots, **keds** (sheep ticks), lice, mites, screw-worms and sheep bots. External parasites can be controlled by making two applications of an effective insecticide that is not harmful to sheep. The two applications should be spaced so that eggs hatched after the first application will not result in egg-laying adults prior to the second application.

DETERMINING THE AGE OF SHEEP BY THEIR TEETH

The age of a sheep can be determined by its teeth. Lambs have four pairs of narrow lower incisors called milk teeth or baby teeth. At approximately a year of age the middle pair of milk teeth is replaced by a pair of larger, permanent teeth. At 2 years, a second is replaced. This process continues until, at 4 years of age, the sheep has all permanent incisors. The teeth start to spread apart and some are lost at about 6–7 years of age.

TABLE 28.1.   Summary of Compounds for Control of Internal Parasites of Sheep

Internal Parasites	Drench	Remarks
Coccidia Coccidiosis	Sulfonamides 1. Sulfamethazine 2. Sulfaguanidine	Relatively effective. Will form crystals in kidney of lamb which can cause death
Liver fluke	Albendazole Carbon tetrachloride	A new drug released from FDA for experimental use in Texas, Oregon, Washington, Idaho, and Louisiana. Toxic—use at recommended doses. Effective only against adult flukes
Lung worm	Levamisol (Tramisol)	Effective only against adult lung worm
Stomach and round worms	Tramisol Phenothiazine Thiabendazole	Safe, effective. Will kill arrested worms Control is much improved with the use of fine particle size Safe, effective
Tape worm Broad Fringe	Lead arsenate Dipenthane 70 (teniatol, teniazine) No approved control	May be mixed with phenothiazine or thiabendazole for control of round worms and tape worms with one drench

When all the permanent incisors are lost, the sheep has difficulty grazing and should be marketed.

SELECTED REFERENCES

Publications

American Feed Industry Association. 1985. *Proceedings 1985 Meetings of the American Feed Industry Association Nutrition Council.* American Feed Industry Association. Virginia, Arlington.

Battaglia, R. A. and Mayrose, V. B. 1981. *Handbook of Livestock Management Techniques.* New York: Macmillan Publishing Co.

Chevelle, N. F. 1977. *Foot Rot of Sheep.* Washington, D.C.: Agric. Res. Serv. Farmers' Bull. 2206.

Ensminger, M. E. and Parker, R. O. 1986. *Sheep and Goat Science.* Danville, IL: Interstate Printers and Publishers, Inc.

Gee, C. K. 1979. *A New Look at Sheep for Colorado Ranchers and Farmers.* Ft. Collins: Colorado State Univ. Exp. Stat. Gen. Series 981.

Gee, C. K. and Madsen, A. G. 1974. *Structure and Operation of the Colorado Lamb Feeding Industry.* Ft. Collins: Colorado State Univ. Expt. Stat. Tech. Bull. 121.

Gee, C. K. and Magleby, R. S. 1978. *Characteristics of Sheep Production in the Western Region.* Washington, D.C.: USDA Econ. Res. Serv. Agric. Econ. Rpt. #345.

Gee, C. K., Magleby, R. S., Nielson, D. B. and Stevens, D. M. 1977. *Factors in the Decline of the Western Sheep Industry.* U.S.D.A. Res. Serv. Agric. Econ. Rpt. #377.

Gee, C. K. and Van Arsdall, R. 1978. *Structural Characteristics and Costs of Producing Sheep in the North-Central States.* USDA Econ. Stat. and Coop. Serv. SCS-19.

Michalk, D. L. 1979. Sheep Production in the United States. *Wool Technology and Sheep Breeding.* March/April 1979.

Nutrient Requirements of Sheep. 1985. Natl. Research Council, Washington, D.C.: National Academy Press.

Scott, G. A. 1983. *The Sheepman's Production Handbook.* Denver, CO: Abegg Printing Co., Inc.

Sheep Housing and Equipment Handbook. 1982. Ames, IA: Midwest Plan Service.

Tomes, G. L., Robertson, D. E. and Lightfoot, R. J. (editors). 1979. *Sheep Breeding.* Boston: Butterworth & Co. Ltd.

Ulman, M. and Gee, C. K. 1975. *Prices and Demand for Lamb in the United States.* Fort Collins: Colorado State Univ. Expt. Sta. Tech. Bull. 132.

Visuals

The Sheep Management Series (separate sound filmstrips covering: *Ewe and Lamb Management; Fitting and Showing Sheep; Docking Sheep; Controlling Internal Parasites of Sheep; Sheep Castration; Basic Sheep Handling Skills.* Vocational Education Productions, California Polytechnic State University, San Luis Obispo, CA 93407.

Production Systems for Sheep, Sheep Reproduction and Management Programs, and *Nutrition and Health in Sheep* (sound filmstrips). Prentice-Hall Media, 150 White Plains Rd., Tarrytown, NY 10591.

Sheep Obstetrics, Assuring Baby Lamb Survival, Castration, Docking and Identification, Raising Orphan Lambs, Marketing Practices, Feeding the Farm Flock, Applying Health Care Practices, Feed and Water Delivery Systems, Sheep Handling, Using Equipment and Sheep Psychology. Videotapes. Agricultural Products and Services, 2001 Killebrew Dr., Suite 333, Bloomington, MN 55420.

CHAPTER 29

Horses and Donkeys

Horses, donkeys (asses and burros), and their crosses, **mules** and **hinnies,** have contributed significantly to civilization. In the past, all of these animals provided a swifter method for humans to move longer distances than was possible on foot, and pack (carrying) animals and draft (pulling) animals moved heavier loads than humans could. Horses and mules provided a means of transporting grains and livestock to marketing centers in the U.S. The railroad tracks and early roads were constructed by using the draft power of horses and mules. Horses, donkeys, and their crosses are still important sources of power throughout the world.

OVERVIEW OF THE INDUSTRY

World

Table 29.1 shows the world numbers of horses (63.9 mil), donkeys (39.9 mil), and mules (15.3 mil). The leading countries are given for each of the three species, with China showing the highest numbers for all three.

Horses have been companions for people since the domestication of the horse. They were important in wars, mail delivery, handling other livestock, farming, forest harvesting, and mining. The horse today is used in shows, racing, for handling livestock, and for companionship and recreation. From young children to older adults, the horse has an appeal to people.

Table 29.1 shows the U.S. ranking second in the world for horse numbers with 10.3 mil. Estimates of horse numbers in the U.S. are less accurate than for meat animals and poultry. Horses have not been included in most of the reports estimating animal numbers.

Figure 29.1 gives numbers of horses, mules, and donkeys where data were taken from the 1982 Census of Agriculture. These numbers tend to underestimate the actual horse numbers; however, the ranking of states should be reasonably accurate. Based on these

TABLE 29.1. World Numbers for Horses, Mules and Donkeys, 1984

	World Total (mil)	Leading Countries (mil head)
Horses	63.9	China (10.8), U.S. (10.3), USSR (5.7), Mexico (5.6), Brazil (5.2)
Donkeys	39.9	China (9.4), Ethiopia (3.9), Mexico (2.8), Philippines (2.7), Iran (1.8)
Mules	15.3	China (4.6), Mexico (3.6), Brazil (1.9), Ethiopia (1.5)

Source: 1984 FAO Production Yearbook.

numbers the leading states in horse numbers are Texas, California, Oklahoma, Kentucky, Missouri, and Ohio. The total mules and donkeys in the U.S. is approximately 32,000.

The American Horse Council estimated 5.25 mil horses in the U.S. in 1986, with Texas (478,000) and California (389,000) the leading states in horse numbers. Most of the 5 mil horses are considered backyard horses.

ORIGIN AND DOMESTICATION OF THE HORSE

Fossil remains show the Eohippus, or dawn horse, a four-toed animal less than a foot high, to be the oldest relative of the horse. This early ancestor was originally a wet-area inhabitant but, through evolutionary changes, it became larger and became the Mesohippus, an animal about the size of a collie dog. Evolutionary changes in the teeth, length of leg, and length of neck made Mesohippus or middle horse capable of foraging and surviving on the prairies in the Great Plains. The third toe grew longer and the other toes disappeared, resulting in a hoof that is characteristic of horses today.

Even though these horses were present some 50–60 million years ago, no horses were present on the continent of North America when Columbus arrived. All horses had completely disappeared. It is assumed that some crossed from Alaska into Sibera, and it is from these animals that horses evolved in Asia and Europe. The draft horses and Shetland ponies developed in Europe while the lighter, more agile horses developed in Asia and the Middle East.

The horse was one of the last farm animals to be domesticated, which occurred about 5,000 years ago. Horses were first used as food, then for war and sports, and also for draft purposes. They were used for transporting people swiftly and for moving heavy loads. Also, horses became important in farming, mining, and forestry.

The donkey was domesticated in Egypt earlier than when domestication of the horse occurred. They apparently descended from the wild ass of Africa. Donkeys and zebras are in the same genus (Equus) but are different species from horses (see Fig. 13.8, Chapter 13). Horses mate with donkeys and zebras but the offspring produced are sterile.

United States

In the early 1900s there were approximately 25 mil horses and mules in the U.S.; however, shortly after World War I horse numbers began a rapid decline. The war stimulated the development and use of motor-powered equipment such as automobiles, trucks, trac-

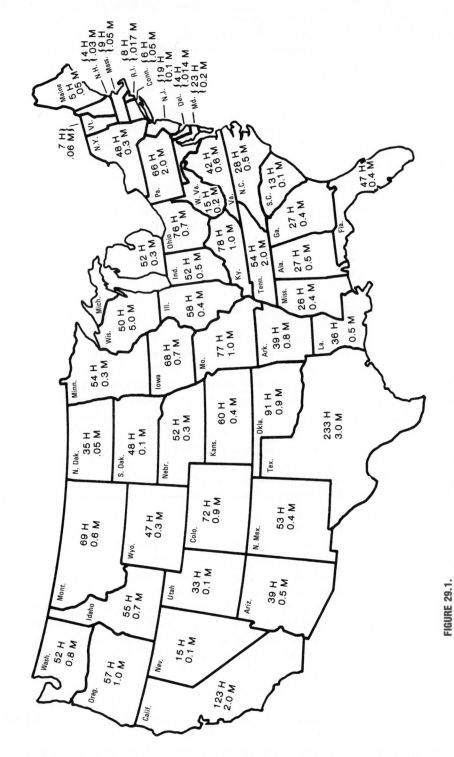

FIGURE 29.1.
Thousands of horses (H) and thousands of mules, burros, and donkeys (M) identified in the 1982 Census of Agriculture. Data courtesy of the U.S. Department of Commerce.

tors, and bulldozers. Railroads were heavily used for transporting people and for moving freight long distances. By the early 1960s, horses and mules had declined to a mere 3 mil.

Recently, horses have increased primarily as an animal used in recreation. With the shorter work week and greater affluence of the working class, there has been more time and money available for recreation. The horse has always been an animal that people enjoy for companionship and exercise. In the mid-1980s, the horse industry employs thousands of people and gives pleasure to millions as well as generating more than 15 bil dollars annually. The money generated by the horse industry includes income for salaries and services, taxes on parimutuel betting, income from breeding, showing and sales of horses, expenditures for feeds and medicines, and investment in the animals, land, and facilities. The horse industry made its greatest growth between 1960 and 1976. Pleasure riding is the main contribution made by horses, though many people think of horse racing as the primary contribution horses make to people.

TERMINOLOGY

A **stallion** is an intact male horse of breeding age, usually over 3 years of age. A **gelding** is a male horse that has been castrated. The young male, less than 3 years old, is called a **colt.** The young female is called a **filly.** Young fillies and colts are collectively called **foals.** A female horse of breeding age is a **mare.**

A close relative of the horse is the donkey, which is also known as an ass or burro. The male ass is referred to as a **jack,** while the female is known as a **jennet.**

A **mule** is produced by crossing a **jackass** with a mare. A **hinny** is produced by crossing a stallion with a jennet. Mules and hinnys are reproductively sterile, although they exhibit normal visual sexual characteristics.

BREEDS OF HORSES

Horses have been used for so many different purposes that many breeds have been developed to fill a specific need. The major breeds of horses and the primary use are listed in Table 29.2. No attempt has been made in Table 29.2 to indicate all the ways each of the breeds are used. For example, the Thoroughbred (Fig. 29.2) is classified primarily as a race horse for long races, but it is used for several other functions.

The "light" breeds of horses provide pleasure for their owners through activities such as racing, riding, and exhibition in shows. Figure 29.3 and Plates N and O (following p. 370) shows examples of these breeds. Quarter Horses are excellent as cutting horses (horses that separate individual cattle out of a herd) and for running short races. Ponies (Plate P, following p. 370) such as the Shetland and Welsh, are selected for their friendliness and safety with children. The American Saddle Horse and the Tennessee Walking Horse have been selected for their comfortable gaits and responsive attitudes. The Palomino, Appaloosa, and Paint horses are color breeds developed for showing and working livestock.

Beauty in color and markings is important in horses used for show and breeding purposes, and color breeds have been developed accordingly. The Appaloosa, for example, has color markings of three patterns: leopard, blanket, and roan. It appears that these color patterns are under different genetic controls. In Appaloosas, Paints, and Palominos, coloration can be adversely affected or eliminated by certain other genes, such as the gene

TABLE 29.2. Characteristics and Uses of Selected Breeds of Horses

Breed	Color	Height in Hands[a]	Weight in lb.	Uses
Riding and Harness Horses				
American Quarter Horse	All colors	14.2–15.2	1,000–1,250	Short racing, showing, stock work
American Saddle Horse	Chestnut, bay, brown, black	15.0–16.0	1,000–1,150	Showing, pleasure riding, 3 and 5 gaited
Arabian	Bay, chestnut, brown, gray, black	14.2–15.2	850–1,000	Pleasure riding, showing
Morgan	Bay, chestnut, brown, black	14.2–15.1	950–1,150	Pleasure riding, driving, showing
Standardbred	Bay, chestnut, roan, brown, black, gray	14.2–16.2	850–1,200	Harness racing
Tennessee Walking Horse	All colors	15.0–16.0	1,000–1,200	Pleasure riding, showing
Thoroughbred	Bay, brown, gray, chestnut, black, roan	15.2–17.0	1,000–1,300	Long racing
Ponies				
Hackney	Bay, chestnut, black, brown	11.2–14.2	450–850	Light harness, showing
Pony of America	Appaloosa	9.2–10.0	300–400	Riding by children, showing
Shetland	Bay, chestnut, brown, black, spotted, mouse			Riding by children, showing
Welsh	Bay, chestnut, black, roan, gray	11.0–13.0	350–850	Riding by children, showing
Draft[b]				
Belgian	Chestnut, roan usually	15.2–17.0	1,900–2,400	Heavy pulling
Clydesdale	Bay, brown, black	15.2–17.0	1,700–2,000	Heavy pulling
Percheron	Black, gray usually	15.2–17.0	1,600–2,200	Heavy pulling
Shire	Bay, brown, usually black	16.2–17.0	1,800–2,200	Heavy pulling
Suffolk	Chestnut	15.2–16.2	1,500–1,900	Heavy pulling
Color Registries				
Appaloosa	Leopard, blanket, roan	14–16	900–1,250	Pleasure riding, showing, stock work
Buckskin	Buckskin, dun, grulla	14–16	900–1,250	Pleasure riding, showing, stock work
Paint	Tobiano, overo	14–16	900–1,250	Pleasure riding, showing, stock work
Palomino	Palomino	14–16	900–1,250	Pleasure riding, showing, stock work
Pinto	Pinto	14–16	900–1,250	Pleasure riding, showing, stock work
White and cremes	White and creme	14–16	900–1,250	Pleasure riding, showing, stock work

[a]Height is measured in inches but reported in "hands." A "hand" is 4 in.
[b]Draft horses are heavy horses used for pulling; other horses are called light horses and are used primarily as pleasure horses.

FIGURE 29.2.
The thoroughbred, Foolish Pleasure, in action. Thoroughbreds are great race horses. Note the extreme muscular development. Courtesy of New York Racing Association, Mr. Louis Weintraub, Photo Communication Company, and The Jockey Club, New York, New York.

for roaning and the gene for graying. The gene for gray can eliminate both colors and markings, as shown by gray horses that turn white with age.

Another important group of horses is the draft horses (Fig. 29.4 and Plate P, following p. 370)—large and powerful animals that are used for heavy work. The Percheron, Shire, Clydesdale, Belgian and Suffolk are examples (Table 29.2).

Popularity of Breeds

Registration numbers is one measure of the popularity of breeds. Table 29.3 shows the breed registration numbers.

BREEDING PROGRAM

Reproduction

Mares of the light breeds reach sexual maturity at 12–18 months of age, whereas draft mares are 18–24 months of age when sexual maturity is reached. Mares come into heat every 21 days during the breeding season if they do not become pregnant. Heat lasts for 5–7 days. Ovulation occurs toward the end of heat. Because of the relatively long duration of heat and because ovulation occurs toward the end of heat, horse owners often delay breeding the mare for 2 days after she has first been observed in heat.

Although about 10% of all ovulations in mares are multiple ovulations, twinning occurs in only about 0.5% of the pregnancies that carry to term. The uterus of the mare apparently cannot support twin fetuses; consequently, most twin conceptions result in the loss of both embryos. The length of gestation is about 340 days (approximately 11 months) with usually only one foal being born. Mares usually come into heat 5 to 12 days following foaling, and fertile matings occur at this heat if the mare has recovered from the previous delivery.

Quarter Horse

Pinto

Appaloosa

American White Horse

Palomino

United States Trotting Horse

FIGURE 29.3.
Some breeds of pleasure horses. Courtesy of American Quarter Horse Association, Amarillo, TX (portrait by Orren Mixer); United States Trotting Association, Columbus, OH; Appaloosa Horse Club, Moscow, ID; Pinto Horse Association, San Diego, CA; Palomino Horse Breeders of America, Mineral Wells, TX (portrait by Orren Mixer); American White Horse Registry, Crabtree, OR.

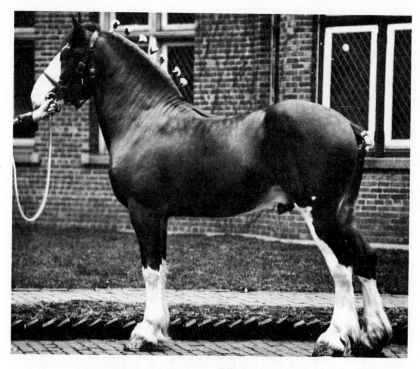

FIGURE 29.4.
A Clydesdale stallion. Courtesy of the USDA.

Selection

Most breeders know that weaknesses exist either in their entire herd or in some individuals of the herd. Breeders generally attempt to find a stallion that is particularly strong in the trait or traits that need strengthening in the herd. Obviously, it is also necessary to be sure that another weakness is not brought into the herd, so it is better to use a stallion with no undesirable traits, even if he is only average in the trait that needs correcting in the herd. To emphasize one trait only for correcting a weakness while bringing another weakness into the herd simply means after 25–30 years of breeding, little or no improvement will be evident.

To have an effective breeding program, it is necessary to be completely objective. There is no place in a breeding program for sympathy toward an animal or for personal pets. Every attempt should be made to see that the environment is the same for all animals in a breeding program. A particularly appealing foal that is given special care and training may develop into a desirable animal. However, many foals less appealing in early life might also develop into desirable animals if given special care and training. A desirable animal that has had special care and training may transmit favorable hereditary traits no better than a less desirable animal that has not had this treatment. Selecting a horse for breeding that has had special care and training may in fact be selecting only for special care and training. Certainly, these environmentally produced differences in horses

TABLE 29.3.   Annual Registration Numbers for Draft, Light Horse, and Pony Breed

| Breed | U.S. Breed Association | | | Canadian Breed Associations |
	1986	1970	Date Association Formed	1985
Draft				
Belgian	4,205	670	1887	1,023
Percheron	1,500	280	1905	582
Clydesdale	200	200	1879	293
Light Horses				
Quarter Horse	153,773	65,326	1956	97
Thoroughbred	49,700	23,201	—	
Arabian	28,283	55,000	1908	1,735
Standardbred	17,637	10,882	—	
Appaloosa	14,551	12,389	1938	—
Paint	11,273	2,258	1956	—
Tennessee Walking	7,900	8,492	1935	—
Morgan	4,329	2,134	1909	682
Palomino	1,500	3,000	1936	10
Ponies				
Shetland	600	1,044	1888	—
Miniature	600	—	1909	—
Welsh	500	889	—	193

Source: Adapted from The 1987 American Horse Industry Directory and the 1985 North American Livestock Census.

are not inherited and will not be transmitted. In fact, it is often wise to select animals that were developed under the type of environment in which they are expected to perform. If stock horses are being developed for herding cattle in rough, rugged country, selection under such conditions is more desirable than where conditions are less rigorous. Horses that possess inherited weaknesses tend to become unsound in rugged environment, and as a result are not used for breeding. Such animals might never show those inherited weaknesses in a less rugged environment.

The ideal environment for most horse-breeding programs is one in which there is plenty of quality forage distributed over an area that requires horses to obtain adequate exercise as they graze. Where the land is limited, highly productive, and valuable, the animals may have to be kept in a small area and forced to exercise a great deal. Forced exercise tends to keep the animals from becoming too fat and gives strength to the feet and legs. Horses need regular, not sporadic, exercise. Regular exercise, even if quite strenuous, is healthy for genetically sound animals and may reveal the weaknesses of those that are not. Strenuous exercise can, however, be harmful to animals that have not exercised for a considerable period to time.

An environment should be provided that will identify horses having genetic superiority for the purposes they are to fulfill. For example, horses that are being bred for endurance in traveling should be made to travel long distances on a regular basis to determine if they can remain sound. Horses bred for jumping should be trained to jump as soon as the animal is physically mature so that those lacking the ability to jump or those that become unsound from jumping can be removed from the breeding program. Draft horses should be trained to pull heavy loads early in life (3 or 4 years of age) to determine their ability to remain sound and their willingness to pull. Performance is needed before animals are used in a breeding program.

Any horse that is unsound should not be used for breeding regardless of the purposes for which the horse is being bred. Such abnormalities as toeing in or toeing out, sickle hocks, cow hocks, and contracted heels will likely lead to unsoundness and difficulties or lack of safety in traveling. Interference and forging actions are extremely objectionable because they can cause the horse to stumble or fall. Defects of the eyes, of the mouth, and of respiration should be selected against in all horses.

CONFORMATION OF THE HORSE

Body Parts

To understand the conformation of the horse, one should be familiar with the body parts (Fig. 29.5). The more detailed skeletal structure of the horse was shown in Chapter 17.

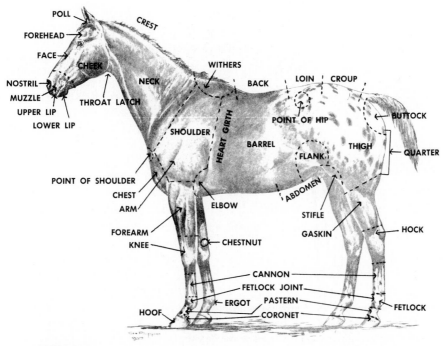

FIGURE 29.5.
The external parts of a horse. Courtesy of Appaloosa Horse Club, Moscow, ID.

Feet and Legs

Major emphasis is placed on feet and legs in describing conformation in the horse—both from identifying correctness and conditions of unsoundness. The old saying, "no feet–no horse" is still considered valid by most horse producers. As feet and leg structure is important, a review of skeletal structure of the front leg and hind leg is given in Figs. 29.6 and 29.7.

Figure 29.8 shows the front legs from a front view. From this view, a vertical line from the point of the shoulder should fall in the centers of the knee, pastern, cannon and foot. Each leg is divided into two equal halves. Deviations from this "ideal position" are shown with the common terminology associated with them.

The front legs from a side view are shown in Fig. 29.9. A vertical line from the shoulder should fall through the center of the elbow and the center of the foot. The angle of the pastern in the "ideal position" is 45°.

Figure 29.10 shows the correct hind leg position from a rear view with less desirable positions of the feet and legs. In the "ideal position," a vertical line from the point of the buttocks should fall through the centers of the hock, cannon, pastern, and foot.

The "ideal position" of the hind legs from a side view is shown in Fig. 29.11. Deviations from the ideal position are shown and described. The "ideal position" is shown where a vertical line from the point of the buttocks touches the rear edge of the cannon from the hock to the fetlock and meets the ground behind the heel.

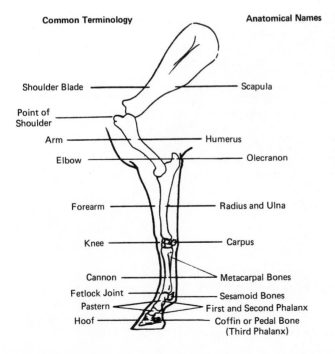

Common Terminology   Anatomical Names

Shoulder Blade — Scapula
Point of Shoulder
Arm — Humerus
Elbow — Olecranon
Forearm — Radius and Ulna
Knee — Carpus
Cannon — Metacarpal Bones
Fetlock Joint — Sesamoid Bones
Pastern — First and Second Phalanx
Hoof — Coffin or Pedal Bone (Third Phalanx)

FIGURE 29.6.
Skeletal front leg with common terminology and anatomical names. Courtesy of Colorado State University Extension Service Publication MOOOOOG.

Common Terminology   **Anatomical Names**

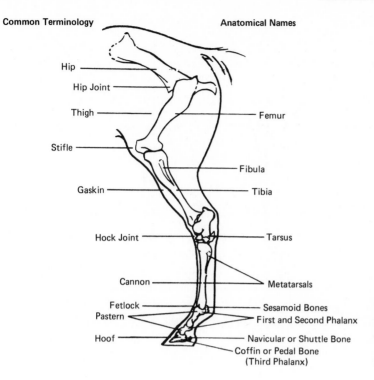

- Hip
- Hip Joint
- Thigh ———————————— Femur
- Stifle
- Fibula
- Gaskin ———————————— Tibia
- Hock Joint ———————————— Tarsus
- Cannon ———————————— Metatarsals
- Fetlock ———————————— Sesamoid Bones
- Pastern ———————————— First and Second Phalanx
- Hoof ———————————— Navicular or Shuttle Bone
  - Coffin or Pedal Bone
  - (Third Phalanx)

FIGURE 29.7.
Skeletal hind leg with common terminology and anatomical names. Courtesy of Colorado State University Extension Service Publication MOOOOOG.

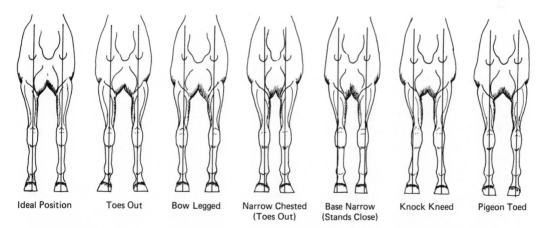

| Ideal Position | Toes Out | Bow Legged | Narrow Chested (Toes Out) | Base Narrow (Stands Close) | Knock Kneed | Pigeon Toed |

FIGURE 29.8.
Correct and faulty conformation of front feet and legs from a front view. Courtesy of Bill Culbertson.

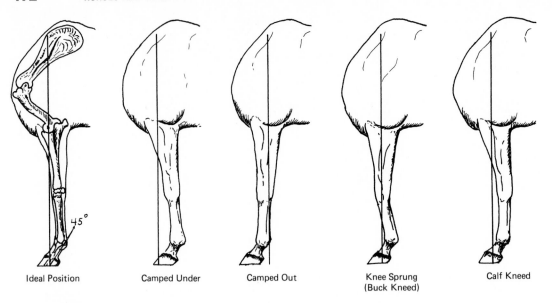

FIGURE 29.9.
Correct and faulty conformation of front feet and legs from a side view. Courtesy of Bill Culbertson.

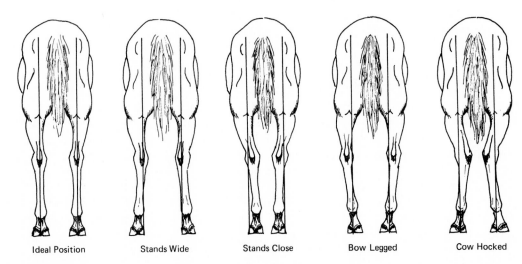

FIGURE 29.10.
Correct and faulty conformation of hind feet and legs as shown in a rear view. Courtesy of Bill Culbertson.

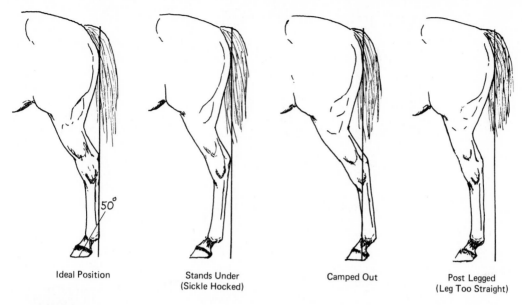

| Ideal Position | Stands Under (Sickle Hocked) | Camped Out | Post Legged (Leg Too Straight) |

FIGURE 29.11.
Correct and faulty conformation of hind feet and legs from a side view. Courtesy of Bill Culbertson.

The Hoof

Care of the horse's feet is essential to keep the horse sound and serviceable. Regular cleaning, trimming, and shoeing (depends on frequency of use) are needed. The hoof will, in mature horses, grow $\frac{1}{4}$–$\frac{1}{2}$ in. per month so trimming is needed every 6–8 weeks.

The external parts of the hoof are shown in Fig. 29.12.

UNSOUNDNESS AND BLEMISHES OF HORSES

Two terms, unsoundness and blemish, are used in denoting abnormal conditions in horses. An **unsoundness** is any defect that interferes with the usefulness of the horse. It may be caused by an injury or improper feeding, be inherited, or develop as a result of inherited abnormalities in conformation. A **blemish** is a defect that detracts from the appearance of the horse but does not interfere with its usefulness. A wire cut or saddle sore may cause a blemish without interfering with the usefulness of the horse.

Horses may have anatomical abnormalities that interfere with their usefulness. Many of these abnormalities are either inherited directly or develop because of an inherited condition. Abnormalities of the eyes, respiratory system, circulatory system, and conformation of the feet and legs are all important.

Some of the major unsoundnesses and blemishes are described, with the location of several of these shown in Fig. 29.13.

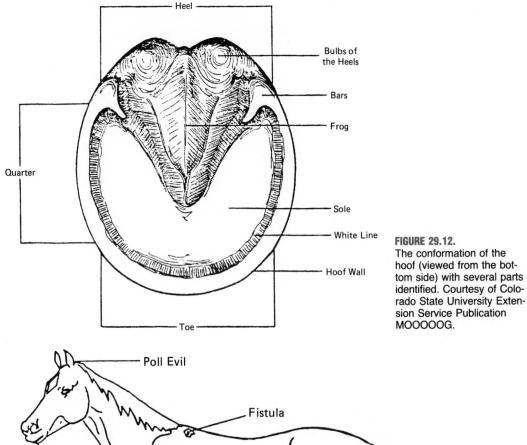

Heel

Bulbs of
the Heels

Bars

Frog

Quarter

Sole

White Line

Hoof Wall

Toe

FIGURE 29.12.
The conformation of the
hoof (viewed from the bottom side) with several parts
identified. Courtesy of Colorado State University Extension Service Publication
MOOOOOG.

Poll Evil

Fistula

Under- or
Overshot Jaw

Sweeney

Hernia

Shoe Boil

Bog Spavin

Bone Spavin

Splints

Stifled

Thoroughpin

Capped Hock

Curbed Hock

Bowed Tendon

Toe Crack

Side Bone

Ringbone

Quarter Crack

FIGURE 29.13.
Location of several potential unsoundnesses and blemishes. Courtesy of Colorado State University Extension
Service Publication MOOOOG.

Bog Spavin

Bog spavin is a soft swelling on the inner, anterior aspect of the hock. Although unsightly, the condition usually does not cause lameness.

Bone Spavin

Bone spavin is a bony enlargement on the inner aspect of the hock. Both bog and bone spavin arise when stresses are applied to horses that have improperly constructed hocks. Lameness usually accompanies this condition, but most animals return to service after rest and in some cases surgery.

Capped Hock

Capped hock is a thickening of the skin at the point of the hock. **Curb** is a hard swelling at the upper-rear of the cannon bone. The plantar ligament becomes inflamed and swollen, usually owing to poor conformation or a direct blow.

Cataract

Cataract is inherited as a dominant trait that could be eliminated if it were not that horses produce several foals before the cataract develops.

Contracted Heels

Contracted heels can be inherited or acquired through improper shoeing and foot care. Contracted heels may result when the frog or cushion of the foot is damaged and shrinks, allowing the heels to come together. The bottom surface of the foot becomes smaller in circumference than at the coronet band.

Cow Hocks

The term cow hocks (Fig. 29.10) indicates that the point of the hocks turn inward. Such hocks are greatly stressed when the horse is pulling, running, or jumping.

Heaves

Heaves is a respiratory disease in which the horse experiences difficulty in exhaling air. The horse can exhale a certain volume of air normally, after which an effort is exerted to complete exhalation. A horse with a mild case of heaves can continue with light work, but horses with moderate to severe heaves have a limited ability to work.

Laminitis

This is also called founder. It is an inflammation of the laminae of the foot causing severe pain and lameness. The front feet are affected more frequently than the hindfeet. Overeating of grain, consumption of large amounts of water by overheated horses, and over-working horses on hard surfaces are frequent causes of laminitis.

Moon Blindness

Moon blindness is periodic ophthalmia in which the horse is blind for a short time, regains its sight, and then again becomes blind for a time. Periods of blindness may initially be spaced as much as 6 months apart. The periods of blindness become progressively closer together until the horse is continuously blind. This condition received the name moon blindness because the trait is first noticed when the periods of blindness occur about a month apart. It was originally thought that the periods of blindness were associated with changes in the moon.

Navicular Disease

Navicular disease is an inflammation of the navicular bone in the front foot. The exact cause is unknown but hard work, small feet, and trimming heels too low may contribute to the development of the disease. Corrective shoeing or surgery (cutting the navicular nerve) are recommended treatments, although the latter may not restore horses to their original usefulness.

Quittor

This is a deep sore that drains at the coronet. The infection resulting from puncture wounds, corns, and other causes, results in severe lameness.

Ring Bone

Ring bone is a condition in which the cartilage around the pastern bone ossifies. Ringbone shows as a hard bony enlargement encircling the areas of the pastern joint and coronet.

Ruptured Blood Vessels

Ruptured blood vessels is a defect of circulation in which the blood vessels in the lungs are fragile and may rupture when the horse is put under the stress of exercising. Some racehorses have been lost owing to hemorrhage from these fragile blood vessels.

Shoe Boil

Shoe boil or capped elbow is a soft swelling on the elbow. Common causes are injury to the elbow while the horse is lying down or injury from a long heel on a front shoe.

Sickle Hocked

Sickle hocked (Fig. 29.11) is a term used when the hock has too much set or bend. As a result, the hind feet are set too far forward. The strain of pulling, jumping, or running is much more severe on a horse with sickle hocks than on a horse whose hocks are of normal conformation.

Sidebones

Sidebones is an abnormality that occurs when the lateral cartilages in the foot ossify. During the ossification process, lameness can occur, but many horses regain the soundness with proper rest and shoeing.

Stifled

A stifled horse is one in which the patella (knee cap in humans) has been displaced. Older horses seldom become sound once they are stifled, while younger horses usually recover.

Stringhalt

This is an involuntary flexion of the hock during movement. It is considered a nerve disorder where surgery can improve the condition.

Sweeny

Sweeny refers to atrophied muscles at any location, although many people refer only to shoulder muscles. In the shoulder sweeny, the nerve crossing the shoulder blade has been injured.

Thoroughpin

A thoroughpin is a soft enlargement between the large tendon (tendon of Achilles) of the hock and the fleshy portion of the hind leg. Stress on the flexor tendon allows synovial fluid to collect in the depression of the hock. Lameness rarely occurs.

Toeing-in and Toeing-out

Toeing-in or pigeon-toed (Fig. 29.8) refers to the turning-in of the toes of the front feet, whereas toeing-out (Fig. 29.8) refers to the turning-out of the toes of the front feet. These conditions influence the way in which the horse will move its feet when traveling. Toeing-out or moving the front feet inward is considered the more serious defect as it can lead to further interference and faults.

Windgalls

These are sometimes referred to as wind puffs or puffs. The fluid sacs around the pastern or fetlock joints are enlarged. They are common, but not serious, in hardworking horses.

GAITS OF HORSES

The major gaits of horses, along with their modifications, are as follows:

1. **Walk** is a four-beat gait in which each of the four feet strikes the ground separately from the others.
2. **Trot** is a rapid diagonal two-beat gait in which the right front and left rear feet hit the ground in unison, and the left front and right rear feet hit the ground in unison. The horse travels straight without weaving sideways when trotting.
3. **Pace** is a lateral two-beat gait in which the right front and rear feet hit the ground in unison and the left front and rear feet hit the ground in unison. There is a swaying from right to left when the horse paces.
4. **Gallop** is the fastest gait with four beats.
5. **Canter** is a fast three-beat gait. Depending on the lead, the two diagonal legs will hit the ground at the same time with the other hind leg and foreleg hitting at different times.
6. **Rack** is a snappy four-beat gait in which the joints of the legs are highly flexed. The forelegs are lifted upward to produce a flashy effect. This is an artificial gait whereas the walk, trot, pace, gallop, and canter are natural gaits. The rack is popular in the showring for speed and animation.
7. **Running Walk** is the fast ground-covering walk of the Tennessee Walking Horse. It is an artificial gait that is faster than the normal walk. The horse moves with a gliding motion as the hindleg oversteps the forefoot print by 12–18 in.

EASE OF RIDING AND WAY OF GOING

When a horse's foot strikes the ground, a large shock is created that would be objectionable to the rider if no shock absorption existed. There are several shock absorbing mechanisms existing in horses' feet and legs. The horse has lateral cartilages on all four feet that expand outward when the foot strikes the ground. This absorbs some of the shock. The pastern on each leg absorbs some of the shock when the foot strikes the ground by bending somewhat. A pastern that is too straight will not absorb much of the shock and one that is too long and weak will let the leg go to the ground. These kinds of a pasterns will soon result in unsound horses. Thus, it is very important that a pastern has the proper slope so it can absorb the optimal amount of shock without the leg going to the ground or causing too much concussion on the joints and ultimately the rider.

The front legs each have two joints that allow movement that absorb shock; the joint between the ulna and the humerus and the joint between the humerus and the scapula. Having movement at these joints results in some absorbing of shock. Also the hind legs have two joints in each that can bend and, thus, absorb some shock. These joints are between the metatarsus and tibia and between the tibia and femur.

If the horse's feet and legs have proper conformation, a pleasant ride can be enjoyed. If there are abnormalities due to inheritance, injury, improper nutrition or disease, the horse will give the rider a less comfortable ride.

Abnormalities in Way of Going

A horse that toes out with its front feet tends to dish or swing its feet inward (wings in) when its legs are in action. Swinging the feet inward can cause the striding foot to strike

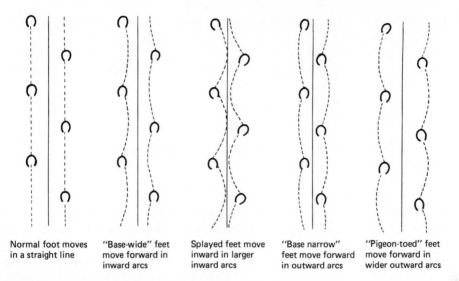

| Normal foot moves in a straight line | "Base-wide" feet move forward in inward arcs | Splayed feet move inward in larger inward arcs | "Base narrow" feet move forward in outward arcs | "Pigeon-toed" feet move forward in wider outward arcs |

FIGURE 29.14.
Path of the feet (way of going) as seen from above, which relates to feet and leg structure. Courtesy of Colorado State University Extension Service Publication MOOOOOG.

the supporting leg so that **interference** to forward movement results. A horse that toes in (pigeon toed) tends to swing its front feet outward, giving a **paddling** action.

Some horses overreach with the hind leg and catch the heel of the front foot with the toe of the hind foot. This action, called **forging,** can cause the horse to stumble or fall.

Figure 29.14 shows the way of going well as the horse moves straight and true, each foot moving in a straight line. The other illustrations show the path of flight of each foot

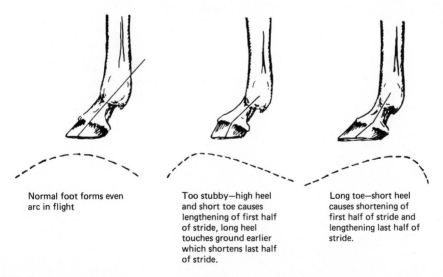

| Normal foot forms even arc in flight | Too stubby—high heel and short toe causes lengthening of first half of stride, long heel touches ground earlier which shortens last half of stride. | Long toe—short heel causes shortening of first half of stride and lengthening last half of stride. |

FIGURE 29.15.
Illustration of how length and slope of the hoof affects way of going. Courtesy of Colorado State University Extension Service Publication MOOOOOG.

when the structure of the foot and leg deviates from the desired norm. How the length and slope of the hoof affects way of going is shown in Fig. 29.15.

Since few horses move perfectly true, it is important to know which movements may be unsafe. A horse which wings in (inteferes) is potentially more unsafe compared with one which wings out (paddles), as the former horse may trip itself.

DETERMINING THE AGE OF A HORSE BY ITS TEETH

The age of a horse can be estimated by its teeth (Figs. 29.16, 29.17, and 29.18). A foal at 6–10 months of age has 24 so-called baby or milk-teeth (12 incisors and 12 molars). The incisors include three pairs of upper and three pairs of lower incisors.

Chewing causes the incisors to become worn. The wearing starts with the middle pair and continues laterally. At 1 year of age, the center incisors show wear; at 1.5 years, the intermediates show wear; and at 2 years, the outer, or lateral, incisors show wear. At 2.5 years, shedding of the baby teeth starts. The center incisors are shed first. Thus, at 2.5 years, the center incisors become permanent teeth; at 4 years, the intermediates are shed; at 5 years, the outer, or lateral, incisors are shed and replaced by permanent teeth.

A horse at 5 years of age is said to have a **full mouth,** because all the teeth are permanent. At 6 years, the center incisors show wear; at 7 years, the intermediates show wear; and at 8 years, the outer, or lateral, incisors show wear. Wearing is shown by a change from a deep groove to a rounded dental cup on the grinding surface of a tooth.

DONKEYS, MULES, AND HINNIES

Mules and hinnies have been used as draft animals in mining and farming operations and as pack animals. When mules were needed for heavy loads, it was important to

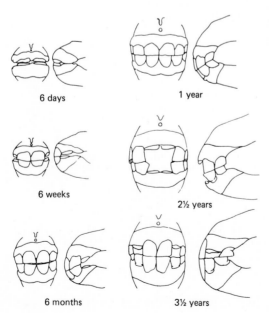

6 days

6 weeks

6 months

1 year

2½ years

3½ years

FIGURE 29.16.
Front and side views of the teeth of the horse from 6 days to 3.6 years of age. The young horse has a full set of "baby," or "milk," teeth by 6 months of age. It starts shedding the baby teeth and developing permanent teeth at 2.5 years of age. From Bone, J. F., *Animal Anatomy and Physiology*, 4th ed., Corvallis: Oregon State University Book Stores, © 1975.

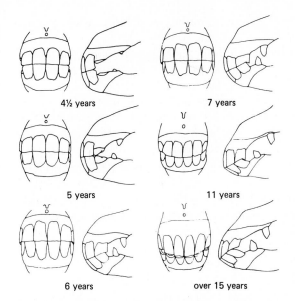

4½ years   7 years

5 years   11 years

6 years   over 15 years

FIGURE 29.17.
Front and side views of the teeth of the horse from
4.5 to 15 years of age. The horse's mouth changes
in shape as it becomes older such that the front
teeth protrude somewhat forward. From Bone, J. F.,
Animal Anatomy and Physiology, 4th ed., Corvallis:
Oregon State University Book Stores, © 1975.

breed mammoth jacks to mares of one of the draft breeds. Crossing of smaller jacks with
mares of medium size produced mules that were useful in mining and farming operations.

Donkeys, mules, and hinnies have some characteristics that make them more useful
than horses for certain purposes. Their sure-footedness makes them ideal pack animals
for moving loads over rough areas. They are rugged and can endure strenuous work. In
addition, they have the characteristic of taking care of themselves. For example, mules
do not gorge themselves when given free access to grain; consequently, they do not

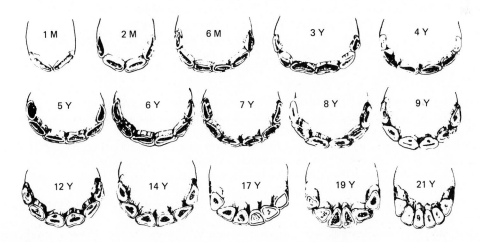

FIGURE 29.18.
Table surfaces of the lower incisors of the horse from 1 month (M) to 21 years (Y) of age. From Bone, J. F.,
Animal Anatomy and Physiology, 4th ed., Corvallis: Oregon State University Book Stores, © 1975.

normally founder from overeating. However, donkeys, mules, and hinnies do not respond to the wishes or commands of humans as well as horses.

Recently, there has been a movement to use small donkeys as pets for children. Small mules from crosses between male donkeys and pony mares have also been popular as children's pets.

The primary reason trucks and tractors replaced horses and mules in farming operations was that work could be done more rapidly; one person could do much more with tractors and trucks than with horses and mules. The high cost of petroleum products (gasoline and diesel fuels) and the possible scarcity of these items may result in more use of horses and mules in farming, logging, and mining operations and a partial return to horse power. It is unlikely that horses and mules will ever replace trucks and tractors, but certain operations on farms or in the timber industry may be done to advantage by horses and mules.

SELECTED REFERENCES

Publications

Blakely, J. 1981. *Horses and Horse Sense: The Practical Science of Horse Husbandry.* Reston, Virginia: Reston Publishing Co., Inc.

Blakely, R. L. 1985. Miniature Horses. *National Geographic* 167:384.

Bone, J. F. 1975. *Animal Anatomy and Physiology.* 4th edition. Corvallis: Oregon State University Book Stores.

Bradley, M. 1981. *Horses. A Practical and Scientific Approach.* Highstown, NJ: McGraw-Hill Book Co.

Evans, J. W., Borton, A., Hintz, H. F. and Van Vleck, L. D. 1986. *The Horse.* San Francisco: W. H. Freeman and Co.

Evans, J. W. 1981. *Horses. A Guide to Selection, Care and Enjoyment.* San Francisco: W. H. Freeman and Co.

Horse Industry Directory. 1987. American Horse Council, 1700 K Street, N.W., Washington, D.C. 20006.

Jones, W. E. 1982. *Genetics and Horse Breeding.* Philadelphia: Lea and Febiger.

Lasley, J. F. 1987. *Genetics of Livestock Improvement.* Englewood Cliffs, NJ: Prentice-Hall.

Rich, G. A. 1981. *Horse Judging Guide.* Colorado State University Ext. Serv. Public. MOOOOOG.

Visuals

The Basic Horsemanship Kit (separate sound filmstrips) covering: *Grooming Your Horse; English Equitation; The Western Equitation Class.* Vocational Education Productions, California Polytechnic State University, San Luis Obispo, CA 93407.

Beginner's Guide to Buying a Horse (sound filmstrip). Vocational Education Productions, California Polytechnic State University, San Luis Obispo, CA 93407.

Horse Breeds and Color Types (slides). Vocational Education Productions, California Polytechnic State University, San Luis Obispo, CA 93407.

Care of Your Horses Feet. SL529-Slide Set. Educational Aids, National 4-H Council 7100 Connecticut Ave., Chevy Chase, MD 20815.

Survival of the Fittest. Two-part 16 mm film (26 min. each). Covers Quarter Horse conformation—the relation of form to function. American Quarter Horse Association, Amarillo, TX 79168.

CHAPTER 30

Feeding and Managing Horses

Horses relate well to people and provide many forms of pleasure. Although many people enjoy horses, most own only a few; however, whether a person keeps only one horse for a pet or operates a large breeding farm, basic horse knowledge is vital. Information on managing horses, such as breeding, feeding, facilities, disease prevention, and parasite control is important.

FEEDS AND FEEDING

Horses eat roughages such as hay, graze on grass and legume pastures, and eat grains of all kinds. Concentrate mixtures containing grains, protein supplements, and vitamin and mineral additives are prepared and sold by commercial feed companies, especially for owners with only one or two horses. Wet molasses can be added to these concentrate mixtures to make sweet feed. The forage and concentrate mixtures may be mixed and pelleted to make a complete, higher-priced convenience feed for horse owners.

Although horses spend less time chewing than ruminants, the normal, healthy horse, with a full set of teeth, can grind grains such as oats, barley, and corn so that cracking or rolling these feeds is unnecessary. Wheat and milo, however, should be cracked to improve digestibility. The stomach is relatively small, composing only 10% of the total digestive capacity. Only a small amount of digestion takes place in the stomach, and food moves rapidly to the small intestine. From 60–70% of the protein and soluble carbohydrates are digested in the small intestine, and about 80% of the fiber is digested in the cecum and colon. The large intestine has about 60% of the total digestive capacity with the colon being the largest component. Bacteria that live in the cecum aid digestion there. Minerals, proteins as amino acids, lipids, and readily available carbohydrates as glucose are absorbed in the small intestine.

Owners of pleasure horses are interested in having them make a desirable appearance and many liberally feed their horses. Perhaps more horses are overfed than are underfed. Also, many people want to be kind to their animals and keep them housed in a box stall when weather conditions are undesirable. This may not be best for the physiological state of the horse. Certainly if any deficiency exists in the feed provided, such a deficiency is much more likely to affect horses that are not running on good pasture, where they can forage for themselves.

Young, growing foals should be fed well to allow for proper growth, but overfeeding and obesity are discouraged. Quality of protein and amounts of protein, minerals, and energy are important for young growing horses. Soybean meal or dried milk products in the concentrate mixture will provide the amino acids and minerals that might otherwise be deficient or marginal in the weanling ration.

Good-quality pasture or hay supplemented with grain can provide the nutrition needed by young horses. An appropriate salt-mineral source should be provided at all times and clean water is essential (Fig. 30.1).

During the first 8 months of gestation, pregnant mares perform well on good pastures or on good-quality hay supplemented with a small amount of grain and an appropriate source of salt-mineral. As the fetus grows during the last trimester of pregnancy, the mare will require more concentrate and less fibrous, bulky hay. Oats, corn, or barley make an excellent feed grain for pregnant mares. Pregnant mares should be in good body condition but avoid obesity as in any horse. Animals used for riding or working, whether they are pregnant or not, need more energy than those not working. Horses being exercised heavily should be fed ample amounts of concentrates to replace the energy used in the work.

Generally, lactating mares require more grain feeding than do geldings, and pregnant or nonpregnant mares. Lactation is the most stressful period for a mare in regards to

FIGURE 30.1.
Horses need a readily available supply of fresh, clean water. Courtesy of Dr. Robert Henderson, Oregon Agricultural Experimental Station.

nutrition. If lactating mares are exercised, they must be fed additional grain and hay to meet the nutrient demand of the physical activity.

Stallions need to be fed as a working horse during the breeding season, but fed a maintenance ration during the nonbreeding season. Feeding good-quality hay with limited amounts of grain is usually sufficient for the stallion.

Feed companies provide properly balanced rations for horse farms of all sizes. An owner who has only one or two horses may find it highly advantageous to use a prepared feed, since it is difficult and laborious to prepare balanced rations for only a few animals. Using commercially prepared feeds or custom-blended rations can prevent nutritional errors, can save time, and is more cost effective with larger farms.

Table 30.1 shows some examples of horse rations. Some comments are given as to when and how these rations should be fed.

MANAGING HORSES

Proper management of horses is essential at several critical times. These critical times are the breeding season, foaling, weaning of foals, castration, and during strenuous work.

Breeding Season

Mares should be **teased** daily with a stallion to determine the stage of her estrous cycle (Fig. 30.2). When the mare is in standing heat (estrus), she is ready to be bred. When the stallion approaches the front of the mare, the mare reacts violently against the stallion if not in heat, but squats and urinates with a winking of the vulva if in heat.

To insure safety for the mare, stallion, and handlers, breeding hobbles (Fig. 30.3) should be fitted to the mare. Cleanliness is paramount for both natural breedings and artificial insemination (AI) programs.

Mares bred naturally should have the vulva washed and dried and the tail wrapped before being served by the stallion (Fig. 30.4). If the mare can be bred twice during heat

FIGURE 30.2.
Chute teasing. Stallion is led along a chute holding several mares. The behavior of each mare will determine if she is in estrus. Courtesy of Colorado State University.

TABLE 30.1. Sample Rations for Horses of Different Ages and In Various Stages of Production

Creep feed for nursing foals. The grain should be fed at the rate of 0.5–0.75 lb of grain/100 lb body weight.

	Percentage in Grain Mix
Corn, rolled or flaked	34.0
Oats, rolled or flaked	34.0
Soybean oil meal	22.0
Molasses	6.0
Dicalcium phosphate	2.0
Limestone	1.5
Trace mineral salt	0.5

Grain mix for weanlings. The grain should be fed at the rate of 0.75–1.5 lb of grain/100 lb body weight. Select the grain according to the type of roughage fed. Allow free choice consumption of either roughage type.

	Percentage in Grain Mix	
	Alfalfa Hay	**Grass Hay**
Corn, rolled or flaked	40	34
Oats, rolled or flaked	40	34
Soybean oil meal	12.0	23.0
Molasses	5.0	5.0
Dicalcium phosphate	2.5	3.0
Limestone	0	0.5
Trace mineral salt	0.5	0.5

Grain mix for yearlings and mares. The grain should be fed at the rate of 0.5–1.0 lb of grain/100 lb of body weight. Select the grain according to the type of roughage fed. Allow free choice consumption of either roughage.

Mares during late gestation and lactation can be fed the same grain mixes as yearlings. The grains should be fed at the rate of 0–0.5 lb/100 lb of body weight. Roughage consumption can vary from 1.5 to 2.5 lb/100 lb of body weight.

	Percentage in Grain mix	
	Alfalfa Hay	**Grass Hay**
Corn, rolled or flaked	46.5	38
Oats, rolled or flaked	46.5	38
Soybean oil meal	0	15.5
Molasses	5	5
Dicalcium phosphate	1.5	2.0
Limestone	0	1.0
Trace mineral salt	0.5	0.5

Grain mix for horses at maintenance, and at work, dry mares during the first 8 months of gestation or stallions. The grain should be fed as needed (to maintain body condition). Roughage consumption will vary from 1.5 to 2.5 lb/100 lb of body weight.

	Percentage in Grain Mix	
	Alfalfa Hay	**Grass Hay**
Corn	46.5	46.5
Oats	46.5	46.5
Molasses	5.0	5.0
Dicalcium phosphate	0	1.5
Monosodium phosphate	1.5	0
Trace mineral salt	0.5	0.5

Courtesy of Ginger Rich, Colorado State University

FIGURE 30.3.
Breeding hobbies. From Battaglia and Mayrose, *Handbook of Livestock Management
Techniques,* Macmillan Publishing Company,
1981.

FIGURE 30.4.
Tail of horse being wrapped prior to breeding. From Battag-
lia and Mayrose, *Handbook of Livestock Management
Techniques,* Macmillan Publishing Company, 1981.

without overusing the stallion, breeding 2 and 4 days after the mare is first noticed in heat is desirable. A mature stallion can serve twice daily over a short time and once per day over a period of 1 or 2 months. A young stallion should be used lightly at about three or fewer services per week. In an AI program the stallion can be collected every other day to cover as many mares as possible depending on the stallions sperm numbers and motility. AI programs allow better management of the stallion.

Foaling Time.   Mares normally give birth to foals in early spring, at which time the weather can be unpleasant. A clean box stall that is bedded with fresh straw should be available. If the weather is pleasant, mares can foal on clean pastures. When a mare starts to foal, she should be observed carefully, but she should not be disturbed unless assistance is necessary. If the head and front feet of the foal are being presented, it should be delivered without difficulty (Fig. 30.5). If necessary, however, a qualified person can assist by pulling the foal as the mare labors. Do not pull when the mare is not laboring and do not use a tackle to pull the foal. If the front feet are presented but not the head (Fig. 30.6), it may be necessary to push the foal back enough to get the head started along with the front feet. Breech presentations can endanger the foal if delivery is delayed; therefore, assistance should be given to help the mare make a rapid delivery if breech presentation occurs. If it appears that difficulties are likely to occur, a veterinarian should be called as soon as possible.

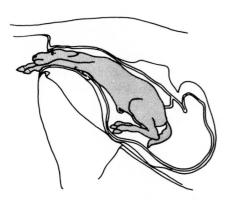

FIGURE 30.5.
Correct presentation position of foal for delivery. From Battaglia and Mayrose, *Handbook of Livestock Management Techniques,* Macmillan Publishing Company, 1981.

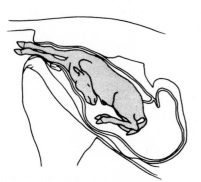

FIGURE 30.6.
Malpresentation; head and neck back. From Battaglia and Mayrose. *Handbook of Livestock Management Techniques,* Macmillan Publishing Company, 1981.

As soon as the foal is delivered, its mouth and nostrils should be cleared of membranes and mucus so it can breathe. If the weather is cold, the foal should be wiped dry and assisted in nursing. As soon as the foal nurses, its metabolic rate increases and helps it stay warm. The **umbilical cord** should be dipped in a tincture of iodine solution to prevent harmful microorganisms from invading the body.

In general, it is highly desirable to exercise pregnant mares up to the time of foaling. Mares that are exercised properly while pregnant have better muscle tone and are likely to experience less difficulty when foaling than those that get little or no exercise while pregnant. During the horse-power era, many mares used for plowing and similar work foaled in the field without difficulty. The foal was usually left with the mare for a few days, after which the foal was left in a box stall while the mare was worked.

Weaning the Foal.   When weaning time arrives, it is best to remove the mare and allow the foal to remain in the surroundings to which it is accustomed. Since the foal will make every attempt to escape to find its mother, it should be left in a box stall or secure and safe fenced lot. Fencing other than barbed wire should be used. Prior to weaning, the foal should become accustomed to eating and drinking on its own. High quality hay or pasture and a balanced concentrate feed should be provided for the foal.

Castration.   Colts that are not kept for breeding can be castrated any time after the testicles have descended, however the stress of castration should not be imposed at weaning time. Some people prefer to delay castration until the colt has reached a year of age while others prefer an earlier time. Genetic potential and nutrition determine mature height and weight and not time of castration.

The colt can be put onto a turntable or manually restrained with ropes for castration. Pain killers and muscle relaxants can be used. The scrotum is opened on each side, by a qualified person, and the membrane around each testicle is split to expose the testicle. The testicle is pulled from the body cavity far enough to expose the cord for clamping. The cord is crushed by the clamp and then severed. The crushing of the cord prevents excessive bleeding. It may be desirable to give the colt an injection of an antibiotic to prevent infection. If the scrotum was cleaned with a mild disinfectant before castration, the wound need not be washed with an antiseptic. Harsh disinfectants should not be applied to the wound. After the colt is castrated, he should be in a small clean pasture for close observation. If fly infestation is a problem, a fly repellant should be applied about the scrotal area.

Horses can be permanently identified by tattooing them on the inside of the upper lip. This does not disfigure the animals and can easily be read by raising the upper lip. Other methods of identification include freeze branding, hot-iron branding, electronic implants, and chestnut implants.

Care of Hard-Working Animals.   Hard-working horses that are sweating and have elevated temperature, pulse, and respiration rates should be washed ("cooled down") before receiving water and feed. The animals should be given only small amounts of water, walked until they have stopped sweating, and their pulse and respiration are within 10% of resting levels. At that time, water, hay, and grain can be given.

Hard-working horses do require extra energy usually in the form of grain, however care must be taken to avoid overfeeding grain to horses. Excess grain and sometimes lush green pasture can cause a horse to founder. Founder can cause death or severe lameness.

HOUSING AND EQUIPMENT

Barn.   Barns should be located on a higher elevation than the surrounding area to assure good drainage. They should be accessible to utilities and vehicles, and preferably have a southeast exposure.

Although many styles of barns exist, one constructed with an aisleway between two rows of stalls provides easy access and efficient use of space. Feed can be placed in the stalls on either side as the feed cart goes down the aisle. If the stalls can be opened from the outside and cleaned with mechanical equipment, labor also is saved. The hay manger in stalls should be constructed at chest-height to the horse. Hay racks placed above the horse's head force the horse to inhale hay dust as they reach for the hay. Health problems can result.

To alleviate mixing hay with grain, a separate grain feeder should be placed several feet from the hay rack. The waterer or water bucket should be located along the outside wall for drainage purposes. The water source should also be placed some distance from the hay and grain so feed will not drop in the water. Electric waterers save labor but are prone to malfunction. Water buckets require more labor but water intake can be easily monitored.

Box Stall.   Box stalls are used for foaling and for the mare and the foal when the weather is severe. The stall should be constructed to allow complete cleaning and proper drainage. Minimum dimensions for a foaling stall are 14 ft. × 14 ft., whereas a regular box stall should be at least 10 ft. × 10 ft.

Feed.   Feed should be stored in an area connected with or adjacent to the barn. Truck access to feed storage area is vital. The floor should be solid (i.e., concrete) and the rodent-proof bins should be easily cleaned. Overhead (loft) storage of hay requires a great deal of labor and expensive construction. Large amounts of hay stored in the barn where the horses are housed increases dust and fire danger.

Fences.   Lot fences should be constructed with wooden or steel posts and cable, or with wooden posts and 2-in. lumber. Barbed wire should not be used because it can cut a horse severely.

Tack Room.   A dry, dust-free tack room is needed to house riding equipment, including saddles, blankets, bridles, and halters. In addition, it may be advisable to have a wash room where horses can be washed.

Chute.   A chute is essential to care for horses that have been injured and need attention. If the horse is in a chute, it is unlikely that the animal can kick the person working around the horse. The chute can be used for AI, teasing, palpation, or treating health problems. The chute is also useful for injections or when blood samples are being taken.

Often, horses strike with a front foot when a needle is inserted for vaccinations or blood samples. A properly constructed chute will protect people as well as the horse.

CONTROLLING DISEASES AND PARASITES

Sanitation is of vital importance in controlling diseases and parasites of horses. Horses should have clean stalls and should be groomed regularly. A horse that is to be introduced into the herd should be isolated for a month to prevent exposing other horses to diseases and parasites.

Horses may have illnesses caused by bacteria or viruses, internal or external parasites, poisonous plants such as Tanzy Ragwort or Brachen fern, inhalation or ingestion of chemicals that have been used about the barn, or receiving an imbalanced ration.

The person who owns and cares for one horse or a dozen should engage the services of a veterinarian in whom the owner has complete confidence. Veterinarians can help prevent diseases as well as treat animals that are diseased. Most veterinarians prefer to assist in preventing health problems rather than treat the animals after they become ill.

Horse manure is an excellent medium for microorganisms that cause **tetanus.** Horses should be given shots to prevent tetanus, which can develop if an injury allows tetanus-causing microorganisms to invade through the skin. Usually two shots are given to establish **immunity,** after which a booster shot is given each year. Because the same microorganisms that cause tetanus in horses also affect humans, those who work with horses should also have tetanus shots.

Strangles, also known as distemper, is a bacterial disease that affects the upper respiratory tract and associated lymph glands. High fever, nasal discharge, and swollen lymph glands are signs of strangles. This disease is spread by contamination of feed and water. Afflicted horses must be isolated and provided clean water and feed. A strangles bacterin is available, but postvaccinal reactions limit its use to stables and ranches in which the disease is endemic.

Brood mares are subject to many infectious agents that invade the uterus and cause **abortion;** examples include **Salmonella** and **Streptococcus** bacteria and the viruses of **rhinopneumonitis** and **arteritis.** Should abortion occur, professional assistance should be obtained to determine the specific cause and develop a preventative plan for the future.

Sleeping sickness, or **equine encephalomyelitis,** is caused by viral infections that affect the brain of the horse. Different types, such as the eastern, western, and Venezuelan are known. Encephalomyelitis is transmitted by vectors such as mosquitoes. It can also be spread by horses rubbing noses together or sharing water and feed containers. This disease is also transmissable to humans. Vaccination against the disease consists of two intradermal injections spaced a week to 10 days apart. These injections should be given in April. The vaccination ensures immunity for only 6 months; therefore, the injections should be repeated each six months when the disease is prevalent.

Influenza is a common respiratory disease of horses. The virus that causes influenza is airborne, so frequent exposure may occur where horses congregate. The acute disease causes high fever and a severe cough when the horse is exercised. Rest and good nursing care for 3 weeks usually gives the horse an opportunity to recover. Severe after-effects are rare when complete rest is provided. Horse owners who plan for shows should vaccinate for influenza each spring. Two injections are required the first year, with one annual booster thereafter.

Horses can become infested with internal and external parasites. Control of internal parasites consists of rotating horses from one pasture to another, spreading manure from stables on land that horses do not graze, and treating infested animals.

Pinworms develop in the colon and rectum from eggs that are swallowed as the horse consumes contaminated feed or water. These parasites irritate the anus, which causes the horse to rub the base of its tail against objects even to the point of wearing off hair and causing skin abrasions. Pinworms are controlled by oral administration of proper vermifuges.

Bots are the larvae stage of the bot fly. The female bot fly lays eggs on the hairs of the throat, front legs and belly of the horse. The irritation of the bot fly causes the horse to lick itself. The eggs are then attached to the tongue and lips of the horse, where they hatch into larvae that burrow into the tissues. The larvae later migrate down the throat and attach to the lining of the stomach, where they remain for about 6 months and cause serious damage. As the chemicals used in the control of bots can be injurious if not properly administered, or given at the proper dosages, professional assistance should be obtained for treatment.

Adult **strongyles** (bloodworms) firmly attach to the walls of the large intestine. The adult female lays eggs that pass out with the feces. After the eggs hatch, the larvae climb blades of grass where they are swallowed by grazing horses. The larvae migrate to various organs and arteries where severe damage results. Blood clots form where arteries are damaged. Clots can break loose and plug an artery. Treatment of bloodworms consists of phenothiazine mixed with the feed or phenothiazine–piperazone mixture administered orally.

Adult **ascaris** worms are located in the small intestine. The adult female produces large numbers of eggs that pass out with the feces. The eggs become infective if they are swallowed when the horse eats them while grazing. The eggs hatch in the stomach and small intestine and the larvae migrate into the bloodstream and are carried to the liver and lungs. The small larvae are coughed up from the lungs and swallowed. When they reach the small intestine, they mature and produce eggs. The same chemicals used for the control of bots are effective in the control of ascarids.

SELECTED REFERENCES

Publications

Battaglia, R. A. and Mayrose, V. B. 1981. *Handbook of Livestock Management Techniques*. New York: Macmillan Publishing Co.

Bradley, M. 1981. *Horses, a Practical and Scientific Approach*. Highstown, NJ: McGraw-Hill Book Co.

Cunha, T. J. 1980. *Horse Feeding and Nutrition*. New York: Academic Press.

Evans, J. W., Borton, A., Hintz, H. F. and Van Vleck, L. D. 1986. *The Horse*. San Francisco: W. H. Freeman and Co.

Lewis, L. D. 1982. *Feeding and Care of the Horse*. Philadelphia: Lea and Febiger.

Nutrient Requirements of Horses. 1978. National Research Council. Washington, D.C.: National Academy Press.

Shidler, R. K. and Voss, J. L. 1984. *Management of the Pregnant Mare and Newborn Foal.* Colorado State Univ. Expt. Sta. Spec. Series 35.

Voss, J. L. and Pickett, B. W. *Reproductive Management of the Broodmare.* Colorado State University.

Visuals

The Basic Horsemanship Kit (separate sound filmstrip) covering: *Horse Safety; Care of the Brood Mare and Foal; Hoof Care.* Vocational Education Productions, California Polytechnic State University, San Luis Obispo, CA 93407.

Breaking and Training the Western Horse (sound filmstrip). Vocational Education Productions, California Polytechnic State University, San Luis Obispo, CA 93407.

CHAPTER 31 ▬▬▬▬▬▬▬▬▬▬▬

The Poultry Industry

The production of poultry and poultry products is an extremely important industry in the world and the United States. World production of poultry meat and eggs is presented in Chapter 5.

The estimated 276 million laying hens that produced 68.5 billion eggs worth $4.0 billion; 4.6 billion broilers worth $2.6 billion; and 207 million turkeys, worth $1.5 billion for a total of 8.1 billion for income from chickens and turkeys in 1986 is a sizeable industry. Figures for geese and ducks are not available, but they contribute less to poultry income than turkeys. Operations that manage poultry are becoming increasingly efficient.

The term **poultry** applies to chickens, turkeys, geese, ducks, pigeons, peafowls, and guineas. The head and neck characteristics which distinguish several of the poultry types are shown in Fig. 31.1. Turkeys have some rather unusual identifying characteristics including a beard, which is a black lock of hair on the upper chest of the male turkey. They also have **carnuncles**—red-pinkish flesh-like covering on the throat and neck, with the **snood** hanging over the beak.

CHARACTERISTICS OF BREEDS

Chickens. Chickens are classified according to class, breed, and variety. A class is a group of birds that has been developed in the same broad geographical area. The four major classes of chickens are American, Asiatic, English, and Mediterranean. A breed is a subdivision of a class composed of birds of similar size and shape. Some important breeds, strains, lines and synthetics are shown in Fig. 31.2. The parts of a chicken are shown in Fig. 31.3. A variety is a subdivision of a breed composed of birds of the same feather color and type of comb.

Factors such as egg numbers, eggshell quality, egg size, efficiency of production, fertility, and hatchability are most important to commercial egg producers. Broiler producers consider such characteristics as white plumage color and picking quality, egg production,

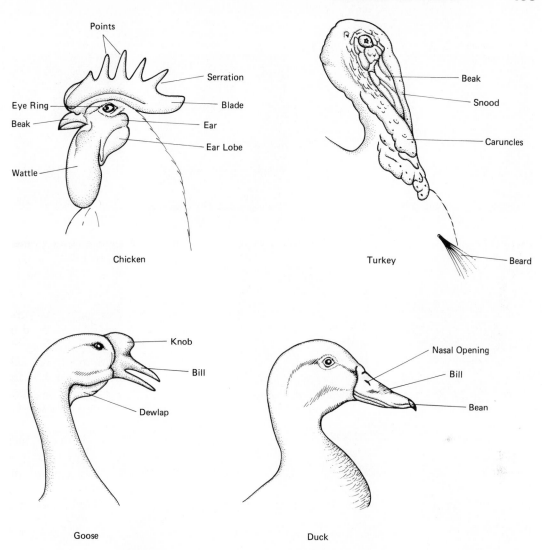

FIGURE 31.1.
Distinguishing characteristics of several different types of poultry.

fertility, hatchability, growth rate, carcass quality, feed efficiency, livability, and egg production. Also, breeds, strains, and lines that cross well with each other are important to broiler breeders.

The breeds of chickens listed in Table 31.1 were developed many years ago. These specific breeds are not easily identified in the commercial poultry industry as the breeds have been crossed to produce different varieties and strains (see Fig. 13.2). Some breeds are exhibited at shows or propagated for specialty marketing and are more novelty than part of today's industry.

PURE BREEDS

EXAMPLES OF CURRENT SYNTHETICS,
STRAINS OR LINES DEVELOPED FROM
ONE OR MORE OF THE PURE BREEDS

Layers (White Eggs)

Leghorn

DEKALB XL-Link

H&N "Nick Chick"

Layers (Brown Eggs)

Rhode Island Red

Plymouth Rock

Hubbard Golden Comet

Meat Type (Broilers)

New Hampshire

Cornish

Indian River Broilers
Male and Female (Variety FS99)

FIGURE 31.2.
Breeds, synthetics, lines, or strains of chickens. Courtesy of Dekalb Poultry Research, H&N International, Hubbard Farms, and Watt Publishing Company.

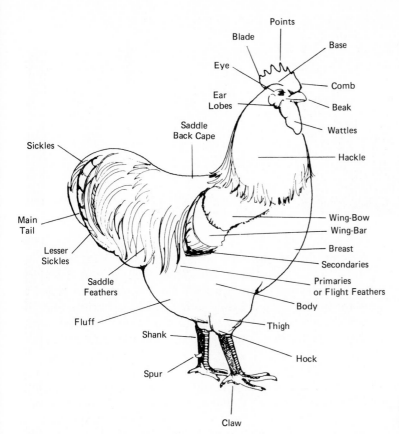

FIGURE 31.3.
The external parts of a chicken.

TABLE 31.1.   Certain Breeds of Chickens and Their Main Characteristics

Breed	Purpose	Type of Comb	Color of Egg
American breeds			
White Plymouth Rock	Eggs and meat	Single	Brown
Wyandotte[a]	Eggs	Rose	Brown
Rhode Island Red	Eggs	Single and Rose	Brown
New Hampshire	Eggs and meat	Single	Brown
Asiatic breeds			
Brahma[a]	Meat	Pea	Brown
Cochin[a]	Meat	Single	Brown
English breeds			
Australorp	Eggs	Single	Brown
Cornish	Meat	Pea	Brown
Orpington[a]	Meat	Single	Brown
Mediterranean breed			
Leghorn	Eggs	Single and Rose	White

[a]These breeds are of minor importance to the U.S. poultry industry.

FIGURE 31.4.
Male and female turkeys that are nearing maturity.
Courtesy of Hubbard Farms.

Turkeys.   There are eight varieties of turkeys but most turkeys grown for meat are white. The eight varieties are: (1) Bronze, (2) Narragansett, (3) White Holland, (4) Black, (5) Slate, (6) Bourbon Red, (7) Small Beltsville White, and (8) Royal Palm. Two types of turkey are commercially important in the United States today: the small White and the large Broad White. Since the 1950s, the emphasis has been to produce more large, white turkeys of the type shown in Fig. 31.4. A corresponding decrease in the number of small whites has occurred.

Ducks.   Many breeds of domestic ducks are known, but only a few are of economic importance. The most popular breed in the United States is the White Pekin. It is most valuable for its meat and produces excellent carcasses at 7–8 weeks of age. Another breed, the Khaki Campbell, is used commercially in some countries for egg production.

Geese.   Several breeds of geese are rather popular. The Embden, Toulouse, White Chinese, and Pilgrim are all satisfactory for meat production. The characteristics preferred by most commercial geese producers include a medium-sized carcass, good livability (low mortality), rapid growth, and a heavy coat of white or nearly white feathers. The Embden and White Chinese are breeds that meet these requirements. One variety of the Toulouse is gray, another variety is buff. The Pilgrim gander is white, the female is grayish. The White Chinese breed is pure white.

CHANGES WITHIN THE POULTRY INDUSTRY

From 1900 until 1940, the primary concerns of the poultry industry in the United States were egg production by chickens and meat production by turkeys and waterfowl. Meat production by chickens was largely a byproduct of the egg-producing enterprises. The broiler industry, as we presently know it, was yet to be established. Egg production was well established near large population centers, but the quality of eggs was often low because of seasonal production, poor storage, and the lack of laws to control grading standards for eggs.

Before 1940, large numbers of small-farm flocks existed, and management practices,

**TABLE 31.2. Egg Production in the Ten
Leading States, 1986**

State	Number of Eggs (millions)
California	7.8
Indiana	5.6
Pennsylvania	4.7
Georgia	4.3
Ohio	3.9
Arkansas	3.7
North Carolina	3.4
Texas	3.4
Alabama	2.7
Florida	2.7

Source: USDA

as applied today, were practically unknown. In the late 1950s, however, the modern mechanized poultry industry of the United States emerged. The number of poultry farms and hatcheries decreased, but the number of birds per installation dramatically increased. Larger cage-type layer operations appeared, and egg-production units grew, with production geared to provide the consumer with eggs of uniform size and high quality. Huge broiler farms, which provided the consumer with fresh meat throughout the year, were established, and large dressing plants capable of dressing 50,000 or more broilers per day were built. Thus, the poultry industry was completely revolutionized.

One of the most striking achievements in the poultry industry is the increased production of eggs and meat per hour of labor. Poultry operations with 1 million birds at one location are not uncommon. Labor reduction has been accomplished by automatic feeding, watering, egg collecting, egg packing, and manure removal.

In this era of rapid expansion, the broiler industry shifted toward the southeastern United States—principally into Alabama, Arkansas, Georgia, Mississippi, and North Carolina. The main egg-production centers became located on the West Coast, in the Southeast, and in the Midwest (Table 31.2).

One dramatic change that occurred in the 1950s and 1960s was **integration,** which began in the broiler industry and is applied, to a lesser degree, in egg and turkey operations. Integration brings all phases of the enterprise under the control of one head, frequently the corporate ownership of breeding flocks, hatcheries, feed mills, raising phase, dressing plants, services, and marketing and distribution of products (Fig. 31.5).

The actual raising of broilers is sometimes on a contract basis with a person who owns houses and equipment and who furnishes the necessary labor. The corporation furnishes the birds, feed, field service, dressing, and marketing. Payment for raising birds is generally made on the basis of a certain price per bird reared to market age. Bonuses are usually paid to those who have done a commendable job of raising birds. An important advantage to this system is that all phases are synchronized so that the utmost efficiency is realized.

The number of laying hens in the United States has actually declined over the years,

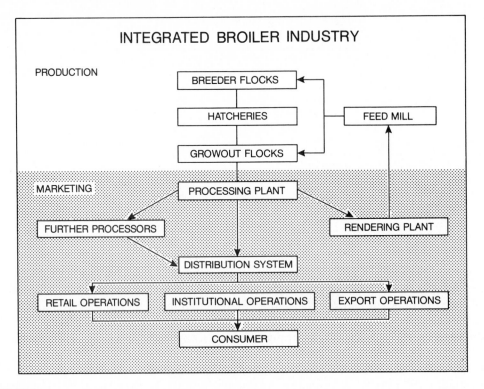

FIGURE 31.5.
Most of the activities in producing and marketing broilers are under the same ownership and management. This integrated system produces broilers very efficiently. Courtesy of CSU Graphics.

but egg production is up because of a dramatic increase in the performance of the individual hen. In 1880, for example, the average laying hen produced 100 eggs per year; in 1950 the average was 175; and in 1986 it was 250!

The USDA estimates that 85% of the market eggs produced in the United States are from large commercial producers (those that maintain one million or more birds, with the largest having 11 million). The 45 largest egg-producing companies in the United States have 97,276,600 layers, which is 35.5% of the nation's total.

The average American eats approximately 250 eggs per year (this figure includes eggs used in all food preparations as well as fresh eggs). The consumer demand for eggs has been decreasing during the last several years. Egg production efficiency has increased but the demand for eggs by consumers has not increased as has the demand for chicken meat.

Turkeys

Great changes have occurred in turkey production in the United States. Some 50 years ago, turkeys were raised in small numbers on many farms. Today, turkeys are raised in large numbers on fewer farms. In 1982, the Turkey Federation reported that there were 2,000 full-time turkey farmers in the United States. This would mean that the average turkey farmer is producing well over 50,000 turkeys per year.

TABLE 31.3. Leading States in Turkey Production, 1985

State	Mil Birds
North Carolina	32.6
Minnesota	30.8
California	20.4
Arkansas	15.0
Virginia	12.1
Missouri	12.0

Source: USDA.

There have also been changes in the eating habits of consumers. Years ago, turkey was considered a holiday meat item; therefore, much of the turkey meat was consumed at the Thanksgiving, Christmas, and Easter seasons. Today, turkey meat is consumed throughout the year in large amounts but it is still consumed in greater quantities during holiday seasons. For example, the estimate of whole-body turkey consumption for Thanksgiving is 45 million, for Christmas is 13 million, and for Easter is 9 million turkeys.

The people of the east and west coast areas consume more turkey per person than people in the midwest. California has the greatest turkey consumption, averaging 16 lb per person per year.

Most of the turkeys produced today are the heavy or large, broad-breasted white type. At one time, the fryer-roaster type turkey (5–9 lb) was popular and bridged the gap between the turkey and broiler chicken. Today the broiler industry has taken over the fryer-roaster market since they can produce fryer-roaster chickens at a much lower price.

The six leading turkey producing states are shown in Table 31.3.

As it takes 16 weeks for turkey hens to reach market weight and 19 weeks for turkey toms to reach market size, and as there is a heavy demand for turkey meat in November and December, eggs are set in large numbers in April, May, and June. Eggs are set the year round because there is a constant though lower demand for fresh turkey meat throughout the year. The normal incubation period for turkey eggs is 28 days.

Table 31.4 shows the changes in turkey production in recent years. During 1980–1986, an increase of 32 million more turkeys produced approximately 700 million more pounds of ready-to-cook product.

Consumption of turkey meat per person has increased from 1.7 lb in 1935 to 6.1 lb in 1960, to 8.0 lb in 1970, and to 12.1 lb in 1986. A great deal of the increase in turkey consumption was stimulated by development of processed products such as turkey rolls,

TABLE 31.4. Changes in Turkey Production from 1980 to 1986

Year	Number of Turkeys (mil)	Live Weight (bil lb)	Ready-to-Cook (bil of lb)
1980	165	3.1	2.3
1986	197	3.9	3.0

Source: USDA.

roasts, pot pies, and frozen dinners. The selling of prepackaged turkey parts in small packages has also increased consumption.

The number of processing plants has decreased but the volume processed per plant has increased greatly. For example, the number of plants was 281 in 1963 and 115 in 1982 but slaughter per plant increased from 4.9 million pounds to 26.8 million pounds. The 20 largest firms operated 45 plants that slaughtered 82% of the turkeys in 1982. In 1983, the eight largest firms processed more turkey than all firms did in 1962. If price of turkey meat is on a constant dollar value of 1967, the cost fell from 77 cents/lb in 1955 to 32 cents in 1982.

Broilers.   Broiler production in the United States has increased steadily since the inception of large-scale operations: about 4.5 billion broilers were produced in 1985, which is more than 100 times the 34 million produced in 1934.

The reasons for the great increase in production and demand for chicken include the short time needed and the economy of producing a broiler. It takes only 47 days and 8 lb of feed to produce a broiler from hatching to slaughter. There is no country in the world that discourages consumption of chicken, but beef consumption is discouraged in India because the cow is considered sacred and pork consumption is discouraged by those of the Islamic faith. Lamb and rabbit are influenced negatively by the name of the meat, causing children to refrain from eating the meat because the name makes them think of the cute little lambs or bunnies.

A very important reason for marked increase in chicken consumption has been the speed of preparing and serving chicken. The Kentucky Fried Chicken Corporation (KFC) led in preparation and serving chicken in a short time and now chicken is rapidly becoming a major item in fast-food service. The KFC restaurants are found in 56 countries and they sell more than 3 billion pieces of chicken a year. In recent years, other fast food restaurants have added more chicken items to their menus.

In 1940 the average American consumed (actually the disappearance rate of retail cuts) 43 lb of beef and 14 lb of chicken but in 1976 the figures were 94 lb of beef and 43 lb of chicken. In 1986 the consumption was 78 lb of beef and 60 lb of chicken. Chicken consumption has increased primarily because of relative cost and some perceived health reasons.

Chicken is a favorite food at home because it can be prepared for eating quickly in several ways—fried, baked, barbequed, shish-ka-bob, nuggets, and others. Chicken goes well with other food—particularly vegetables.

Chicken can be produced in less time and at lower cost than any other meat. The broiler industry is now one of the most efficient industries that exists and it is the most efficient in meat production.

BREEDING POULTRY

Turkeys

There is a definite breeding cycle in the production of baby poults for distribution to growers who raise them to market weight. The pedigree flocks (generation #1) or parent

FIGURE 31.6.
A flock of young, pullet turkey hens in the breeding program. Courtesy of Hubbard Farms.

stock are pure lines. The pure lines are crossed one with another to produce generation #2. Generation #2 lines are crossed one with another to produce generation #3. Eggs from these line crosses are selected into female or male lines and sent to hatcheries. Males from the male lines are selected for such meat traits as thicker thighs, meatier drumsticks, plumper breasts, rate of growth, and feed efficiency. The females of the female line are selected for greater fertility, hatchability, egg size, and meat conformation. The male line males are crossed with female line females and the eggs produced are hatched and the poults are sold for the production of market turkeys.

Laying hens are approximately 30 weeks of age when they reach sexual maturity (Fig. 31.6). They are put under controlled lighting to stimulate them to start laying. The average hen usually lays for about 25 weeks and produces 88–93 eggs. The hen is considered "spent" at the end of the 25-week laying cycle. Most of them are marketed at that time for meat. Hens can be molted and stimulated to lay another 25-week cycle. They require 90 days for molting and produce only 75–80 eggs during this later laying cycle, therefore one needs to compare the costs of growing new layers with that of recycling the old ones before deciding if it is economical to recycle old layers. The "spent" hens usually go into soup products.

Turkey eggs are not produced for sale for eating because it costs much more to produce food from turkey eggs than from chicken eggs. It would cost about 50 cents each to produce turkey eggs for human consumption.

All turkey hens are artificially inseminated to obtain fertile eggs. As turkeys have been selected for such broad breasts and large mature size, it is difficult for toms to mate successfully with hens. Semen is collected from the toms and used for inseminating the hens. Usually 1 tom to 10 hens is sufficient for providing the semen needed.

The present-day turkey is much different from the turkeys of many years ago. They grow more rapidly, convert feed into meat more efficiently, have more edible meat per unit of weight, require less time to reach market, and most turkeys grown today are white feathered so there are no dark pin feathers to discolor the carcass. It requires only 2.8 lb feed for every pound of gain with present-day turkeys (Fig. 31.7).

FIGURE 31.7.
World's largest turkey feed conversion facility used to accurately measure individual feed efficiency of more than 3,000 female line pedigree candidate toms per year. This company also tests individual feed conversion on approximately 7,400 other toms for potential use in their breeding program. Courtesy of Nicholas Turkey Breeding Farms.

Chickens

Sophisticated selection and breeding methods have been developed in the United States for increased productivity of chickens for both eggs and meat. The discussion to follow is directed primarily at improvement of chickens; however, methods described for chickens can be applied to improve meat production in turkeys, geese, and ducks.

Early poultry breeding and selection concentrated on qualitative traits which are, from a genetics standpoint, more predictable than quantitative traits. Qualitative traits, such as color, comb type, abnormalities and sex-linked characteristics are important; however, quantitative traits, such as egg production, egg characteristics, growth, fertility, and hatchability, are economically more important today.

Quantitative traits are more difficult to select for than are qualitative traits because the mode of inheritance is more complex and the role of the environment is greater. Quantitative traits differ greatly as to the amount of progress that can be attained through selection. For example, increase in body weight is much easier to attain than increase in egg production. Furthermore, a relationship usually exists between body size and egg size. Generally, if body size increases, a corresponding increase in egg size will occur. Most quantitative traits of chickens fall in the low to medium range of heritability, whereas qualitative traits fall in the high range.

Progress in selecting for egg production has been aided by the trapnest, a nest equipped with a door that allows a hen to enter but prevents her from leaving. The trapnest enables accurate determination of egg production of individual hens for any given period and helps to identify and eliminate undesirable egg traits and broodiness (the hen wanting to set on eggs to hatch them). Furthermore, it allowed the breeder to begin pedigree work within a flock.

The two most important types of selection applied primarily to chickens but to a lesser extent to other poultry today are mass selection and family selection. In mass selection, which is the older method, the best-performing males are mated to the best-performing females (Fig. 31.8). This program is very effective in improving traits of high heritability. Family selection is a system whereby all offspring from a particular mating are designated as a family, and selection and culling within a population of birds are done on the performance level of the entire family. This type of selection is most adaptable to selection for traits of low heritability. This is not to say that progress for traits of high heritability cannot be made by using family selection—quite the opposite is true. However, mass selection is very effective for traits of high heritability and certainly easier.

Progress through selection is rather slow for the reproductive traits because they are lowly heritable. Most breeding programs are geared to improvement of more than one trait at a time. For example, to improve egg-laying lines, it might be necessary to attempt simultaneously to improve egg numbers, shell thickness, interior quality of the egg, and feed efficiency.

Outcrossing. Defined as mating unrelated breeds or strains, outcrossing is probably more adaptable to the modern broiler industry than to the egg-production industry, although it can be used to improve egg production. Numerous experiments have established fairly accurately which breeds or strains will cross well with each other. Knowledge today is sophisticated to the extent that breeders know which strain or line to use as the male line and which to use as the female line.

Crossing Inbred Lines. Development of inbred lines for crossing is used largely for increasing egg production. Inbreeding is defined as mating related individuals or as mating individuals more closely related than the average of the flock from which they came. The mating of brother with sister is the most common system of inbreeding for poultry; how-

FIGURE 31.8.
A broiler breeding pen containing 1 male and 12 females. This is a typical breeding unit. Courtesy of Hubbard Farms.

ever, mating parents with offspring gives the same results. Both of these types of inbreeding increase the degree of inbreeding at the same rate per generation.

Hybrid chickens are produced by developing inbred lines and then crossing them to produce chickens that exhibit hybrid vigor. The technique employed in production of hybrid chickens is essentially the same as that used in development of hybrid plants. The breeder works with many egg-laying strains or breeds while producing hybrids. The strains or breeds are inbred (mostly brother or sister) for a number of generations, preferably five or more. In the inbreeding phase, many undesirable factors are culled out—selection is most rigid at this stage.

In all phases of hybrid production, the strains are completely tested for egg production and for important egg-quality traits. After inbreeding the birds for the necessary number of generations, the second step, crossing the remaining inbred lines in all possible combinations, is initiated. These crossbreds are tested for egg production and for egg-quality traits. Crossbreds that show improvement in these traits are bred in the third phase, in which the best test crosses are crossed into three-way and four-way crosses in all possible combinations. Most breeders prefer the four-way cross. The offspring that results from three-way and four-way crosses are known as hybrid chickens.

Inbreeding is an extremely costly way to produce chickens that lay eggs of superior quality and quantity, because inbreeding usually decreases reproductive performance and survivability. Many strains do not survive the inbreeding phase. Events in this breeding scheme are timed to provide the commercial producer with chickens that will give maximum production at a definite point (usually after five or six generations of inbreeding). It is imperative that the breeder carry a control population for each strain to enable evaluation of the inbred lines and also to have birds from which to start new inbred lines as needed.

Strain Crossing.   Strain crossing is more easily done than inbreeding and crossing inbred lines because it entails only the crossing of two strains that possess similar egg-production traits. It is definitely a type of crossbreeding if the two strains are not related to each other.

Strain crossing has two definite advantages over inbreeding in that it is easy to do and is relatively inexpensive. In general, the day-old female chick produced through strain crossing costs less than the day-old hybrid chick.

The computer is used largely in chicken breeding to help select breeding stock. An overall merit index is computed using data on heritability of each trait considered, the relative economic importance of each trait, and genetic correlations among the traits. Birds having the highest index are used for breeding.

In production of broiler or layer hybrid chicks, the computer is used to determine which lines or strains are most likely to give chickens that are superior. In broiler hybrids, hatchability, growth rate, economy of feed use, and carcass desirability are important. In layer hybrids, hatchability, egg production, egg size, egg and shell quality and livability are important.

Practically all line-cross layers are now free from **broodiness** because attempts have been made to eliminate genes for broodiness, and records are available on which lines produce broody chicks when crossed. Lines that produce broody chicks in a line cross are no longer used for producing commercial layers.

SELECTED REFERENCES

Publications

Eggs, Chickens, and Turkeys. 1986. Washington, D.C.: USDA.

Lasley, F. A., Henson, W. L. and Jones, H. B., Jr. 1984. *The U.S. Turkey Industry.* Agric. Econ. Rpt. 525. USDA, ERS.

Moreng, R. E. and Avens, J. S. 1985. *Poultry Science and Production.* Reston, VA: Reston Publishing Co., Inc.

Neshlim, M. C., Austic, R. E., and Card, L. E. 1979. *Poultry Production.* 12th edition. Philadelphia: Lea and Febiger.

Plunkett, James. 1980. Four-house, 80,000-hen complex managed by a single employee. *Poultry Digest* 39:64, 68.

Poultry and Egg Situation. 1986. Washington, D.C. USDA

Visuals

Anatomy of the Fowl (sound filmstrip). Vocational Education Productions, California Polytechnic State University, San Luis Obispo, CA 93407.

An Introduction to the Poultry Industry (77 slides and audiotape; 12 min.), Poultry Science Department, Ohio State University, 674 W. Lane Ave., Columbus, OH 43210.

CHAPTER 32

Managing Poultry

The success of modern poultry operations depends on many factors. The hatchery operator must care for breeding stock properly so that eggs of good quality are available to the hatchery and must incubate eggs under environmental conditions that ensure the hatching of healthy and vigorous birds.

INCUBATION MANAGEMENT

Each type of poultry has an incubation period of definite length (Table 32.1), and incubation management practices are geared to the needs of the eggs throughout that time. This discussion on incubation management applies specifically to chickens. Although the principles of incubation management generally apply to other types of poultry, specific information should be obtained from other sources.

All modern commercial hatcheries have some form of forced-air incubation system. Today's commercial incubator has a forced-air (fan) system that creates a rather uniform environment inside the incubator. Forced-air incubators are available in many sizes. All sizes are equipped with sophisticated systems that control temperature and humidity, turn eggs, and bring about an adequate exchange of air between the inside and outside of the incubator. The egg-holding capacity of forced-air incubators ranges from several hundred to 100,000 or more (Fig. 32.1).

Temperature, humidity, position of eggs, turning of eggs, oxygen content, carbon dioxide content, and sanitation must be regulated.

Temperature.   Proper temperature is probably the most critical requirement for successful incubation of chicken eggs. The usual beginning incubation temperature in the forced-air incubator is 99.5–100.0°F. The temperature should usually be lowered slightly (by 0.25–0.5°F) at the end of the fourth day and remain constant until the eggs are to be transferred to the hatching compartments (approximately 3 days before hatching). Three days before hatching, the temperature is again lowered, usually by about 1.0–1.5°F. Just

TABLE 32.1.   Incubation Times for Various Birds

Type	Incubation Period (days)
Chicken	21
Turkey	28
Duck	28
Muscovy duck	35–37
Goose	28–34
Pheasant	23–28
Bobwhite quail	23
Coturnix quail	17
Guineas	28

before hatching, the chicks change from embryonic respiration to normal respiration and give off considerable heat, which results in a high incubator temperature.

With reference to temperature, there are two especially critical periods during incubation—the first through the fourth day, and the last portion of incubation. Higher-than-optimum temperatures usually speed the embryonic process and result in embryonic mortality or deformed chicks at hatching. Lower-than-optimum incubation temperatures usually slow the embryonic process and also cause embryonic mortality or deformed chicks.

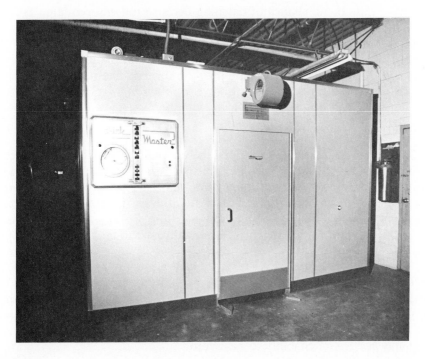

FIGURE 32.1.
Large commercial room-type incubator. Courtesy of Chick Master Incubator Co., Medina, OH.

Humidity.   A relative humidity of 60–65% is needed for optimum **hatchability.** All modern commercial forced-air incubators are equipped with sensitive humidity controls. The relative humidity should usually be raised slightly in the last few days of incubation. It has been shown that when the relative humidity is close to 70% in the last few days of incubation, successful hatching is greater. Providing optimum relative humidity is essential in reducing evaporation from eggs during incubation.

Position of Eggs.   Eggs hatch best when incubated with the large end (the area of the space called the "air cell") up; however, good hatchability can also be achieved when eggs are set in a horizontal position. Eggs should never be set with the small end up, because a high percentage of the developing embryos will die before reaching the hatching stage. The head of the developing embryo should develop near the air cell. In the last few days of incubation, the eggs are transferred to a different type of tray—one in which the eggs are placed in a horizontal position.

Shortly before hatching, the beak of the chick penetrates the air cell. The chick is now able to receive an adequate supply of air for its normal respiratory processes. The horny point (egg tooth) of the beak eventually weakens the eggshell until a small hole is opened. The egg is now said to be "pipped." The chick is typically out of the shell within a few hours. The hatching process varies considerably among different species.

Turning of Eggs.   Modern commercial incubators are equipped with time-controlled devices that permit eggs to be turned periodically. Eggs that are not turned enough during incubation have little or no chance of hatching because the embryo often becomes stuck to the shell membrane.

Commercial incubators are equipped with setting trays or compartments that allow eggs to be set in a vertical position. They also have mechanisms that rotate these trays so that the chicken eggs rotate 90° each time they are turned. The number of times the eggs are turned daily is most important. If one were to rotate the eggs once or twice daily, the percent of hatchability would be much lower than if the eggs were rotated five or more times daily.

Oxygen Content.   The air surrounding incubating eggs should be 21% oxygen by volume. At high altitudes, however, the available oxygen in the air may be too low to sustain the physiological needs of developing chick embryos, many of which, therefore, die. Good hatchability can often be attained at high altitudes if supplemental oxygen is provided. This is necessary for commercial turkey hatching, but hatchability can be improved by selecting breeding birds that hatch well in environments in which the supply of oxygen is limited.

Carbon Dioxide Content.   It is vital that the incubator be properly ventilated to prevent excessive accumulation of carbon dioxide (CO_2). The CO_2 content of the air in the incubator should never be allowed to exceed 0.5% by volume. Hatchability is lowered drastically if the CO_2 content of air in the incubator reaches 2.0%. Levels of more than 2.0% almost certainly reduce hatchability to near zero.

Sanitation.   The incubator must be kept as free of disease-causing microorganisms as possible. The setting and hatching compartments must be thoroughly washed or steam-cleaned and fumigated between settings. In many cases, they should be fumigated more than once for each setting. An excellent schedule is to fumigate eggs immediately after they have been set and again as soon as they have been transferred to the hatching compartment. A good procedure for fumigation is:

1. Prepare 1.5 ml of 40% formalin for each cubic foot of incubator space.
2. Prepare 1.0 g of potassium permanganate for each cubic foot of incubator space.
3. Set temperature of setting compartment or hatching compartment at proper reading (99.5–100.0°F, and 98.5–99.0°F, respectively).
4. Turn on fan.
5. Close vents.
6. Set relative humidity at proper reading (65%).
7. Place potassium permanganate in open container on compartment.
8. Pour formalin into potassium permanganate container and close incubator door for approximately 20 min.
9. Open door for 5–10 min or turn on exhaust system to allow formaldehyde gas to escape.
10. After gas has escaped, remove potassium permanganate container, close door, and continue normal operating procedures.

Candling of Eggs.   In maintaining a healthy and germ-free environment, eggs must be candled (examined by shining a light through each egg to see if a chick embryo is developing) at least once during incubation so that infertile and dead-germ eggs (eggs containing dead embryos) can be identified. Many operators candle chicken eggs on the 4th or 5th day and turkey and waterfowl eggs on the 7th to 10th day. Some operators candle eggs a second time when transferring eggs to the hatching compartment.

MANAGING YOUNG POULTRY

The main objective in managing young poultry is to provide a clean and comfortable environment with sufficient feed and water (Fig. 32.2).

House Preparation.   The brooder house should be thoroughly cleaned, **disinfected,** and dried several days before it receives young birds. All necessary brooding equipment should have been tested ahead of time and be in the proper place. One must make sure the brooders are working properly and *check thermostats and micro-switches!* Such advance preparation is probably as essential to success as any management practice applied to young birds. It is imperative that the disinfectant have no effect on the meat or eggs produced by birds. Food and Drug Administration regulations that govern the use of disinfectants should be strictly observed.

Litter.   Litter should be placed in the house when equipment is checked. Some commonly used substances are planer shavings, sawdust, wood chips, peat moss, ground corn cobs, peanut hulls, rice hulls, and sugar cane fiber. The entire floor should be covered by at least 2 in. of litter, which must be perfectly dry when the birds enter.

The primary purpose of litter is to absorb moisture. Litter that gets extremely damp should be replaced. Some operators add small amounts of litter as the birds grow.

FIGURE 32.2.
A commercial poultry house containing young broilers. Environmental conditions are well controlled for these many hundred birds. Courtesy of Hubbard Farms.

Floor Space.   The availability of too much or too little floor space can affect growth and efficiency of production adversely. If floor space is insufficient, young birds have difficulty finding adequate feed and water. This could lead to feather picking and actual cannibalism. Too much space can cause the birds to become bored, which can cause problems similar to those caused by overcrowding.

The amount of space required varies with the type and age of bird. Regardless of type, each bird should have at least 7–10 square in. of space around and under the hover (a metal canopy that covers most brooder stoves). Chicks of the egg-producing type and birds of similar size fare well for 5–6 weeks on 70 square in. of floor space. Between the time these birds are 6 and 10 weeks old, the space per bird must be increased to nearly 145 square in.

Broilers, turkeys, and waterfowl should have an area of 110 to 145 square in. for at least their first 8 weeks. In certain situations, such birds could be allowed as much as 145–220 square in. before they are 8 weeks old.

Feeder Space.   Chick-trough-feeders should be placed within the hover guard area or partially under the hover (0.8 in./chick or 80 in./100 chicks to 6 weeks of age). One should be sure the 10 watts attraction lights under the hover are working. One linear inch of feeder space per bird is sufficient for most species from the age of 1 day to 3 weeks; 2 in. for 3–6 weeks; and 3 in. beyond 6 weeks. Quail require less space, whereas turkeys, geese, and ducks require slightly more. Most feeders are designed so that birds can eat on either side of them.

The operator can determine how much feeder space is needed through observation. If relatively few birds are eating at one time, the feeding area is probably too large. If too many are eating at once, the area is probably too small. Young birds are usually fed automatically after the brooder stove enclosure is removed.

Water Requirements.   Three fountain-type waterers should be placed within the hover guard area (one fountain-type waterer/100 chicks). Generally, two 1-gallon-size water fountains are adequate for 100 birds that are 1-day old. More waterers can be added as necessary. Most commercial producers switch from water fountains to an automatic watering system when birds are put on automatic feeding. If trough-type waterers are used, allow at least 1 in. of water space per bird until 10–12 weeks of age. More space might be needed thereafter. Trough-type waterers designed so that birds can drink from both sides are available. A space of at least 0.5 in. per bird is needed if pan-type waterers are used.

Lighting Requirements.   The amount of light provided by the operator varies according to the type of bird being raised. No artificial lights are needed for chicks hatched during April through July. The time clock is set for 12 hours (5 A.M. to 5 P.M.) so lights are available when personnel are working in the area. A step-down lighting program is used for chicks hatched August through March. Many broiler producers use a 24-hour light regime, whereas others use systems such as 20 hours of light alternating with 4 hours of darkness. Growing pullets do not require as much light per day as broilers because the producer desires that future layers not reach sexual maturity too quickly. Generally, about 8–10 hours of light are sufficient for future layers until they are about 12–14 weeks old.

A system of lighting that takes into account the age of the bird, the time of year in which the bird was hatched (seasonal effect), and the type of housing (windowed or windowless) should be used. Pullets should be on a constant day-length regime or a decreasing light regime at 10 weeks of age to prevent early sexual maturity. In most cases, some type of dim light is provided to prevent young chicks from piling up in periods of darkness. After birds are about 1 week old, some producers substitute light bulbs of relatively low intensity (15–25 watts) for a bulb of higher intensity (40 watts is standard for the 1st week).

Other Management Factors.   Young birds should have access to the proper feed as soon as they are placed in the brooder house. It is important that birds such as turkey poults start eating feed early. Some young poults die for lack of feed, although feed is readily available because they have not learned to eat it.

Young birds are commonly debeaked sometime between their first day of life and several weeks of age. Debeaking (also called beak trimming) is done with a machine by searing off approximately one-half of the upper mandible and removing a small portion of the lower one. Care should be taken to avoid searing the bird's tongue while debeaking. Debeaking is done to prevent birds from picking feathers from other birds; it prevents cannibalism.

Birds should be vaccinated and given health care according to a prescribed schedule; a veterinarian should be consulted.

MANAGING 10- TO 20-WEEK-OLD POULTRY

Management of poultry from approximately 10 weeks to 20 weeks of age is quite different from managing younger birds. Improper practices in this most critical period could adversely affect subsequent production.

Confinement rearing is used by commercial producers of replacement birds, those birds that will be kept for egg production. In the three basic systems of confinement rearing, birds are raised on solid floors, slatted floors, or wire floors or cages.

Birds raised on floors in confinement should have from 1–2 square feet of floor space per individual. The amount of space required depends on the environmental conditions and also on the condition of the house. Caged birds should be allowed 0.5 square feet to not more than 0.75 square feet.

Replacement chickens raised in confinement are generally fed a completely balanced growing, or developing, ration that is 15–18% protein. Most birds are kept on a full-feeding program, but in certain conditions some producers restrict the amount of feed provided. Restricted feeding can take different forms: total feed intake, protein intake, or energy intake may be limited. The main purposes of a restricted feeding program are to slow growth rate (thus delaying the onset of sexual maturity and reducing the number of small eggs produced) and to lower feeding cost. Feed intake must not be restricted whenever a restricted lighting system (which is also used to delay sexual maturity) is in effect.

Automatic feeding and watering devices are used in most confinement operations. Watering- and feeding-space requirements are practically the same as those for birds that are 6–10 weeks old. As long as birds are not crowding the feeders and waterers, there is no particular need to increase the feeding and watering space.

The lighting regime used in confinement rearing is very important. Birds raised in window-type housing receive the normal light of long-day periods unless the house is equipped with some type of light-check. In short-day periods, supplemental lighting can be used to meet the requirements of growing chickens. Replacement chickens should receive approximately 14 hours of light daily up to 12 weeks of age. To delay sexual maturity, the amount should then be reduced to about 8–9 hours daily until the birds have reached 20–22 weeks of age. The light is then either abruptly increased to 16 hours/ day, or it is increased by 2–3 hours with weekly increments of 15–20 min then added until 16 hours of light per day are reached.

Regardless of which system replacement pullets are reared under, they should be placed in the laying house when approximately 20 weeks of age so they can adjust to the house and its equipment before beginning to lay.

MANAGEMENT OF LAYING HENS

Requirements of laying hens for floor space vary from 1.5 to 2 square feet per bird for egg-production strains and from 2.5 to 3.5 square feet for dual-purpose and broiler strains. Turkey breeding hens require 4–6 square feet, but less area is needed if hens are housed in cages. If turkey hens have access to an outside yard, 4 square feet of floor space is best. Game birds such as quail and pheasants require less floor space than egg-production hens. In general, 3 square feet or more of floor space is adequate for ducks, whereas geese need approximately 5 square feet.

Breeder hens are housed on litter or on slatted floors while birds used for commercial egg production are kept in cages (Fig. 32.3).

Floor-type houses for breeder hens usually have 60% of the floor space covered with slats. Slatted floors are usually several feet above the base of the building and manure is often allowed to accumulate for a rather long time before being removed. Many slatted-floor houses are equipped with mechanical floor scrapers that remove the manure period-

FIGURE 32.3.
Composite pictures of an automated commercial cage house for laying hens. Feeding, watering, egg gathering (vertical elevator at end of row), and manure removal are done automatically.

ically. Frequent removal of manure lessens the chance that ammonia will accumulate.

Gathering of eggs in floor-type houses can be done automatically if some type of roll-away nesting equipment is used. In houses equipped with individual nests, the eggs are gathered manually. Some operators prefer colony nests (an open nest that accommodates several birds at a time).

Some type of litter or nesting material must be placed in the bottom of individual and colony nests if eggs are to be gathered manually. Birds should not be allowed to roost in the nests in darkness because dirty nests result.

Cage operations are used by most commercial egg-producing farms. Young chickens may be brooded in colony cages. (Fig. 32.4) whereas laying hens may be housed in triple-decked laying cages. The individual cages range in size from 8 to 12 in. in width, 16 to 20 in. in depth (front to back), and 12 to 15 in. in height. The number of birds housed per unit varies with the operator. Density of the cage population is an important factor in production.

All cages are equipped with feed troughs that usually extend the entire length of the

FIGURE 32.4.
Rearing pullets in cages. Courtesy of J.F. Stephens.

cage. Some troughs are filled with feed manually, others automatically. It is sometimes advantageous to dub (remove the combs and wattles) caged layers so they can obtain feed from automatic feeders by reaching the head through the openings.

Several types of watering devices (troughs, "nipple" type, or individual "cup-type" waterers) are available for use in individual and colony cages. The watering system should be equipped with a metering device that makes it possible to medicate the birds quickly when necessary by mixing the exact dosage of medicant required with the water.

The arrangement of cages within a house varies greatly. Some producers have a "step-up" arrangement, such as a double row of cages at a high position with a single row at a low position on either side. Droppings from birds caged in the double row fall free of the birds in the single row. An aisle approximately 3 feet wide is usually between each group of cages. Rather than aisles, some systems have a movable ramp that can travel above the birds, from one end of the house to the other. In other systems, several double cages are stacked on top of each other.

In cage operations, manure is typically collected into pits below the cages and removed by mechanical pit scrapers or belts. Manure may also drop into pits that are slightly sloped from end to end and partially filled with water. The manure can be flushed out with the water into lagoons.

HOUSING POULTRY

Factors such as temperature, moisture, ventilation, and insulation are given careful consideration in planning and managing poultry houses.

Temperature. Most poultry houses are built to prevent sudden changes in house temperature. A bird having an average body temperature of 106.5°F usually loses heat to its environment except in extremely hot weather. Chickens perform well in temperatures between 35° and 85°F, but 55°–75°F seems optimal.

House temperature can be influenced by such factors as the prevailing ambient temperature, solar radiation, wind velocity, and heat production of the birds. A four-and-a-half-pound laying hen can produce nearly 44 **British thermal units** (BTU) of heat per hour

(1 BTU is the quantity of heat required to raise the temperature of one pound of water 1°F at or near 39°F). The amount of BTU produced varies with activity, egg-production rate, and the amount of feed consumed. Heat production by birds must be considered in designing poultry houses (about 40 of the 44 BTU produced per hour by a laying hen are available for heating). Although heat produced by birds can be of great benefit in severe cold, the house must be designed so that excess heat produced by birds in hot weather can be dissipated.

Moisture.   Excess house moisture, especially in cold weather, can create an environment that can be extremely uncomfortable, can lead to a drop in production by laying birds, and, if allowed to continue for long, can cause illness.

Much moisture in a poultry house comes from water spilled from waterers, water in inflowing air, in water vapor from the birds themselves, and in their droppings. Every effort should be made, of course, to minimize spillage from watering equipment. The amount of moisture in a poultry house can be reduced by increasing the air temperature or by increasing the rate at which air is removed. Because air holds more moisture at high temperatures than at low temperatures (Table 32.2), raising the air temperature will cause moisture from the litter to enter the air and thus be dissipated. Air temperature in the house can be increased by retaining the heat produced by the birds themselves and by supplemental heat.

If incoming air is considerably colder than the air in the house, it must be warmed or it will fail to aid in moisture removal. In cold weather, most exhaust ventilation fans are run slower than normal, so most moisture removal is than accomplished by increasing the air temperature in the house.

Ventilation.   A properly designed ventilation system provides adequate fresh air, aids in removing excess moisture, and is essential in maintaining a proper temperature within the house. Type and amount of insulation and heat produced by the birds themselves must be considered in planning for ventilation.

Ventilation is accomplished by positive pressure in which air is forced into the house

TABLE 32.2.   The Water-Holding Capacity of Air at Various Temperatures

Temperature (°F)	Pounds of Water per 1,000 lb. of Dry Air
50	7.62
40	5.20
30	3.45
20	2.14
10	1.31
0	0.75
−10	0.45

Source: Meyer, V. M., and Walther, P. *Ventilate Your Poultry House: For Clean Eggs, Healthy Hens, More Profit.* Ames: Iowa State University Cooperative Extension Service Pamphlet 292. 1963.

FIGURE 32.5.
A battery of fans used to ventilate a large poultry house. Courtesy of Big Dutchman, a division of U.S. Industries, Atlanta, GA.

to create air turbulence or by negative pressure in which air is removed from the house by exhaust fans (Fig. 32.5). The positive pressure system is accomplished by having fans in the attic so that air is forced through holes. The negative pressure system is accomplished by locating exhaust fans near the ceiling. Some houses are ventilated so that the air is fairly warm and dry by having open walls. Also, when outside air is cold and dry, air can enter at low portions of the house and leave through vents near the ceiling as it warms. The heat created by the chickens will warm the colder air, which will then take moisture from the house.

Two integral and necessary parts of the most common ventilation systems are the exhaust fan and the air-intake arrangement. The number of exhaust fans varies with size of the house, number of birds, and capacity of the fans (Fig. 32.5). Most fans are rated on the basis of their cubic feet per minute (cfm) capacity, that is, how many cubic feet of air they move per minute. Fans in laying houses should operate at 4–4.5 cfm per bird when the temperature is moderate. In summer, cfm per bird could be as high as 10.

Fans also have a static pressure (water pressure for low house temperatures and a high speed for warm house temperatures). Others operate at only one speed; the amount of air removed is controlled by shutters that open so that more air can be removed when the temperature of the house exceeds the thermostatically fixed temperature. These shutters close at lower temperatures.

An air-intake area must be provided for the ventilation system. It is usually a slotted area in the ceiling or near the top of the walls. There should be at least 100 square in. of inlet space for each 400 cfm of fan capacity.

Insulation.   Because energy needs are becoming ever more critical, the insulation of poultry houses of the future must be superior to that of the past. Sudden temperature changes inside the house must be avoided, especially for young birds. Houses in cold and windy areas require better insulation than those in milder climates, but good insulation is also essential in areas in which outside temperatures are high. The insulation ability of any material is measured by its R value. This value is based on the material's ability to limit heat loss, as expressed in BTUs. The better the insulating material, the higher its R value. For example, if the outside temperature is 11°F cooler than the inside, a material having an R value of 11 loses approximately 1 BTU/hour to the outside for each square foot of area. A material having an R value of 20 loses 1 BTU/hour to the outside for each square foot of area when the outside temperature is 20°F cooler than the inside.

By knowing how much heat the birds generate, local variations in ambient temperature, expected wind velocities, and relative humidities experienced, the appropriate R value can be established. In many areas of the United States, especially in the cold regions, wall insulation in poultry houses should have an R value of 15; ceiling insulation, at least 20 (heat loss through the floor is negligible). In warmer areas, wall insulation should have an R value of 5–10; ceiling insulation, 10–15.

When computing the R value for a particular area of the house (walls, for example), resistance of the outer surface, insulation, air space between the studding, inner wall material, resistance of the inner surface and, if windows are present, presence of glass, must all be considered. Each of these factors has an R value, and the total of these values establishes the R value for that area of the house. A building materials dealer can furnish the R value ratings of these factors.

Because too much moisture may accumulate in the house, especially in cold weather, some type of vapor barrier should be present in the walls and ceiling. This barrier can be a part of the insulation itself (foil backing) or separate. It should be placed between the insulation and the inside wall. The foil-back portion of the insulation should be placed next to the inside wall.

FEEDS AND FEEDING

Rations fed to poultry today are complex mixtures that should include all ingredients, in a balanced proportion, that have been found to be necessary for body maintenance, for maximum production of eggs and meat, and for optimum reproduction (**fertility** and hatchability). Poultry nutritionists are constantly searching for and testing new feedstuffs that, when incorporated into diets, will permit better efficiency of production. Feeding can be expensive, because an adequate intake of energy, protein, minerals, and vitamins is essential.

The computer is used in formulating least-cost rations that meet the nutritional requirements of poultry. Data input that must be provided the computer includes constraints that depend on age of birds and their productivity, cost of available ingredients, ingredients desired in the ration, composition of the nutritional materials, and requirements of the chickens to be fed.

The constraints set minimum and maximum percentages. For example, an upper limit on the amount of fiber that could be in the ration must be established because birds cannot use fiber effectively. Likewise, a low or minimal level of soybean meal is important if it is to provide the amino acid needs.

The large computers are expensive and only large poultry operations can afford them; however, the microcomputers are now designed and ready for use in computing least-cost rations. Most poultry producers can afford a microcomputer.

Energy requirements are supplied mainly by cereal grains, grain by-products, and fats. Some important grains are yellow corn, wheat, sorghum grains (milo), barley, and oats. Most rations contain rather high amounts of grain (60% or higher, depending on the type of ration). A ration containing a combination of grains is generally better than a ration having only one type. Animal fats and vegetable oils are excellent sources of energy. They are usually incorporated into broiler rations or any high-energy rations.

Protein is so highly essential that most commercial poultry feeds are sold on the basis of their protein content. The types of amino acids present determine the nutritional value of protein (see Chapter 14). Excellent protein can be derived from both plants and animals. Most rations contain both plant and animal protein so that each source can supply amino acids that the other source lacks. The most common sources of plant protein are soybean meal, cottonseed meal, peanut meal, alfalfa meal, and corn gluten meal. Cereal grains contain insufficient protein to meet the needs of birds. The best sources of animal protein are fish meal, milk by-products, meat by-products, tankage, blood meal, and feather meal. The protein requirements of birds vary according to the species, age, and purpose for which they are being raised. The protein requirements of certain birds are shown in Table 32.3. Where variable values are listed, the highest level is to be fed to the youngest individuals.

Whatever ration is adequate for turkeys is generally suitable for game birds. Some game bird producers feed a turkey ration at all times. Others feed a complete game bird ration. The actual nutrient requirements of game birds are not as well known as are the requirements of chickens and turkeys.

TABLE 32.3. Protein Requirements of Poultry

Type	Age (weeks)	Percent Protein Required in Diets
Chickens		
Broilers	0–6	20.0–23.0
Replacement pullets	0–14	15.0–18.0
Replacement pullets	14–20	12.0
Laying and breeding hens		14.5
Turkeys		
Starting	0–8	26.0–28.0
Growing	8–24	14.0–22.0
Breeders		14.0
Pheasant and Quail		
Starting and growing		16.0–30.0
Ducks		
Starting and growing		16.0–22.0

Source: Nutrient Requirements of Poultry. 8th ed. Washington, D.C.: National Academy of Sciences, National Research Council, 1984.

A considerable number of minerals are essential, especially calcium, phosphorus, magnesium, manganese, iron, copper, zinc, and iodine. Calcium and phosphorus, along with vitamin D, are essential for proper bone formation. A deficiency of either of these elements can lead to a bone condition known as rickets. Calcium is also essential for proper eggshell formation.

The amount of calcium required varies somewhat with age, rate of egg production, and temperature. Chickens and turkeys up to 8 weeks of age require 0.8–1.2% of calcium in their diets. Calcium can be reduced from 0.6 to 0.8% in 8- to 16-week-old birds. Laying hens require at least 3.4% calcium. The amounts of phosphorus required for chickens of different ages are as follows: 0–6 weeks, 0.4%; 6–14 weeks, 0.35%; and mature, 0.32%. Turkeys require from 0.3 to 0.6% phosphorus; game birds, approximately 0.5%. Requirements for other minerals vary greatly depending on the stage of growth and production.

The vitamins that are most important to poultry are A, D, K, and E (fat soluble), and thiamin, riboflavin, pantothenic acid, niacin, vitamin B_6, choline, biotin, folacin, and vitamin B_{12} (water soluble).

SELECTED REFERENCES

Publications

Day, E. J. 1980. Microcomputers: Ready for least-cost ration formulation. *Feedstuffs* 52:1, 50, 52.

Hicks, F. W. 1974. *Farm Flock Management Guide.* University Park: Pennsylvania State University Extension Bulletin 4-663.

Moreng, R. E. and Avens, J. S. 1985. *Poultry Science and Production.* Reston, VA: Reston Publishing Co., Inc.

Nakaue, H. S. 1975. *Fundamentals of Computer Ration Formulation.* Official Proceedings, 10th Annual Pacific Northwest Animal Nutrition Conference, pp. 31–36.

Nesheim, M. C., Austic, R. E., and Card, L. E. 1979. *Poultry Production.* 12th edition. Philadelphia: Lea and Febiger.

Nutrient Requirements of Poultry. 1984. National Research Council. Washington, D. C.: National Academy Press.

Visuals

How to Do a Poultry Autopsy (silent filmstrip) and *Small Flock Brooding Techniques* (sound filmstrip). Vocational Education Productions, California Polytechnic State University, San Luis Obispo, CA 93407.

Hatchery Operations (80 slides and audio tape; 18 min.). Poultry Science Department, Ohio State Univ., 674 W. Lane Ave., Columbus, OH 43210.

CHAPTER 33 ▰▰▰▰▰

Goat Breeding, Feeding, and Management

Geological records reveal that goats existed five million years ago and bones of goats closely associated with people have been found dating back 12,000 years. It appears that people at that time existed in small groups, moving to where food and water could be found for them and the goats. The goats were particularly important to people because they provided food (milk and meat), skins for making certain items of clothing, and mohair for many uses.

Goats thrive on **browse** (brushy plants) and **forbs** (broad leaf plants) but also eat grass extensively, depending on species and stage of maturity. Choice of plants and plant parts is important to goats because they are more likely to select the more nutritious parts, and because of their browsing over a wide territory they are less likely to suffer from internal parasites.

Goats have a large impact on the economy and food supply for people of the tropical world. In many countries, people consume more goats' milk than cows' milk, and goats are an important source of meat. In the U. S., the consumption of both goat milk and meat is increasing, and the production of mohair is an important industry in Texas.

There are four major kinds of domesticated goats: the dairy goat, which is used largely for the production of milk and to a lesser extent for meat; the Angora goat, which is used mainly for the production of mohair and to a lesser extent for brush clearance and the production of meat; the meat goat, which is used for brush clearance and for the production of meat and in some parts of the world for skins and fine leather; and the Pygmy goat, which is used as a laboratory ruminant animal and pet in the U. S. but is an important disease resistant meat and milk producer in Africa and other countries.

In some parts of the world, goat meat production is extremely important and, in fact, meat goats outnumber dairy goats worldwide. Goat numbers in the U. S. are modest compared to cattle and sheep because of the highly specialized and effective dairy industry built around the dairy cow, as well as the specialized beef cattle and sheep industries. Yet in some parts of the world, including Mediterranean countries, France and Norway, large quantities of dairy products are produced from goats.

Milk produced by dairy goats differs from cows' milk in that all carotene has been converted into vitamin A in goat milk. The type of curd formed from goat milk is different from the curd from cow milk because of differences in the major caseins; milk fat in goat milk is in smaller globules than in cow milk, does not rise or coalesce as readily and contains much more short-chain fatty acids. Goat milk is more readily digested and assimilated by people and animals because of these differences. The dairy goat is a desirable animal for providing milk for the family because it is small and less expensive to feed than a cow. Goats can consume large quantities of browse, which is not very palatable to cattle.

IMPORTANCE OF GOATS

The number of goats in the different countries of the world presented by FAO Production Yearbook, (1984) is the best existing record but in some developing countries this information is difficult to obtain. There are many families that have a few goats for the production of milk, meat, and hides for family use. These goats live primarily on browse, and the numbers vary from day to day because of death losses and slaughter. World goat numbers are presented in Table 33.1. Most goats are found in developing countries in which they contribute greatly to the needs of the people in these countries.

One major by-product of goats is skins. In areas with extensive goat numbers for milk

TABLE 33.1.   Number of Goats in the World, Different Areas of the World, and Developed and Developing Nations for 1984

Area or Country	Number of Goats (mil)
Africa	150.9
Nigeria	26.0
Ethiopia	17.2
Asia	255.2
India	80.8
China	68.2
Europe	12.5
Greece	4.6
Spain	2.4
Oceania	0.5
North and Central America	13.9
Mexico	10.0
U.S.	1.4
South America	19.8
Brazil	8.5
Bolivia	3.7
Developed countries	26.9
Developing countries	432.7
World	459.6

Source: 1984 FAO Production Yearbook

TABLE 33.2. **Production of Fresh Goatskins for the Five Leading Countries in 1984**

Country	Goatskins (mil lb)
World total	854.8
India	161.8
China	129.5
Pakistan	74.8
Nigeria	46.6
Bangladesh	32.1

Source: 1984 FAO Production Yearbook

and meat production, hides become important. The five leading countries in goatskins in 1984 are given in Table 33.2. These five countries produce 52% of the goatskins produced in the world.

The four leading countries in the production of goat meat in 1984 are presented in Table 33.3. These four countries produce more than 75% of the world's goat meat.

PHYSIOLOGICAL CHARACTERISTICS

Goats have many characteristics that are similar to sheep. Some of the production traits of goats are summarized in Table 33.4.

CHARACTERISTICS OF BREEDS OF DAIRY GOATS

The external parts of a goat are shown in Fig. 33.1. The dairy goat and dairy cow are about equal in efficiency of converting feed into milk, even though the dairy goat produces much more milk in relation to its size and weight then does the dairy cow. Feed needed for maintenance is higher per unit of body weight for goats than for dairy cows.

TABLE 33.3. **Goat Meat Production for the Four Leading Countries for 1984**

Country	Amount of Goat Meat (mil lb)
World Total	2,200
China	700
Pakistan	500
Nigeria	300
Turkey	200

Source: 1984 FAO Production Yearbook

TABLE 33.4.   Production Characteristics of Goats

Trait	Time
Gestation length	144–155 days (average 150 days)
Length of estrous cycle	15–18 days (average 16 days)
Length of estrus	1–3 days (average 2 days)
Age at puberty	120 days to over a year[a]
Normal breeding season	Late summer or early fall early fall or late winter[b] (Breeds near the equator are less seasonal)
Size of kids at birth	1.5–11.0 lb[c] (average 5.5 lb.)
Adult size	
Does	40–190 lb. (average 130 lb.)
Bucks	50–300 lb.[d] (average 160 lb.)

[a] Age of puberty depends upon the breed and nutrition. Pygmy goats tend to reach puberty at an early age. Some kid at 9 months of age.

[b] Normal breeding season depends on the breed. Pygmy does are capable of breeding over a long breeding season but are more fertile when bred in the fall or early winter. Angora goats usually have a highly restrictive breeding season during the fall.

[c] Pygmy and Angora are smaller at birth than dairy goat kids.

[d] Pygmy goats are the smallest, Angora goats are intermediate, and dairy goats are the largest. Bucks of all breeds are larger than corresponding does.

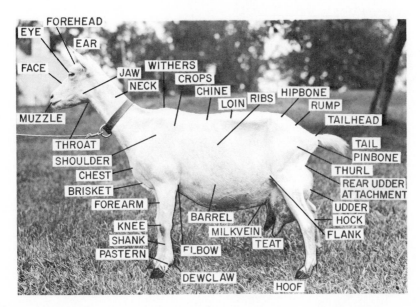

FIGURE 33.1.
The external parts of the dairy goat. Courtesy of Mr. R.F. Crawford and Mr. Ted Edwards, Emerald Dairy Goat Association Newsletter. Photograph by Mr. Ole Hoskinson.

Gall (1981) compared the efficiency of Holstein dairy cows with Alpine Beetel dairy goats. The cows weighed 1325 lb and produced 13,260 lb milk whereas the dairy goats weighed 108 lb and produced 1083 lb of milk per lactation. The dry matter intakes per lactation and per liter of milk for the Holstein cows were 13,300 lb and 2.3 lb per quart of milk and for the goats 1173 lb and 2.4 lb per quart of milk.

The six major dairy goat breeds registered in the United States are Toggenburg, Saanen, Alpine, Oberhasli, Nubian, and LaMancha, all of which are capable of high milk productivity. These breeds are shown in Figs. 33.2–33.7.

The Toggenburg breed is medium in size, is vigorous, and produces much milk (Fig. 33.2). The highest record of milk production for one lactation by a U.S. Toggenburg female is 5,750 lb. The Saanen is a large all white breed, capable of producing the most milk (Fig. 33.3). A world record milk production for one lactation by a Saanen female is 7,713 lb. The Alpine breed is a large, somewhat rangy goat (Fig. 33.4). Alpines also produce much milk; the record for one lactation of a female is 5,729 lb.

The Nubian is a tall proud-looking goat breed that differs from the other five breeds in being a better meat producer and having long, wide, pendulous ears and a Roman

FIGURE 33.2.
A Toggenburg doe. Note the fore attachment of the udder and the well-placed teats. Courtesy of the *Dairy Goat Journal.*

FIGURE 33.3.
A Saanen doe. Courtesy of Nancy Lee Owen.

FIGURE 33.4.
A French Alpine doe. Courtesy of Mrs. Eva Rappaport.

nose (Fig. 33.5). The U. S. record milk production of a Nubian female is 4,420 lb, which is lower than for the three Swiss breeds previously described. However, Nubian milk is distinctive for its higher milk fat content (7.4% vs. 3.5% for Swiss breeds).

The LaMancha goat breed of California origin (Fig. 33.6) is smaller and differs from all other breeds in having almost no external ears ("gopher" ears) or extremely short ears ("elf" or "cookie" ears). Milk production is comparatively lower. Records for one lactation of up to 4510 lb have been recorded. Fewer females of this breed and the Oberhasli breed have been officially tested for milk production. Therefore the breed's true capacity has yet to be demonstrated.

The Oberhasli (Fig. 33.7) is a medium-sized dairy goat breed, also of Swiss origin, with usually solid red or black colors. Production records of up to 3300 lb milk have been reported.

Gall (1981) reported the following annual average milk production records from the major dairy goat breeds in the U. S. (only goats having more than five lactations during 276 to 305 days were used): Toggenburg, 1940 lb; Saanen, 2035 lb; Alpine, 2024 lb; Nubian, 1662 lb; and LaMancha, 1719 lb.

FIGURE 33.5.
A Nubian doe. Note the ear size and shape. Courtesy of Cindy Schneider.

FIGURE 33.6.
A LaMancha doe. Note the earless condition. This individual is homozygous for earlessness. Courtesy of Cindy Schneider.

FIGURE 33.7.
An Oberlasli doe. Courtesy of Mary Slabach.

CARE AND MANAGEMENT

Simple housing for goats is adequate in areas in which the weather is mild because goats do best when they are outside on pasture or in an area in which they can exercise freely. If the weather is wet at times, an open shed with a good roof is necessary. Lactating goats can be fed in an open shed in stormy weather and taken to the milking parlor for milking. If the weather is severely cold, an enclosed barn with ample space is needed. At least 20 square feet per goat is needed if goats are to be housed in a barn, but 16 square feet per goat may suffice in open sheds.

The use of several small pastures permits goats to be moved about, thus increasing grazing efficiency and reducing the risk of parasite infestation. Fences must be properly constructed with woven wire and 6-in. stays. Fencing for goats is different from fencing for cows as a woven-wire fence is best for containing goats, which can climb a rail fence. In addition, a special type of bracing at corners of the fence is necessary because a goat can walk up a brace pole and jump over the fence. With electric fences two electric wires should be used—one located low enough to keep goats from going underneath and one high enough to keep them from jumping out. However, an electric fence may not contain bucks of any breed, because they might go through the fence in spite of the shock.

Several pieces of equipment are needed for normal care: a tattoo set, hoof trimmers, a hoof knife, a grooming brush, an emasculator, and a balling gun for administering boluses (large pills for dosing animals).

All breeds of dairy goats are available in polled strains. The polled condition has advantages because a polled goat is less likely to catch its head in a woven-wire fence than is a horned goat. Horned animals should be disbudded during the first week following birth by use of an electric dehorning iron. Goat breeders were plagued by the genetic linkage between hermaphroditism and polledness. However, polled goats with good fertility can be found.

Goat's feet should be kept properly trimmed to prevent deformities and footrot. A good pruning shear is ideal for leveling and shaping the hoof, but final trimming can be done with a hoof knife. Footrot should be treated with formaldehyde, cooper sulfate, or iodine. Affected goats are best isolated from the others and placed on clean, dry ground after treatment for footrot.

All kids should be identified by a tattoo in the ear, except for LaMancha and Pygmies, where identification is in the tail-web. Ear tags are also useful and are easy to read. The tattoo serves as permanent identification in case an ear tag is lost. All registered goats must be tattooed because none of the goat registry associations accept ear-tag identification.

Male kids not acceptable for breeding should be castrated. This can be done by constricting the blood circulation to the testicles by use of a rubber band elastrator, or by surgically removing the testicles. Some breeders crush the cords to the testicles with an emasculator without surgery. After the blood supply to the testicles is discontinued, they will usually atrophy but exceptions occur. After using the emasculator, the blood circulation to the testicles may continue and, although the male is sterile, he continues to produce testosterone and has normal sex drive. A problem with the elastrator is that tetanus can occur when the tissue dies below the elastrator band. Surgical removal of the testicles creates a wound that attracts flies in the warm season. A fly repellant should be applied around the wound. It is recommended that a veterinarian castrate older goats. Whatever method, caution must be used to prevent infection, gangrene, and other complications.

It is important that the udder of dairy goats have strong fore and rear attachments and that the two teats be well spaced. Goats are milked by hand (Fig. 33.8) or by milking

FIGURE 33.8.
Hand milking the goat. Courtesy of G.F.W. Haenlein, University of Delaware.

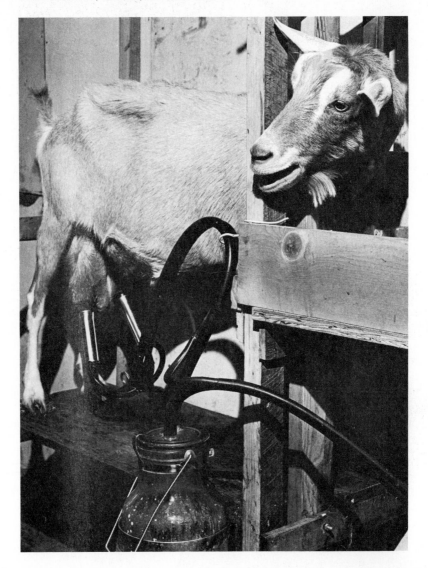

FIGURE 33.9.
A goat being milked by machine in a stanchion at a convenient level for milking. Courtesy of Mrs. Eva Rappaport.

machines. The milking stanchion should be elevated to place the goat at a convenient level for a person doing the milking (Fig. 33.9). Clean, sanitary conditions of the milking stanchion, the milking machine, the goat, and her udder and teats are very important prior to milking. After milking, the teats should be dipped in a weak iodine solution.

Considerations of special importance in managing dairy goats are care at time of breeding, kidding, and feeding.

Time of Breeding

The female goat comes into heat at intervals of 15–18 days until she becomes pregnant. Young does can be bred first when they weigh 85–95 lb. at about 6 to 10 months of age, which means that does in good general condition may be bred to have kids at 1 year of age. Goats are seasonal breeders, with the normal breeding season occurring in September, October, and November. If no effort is made to breed does at other times, most of the young will be born in February, March, or April. Normal goat lactations are 7–10 months in duration. A period of at least 2 months when no goats are lactating might occur, but with staggered breeding, the kidding and milking seasons can be extended. Housing goats in the dark for several hours each day in the spring and summer months, i.e., to simulate onset of shorter days, causes some to come into estrus earlier than usual. Conversely, artificial additional light in the goat barn may delay estrus in the autumn.

One service is all that is needed to obtain pregnancy, but it is generally wise to delay breeding for a day after the goat first shows signs of heat. The doe stays in heat from 1–3 days, but the optimum period of standing heat may last only a few hours at the end of estrus. A female in heat is often noisy and restless, and her milk production may decrease sharply. She may disturb other goats in the herd, so it is advisable to keep her in a separate stall until she goes out of heat.

Male goats during the breeding season usually have a rank odor, especially when confined to a pen. Housing the males downwind apart from the does will reduce some of the nuisance of male odors. Cleanliness of the male's long beard and shaggy hair helps reduce "bucky" odors.

Careful records should be kept showing breeding dates and the bucks used. From breeding records, it is possible to calculate when does should deliver kids as the gestation period is about 150 days. Detailed breeding records are essential for knowing ancestors, which in turn helps in the selection of superior males to produce especially desirable offspring.

Delivery of twins is common, and delivery of triplets occurs on occasion, particularly among mature does. More males are conceived and born than females. The ratio is approximately 115:100. Mature does average two kids whereas younger does average 1.5. Kids at birth weigh 5–8 lb, but singles may be heavier than twins or triplets, and males are usually heavier than females.

Time of Kidding

A doe, almost ready to deliver young, should be placed in a clean pen that is well-bedded with clean straw or shavings. The doe should have plenty of clean water and some laxative feed (such as wheat bran) as well as fresh soft legume hay. She should be carefully observed but not disturbed unless assistance is necessary, as indicated by excessive straining for 3 hours or more. If the kid presents the front feet and head, delivery should be easy without assistance. If only the front feet, but not the head, are presented, the kid should be pushed back enough to bring the head forward in line with the front feet. Breech presentations are not uncommon, but these deliveries should be rapid to prevent the kid from suffocating. If assistance is needed, a qualified attendant with small well-lubricated hands should pull when contractions occur. Gentleness is essential, and harsh or ill-timed pulling can cause severe internal damage.

As soon as the kid arrives, its mouth and nostrils should be wiped clean of membranes and mucus. In cold weather, it is necessary to take the newborn to a heated room for drying. A chilled kid must be helped to regain its body temperature with a heat lamp or maybe even by immersing its body up to the chin in water that is as hot as can be tolerated when the attendant's elbow is immersed in it for 2 min. Kids should be encouraged to nurse as soon as dry. Nursing helps the newborn kid to keep warm.

Difficulty in kidding may be caused by diseases and internal parasites. It is advisable to use disposable rubber or plastic gloves when assisting with delivery and to wash and disinfect everything diligently. Some diseases (including brucellosis) are transmissable to people; however, brucellosis has not been recently reported in U. S. goats.

Feeding

Because goats are ruminants, they can digest roughage effectively. However, types and proportions of feed should be related to the functions of the goats. For example, dry (nonmilking) does and bucks that are not actively breeding perform satisfactorily on ample browse, good pasture, and good-quality grass and legume hay. If grass is short and hay is of poor quality, and goats are milking well, feeding of supplemental concentrates is necessary. Overfeeding of supplemental concentrates can cause diarrhea or obesity, which interferes with reproduction and subsequent lactation.

Heavily lactating does and young does that must grow while lactating should be given good-quality hay *ad libitum* and concentrates liberally. Good-quality pasture should supplement hay well, but even then lactating goats require some roughage in the form of hay or other source of long fiber to prevent scouring.

Pregnant does should be fed to gain weight to assure adequate nutrition of the kids. A doe should be in good flesh but not fat when she kids, because she draws from her body reserves for milk production.

Grains such as corn, oats, barley, and milo may be fed whole because goats crack grains by chewing. If protein supplements are mixed with the grain or if the feed mix is being pelleted, the grain may be rolled, cracked or coarsely ground. Some people prefer to mix the hay and grain and prepare the ration in a total pellet form. There is added expense in pelleting, but much less feed is wasted.

In some areas, deficiencies in minerals such as phosphorus, selenium and iodine may exist. The use of iodized or trace-mineralized salt along with dicalcium phosphate or steamed bone meal usually provides enough minerals if legume hay or good pastures are available. Calcium is usually present in sufficient quantity in legume hay and phosphorus is adequately provided in grain feeds. To be safe, it is best to provide a mixture of trace-mineralized salt and bone meal free choice.

Young, growing goats and lactating goats need more protein than do bucks or dry does. A ration containing 12–15% protein is desirable for bucks and dry does, but 15–20% protein may be better for young goats and for does that are producing much milk.

Generally, kids must be allowed to nurse their dams to obtain the first milk, colostrum. After 3 days, kids may be removed from their mothers and given milk replacer by means of a lamb feeder or hand-held bottle until they are large enough to eat hay and concentrates. It is advisable to encourage young kids to eat solid feed at an early age, by 2–3 weeks, by having leafy legume hay and palatable fresh concentrates such as rolled grain available at all times. Solid feeds are less expensive than milk replacers, and when the kids can do well on solid feeds, milk replacers should not be fed. Small kids need concentrates until their rumens are sufficiently developed to digest enough roughage to meet all their nutritional needs. At 5–6 months of age, young goats can do well on good pasture or good quality legume hay alone.

CONTROLLING DISEASES AND PARASITES

Major diseases affecting goats are Johnes disease, caseous lymphadenitis, caprine pleuropneumonia, ecthyma, enterotoxemia, goat pox, herpesvirus, Pasteurella hemolytica, tetanus, and viral leukocencephalomelitis. As goats are highly resistant to bluetongue, there is no reason for alarm even if sheep having bluetongue are grazing in the same pasture.

Johnes disease also affects sheep and it is a serious health problem in goats. Affected goats become unthrifty, emaciated, and unproductive.

A troublesome disease is caseous lymphadenitis. It is characterized by nodules in the lymph area (throat). When one of the infected nodes is opened, a thick, caseous material is exposed. The disease is difficult to cure or to control. One should check animals in a herd from which breeding stock is considered for purchase and not buy animals from a herd that has infected animals. Also an isolation program is desirable for purchased animals to see if they develop the disease. Infected animals should be isolated, then infected nodes are opened and flushed with H_2O_2.

Caprine pleuropneumonia results in high mortality. A veterinarian should help control this disease.

Contagious ecthyma is also called soremouth. It is contagious to sheep and humans. Animals that recover from it are immune for several years. A live vaccine can be used on kids on premises that are infected.

Enterotoximia is caused by clostridium perfringes types C and D. Animals of all ages are susceptible. C. perfringes antitoxin given intraveneously or subcutaneously will give dramatic response. Toxoid vaccination given to month-old kids followed by a second dose in 2 weeks and booster doses each year will control the disease.

Goat pox causes lesions on mucous membranes and skin. It is prevalent in North Africa, middle East, Australia, and Scandinavia but not in the U. S.

Herpesvirus is serious in neonatal kids.

Young kids are highly susceptible to Pasteurella hemolytica. They can be treated with penicillin with successful results.

Tetanus is caused by *Clostridum tetani,* which infects wounds and often causes death. The causitive organism is prevalent in soil contaminated with horse feces. Two doses of toxoid to goats over 1 month of age and then an annual booster injection will control the disease.

Mastitis is an inflammation of the udder predisposed by bruising, lack of proper sanitation, and improper milking. The milk becomes curdled and stringy. It is advisable to engage the services of a veterinarian for treating severe mastitic cases. An udder may be treated with a suitable antibiotic by sliding a special dull plastic needle up the teat canal into the udder cistern and depositing the antibiotic, or by systemic antibiotic treatment intramuscularly or intravenously. Hot packs may help reduce the edema and in severe cases frequent milking will be necessary. Dry goat treatments can be effective prevention of mastitis.

Brucellosis is transmissable to humans. All goats should be tested for brucellosis, and reactor animals or suspects must be slaughtered.

Footrot occurs when goats are kept on wet land. It is contagious, but the bacterial organism causing it does not live long in the soil. Any animal to be introduced into a herd should be isolated for several days to see if footrot is present. Severe hoof trimming of infected feet, followed by treating with formaldehyde (as was described in Chapter 28) cures this disease. Generally, dairy goats dislike wet land and footrot is not a common problem in U. S. dairy goats.

Ketosis, or pregnancy disease, rarely occurs in goats. Affected goats are in pain and cannot walk. Exercising goats that are pregnant, reducing the amount of feed in the latter part of the pregnancy to avoid fat conditions and being certain that they have clean fresh water at all times helps prevent ketosis.

Milk fever can, but rarely, occurs in goats that are lactating heavily. Some heavy-milking goats may deplete their calcium stores. They cannot stand and may die of progressive paralysis. An intravenous injection of calcium gluconate results in rapid recovery. The goat may be up and normal in less than an hour after a calcium gluconate injection.

Goats are subject to many severe external and internal parasites. The external parasites include **lice** and **mites.** The animals may be sprayed, dipped, or dusted with an appropriate insecticide similar to those used on dairy cows. Dairy feed stores are usually good sources of materials and advice for treatment.

Internal parasites of goats include **stomach worms** and **coccidia.** The same treatment for stomach worms that is effective in sheep or horses can be used. **Coccidiosis,** which is caused by a protozoan organism in the intestinal tract, is found mostly in young goats. **Drenching** the animal with sulfa drugs or certain antibiotics helps control coccidiosis. Routine treatment with thibenzole or phenothiazine, or an alternation of these each six months, helps control internal parasites. Professional help in administering these drugs is advisable.

THE ANGORA GOAT

Angora goats (Fig. 33.10) are raised chiefly to produce mohair, which is made into fine clothing. These goats are produced largely in the Southwest (Texas has the most Angora goats of any state), the Ozarks of Missouri and Arkansas, and the Pacific Northwest. They do best on browse, so they are most commonly kept on rough, brushy areas. They are not as well adapted for intensive grazing pastures, because they prefer broadleaf plants over some grasses, and they are more likely to become parasitized when grazing near the soil level in overgrazed pastures.

Good Angora goats are shorn twice a year. Some have a tendency to shed their mohair in the spring, but selection against this trait has been successful. Some Angora goats are not shorn for 2–3 years, which allows the mohair to grow to lengths of 1–2 feet. This special mohair brings a high price. It is used for making wigs and doll hair. Shearing is

FIGURE 33.10.
An Angora goat showing a full fleece of mohair. Angora goats are useful for meat production, brush clearing, and mohair production. Courtesy of Mr. and Mrs. Don F. Kessi.

usually done in spring. The clip from does weighs 4–6 lb, whereas bucks and wethers may shear 5–8 lb. Highly improved goats often shear 10–15 lb of mohair per year. Three classes of mohair are produced: the tight lock, which has ringlets the full length of the fibers; the flat lock, which is wavy; and the fluffy, or open, fleece. The tight-lock fleeces are fine in texture but low in yield. The flat lock lacks the fineness of the tight lock but is satisfactory for making cloth and its yield is high. The fluffy fleece is of low quality, often quite coarse, and is easily caught in brush and lost.

Fleeces should be sorted at shearing time and coarse fleeces and those with burs should be kept separate from the good fleeces. The clip from each fleece should be tied separately so that it can be properly graded.

Fertility can be low in Angora goats and losses of kids on the open range or in scattered timber areas of the west coast states are often high, especially due to predators such as coyotes, bobcats, and wild dogs. Generally, the kid crop that is raised to weaning is about 80% of the number conceived as estimated on the basis of the number of does bred. Losses of 20% or more may occur when proper care is lacking at the time kids are born.

Angora goats have straighter horns than dairy goats. When they are in good condition on the range, mature bucks and wethers may weigh 150–200 lb. Mature does weigh 90–110 lb, but well-fed show animals usually weigh more. Kids are usually weaned at 5 months of age, which allows the doe to gain in weight at breeding time. Breeding over a period of days is usually done in late September and October. Since the gestation is 147–152 days, this time of breeding results in the kids arriving in the spring after the weather has moderated. Kidding may take place in open sheds when the weather is severe or outside when the weather is favorable. Young does are usually first bred to kid at the age of 2 years, but, if they are properly developed, they may be bred to kid when they are yearlings.

In some areas, Angora, Spanish meat goats or cull dairy goats are used to kill brush. One method is to stock the area heavily with goats and give them little feed as long as their health is not impaired. Keeping the goats hungry for green feed causes all brush to be destroyed in 2 years because the goats nip off the buds of new growth, thus starving the plant. It usually requires 2 years to starve hardy plants, but some of the less hardy ones will succumb the first year of starvation. Another method is to run goats along with cattle for 6–10 years. The goats tend to eat leaves of brushy plants while the cattle consume the grasses.

Angora goats need a shelter in rainy periods or wet snows, but an open shed is sufficient. The long mohair coats of Angoras keep them warm during cold weather, but they may nevertheless crowd together and smother in severe, cold snowstorms if no shelter is provided.

Angora goats are subject to the same parasites and diseases as dairy goats, and the same treatments are effective. In most respects, the care given Angora goats resembles the care given sheep rather than that given dairy goats.

THE MEAT GOAT

Meat goats are found mainly in the southwest and western states but are raised primarily in Africa and the Middle East where their main uses are for meat and skins. In Africa there has been a preference for a small, well-muscled, agile goat—the pygmy. These goats can obtain most of their feed from brushy plants and low trees. As many of the areas away from cities and towns are lacking in refrigeration, small goats that are slaughtered can be consumed by the family before the meat spoils.

Goats of any kind are usually run with sheep outside of the U. S. and this practice may be advantageous in the U. S. Goats normally eat more browse and forbs than sheep.

Rams should never be run with doe goats and buck goats should never be run with ewes. These animals will mate when the females come in heat (estrus) and may conceive, but the pregenancy will usually terminate at about 3 or 4 months. There have been rare cases of live sheep × goat crosses delivered alive at term.

THE PYGMY GOAT

Pygmy goats were developed in Africa, and those brought to the U. S. came from the Cameroon area to the Caribbean and then imported to the U. S. Early pygmy goats were brought to the U. S. as exotic zoo animals.

FIGURE 33.11.
An Agouti Pygamy goat can be a useful 4-H project animal. Courtesy of Deana Mobley.

Zoos not only sold pygmy animals to other zoos but also to research scientists. A colony of pygmy goats was established in Oregon in 1961 and several other research herds also were established, including herds at the University of California. Because Pygmies are small in size, the cost of keeping them is minimal. Pygmy goats are also raised for meat and milk.

Breeders of Pygmy goats enjoy having shows and showing their goats just like dairy goat and Angora goat breeders. There are over 50 Pygmy goat shows during the year where a total of several hundred goats are shown.

Pygmy goats vary in color from black to white with a dorsal stripe down the back and dark on the legs. On all Pygmy goats except black, the muzzle, forehead, eyes, and ears are accented in tones lighter than the dark portion of the body.

The Pygmy goat also is becoming popular as an animal for 4-H projects (Fig. 33.11). They are small and do not require a lot of space or feed. All goats are very friendly animals, which makes them ideal for young people to handle, and they are easily trained.

SELECTED REFERENCES

Publications

Coop, I. E. editor. 1982. A. Niemann-Sorensen, D. E. *Sheep and Goat Production. C. Production System Approach.* World Animal Science. Elsevier Scientific Publishing Company. New York, N. Y.

Dairy Goat Journal Publishing Co. 1982. *Proceedings of the Third International Conference on Goat Production and Diseases.* Dairy Goat Publishing Co. Scottsdale, Arizona.

Gall, C. 1981. *Dairy Goats.* London: Academic Press.

Haenlein, G. F. W. and D. L. Ace, (Editors). 1984. *Extension Goat Handbook.* Washington, D. C.: USDA Extension Service.

Haenlein, G. F. W. 1981. Dairy goat industry in the United States. *J. Dairy Sci.* 64:1288–1304.

Hale, L. and E. Kritzman, 1982. *Pygmy Goats: The Best of Memo.* National Pygmy Goat Association. Vashon Island, WA: Island Industries, Inc.

International Symposium: Dairy Goats. 1980. *J. Dairy Sci.* 63:1591–1781.

Nutrient Requirements of Goats. 1981. National Research Council. Washington, D. C.: National Academy Press.

Visuals

Livestock Judging, Slide Set No. 7 (Major breeds of goats illustrated). Vocational Education Productions, California Polytechnic State University, San Luis Obispo, CA 93407.

The Dairy Goat Kit (Sound filmstrips on *Dairy Goat Management* and *Fitting and Showing Dairy Goats*). Vocational Education Productions, California Polytechnic State University, San Luis Obispo, CA 93407.

Introduction to Dairy Goat Production, Facilities, Breeding/Prekidding Management/Kidding Management and *Care of Kids: Birth to Weaning, Maintaining Herd Health/Milking Management, Marketing/Obtaining Financing,* and *Dual-Purposes Goat* (videotapes). Agricultural Products and Services, 2001 Killebrew Drive, Suite 333, Bloomington, MN 55420.

CHAPTER 34

Animal Behavior

Ethology is the scientific study of an animal's behavior in response to its environment. Only within the past 30 years has animal behavior been accepted as a bona fide discipline within colleges of agriculture and veterinary medicine. A knowledge of animal behavior is essential to understand the whole animal and its ability to adapt to various management systems imposed by livestock and poultry producers. The value of horses can be increased when horse trainers apply their knowledge of horse behavior.

Animal behavior is a complex process involving the interaction of inherited abilities and environmental experiences to which the animal is subjected. Behavioral changes enable animals to adjust to some external or internal change of conditions, improve their chances of survival, and serve mankind. Also, producers can understand patterns of behavior to manage and train animals more effectively and efficiently.

Instinct (reflexes and responses) is what the animal has at birth. All mammals, at birth, have the instinct to nurse even though they must first learn the location of the teat. Shortly after hatching, chicks begin pecking to obtain feed.

Habituation is learning to respond without thinking. Response to a certain stimulus is established as a result of habituation. **Conditioning** is learning to respond in a particular way to a stimulus as a result of **reinforcement** when the proper response is made. Reinforcement is a reward for making the proper response. Trial and error is the performance of different responses to a stimulus until the correct response is performed, at which time the animal receives a reward. For example, newborn mammals soon become hungry and want to nurse. They search for some place to nurse on any part of the mother's body until they find the teat. This is trial and error until the teat is located; then, when the young nurses, it receives milk as its reward. Soon they learn where the teat is located and find it without having to go through trial and error. Thus, the young has become conditioned in nursing behavior through reinforcement.

Reasoning is the ability to respond correctly to a stimulus the first time that a new situation is presented. **Intelligence** is the ability to learn to adjust successfully to certain situations. Both short-term and long-term memory are part of intelligence.

MAJOR TYPES OF BEHAVIOR

Farm animals exhibit several major types of behavior: (1) sexual, (2) maternal, (3) communicative, (4) social, (5) feeding, (6) eliminative, (7) shelter-seeking, (8) investigative, (9) allelomimetic, and (10) maladaptive. Some of these types of behavior are interrelated, although they will be discussed separately. It is not the intent of this chapter to describe, in detail, the different behavior patterns for all farm animals. The major focus is to identify behavioral activities which most significantly affect animal well-being, productivity, and profitability. By understanding animal behavior, producers can plan and implement more effective management systems for their animals.

SEXUAL BEHAVIOR

Observations on sexual behavior of female farm animals are useful in implementing breeding programs. Cows that are in heat, for example, allow themselves to be mounted by others. Producers observe this condition of "standing heat" or estrus to identify those cows to be **hand-mated** or bred artificially. Ewes in heat are not mounted by other ewes, but vasectomized rams can identify them.

Males and females of certain species produce **pheromones,** which are chemical substances which attract the opposite sex. Cows, ewes, and mares may have pheromones present in vaginal secretions and in their urine when they are in heat. Bulls, rams, and stallions will smell the vagina and the urine using a nasal organ which can detect pheromones. A common behavioral response in this process is called **flehmen,** where the male animal lifts its head and curls the upper lip.

It appears that in a sexually active group of cows, the bull is attracted to a cow in heat most often by visual means (observing cow-to-cow mounting) rather than by olfactory clues. The bull follows a cow that is coming into heat, smells and licks her external genitalia, and puts his chin on her rump. When the cow is in standing heat, she stands still when the bull chins her rump. When she reaches full heat, she allows the bull to mount.

When females are sexually receptive, they will usually seek out a male if mating has not previously occurred. Females are receptive for varying lengths of time; cows are usually in heat for approximately 12 hours, whereas mares show heat for 5 to 7 days with ovulation occurring during the last 24 hours of estrus.

Vigorous bulls breed females several times a day. If more than one cow is in heat at the same time, bulls tend to mate with one cow once or a few times and then go to others. Other bulls may become attached to one female and ignore the others which are also in heat.

The ram will chase a ewe that is coming into heat. The ram champs and licks, puts his head on the side of the ewe, and strikes with his foot. When the ewe reaches standing heat, she stands when approached by the ram.

The buck goat does considerable snorting and blowing when he detects a doe in heat. The doe shows unrest and may be fought by other does. Mating in goats is similar to that in sheep.

The boar does not seem to detect a sow that is in heat by smelling or seeing. If introduced into a group of sows, a boar will chase any sow in the group. The sow that is in heat will seek out the boar for mating, and when the boar is located she stands still and flicks her ears. Boars produce pheromones in the saliva and preputial pouch, which attracts sows and gilts in estrus to the boars. Ejaculation requires several minutes for boars in contrast to an instantaneous ejaculation by rams and bulls.

The sequence of events in estrus detection in horses appears to be similar to swine (previously described). The stallion approaches a mare from the front and a mare not in heat runs and kicks at the stallion. When the mare is in standing heat she stands, squats somewhat, and urinates as he approaches. Her vulva exhibits **"winking"** (opens and closes) when she is in heat.

In chickens and turkeys, a courtship sequence between the male and female usually takes place. If either individual does not respond to the other's previous signal, the courtship does not proceed further. After the courtship has developed properly, some females run from the rooster, who chases them until they stop and squat for mating. The male chicken or turkey stands on a squatting female and ejaculates semen as his rear descends toward the female's cloaca. Semen is ejaculated at the cloaca and the female draws it into her reproductive tract.

Male chickens and turkeys show a preference for certain females and may even refuse to mate with other females. Likewise, female chickens and turkeys may refuse to mate with certain males. This is a serious problem when pen matings of one male and 10–15 females are practiced. The eggs of some females may be infertile. This preferential mating is a greater problem in chickens as AI is the common breeding practice in turkeys.

Little relationship appears to exist between sex drive and fertility in male farm animals. In fact, some males that show extreme sex drive have reduced fertility because of frequent ejaculations that result in semen with reduced sperm numbers.

Research studies show that many individual bulls have sufficient sex drive and mating ability to fertilize more females than are commonly allotted to them. An excessive number of males, however, are used in multiple-sire herds to offset the few which are poor breeders and to cover for the social dominance which exists among several bulls running with the same herd of cows. The bull may guard a female which he has determined is approaching estrus. His success in guarding the female or actually mating with her is dependent on his rank of dominance in a multiple-sire herd. If low fertility exists in the dominant bull or bulls, then calf crop percent will be seriously affected even in multiple-sire herds.

Tests have been developed to measure libido and mating ability differences in young bulls. While behavioral differences are evident between different bulls, these tests, based on pregnancy rates, have not proven accurate for use by the beef industry.

Bulls, being raised with other bulls, commonly mount one another, have a penal erection and occasionally ejaculate. Individual bulls can be observed arching their back, thrusting their penis toward their front legs, and ejaculating.

Bulls can be easily trained to mount objects which provide the stimulus for them to experience ejaculation. AI studs commonly use restrained steers for collection of semen. Bulls soon respond to the artificial vagina, when mounting steers, which provides them with a sensual reward.

Mating behavior has an apparent genetic base as there is evidence of more frequent mountings in hybrid or crossbred animals.

Some profound behavior patterns are associated with sex of the animal and changes resulting from castration. This verifies the importance of hormonal directed expression of behavior. Intact males have more aggressive behavior, while castrates are more docile after losing their source of male hormone.

MATERNAL BEHAVIOR

There is evidence that more cows calve during periods of darkness than during daylight hours. The calving pattern, however, can be changed by when cows are fed. Cows which are fed during late evening will have a higher percentage of their calves during daylight hours.

When the young of cattle, sheep, goats and horses are born, the mothers clean the young by licking them. This stimulates blood circulation and encourages the young to stand and to nurse. Sows do not clean their newborn, but encourage them to nurse by lying down and moving their feet as the young approach the udder region. They thus help the young to the teats.

Most animal mothers tend to fight intruders, especially if the young squeal or bawl. Often cows, sows, and mares become very aggressive in protecting their young shortly after parturition. Serious injury can occur to producers who do not use caution with these animals.

Strong attachments exist between mother and newborn young, particularly between ewe and lamb and cow and calf. Beef cows diminish their output of milk about 100–120 days after birth of young, and ewes do the same after 60–75 days. This reduction in milk forces the young to search for forage, the consumption of which stimulates rumen development. It is at this time that care-giving by the mother declines.

If young pigs have a high-energy feed available at all times, they nurse less frequently. Without a strong stimulus of nursing, sows reduce their output of milk. Some sows may wean their pigs early and show little concern for them a few days after they are weaned. Pigs are usually weaned by producers at 21–35 days of age.

COMMUNICATION BEHAVIOR

Communication exists when some type of information is exchanged between individuals. This may occur with the transfer of information through any of the senses.

A distress call, involving a different type of sound, occurs from either the female or her young when they become separated (Fig. 34.1). Young animals cry for help when disturbed or distressed. Lambs bleat, calves bawl, pigs squeal, and chicks chirp. Even adult animals call for help when under stress. The female and her offspring may recognize each other's vocal sound; however, it appears the most effective way the dam recognizes her offspring is by smell. The young usually nurses with its rear end towards the female's head. This allows dams to smell their offspring and decide to accept or reject it. A rejected young animal is usually bunted with the head of the dam and kicked with the rear legs when it attempts to nurse. The young animals are less discriminate in their nursing behavior than are their dams.

FIGURE 34.1.
A calf separated from its dam has a distict bawl, which communicates distress or dissatisfaction. Courtesy of the American Hereford Association.

Females will more easily adopt other young through transfer of the odor of one young animal to another. Cows have fostered several calves, by removing their own calves at birth, then smearing the foster calves with amniotic fluid previously collected from the second "water bag."

Many farm animals learn to respond to vocal calls or whistles of the producer who wants the animals to come to feed. The animals soon learn that the stimulus of the sound is related to being fed.

The bull vocally communicates his aggressive behavior challenge to other bulls and intruders to his area through a deep bellow. This bellow and aggressive behavior is under the control of the male hormone (testosterone), as the castrated male seldom exhibits similar behavior.

The bull also issues calls to cows and heifers, especially when he is separated from them, but they are within his sight.

Cattle are especially perceptive in their sight as they have 310 to 360 degree vision. This affects their behavior in many ways, for example, as they are approached from different angles and as they are handled through various types of facilities.

SOCIAL BEHAVIOR

Social behavior includes behavior activities of fight and flight and those of aggressive and passive behavior when an animal is in contact (physically) with another animal or with livestock and poultry producers.

Interaction with Other Animals

Unless castrated when young, the males of all farm animals fight when they meet other unfamiliar males of the same species. This behavior has great practical implications for management of farm animals. Male farm animals are often run singly with a group of females in breeding season, but it is often necessary to keep males together in a group at times other than the breeding season. The typical producer simply cannot afford to provide a separate lot for each male.

Bulls and other males may exert prolonged physical activity when fighting and thus generate much heat. Therefore, bulls and other potential fighting males should be put together either early in the morning or late in the evening when the environmental temperature is lower than at midday. Often fighting can be reduced when male animals are mixed and put into a new environment.

If a mature bull is put with young bulls, fighting usually occurs because the younger ones concede to the mature one. After the fighting is over, bulls may start mounting one another (one form of homosexuality).

Cows, sows, and mares usually develop a peck order, but fight less intensely than males. Sows that are strangers to each other sometimes fight. Ewes seldom, if ever, fight, so ewes that are strangers can be grouped together without harm.

Some cows will withdraw from the group to find a secluded spot just before calving. Almost all animals will withdraw from the group if they are sick.

Early and continuous association of calves is associated with greater social tolerance, delayed onset of aggressive behavior, and relatively slow formation of social hierarchies.

Status and social rank typically exist in a herd of cows, with certain individuals dominating others who become submissive. The presence or absence of horns are important in determining social rank, especially when strange cows are mixed together. Also, horned cows will usually outrank polled or dehorned cows where close contact is encountered, such as at feedbunks or on the feed ground.

Large differences in age, size, strength, genetic background, and previous experience have powerful effects in determining social rank. Once the rank is established in a herd, it tends to be consistent from one year to the next. There is evidence that genetic differences exist for social rank.

Animals fed together will consume more feed than when fed individually. This competitive environment evidently is a stimulus for greater feed consumption. Dairy calves, separated from their dams at birth, appear to gain equally well whether fed milk in a group or kept separate. There is, however, evidence that they learn to eat grain earlier when group fed compared with being individually fed. Cattle individually fed in metabolism stalls will consume only 50–60% the amount of feed if the animals are group fed.

When fed in a group of older cows, 2-year-old heifers have difficulty getting their share of supplemental feed (Table 34.1). Two-year-old heifers and 3-year-old cows can be fed together without there being a significant age effect in competition for supplemental feed. These behavior differences no doubt explain some of the nutrition, weight gains,

**TABLE 34.1.   Weight Changes of 2-Year-Old Heifers Fed
Separately As a Group or Together with Older Cows**

Treatment	Weight Change
Pastured and fed with older cows	25 lb loss
Pastured and fed separately	46 lb gain

Source: Wagnon 1965.

and postpartum internal relationships that are age related when cows of all ages compete for the same supplemental feed.

Dominant cows, raised in a confinement operation, will usually consume more feed and wean heavier calves. More submissive cows will wean lighter calves (25%), and fewer of them will be pregnant compared with more aggressive cows. Figure 34.2 shows a 1-hour feeding pattern for two cows of different social rankings.

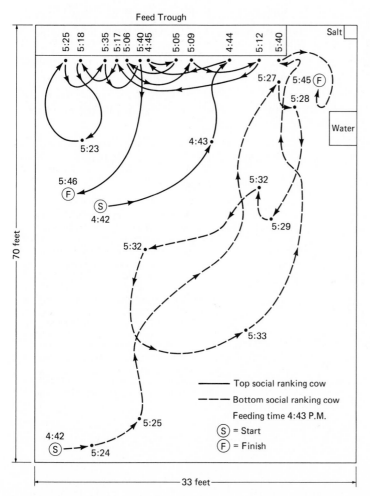

FIGURE 34.2.
One hour feeding pattern for two cows of different social rankings. Adapted from Schake and Riggs, 1972.

Interactions with Humans

Producers consider the disposition or temperament of animals ranging from docile to wild or "high strung." This evaluation is usually made when the animals are being handled through various types of corrals, pens, chutes, and other working facilities. The typical behavior exhibited by animals with poor dispositions is one of fear or an aggressive fighting or kicking behavior.

There is evidence that farm animals develop good or poor dispositions from the way they have been treated or handled, although there is evidence that disposition has an inherited basis. A few heritability estimates for disposition are in the medium-to-high category, indicating the trait would respond to selection. Some producers cull or eliminate animals with poor dispositions from their herds and flocks because of potential personal injury and economic losses among broken fences, facilities, and also to reduce the excitability of other animals.

Behavior During Handling and Restraint

Most animals are handled and restrained several times during their lifetime. Ease of handling will depend largely on their temperament, size, previous experience, and design of the handling facilities. Understanding animal behavior can assist in preventing injury, undue stress, and physical exertion for both animals and producers. An example is knowing how to approach animals so they will respond to how the producer prefers the animals to move (Fig. 34.3). Most animals have a flight zone. When a person is outside this zone the animals will usually exhibit an inquisitive behavior. When a person moves inside the flight zone the animal will usually move away.

Blood odor appears to be offensive to some animals; therefore, the reduction or elimination of such odors may well encourage animals to move through handling facilities

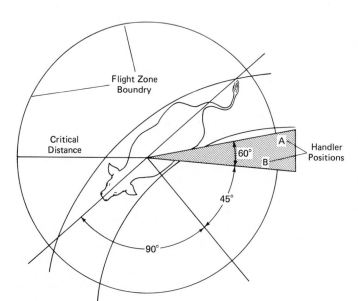

FIGURE 34.3.
Handler positions for moving cattle. Positions A and B are the best places for the handler to stand. The flight zone is penetrated to cause the animal to move forward. Retreating outside the flight zone will cause the animal to stop moving. Handlers should avoid standing directly behind the animal because they will be in the animal's blind spot. If the handler gets in front of the line extending from the animal's shoulder, the animal will back up. The solid curved lines indicate the location of the curved, single-file chute. Courtesy of Temple Grandin.

FIGURE 34.4.
Shadows that fall across a chute can disrupt the handling of animals. The lead animal will often balk and refuse to cross the shadows. Courtesy of Temple Grandin.

with greater ease. Animals are easily disturbed by loud or unusual noises such as motors, pumps, and compressed air.

With their 310–360 degree vision, cattle are sensitive to shadows and unusual movements observed at the end of the chute or outside the chutes (Fig. 34.4). For these reasons, cattle will move with more ease through curved chutes with solid sides (Fig. 34.5).

Round pens, having an absence of square corners, will handle cattle that are more excitable with less injury occurring to them.

Some breeders claim that AI facilities should permit beef cows to be handled quietly and carefully, not using facilities where cows have previously felt pain. They state that pregnancy rates will be higher. This sounds logical, but research needs to substantiate these claims.

FEEDING BEHAVIOR

Ingestive behavior is exhibited by farm animals when they eat and drink. Rather than initially chewing their feed thoroughly, ruminants swallow it as soon as it is well lubri-

FIGURE 34.5.
Animals move easier through curved chutes with solid sides. Courtesy of Temple Grandin.

cated with saliva. After they have consumed a certain amount, they ruminate (regurgitate the feed for chewing). Cattle graze for 4–9 hours/day and sheep and goats graze for 9–11 hours/day. Grazing is usually done in periods, followed by rest and **rumination.** Sheep rest and ruminate more frequently when grazing than do cattle—cattle ruminate from 4 to 9 hours/day; sheep from 7 to 10 hours/day. A cow may regurgitate and chew between 300 and 400 **boluses** of feed per day; sheep between 400 and 600 boluses per day.

Under range conditions, cattle will usually not go more than 3 miles away from water, whereas sheep may travel as much as 8 miles a day. When cattle and sheep are on a large range, they tend to overgraze near the water area and avoid grazing in areas far removed from water. Development of water areas, fencing, placing of salt away from water, and herding the animals are management practices to assure a more uniform utilization of the range forage.

Grazing Behavior

Cattle, horses, and sheep have palatability preferences for certain plants and many have difficulty changing from one type or types of plants to other types. Most animals have preference to graze lower areas, especially if they are near water. These grazing behaviors tend to cause overgrazing in certain areas of the pasture and reduce weight gains.

Age of cow and weather affect the typical behavior of cows grazing native range during the winter (Table 34.2). At the Range Research Station at Miles City, Montana, cows grazed less as temperatures dropped below 20°F, and at a −30°, 3-year-olds grazed approximately 2 h less than 6 year olds. Also, with colder temperatures, cows wait longer before starting to graze in the morning. At 30°, cows started grazing between 6:30 and 7:00 A.M., but at −30°, they waited until about 10 A.M.

TABLE 34.2. Activities of Cows Grazing on Winter Range

Activity	Hours
Grazing	9.45
Ruminating	
Standing	0.63
Lying	8.30
Idle	
Standing	1.11
Lying	3.93
Traveling	0.58
Total	24.00

Eliminative Behavior

Cattle, sheep, horses, goats, and chickens void their feces and urine indiscriminately. Hogs, by contrast, defecate in definite areas of the pasture or pen. Ease of cleaning swine pens can be planned by knowing defecation patterns of the pigs.

Cattle, sheep, goats, and swine usually defecate while standing or walking. All these animals urinate while standing, but not usually when walking. Cattle defecate 12–18 times a day; horses 5–12 times. Cattle and horses urinate 7–11 times per day. Animals on lush pasture drink less water than when they consume dry feeds; therefore, the amount of urine voided may not differ greatly under these two types of feed conditions.

All farm animals urinate and defecate more frequently and void more excreta than normal when stressed or excited. They often lose a minimum of 3% of their liveweight when transported to and from marketing points. Much of the **shrink** in transit occurs in the first hour, so considerable weight loss occurs even when animals are transported for short distances. Weight loss can be reduced by handling animals carefully and quietly, avoiding any excessive stress or excitement of the animals.

SHELTER-SEEKING BEHAVIOR

Animal species vary greatly in the degree to which they seek shelter. Cattle and sheep seek a shady area for rest and rumination if the weather is hot, and pigs try to find a wet area. When the weather is cold, pigs crowd against one another when they are lying down to keep each other warm. In snow and cold winds, animals often crowd together. In extreme situations, they pile up to the extent that some of them smother. Unless the weather is cold and windy, cattle and horses often seek the shelter of trees when it is raining. This may be hazardous where strong electrical storms occur because animals under a tree are more likely to be killed by lightning than those in the open.

INVESTIGATIVE BEHAVIOR

Pigs, horses, and dairy goats are highly curious and will investigate any strange object. They usually approach carefully and slowly, sniffing and looking as they approach. Cattle

FIGURE 34.6.
Cattle expressing investigative behavior as a producer is approaching them. Courtesy of R.W. Henderson.

also do a certain amount of investigating (Fig. 34.6). Sheep are less curious than some other farm animals and are more timid. They may notice a strange object, become excited, and run away from it.

ALLELOMINETIC BEHAVIOR

Animals of a species tend to do the same thing at the same time. Cattle and sheep tend to graze at the same time and rest and ruminate at the same time. On a cattle ranch, cows from over the range gather at the watering place at about the same time each day because one follows another. This behavior is of practical importance because the producer can then observe the herd or flock with little difficulty, notice anything that is wrong with an animal, and have the animal brought in for treatment. If one is artificially inseminating beef cattle, the best time to locate range cows in heat is when they gather at the watering place. This type of behavior is useful in driving groups of animals from one place to another.

MALADAPTIVE OR ABNORMAL BEHAVIOR

Animals that cannot adapt to their environment may exhibit inappropriate or unusual behavior.

Some animals under extensive management systems, such as poultry and swine, are often kept in continuous housing to reduce costs of land and facilities. Frequently, both chickens and swine will resort to cannibalism, which may lead to death if preventive measures are not taken. Some swine producers remove the tails of baby pigs to prevent tail chewing, because tail chewing can cause bleeding, and whenever bleeding occurs, the pigs are likely to become cannibalistic.

Some male uncastrated animals raised with other males will masturbate and demonstrate homosexuality. In the latter situation, males will mount other males attempting to breed them. Some of the more submissive males may have to be physically separated from the more aggressive males to prevent injury or death.

The **buller-steer syndrome** is exhibited in steers that have been castrated before puberty. This demonstrates ʾa masculine behavior other than testosterone origin. Certain steers (bullers) are more sexually attractive for other steers to mount. As one steer mounts a buller, then other steers are attracted to do the same. Thus the activity associated with the buller-steer syndrome can cause physical injury, a reduction in feedlot gains, and additional labor and equipment expense as bullers are usually sorted into separate feedlot pens. Some feedlots experience 1–3% of their steers as bullers.

Growth implants and reduced pen space have been cited as reasons for an increased incidence of buller-steers. Some behaviorists cite evidence of similar homosexual behavior of males in free-ranging natural environments.

SELECTED REFERENCES

Publications

Arave, C. W. and Albright. J. L. 1981. Cattle behavior. *J. Dairy Sci.* 64:1318.

Craig, J. V. 1981 *Domestic Animal Behavior.* Englewood Cliffs, NJ: Prentice-Hall.

Curtis, S. E. and Houpt, K. A. 1983. Animal ethology: Its emergence in animal science. *J. Anim. Sci.* 57(Suppl.2):234.

Fraser, A. F. (Editor). 1985. *Ethology of Farm Animals. A Comprehensive Study of the Behavioral Features of Common Farm Animals.* New York: Elsevier Publishing Co., Inc.

Grandin, T. "Handling facility" fitness. *Managing in Changing Times.* IMC Conf. Proc., Sept. 16, 1986, Saskatoon, Saskatchewan, Canada.

Hart, B. L. 1985. *The Behavior of Domestic Animals.* New York: W. H. Freeman.

Houpt, K. A. 1985. Behavioral problems in horses. *American Association of Equine Practitioners* 31:113.

Houpt, K. A. and Wolski, T. R. 1982. *Domestic Animal Behavior for Veterinarians and Animal Scientists.* Ames, IA: The Iowa State University Press.

Kilgour, R. and Dalton, C. 1984. *Livestock Behavior: A Practical Guide.* Boulder, CO: Westview Press.

Signoret, J. P. (Editor). 1982. *Welfare and Husbandry of Calves.* Boston: Martinus Nijhoff Publishers.

Visuals

Safety in Handling Livestock (sound filmstrip). Vocational Education Productions. California Polytechnic State University, San Luis Obispo, CA 93407.

CHAPTER 35

Animal Welfare and Animal Rights

This chapter is approached differently from other chapters in this book as more questions are posed than answers provided. Many philosophical issues are presented because the scientific basis, which supports most of the other material in this book, is limited or nonexistent.

Although all may not agree, it seems logical to have agricultural, preveterinary medicine, and other students understand the issues associated with animal welfare and animal rights. Only by understanding concerns and challenges can people work together in solving problems.

Some individuals do not differentiate between animal welfare and animal rights whereas other individuals make a sharp distinction between the two. Those who make the distinction say they are supportive for animal well-being or animal welfare but many animal rights issues are not acceptable to them. At the risk of some overlap, animal welfare and animal rights will be discussed in separate sections. This discussion will concentrate primarily on farm animals identified in other chapters of the book.

ANIMAL RIGHTS

Should animals have the same rights as humans? There are those individuals who feel that human's involvement with animals has deprived animals of social interaction. Livestock and poultry producers believe they have the well-being of animals at heart, thus many accept the concept of animal welfare, otherwise the animals would not be productive. Individuals who take a strong stand on animal rights are labeled as extremists by livestock producers.

The issues of animal rights center around intensive animal production practices—e.g., battery caged systems for laying hens; veal calves in single crates; and continuous tethering (tying) of sows in confinement. Typical participants in this controversy are vegetarians, livestock producers, philosophers, lawyers, and others who just associate themselves with "a cause." Sometimes it is individuals who put themselves in the middle of these controversies who materially gain the most—the writers, those who establish new organizations, and others.

Are many of our farm animals deprived? How does one assess deprivation? In one sense the domestication of animals resulted in animals being deprived of some opportunities they had before domestication. It also provided them with a better year-round feed supply and a longer, productive life.

But is intensive management or confinement raising (animal rightists call it factory farming) too depriving? Intensive management systems typically control environment factors (temperature, humidity, and a steady supply of feed and water) more completely. However, these intensive systems may also result in stress, which results in behavioral problems and reduced longevity.

These intensive production operations are economically driven by a consuming public and government who demands high quality foods at the lowest possible cost. If producers return to less efficient production methods who pays for the inefficiency? Not the consumer because the producer cannot pass on the higher production costs. If supply becomes low because of lower production, then price and profitability might increase. But who decides who will lower production or go out of business?

The factory farming of animals (and crops) allowed an agricultural producer to feed many more people—75 more people per farmer from 1900 to 1985. This increased efficiency has allowed other people (or deprived them?) to move from the farm to the city and produce other goods and services for the benefit of mankind. Is the factory worker (confined to an assembly line) who lives in a high-rise apartment complex deprived as some confinement-raised animals? Probably not to the same extent as there is more freedom of choice for the factory worker but there are some similarities considering the pollution of the big cities and a potentially stressful factory environment. Economics has been a driving force in determining the current lifestyles of humans and farm animals in the U. S.

Have we become conditioned in our perception that animals are more human than they really are? Walt Disney conveyed to millions of people through animated motion pictures that animals could think, feel, and communicate on the same level as humans. The toy and book industries have conveyed the same message.

ANIMAL WELFARE

Livestock producers have had concern for the welfare of their animals for centuries. The following biblical verse conveys this thought:

"Be thou diligent to know the state of thy flocks and look well to thy herds." Proverbs 27:23

Animals were domesticated to give nomadic people a more consistent supply of food, companionship, and draft animals for transportation and power. People soon learned that the productive response of animals was greater when animals were given proper care.

Today nutrition, health, and management needs of animals are well known and scientifically based. There is evidence that many animals in the U. S. receive a more nutritious diet than many humans. The development of the veterinary medical profession has provided on-farm services, health clinics, and hospital care, in many cases equal to human health-care services.

Scientific Basis

Many of the issues of animal welfare (rights) are based on emotions and personal beliefs. For example, some people have a personal value system that objects strongly to killing animals for any reason. They would eliminate the killing and suffering of animals to the limit of human capability. There can be little or no scientific information that can resolve differences in personal beliefs or personal value systems. Research can relate to identifying measures of well-being of food-producing animals.

It is not the intent of this chapter to make an exhaustive coverage of research data that relates to the welfare of farm animals. Several points are made to introduce this topic. Readers should refer to the Selected References at the end of this chapter for more extensive reading. We can expect to see numerous research papers during the next few years more clearly identifying (or confusing) parameters of animal well-being.
Items to consider:

1. The principal criteria used by producers and animal scientists as indexes of welfare in food producing animals have been growth rate, feed efficiency, reproductive efficiency, mortality, and morbidity. These are only indicators of animal well-being as there is not total agreement on how to define well-being. In intensive production systems, the effect of stress on longevity is not critically evaluated as these animals have rapid turnover in the production systems.

2. Gonyou (1986) reviewed the research data on comfort and well-being in farm animals in terms of physical damage, physiological responses, and behavior. Some of his conclusions were:

a. There is no universal measure for comfort and well-being and it is unlikely to be discovered.
b. Without a universal measure of comfort and well-being, an accurate assessment is not possible when attempting to compare different management systems.
c. A less publicized but often productive approach is to identify a problem in a specific system and reduce or eliminate that problem. Tauson (1980) evaluated physical damage in feather density, foot damage, claw length, and throat blisters in hens in battery cages. Feather density was improved by using solid partitions between cages; plastic-coated mesh floors reduced foot damage, and horizontal, as opposed to vertical, wires in the area of the feeders reduced throat blisters. Commercial cages combining these features result in less physical damage and represent an improvement in comfort and well-being.
d. Progress is being made in each of these three areas discussed. More importantly, the comfort and well-being of animals has been improved in some areas and other attempts are in progress.

ORGANIZATIONS

Animal Rights (Welfare)

There are reportedly 200 serious animal welfare groups and organizations with a like number of informal groups. They have a financial base estimated at more than $150 million. One long-established animal rights organization claims an endowment of $40 million.

Noted celebrities associate their identity and vocal support for the cause of animal rights. A growing welfare movement claims a following of 2–5 million people with 85% of their followers being women. This movement has attracted leaders who study the system and know how to work it.

There are several different approaches taken by various animal welfare and animal rights organizations. Some take the stand that it is okay to exploit animals if they are humanely treated. Another group is more radical and believes animals should not be exploited in any way for 'any reason. Dr. Michael Fox, Director of the U. S. Humane Society's Institute for the Study of Animal problems, who is one of the most widely quoted individuals associated with the animal welfare movement, appears to take a stand between the two extreme approaches previously mentioned.

Factory farming is one of the main targets for several animal rights organizations. Some extreme groups, Food Animal Concerns Trusts (FACT) and The Farm Animal Reform Movement (FARM), are both urging the reduction and eventual elimination of animals for food. One of their approaches is to publicize how destructive the consumption of animal products is to human health.

Agricultural Groups

Farm Animal Welfare Coalition (FAWC), an organization of agricultural groups, has been formed to provide animal producers and animal industry organizations a clear, single voice on animal welfare issues. Individual producers have been encouraged to avoid confrontations with animal rights activists as it only gives activists the media coverage they desire to have. Activists have been quoted as saying that to achieve public awareness they need mass demonstrations and media headlines.

IN RETROSPECT

The current issues of animal welfare and animal rights causes individuals to be polarized in their views and actions. Is it possible that some individuals who represent the extremes of these issues may not have the best interests of both human beings and farm animals?

"Moderation," "balance," or "optimums" appear to be a logical compromise in considering the well-being of humans and farm animals. It is not logical for human beings to be abusive to one another or to other animals. Humans should not be motivated only by selfish or profit-oriented interests. Humans have the responsibility to be good "caretakers" and stewards of the earth's resources. The replenishing of the earth for future generations is a desirable humanitarian approach.

On the other hand, vegetarians should be respected for their freedom of choice. However, they should not impose their views on others. Those who advocate strict population control may well be selfishly denying other potential human beings the opportunity for earthly experiences. Progress in the efficient production of food has given a reasonable basis for an increased human population in which more individuals can experience "the good life." There is evidence for much higher levels of food production in the world if the problems of selfishness, poor land utilization, and unstable governments can be overcome.

Animal rightists, livestock producers, and other concerned individuals should work together to fund and support bonafide research and investigation into those matters of greatest concern. Scientifically based truths rather than emotionally based opinions can provide a more rational basis for determining the combined well-being of humans and their farm animals.

SELECTED REFERENCES

Publications

Fox, M. W. 1984. *Farm Animals. Husbandry, Behavior, and Veterinary Practice* (Viewpoints of a Critic). Baltimore: University Park Press.

Gonyou, H. W. 1986. Assessment of comfort and well-being in farm animals. *J. Anim. Sci.* 62:1769.

Hart, B. J. 1985. The behavior of domestic animals. Chapter 10. *Contributions of Behavioral Science to Issues in Animal Welfare.* New York: W. H. Freeman and Co.

Moreng, R. E. and Avens, J. E. 1985. Poultry science and production. Chapter 11. *Agricultural Animal Welfare.* Reston, Virginia: Reston Publishing Co.

Regan, T. 1983. *The Case for Animal Rights.* Berkeley: University of California Press.

Rollin, B. E. 1981. *Animal Rights and Human Morality.* Buffalo, N. Y.: Prometheus Books.

Scientific Aspects of the Welfare of Food Animals. 1981. Report No. 91 Council for Agri. Sci. and Tech. (CAST), Ames, Iowa.

Smidt, D. 1983. *Indicators Relevant to Farm Animal Welfare.* Boston: Martinus Nijhoft Publishers.

Tannenbaum, J. 1986. Animal rights: Some guideposts for the veterinarian. *J. Am. Vet. Med. Assoc.* 188:1258.

Tauson, R. 1980. Cages: How could they be improved? In: R. Moss (ed). The laying hen and Its environment. *Curr. Top. Vet. Med. Anim. Sci.* 8:269–299.

Visuals

The Animal Film: (136 minutes, pro-animal rights). 1981. The Cinema Guild, 1697 Broadway, New York, NY 10019.

Facing the Animal Rights Challenge (7 min. slide/tape; how producers should deal with the activist movement). Hugh Johnson, American Farm Bureau Federation, 225 Touhy Ave., Park Ridge, IL 60068.

CHAPTER 36

Making Effective Management Decisions

Effective management of livestock operations implies that available resources are used to maximize net profit while the same resources are conserved or improved. Available resources include fixed resources (land, labor, capital, and management) and renewable biological resources (animals and plants). Effective management requires a manager who knows how to make timely decisions based on a careful assessment of management alternatives. Modern technology is providing useful tools to make more rapid and accurate management decisions.

Previous chapters have shown how biological principles determine the efficiency of animal production. It is important to identify other resources which, combined with the efficiency of animal production, will determine the profitability of an operation.

THE MANAGER

The manager is the individual responsible for planning and decision making. The management process in simple form is to plan, act, and evaluate. This process is described in more detail in Fig. 36.1.

It is imperative that livestock and poultry producers manage their operations as business, and not just as a way of life. Current economic pressures associated with keen competition are forcing more producers to manage their operations as business enterprises.

The manager may be an owner-operator with minimum additional labor, or the manager of a more complex organizational structure involving several other individuals or businesses (Fig. 36.2). An effective manager, whether involved in a one-person operation or a complex organizational structure, needs to (a) be profit-oriented, (b) identify objectives of the business and form both short-term and long-range goals to achieve those objectives, (c) keep abreast of the current knowledge related to the operation, (d) know how to use time effectively, (e) attend to the physical, emotional, and financial needs of those employed in the operation, (f) incorporate incentive programs to motivate employ-

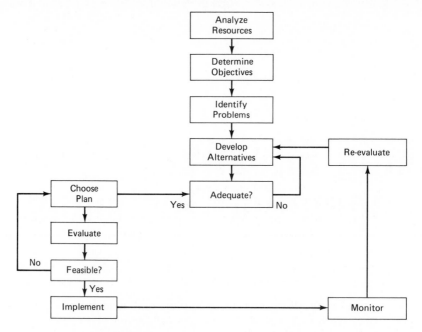

FIGURE 36.1.
Major component parts of the planning process.

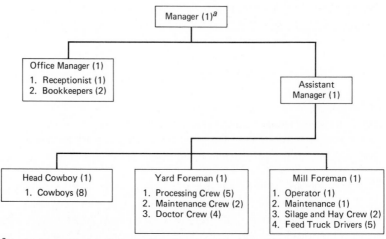

[a]Number in parenthesis indicates number of employees in each position

FIGURE 36.2.
Organizational structure of a large commercial feeding operation.

ees to perform at their full capacity each day, (g) have honest business dealings, (h) effectively communicate responsibilities to all employees and make employees feel they are part of the operation, (i) know what needs to be done and at what time, (j) be a self-starter, (k) set priorities and allocate resources accordingly, (l) remove or alleviate high risks, and (m) manage oneself so that others can see and follow a good example.

FINANCIAL MANAGEMENT

Costs, returns, and profitability of a livestock operation can only be assessed critically with a meaningful set of records. A financial record system should include a financial statement, a yearly budget of anticipated income and expenses, adequate income and expense records to complete the IRS tax forms (primary form shown in Fig. 36.3), and a cash-flow statement that shows the anticipated income and expenses by months. The latter determines if operating money is adequate. Knowing this could save interest for the business because the manager could borrow and repay at certain critical times of the year. Cash-flow statements also identify growth-limiting factors in the operation and determine whether or not a particular investment decision should be accepted or rejected.

Profits should be determined by evaluating all cash costs of the operation, all inventory increases or decreases, and the value of opportunity costs. Opportunity costs represent the returns that would be forfeited if debt-free resources (such as owned land, livestock, and equipment) were used in their next best level of employment; for example, the value of pastureland if it were leased, returns if all capital represented by equipment and livestock were invested in a certificate of deposit (or similar investment), and returns if all family labor and management were utilized in other employment. By calculating this, a price can be established for living on a farm or ranch, and a goal established for increasing profits.

Credit and money management become crucial during periods of inflation, high interest rates, and relatively low livestock prices. Prudent use of credit can enable a livestock operation to grow more rapidly than it could through the use of reinvested earnings and savings, so long as borrowed funds return more over time than they cost. Thus, farmers and ranchers have to look to credit as a financial tool and learn to use it effectively.

The net cost of credit may be less than the interest rate depending on the taxable income bracket (Table 36.1).

TABLE 36.1. Net Cost of Borrowed Dollars

Income Tax Bracket (%)	Annual Rate of Interest Loan (%)				
	6	8	10	12	14
	Net rate of interest after tax saving (%)				
19	4.86	6.48	8.10	9.72	11.34
25	4.50	6.00	7.50	9.00	10.50
32	4.08	5.44	6.80	8.16	9.52
39	3.66	4.88	6.10	7.32	8.54
48	3.12	4.16	5.20	6.24	7.28

SCHEDULE F
(Form 1040)

Department of the Treasury
Internal Revenue Service

Farm Income and Expenses

▶ Attach to Form 1040, Form 1041, or Form 1065.
▶ See Instructions for Schedule F (Form 1040).

OMB No. 1545-0074

1981
16

Name of proprietor(s)	Social security number
JAMES A. BROWN	579 28 6685

Farm name and address ▶ JAMES A. BROWN, RR #1, Box 25
HOMETOWN YOUR STATE 02115

Employer identification number
7 1 9 3 6 7 9 7 4

Part I — Farm Income—Cash Method

Do not include sales of livestock held for draft, breeding, sport, or dairy purposes; report these sales on Form 4797.

Sales of Livestock and Other Items You Bought for Resale

a. Description	b. Amount	c. Cost or other basis
1 Livestock ▶ HEIFERS BOUGHT FOR RESALE	6,900	4,000
2 Other items ▶		
3 Totals	6,900	4,000

4 Profit or (loss), subtract line 3, column c, from line 3, column b . . . . . ▶ **2,900**

Sales of Livestock and Produce You Raised and Other Farm Income

Kind	Amount
5 Cattle and calves	3,316
6 Sheep	
7 Swine	
8 Poultry	
9 Dairy products	91,454
10 Eggs	
11 Wool	
12 Cotton	
13 Tobacco	
14 Vegetables	
15 Soybeans	
16 Corn	
17 Other grains	
18 Hay and straw	2,266
19 Fruits and nuts	
20 Machine work	1,258
21 a Patronage dividends . [272]	
b Less: Nonincome items [22]	
c Net patronage dividends	250
22 Per-unit retains	
23 Nonpatronage distributions from exempt cooperatives . .	
24 Agricultural program payments: **a** Cash . .	80
b Materials and services	324
25 Commodity credit loans under election (or forfeited) .	550
26 Federal gasoline tax credit	100
27 State gasoline tax refund	
28 Crop insurance proceeds	624
29 Other (specify) ▶ FAIR PRIZES	700
30 Add amounts in column for lines 5 through 29 .	100,922
31 Gross profits* (add lines 4 and 30) . . ▶	103,822

Part II — Farm Deductions—Cash and Accrual Method

Do not include personal or living expenses (such as taxes, insurance, repairs, etc., on your home), which do not produce farm income. Reduce the amount of your farm deductions by any reimbursement before entering the deduction below.

Items	Amount
32 a Labor hired	4,438
b Jobs credit	
c WIN credit	
d Total credits	
e Balance (subtract line 32d from line 32a) . .	4,438
33 Repairs, maintenance . .	3,236
34 Interest	4,124
35 Rent of farm, pasture . .	1,200
36 Feed purchased	24,182
37 Seeds, plants purchased .	2,286
38 Fertilizers, lime, chemicals .	4,498
39 Machine hire	1,200
40 Supplies purchased . . .	3,204
41 Breeding fees	1,078
42 Veterinary fees, medicine .	2,020
43 Gasoline, fuel, oil . . .	3,070
44 Storage, warehousing . .	
45 Taxes	3,050
46 Insurance	2,106
47 Utilities	2,050
48 Freight, trucking . . .	2,056
49 Conservation expenses . .	4,580
50 Land clearing expenses . .	
51 Pension and profit-sharing plans	
52 Employee benefit programs other than line 51 . . .	
53 Other (specify) ▶	
ADVERTISING	468
FINANCIAL RECORDS	166
FARM ORG. DUES	350
DEATH LOSS - HEIFER BOUGHT FOR RESALE	450
54 Add lines 32e through 53 .	66,812
55 Depreciation (from Form 4562)	8,483
56 Total deductions (add lines 54 and 55) . . . . ▶	75,295

57 Net farm profit or (loss) (subtract line 56 from line 31). If a profit, enter on Form 1040, line 18, and on Schedule SE, Part I, line 1a. If a loss, go on to line 58. (Fiduciaries and partnerships, see the Instructions.) . . . . | **57** | **28,527**

58 If you have a loss, do you have amounts for which you are not "at risk" in this farm (see instructions)? . . . ☐ Yes ☐ No
If you checked "No," enter the loss on Form 1040, line 18, and on Schedule SE, Part I, line 1a.

*Use amount on line 31 for optional method of computing net earnings from self-employment. (See Schedule SE, Part I, line 3.)

For Paperwork Reduction Act Notice, see Form 1040 Instructions.

FIGURE 36.3.
Internal Revenue Service form showing income and expense items for which documented records must be kept.

INCOME TAX CONSIDERATIONS

In addition to keeping accurate records for income tax purposes, there are some other management considerations regarding taxes that are important. A few of these are briefly mentioned and these should be more critically evaluated, preferably with a tax accountant.

1. A spouse can be paid a salary. This may save social security taxes, enhance retirement and estate planning, and provide eligibility for child-care tax credit. It may have the disadvantage of reducing social security benefits.
2. An investment credit can be taken. A 10% deduction can be used on the amount paid for new or remodeled livestock confinement buildings, milking parlors, other single purpose livestock structures, fences, paved roads, water wells, and drainage tiles.
3. Depreciation can be taken on breeding livestock, barns, sheds, grain storage bins, corrals, fences, irrigation wells and pumps, silos, farm lighting systems, and several others. For example, a dairy cow purchased for $1000 could be depreciated $200 per year over the expected productive lifetime of 5 years. Her slaughter value would have to be claimed as income when sold.
4. Children working on a family farm can be paid up to $3,300 per year without having to pay any federal income tax.
5. Mileage costs or actual expenses of operating and maintaining the farm share of the automobile or pickup can be deducted.

ESTATE AND GIFT TAX PLANNING

Many livestock operations have a large amount of debt-free capital invested in land, livestock, buildings, and equipment. Adequate knowledge and proper planning is necessary for farmers and ranchers so they may pass on viable economic units to their heirs.

In 1981, the Economic Recovery Tax Act made major revisions to the previous tax code. When the new estate tax amendments are fully phased in and implemented, a husband and wife who own farm or ranch property other than in joint tenancy will be able to transfer $2.7 million of this property (fair market value) to their children free of federal estate taxes. Also under the new law, each parent will be permitted to transfer $10,000 to each child and his or her spouse (up to $40,000 for each child and spouse, to a maximum of $160,000 per parent couple of property annually) without incurring any federal gifts tax.

Livestock producers should review tax plans—estate, gift, and income—in light of the amendments contained in the 1981 tax bill, and consult with the professionals who work with them.

COMPUTERS

Livestock operators need to keep a voluminous amount of information for financial and production records, inventories, and to more critically assess management alternatives. Excellent managers, with conventional records systems, often have an inate ability to "use their heads" to make good management decisions with a high degree of accuracy. Evidently they have a unique mind (the most complex computer) that functions exceptionally well in assimilating, storing, and recalling useful information.

Small computers (microcomputers) are becoming more frequently used as a management tool by farmers and ranchers. Producers must recognize what computers can and cannot do, at the present time and in the future. The technology of computers is changing rapidly; therefore, producers must be ready to accept the frustration of change if they become involved with computers.

Every computer, regardless of size and shape, is nothing more than a fast adding machine and an electronic filing cabinet. A microcomputer system consists primarily of a system of hardware and software. The hardware for microcomputers (Fig. 36.4) consists of a terminal (a typewriter keyboard and an optional television screen), a printer, data storage devices (which are the filing cabinets), main memory, and a central processing unit (CPU), which is the adding machine. The data storage devices can be either magnetic type (similar to cassette tapes), floppy discs (look like 45 rpm records), or hard discs (large storage capacity in one unit). The devices that read and write the information on these discs are called "disc drives" or just "drives."

Software are the programs that make the hardware function. Each program typically consists of thousands of minutely detailed instructions that tell the computer exactly what data to manipulate mathematically, print, store, or display.

Good software (programs) are the most limiting factor in microcomputer utilization by livestock producers. Software developed for one microcomputer may not function on a different brand machine. Computer utilization is expected to increase rapidly as more compatible programs are developed. The four ways producers can obtain software are: (a) purchase a complete commercial package, (b) hire someone to do the programming, (c) learn to program the computer themselves, or (d) obtain it from their land grant university.

FIGURE 36.4.
Microcomputers are becoming more commonly used by livestock producers. Courtesy of *BEEF*.

The three major areas where microcomputers have the largest potential for livestock producers are: (a) business accounting, (b) herd performance, and (c) financial management. A computer to handle these three functions would need a relatively large memory system and may cost several thousand dollars. Some producers have access to computer terminals with a telephone hook-up to a large main-frame computer. An example of such a system is AGNET (Agricultural Computer Network), which functions through the joint efforts of the Agricultural Extension Services of Wyoming, Nebraska, South Dakota, North Dakota, and Washington. An example of one of approximately 250 AGNET programs is COWCOST, which evaluates the cost and returns for beef cow-calf enterprises.

Computers do not simplify a livestock business or cut overhead costs. They can make a business more complex and more costly. If wisely acquired and properly utilized, a computer can increase profits, but usually not by reducing expenses. The computer will not solve the problem of a poorly organized operation. Well-planned and efficient office management practices must exist and function smoothly before the addition of a computer will do anything but aggravate the organizational problem of the business. The computer cannot assist the producer in managing the business effectively unless the computer receives all the required data in precisely the prescribed manner. Unless the computer can provide the information needed by producers at the appropriate time and in an easily understood form, it should not be part of a management program, regardless of price.

A producer should follow six steps before deciding to purchase a computer. They are: (a) determine what important management decisions need to be made and what information is needed to effectively make these decisions, (b) see what programs are available to meet the management information needs, (c) decide what are the hardware needs to utilize the needed software, (d) contact local computer hardware dealers and evaluate the various alternatives and service, (e) estimate the cost/benefit ratio of the proposed management information system, and (f) make the final decision after evaluating the experiences of other producers who are using computers in their operations.

MANAGEMENT SYSTEMS

Management systems analysis provides a method of systematically organizing information needed to make valid management decisions. It permits variables to be more critically assessed and analyzed as to their contribution to the desired end point, which is usually net profit. Individuals who have been educated to think broadly in the framework of management systems can often make valid management decisions without the use of data processing equipment. A pencil, hand calculator, and a well-trained mind are the primary components required for making competent management decisions. Without question, however, the use of the computer in synthesizing voluminous amounts of information enhances the management system.

Resources are different for each operation; no fixed recipe exists for successful livestock production. The uniqueness of each operation results from variables such as different levels of forage production, varying marketing alternatives, varying energy costs, different types of animals, environmental differences, feed nutrients at various costs, varying levels of competence in labor and management, and others to numerous to mention. All of these variables and their interactions pose challenges to the producer who needs to combine them into sound management decisions for a specific operation.

It is a common practice to increase or maximize production of animals by using known biological relationships. Animal production typically has been maximized without careful consideration of cost-benefit ratios and how increased productivity relates to land, feed, and management resources. However, recognition is now being given to the need for optimization rather than maximization of animal productivity. Thus, there is an increased interest in management systems, which attempt to optimize production with net profit being the primary (and possibly the only) goal involved. As a result, some valid biological relationships will not be useful or applicable because an economic analysis will prevent their inclusion in sound management decisions. For example, it is a well-known biological fact that calves born earlier in the calving season will have heavier weaning weights. Because of this relationship, some ranchers move the calving season to a period earlier in the year without a careful economic assessment. For example, an economic evaluation for a specific ranch demonstrated that by changing the calving season to match forage availability (in this case about 30 days to grazable forage), profits were increased significantly. Changing calving season alone was estimated to increase the ranch's carrying capacity by 30–40%, in terms of animal units. In addition, the annual cow cost was reduced 18% per cow because of reduction in winter feed requirements.

In dairy production, there is often a difference between maximum milk production and optimum milk production if producers want to maximize profit. The extra feed and management needed to maximize milk production may cost more than the value of the added milk.

During the past several years, research workers have increasingly applied management systems to livestock management. After a system is conceptualized, it is usually described by a set of mathematical equations and called a model. Models are constructed to simulate real-life situations, and attempts are made to validate them on that basis. Obviously, the models are no better than the data used to construct the models, and output data depend on the validity of the input data.

The farmer and rancher are limited in utilizing the computer in management systems because of the lack of software. Few programs being used combine production data with an economic assessment. However, there is increased discussion about and movement toward making more management system software available to producers. Several projections indicate that programs will be used commonly in the next few years. Producers have to make management decisions now. To be competitive, livestock operations will have to be operated more like a business. This means keeping useful and accurate financial and production records to be used in making management decisions. The computer will become more important as a tool to improve the effectivness of the management-decision process. However, until more usable computer programs become available, producers must use the traditional tools (pencil, hand calculator, and a keen mind) more effectively than in the past.

SELECTED REFERENCES

Publications

Bourdon, R. 1985. The systems concept of beef production. Beef Improvement Federation Fact Sheet.

Fitzhugh, H. A. 1978. Bioeconomic analysis of ruminant production systems. *J. Anim. Sci.* 46:797.

Gosey, J. Matching genetic potential to feed resources. Cornbelt Cow-calf Conference, Feb. 25, 1984.

Harl, N. 1980. *Farm Estate and Business Planning*. 6th Edition. Skokie, IL: Century Communications, Inc.

Hughes, H. 1981. The Computer Explosion. Proceedings, The Range Beef Cow, A Symposium on Production, VII. Rapid City, SD, December.

Hughes, H. 1981. Six steps to take in making a decision to buy a computer. *BEEF*, December.

Killcreas, W. E., and Hickel. R. 1982. A Set of Microcomputer Programs for Swine Record Keeping and Production Management. Miss. Agric. and Forestry Expt. Stat. AEC Tech. Public. No. 35.

Klinefelter, D. A., and Hottel, B. 1981. Farmers' and Ranchers' Guide to Borrowing Money. Texas Agric. Expt. Stat. MP-1494.

Luft, L. D. Sources of credit and the cost of credit. Great Plains Beef Cattle Handbook, GPE-4351 and 4352.

Maddux, J. 1981. The Man in Management. Proceedings, The Range Beef Cow, A Symposium on Production, VII. Rapid City, SD, December.

Miller, W. C., Brinks, J. S. and Greathouse, G. A. 1985. A systems analysis model for cattle ranch management. *J. Anim. Sci.* 61(Suppl. 1):177.

Notter, D. R., Sanders, J. O., Dickerson, Smith, G. R., and Cartwright, T. C. 1979. Simulated efficiency of beef production for a midwestern cow-calf-feedlot management system. III. Cross-breeding systems. *J. Anim. Sci.* 49:92.

Ott, G. Planning for profit with partial budgeting. Great Plains Beef Cattle Handbook, GPE-4551.

Simms, D. D. and Rush, I. 1986. Beef cattle software. Kansas State University and University of Nebraska.

Sobba, A. Focusing on the new tax code: changes that will affect cattlemen. NCA Beef Business Bulletin, July 22, 1986.

Trede, L. D., Boehlje, M. D. and Geasler, M. R. 1977. Systems approach to management for the cattle feeder. *J. Anim. Sci.* 45:1213.

Trede, L. D. Calculators, computers and beef cows. Cornbelt Cow-calf Conference.

Wilt, R. W. Jr., and Bell, S. D. 1977. Use of Financial Cash Flow Statements as a Financial Management Tool. Ala. Agric. Expt. Sta. Bull.

Visuals

Farm and Ranch Management and *Agricultural Credit*. Separate sound filmstrips. Vocational Education Productions, California Polytechnic State University, San Luis Obispo, CA 93407.

Computer Software

Swine Management Series One and *Swine Record Keeping*. Vocational Education Productions, California Polytechnic State University, San Luis Obispo, CA 93407.

CHAPTER 37

Careers and Career Preparation in the Animal Sciences

Millions of domestic animals that provide food, fiber, and recreation for mankind create many and varied types of career opportunities. A placement survey of approximately 25,000 agriculture graduates of state colleges and universities is shown in Tables 37.1 and 37.2. Animal sciences placement data follow a similar pattern because they compose a high percent of the graduates shown in these tables.

Beef cattle, dairy cattle, horses, poultry, sheep, and swine are the animals of primary importance in animal science curricula, whereas reproduction, nutrition, breeding (genetics), meats, and live animal appraisal are the specialized topics typically covered in animal science courses. A few animal science programs include studies of pet and companion animals. Most college and university curricula in the animal sciences are designed to assist students in the broad career areas of production, science, agribusiness, and the food industry. The major careers in each of these areas are shown in Table 37.3.

TABLE 37.1. Postgraduation Activities of Bachelor Degree Recipients in Agriculture

Postgraduation Activity	Percent of Graduates
Agribusiness (industry)	35
Graduate and professional study	20
Farming and ranching	20
Education, including extension	5
Government (national, state, and local)	10
Miscellaneous (not placed or not seeking employment)	10

Source: National Association of State Universities and Land Grant Colleges, 1985.

TABLE 37.2.   Average Starting Salaries of Agriculture Graduates from State Universities

Graduate	Average Annual Starting Salary		
	1975	1980	1985
Bachelor of Science (B.S.)	$ 9,792	$13,848	$17,585
Master of Science (M.S.)	11,748	17,378	22,020
Doctor of Philosophy (Ph.D.)	15,696	22,472	29,448

Source: National Association of State Universities and Land Grant Colleges

TABLE 37.3.   Animal Sciences Careers in Production, Science, Agribusiness, and Food Industry

Animal Sciences			
Production	Science	Agribusiness	Food Industry
Feedlot positions	Graduate school for Master of Science (M.S.) and Doctor of Philosophy (Ph.D) degrees	Sales and management positions with feed companies, packing companies, drug and pharmacy companies, equipment companies, etc.	Food processing plants
Livestock production operations (beef, dairy, swine, sheep and horses)			Food ingredient plants
			Food manufacturing plants
Ranch positions	Research (university or industry) in nutrition, reproduction, breeding and genetics, products, and production management		Government—protection and regulatory agencies
Breed associations		Livestock publications	
AI studs and breeding		Advertising and promotion	Government—Department of Defense (food supply and food service)
Livestock buyers for feeders and packers			
County extension agents	University or college teaching	Finance (PCAs, banks, etc.)	Government—Department of Agriculture (Research and Information)
Meat grading and handling distribution	University extension and area extension	Public relations	
		Meat grading (federal government)	
Marketing (auctions, Cattle Fax, livestock sales management, etc.)	Management positions in industry	International opportunities	University research, teaching, and extension
	Government work	Graduate school for Masters in Business Administration (M.B.A.)	Positions in food companies
International opportunities	International opportunities		
Livestock and meat market reporting (government)		Foreign agriculture	Research and development with food companies
	Veterinary school for Doctor of Veterinary Medicine (D.V.M.)	Technical sales and service	
Riding instructors	Private practice	Positions in poultry production units	
Feed manufacturing	Consulting		
Companion animals	University teaching and research	Companion animals	
Boarding			
Breeding	Meat inspection		
Training	Vet assistant		
Humane Society	Companion animal research		

PRODUCTION

Many individuals are intrigued with animal production because of the possibility of being their own boss and working directly with animals. (Fig. 37.1). These ambitions may be somewhat idealistic, because the financial investment required in land is typically hundreds of thousands of dollars. Beef cattle, dairy, poultry, sheep, and swine production operations continue to become fewer in number, more specialized, and larger in terms of the amount of capital invested. Those individuals who find a career in the production area usually have a family operation to which they can return. A few others have the needed capital to invest.

Some students who seek employment in animal production are disillusioned with the base salary, benefits, and working hours when production work is compared with agribusiness careers. However, some graduates are anxious and willing to obtain experience, prove they can work, and eventually locate permanent employment in production operations. Some producers of cattle, sheep, swine, and horse operations allow an employee to buy into an operation on a limited basis or own some of the animals after working a year or two.

Certain careers require education and experience in the production area even though the individual will not be producing animals directly. Career opportunities such as breed associations and publications, fieldmen, extension agents, and livestock marketing require an understanding of production as one works directly with livestock producers.

SCIENCE

An animal science student who concentrates heavily in science courses is usually preparing for further academic work with a goal of achieving one or more advanced degrees. A minimum grade average of B is usually required for admission into graduate school or a professional veterinary medicine program. Advanced degrees are usually obtained after entrance into graduate school or after being admitted to a professional veterinary medicine program where a Doctor of Veterinary Medicine (D.V.M.) degree is awarded. Table 37.4 shows where graduates of Veterinary medical colleges are employed and their average starting salary. The two most employment areas in 1985 were small animal practice and mixed animal practice.

The income of D.V.M.s after several years of private practice is given in Table 37.5. Actual and real incomes for 1977 and 1983 are compared. Even though the actual income is higher in 1983 compared with 1977, the purchasing power (real income) was higher for 1977 than 1983.

The Master of Sciences (M.S.) and Doctor of Philosphy (Ph.D.) are the advanced degrees received by the science-oriented student. Although the 2-year college certificate or Bachelor of Science (B.S.) degree allows breadth of education, advanced degrees are generally directed to a specialization. These specialized animal science areas are typically in nutrition, reproduction, breeding (genetics and statistics), animal products, and, less frequently, in production and management systems (Figs. 37.2–37.5).

AGRIBUSINESS

Although the number of individuals employed in livestock and poultry production has decreased, the number of individuals and businesses serving producers has greatly

FIGURE 37.1.
There are numerous career opportunities in the production of the various species of farm animals. Some students interested in the production area will not be directly involved in animal production. These students will work for companies, associations, or groups closely associated with the animal producers. Courtesy of Institute of Agriculture and Natural Resources, University of Nebraska; lower left photo courtesy of University of Illinois.

TABLE 37.4. Types of Employment and Starting Incomes of a Large Sample of 1985 U.S. Veterinary Medical College Graduates

Employment Area	Percentage of Graduates	Average Starting Salary (No.)	Average Additional Compensation Expected in 1st Year (No.)
Large animal exclusive	1.0	$20,857(7)	$2,416(6)
Large animal predominant	8.7	20,640(68)	2,205(38)
Mixed animal	14.8	20,036(106)	2,108(58)
Small animal predominant	12.0	20,981(91)	2,434(55)
Small animal exclusive	35.6	21,427(262)	2,596(149)
Equine predominant	4.2	20,710(30)	4,768(11)
University	1.4	14,171(7)	—
Federal government	0.2	22,000(2)	—
Military service	1.5	26,560(10)	3,333(3)
State/local government	0.2	20,000(1)	—
Industry/commercial	0.2	29,500(2)	—
Not-for-profit organization	0.1	—	—
Advanced study	12.8	14,199(87)	2,820(15)
Self-employed	6.4	21,016(30)	5,000(10)
Other	0.8	20,684(7)	3,250(3)
	100	$20,164(710)	$2,602(348)

Source: JAVMA 188:193

increased. Positions in sales, management, finance, advertising, public relations, and publications are prevalent in feed, drug, equipment, packing, and livestock organizations.

Agribusiness careers that relate to the livestock industry require a student to have a foundation of knowledge of livestock along with an excellent comprehension of business, economics, computer science, and effective communication. A graduate should understand people, know how to communicate with people, and enjoy working with people.

TABLE 37.5. Money and Real Professional Incomes of Private Practice Owners in the U.S., 1977–1983

Type of Practice	Money Income		Real Income (1983 dollars)	
	1977	1983	1977	1983
Large animal exclusive	$30,048	$46,553	$49,421	$46,553
Large animal predominant	30,663	47,148	50,432	47,148
Mixed animal	27,547	38,576	45,308	38,376
Small animal predominant	29,908	38,255	49,191	38,235
Small animal exclusive	32,635	44,905	53,676	44,905
Equine	40,249	56,086	66,199	56,086
All practices	$31,613	$44,249	$51,995	$44,249
Number D.V.M.s surveyed	(2,309)	(1,078)		

Source: Adapted from JAVMA 188:1195

FIGURE 37.2.
Many students concentrating in the sciences are preparing themselves for entrance into graduate school. There they will complete M.S. and Ph.D. degrees, which are primarily research-oriented degrees. A few examples are depicted here. Courtesy of Institute of Agriculture and Natural Resources, University of Nebraska photo; lower left photo courtesy of West Virginia University.

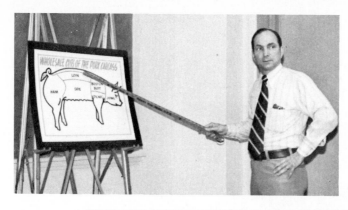

FIGURE 37.3.
A Ph.D. is usually required for teaching in a college or university. Courtesy of Oklahoma State University.

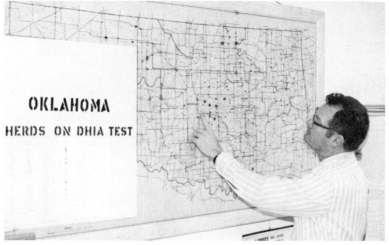

FIGURE 37.4.
University extension personnel need a Ph.D. degree so they can interpret the current research and apply it in off-campus educational programs. Courtesy of Oklahoma State University.

FIGURE 37.5.
Students will usually have a B.S. degree before applying for admission to the College of Veterinary Medicine. Students with high academic performance in the science aspect of animal science are typically well prepared for completing the D.V.M. degree. Courtesy of Colorado State University.

Extracurricular activities that give students experience in leadership and working with people are invaluable for meaningful career preparation.

Some animal science students who concentrate heavily in business and economics courses will pursue an advanced degree. This is typically a Masters in Business Administration (M.B.A.). A combination of a B.S. degree in Animal Science with an M.B.A. is an excellent preparation for advancement in many livestock oriented businesses.

FOOD INDUSTRY

Red meats, poultry, milk, and eggs, the primary end products of animal production provide basis nutrition and eating enjoyment for millions of consumers. Career opportunities are numerous in providing one of mankind's basic needs—food. Processing, packaging, and distribution of food are important components of the food production chain. New food products are continually being produced, and new methods of manufacturing and fabricating food are being developed. Electrical stimulation of carcasses for increased tenderization, vacuum packaging of primal cuts for a longer shelf life, and preparation of convenience products are examples of recent innovations in meat processing. A vital and continuing challenge over the next several decades will be to provide a food supply that is nutritious, safe, convenient, attractive, and economical while providing the desired eating satisfaction to the consumer (Fig.37.6).

INTERNATIONAL OPPORTUNITIES

Much has been written and said during the past decade of the challenge to provide adequate nutrition to an ever-expanding world population. Many countries have tremendous natural resources for expanded food production but lack technical knowledge and adequate capital to develop these resources. Federal government programs, designed to help foreign countries help themselves, offer several career opportunities in organizations such as the Peace Corps, Vista, and USAID. Many of these opportunities of assisting people in agricultural production in developing countries are open to individuals educated and experienced in animal sciences.

The Foreign Agricultural Service of the USDA employs individuals in animal economics, marketing, administration, as attachés and as international secretaries. Animal science students take several courses in economics, marketing, foreign languages, and business administration if they wish to qualify for these positions.

Certain private industries offer opportunities in animal production and related businesses in foreign countries. Multinational firms that develop livestock feed companies, drug and pharmaceutical companies, companies that export and import animals and animal products, and consulting companies are some examples.

MANAGEMENT POSITIONS

Management and administrative positions exist in production, science, agribusiness, and food industry areas of animal science. Because management generally implies less work and more pay than other positions that require more physical labor, young men and women typically desire management positions. Nevertheless, although management and

FIGURE 37.6.
The demand for animal products creates many industries that support the animals and process the products for human consumption. Courtesy of the University of Nebraska; Colorado State University; and Oklahoma State University.

administrative salaries are usually higher, the work load and pressures are usually much greater.

Many college graduates have the impression that once the degree is in hand, it automatically qualifies them for management positions. In actuality, management positions are usually earned based on an individual's proven ability to work with people, solve problems, and make effective decisions. After being employed for a few years, individuals having previously learned the academic principles must also experience the component parts of an operation or business. For example, presidents or vice-presidents of feed companies have typically started their initial careers in feed sales. Most successful managers have had a continual series of learning experiences since the end of their formal education. Managers need to understand all aspects of the business they are managing, particularly the products that are produced and sold.

WOMEN IN ANIMAL SCIENCE CAREERS

Women are pursuing and finding job opportunities in all areas of the animal sciences. The number of women majoring in animal science has been growing steadily in most colleges and universities. In the early 1970s, the number of females majoring in animal science was 20–25% of the total majors. Currently nearly one-half of the animal science majors are women (Fig. 37.7).

This trend is not so much a part of the Women's Liberation Movement as it is an equal opportunity crusade. City-raised women, as well as those raised on a farm, are looking for career opportunities in which they can work with animals or some business associated with livestock and poultry. Many agricultural employers, especially in the agribusinesses, are now actively recruiting women with animal science degrees.

During the 1970s, women found career opportunities that many felt were reserved for men only. Although some men expressed fear and had serious reservations, many of these women proved they could perform competently. It does, however, appear that some employers in private production agriculture will still be reluctant in hiring women on the same basis as men. Nevertheless, women interested in animal science careers should prepare themselves adequately in both education and experience. Challenging career opportunities await both women and men who are prepared and are willing to work hard.

Recent observations in a California study showed that women graduates in agriculture were at a disadvantage in salary and job status when compared with men graduating with similar degrees. However, a Colorado State University survey of its women graduates in animal sciences since 1979, revealed that many of them were in high-paying careers. This survey also demonstrated that women desiring employment in the production area had struggled in identifying meaningful employment. The agribusiness area (e.g., Farmer's Home Administration, feed companies, food processing companies, and drug and pharmaceutical companies) had provided attractive careers for women graduates, as several reported annual salaries above $20,000.

FUTURE CAREER OPPORTUNITIES

A 1985 USDA national assessment report projected the employment opportunities for food and agricultural sciences graduates through 1990. The estimated number of annual

A

B

C

D

E

FIGURE 37.7.
Women are seeking and finding career opportunities in the animal sciences. Some examples shown are: (A) Teaching equitation; (B) veterinarian's assistant; (C) feedlot manager; (D) researcher; (E) photographer. Courtesy of Colorado State University; University of Illinois; Office of Agricultural Communications; *BEEF;* University of Wyoming; *The American Hereford Journal;* and Oklahoma State University.

NO. POSITIONS (ANNUALLY)

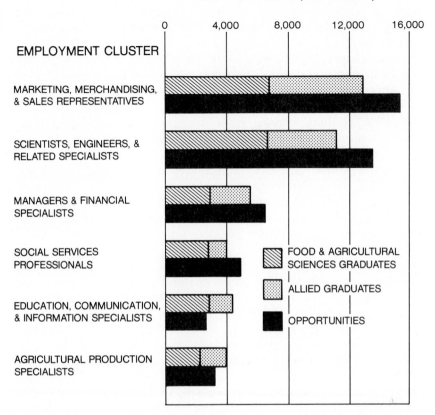

FIGURE 37.8.
Employment opportunities and available graduates through 1990. Courtesy of the USDA and CSU Graphics.

job openings are 48,000 with approximately 44,000 qualified college graduates available each year. The supply of graduates is expected to be approximately 10% less than the available job openings.

Figure 37.8 shows the employment opportunities and the available graduates for six different career areas. The USDA report has the following summary:

> In summary, basic plant and animal research, food and fiber processing, and agribusiness management and marketing are expected to provide the most significant employment opportunities for college graduates with expertise in agriculture, natural resources, and veterinary medicine through 1990. In contrast, college graduates seeking positions in production agriculture, education, and communications will encounter strong competition for somewhat limited employment opportunities within the U.S. food and agricultural system during the next five years.

CAREER PREPARATION

An education should provide opportunities for individuals to earn a living, continue learning, and live a full, productive life even beyond the typical retirement years. A broadbased education is important because an individual may be preparing for a career that does not exist at the present time. Change has occurred and will no doubt continue to occur, so a person needs to be flexible and adaptable and must be willing to continue learning throughout life.

Occasionally an individual chooses a career for which he or she has little real talent or for which she or he is not qualified or properly motivated. These individuals may spend years in frustrating preparation or achieving partial success when they could have been outstanding in another area. In animal sciences, students preparing for acceptance into veterinary medicine, graduate school, or another professional school seem to experience this frustration the most. Individuals are capable of success in several different careers if they receive the appropriate education and experience. A career choice should be consistent with a person's interest, desire, and motivation. A career goal should be established with flexibility for change if it proves to be unrealistic.

After tentatively choosing a career area, it should be pursued vigorously with sustained personal motivation. Students should write and visit with individuals currently working in the student's chosen career area. Every personal visit or written letter should request another lead or source of further information. Students should continually evaluate the reality of a chosen career. They should think realistically about the facts rather than getting caught up in the glamor, glory, or imaginations they have perceived about a career choice. Career interest tests can be helpful in identifying areas to pursue in more detail. However, these tests are not infallible. They are only helpful tools to give direction but they are not the final answer. Students should always temper the results with additional information from several sources before making final decisions.

A nationwide survey conducted by an American Society of Animal Science committee identified the primary concerns that employers had experienced with animal science graduates as the graduates started their first job. In descending order of importance, these concerns were: lack of practical knowledge and experience, lack of communicative skills, lack of drive and initiative, overly high expectations, failure to assume responsibility, and lack of business knowledge. Students preparing for an animal science career should obtain knowledge and experiences that will allow them to overcome these common deficiencies.

Employers hire people, not a college major or area of study. The classes a student takes in college are only part of the necessary preparation. Therefore, in addition to academic preparation, students should develop their marketable personal assets such as leadership, problem solving, ability to communicate and work with people, and a desire to consistently work hard. A person develops many of these personal characteristics through experience, part-time and summer jobs, and extracurricular activities. Internships are especially valuable in obtaining exposure to a career area and developing personal abilities and skills through practical experience. A combination of useful courses, practical experience, effective communication, and excellent work skills is the best assurance of finding and keeping a meaningful career in the animal sciences.

SELECTED REFERENCES

Careers in the Meat Industry. American Meat Science Association, 444 N. Michigan Ave., Chicago, IL 60611.

Coulter, K. J. Stanton, M. and Goecker. 1986. Employment Opportunities for College Graduates in the Food and Agricultural Sciences. USDA, Higher Education Program.

Employer Reactions to Animal Science Graduates and Their Qualifications. Report II. National Survey by the ASAS Committee on Future Career Opportunities for Animal Science Graduates. July 29, 1979.

Survey of Agricultural Degrees Granted and Post Graduation Activities of Graduates. National Association of State Universities and Land Grant Colleges. July 1980.

Wise, J. K. 1986. Employment, starting salaries and educational indebtedness of 1985 graduates of U.S. Veterinary Medical Colleges, *JAVMA* 188:193.

Wise, J. K. 1986. Real starting salaries of US Veterinary Medical college graduates, 1978–1985. *JAVMA* 188:427.

Wise, J. K. 1986. Real professional income of US Veterinarians. *JAVMA* 188:1195.

Wood, J. B., Dupre, D. H., and Thompson, O. E. *Women in the Agricultural Labor Market*. Sept.–Oct. 1981, p. 16.

GLOSSARY

abomasum The fourth stomach compartment of ruminant animals that corresponds to the true stomach of monogastric animals.

abortion Delivery of fetus between conception and a few days before normal parturition.

abscess Localized collection of pus in a cavity formed by disintegration of tissues.

absorption The passage of liquid and digested (soluble) food across the gut wall.

accessory organs The seminal vesicles, prostate, and Cowper's glands in the male. These glands add their secretions to the sperm to form semen.

accuracy (ACC) Correlation between an animal's actual breeding value (unknown genetic make-up) and an estimated breeding value.

adipose Fat cells or fat tissue.

ad libitum Free choice; allowing animals to eat all they want.

afterbirth The membranes, attached to the fetus, which are expelled after parturition.

AI Abbreviation for artificial insemination.

air dry Refers to feeds in equilibrium with air; they would contain approximately 10% water (90% dry matter).

albumen The white of an egg.

alimentary tract Passageway for food and waste products through the body.

alleles Genes occupying corresponding loci in homologous chromosomes that affect the same hereditary trait but in different ways.

allelomimetic behavior Doing the same thing. Animals tend to follow the actions of other animals.

alveolus *(plural, alveoli)* A hollow cluster of cells. In the mammary gland, these cells secrete milk.

animo acid An organic acid in which one or more of the hydrogen atoms has been replaced by the amino group ($-NH_2$). Amino acids are the building blocks in the formation of proteins.

amnion A fluid-filled membrane located next to the fetus.

ampulla The dilated or enlarged upper portion of the vas deferens in bulls, bucks, and rams, where sperm are stored for sudden release at ejaculation.

anabolic A constructive, or "building up", process.

anatomy Science of animal body structure and the relation of the body parts.

androgen A male sex hormone, such as testosterone.

anemia Deficiency of hemoglobin, usually accompanied by reduced number of red blood cells. Usually results from an iron deficiency.

anestrous Period of time when female is not in estrus; the nonbreeding season.

ante mortem Before death.

anterior Situated in front of, or toward the front part of, a point of reference. Toward the head end of an animal.

anthelmintic A drug or chemical agent used to kill or remove internal parasites.

antibiotic A product produced by living organisms, such as yeast, which destroys or inhibits the growth of other microorganisms, especially bacteria.

antibody A specific protein molecule that is produced in response to a foreign protein (antigen) that has been introduced into the body.

antigen A foreign substance which, when introduced into the blood or tissues, causes the formation of antibodies. Antigens may be toxins or native proteins.

antiseptic   A chemical agent used on living tissue to control the growth and development of microorganisms.

artery   Vessel through which blood passes from the heart to all parts of the body.

arteriosclerosis   A disease resulting in the thickening and hardening of the artery walls.

artificial insemination   The introduction of semen into the female reproductive tract (usually the cervix or uterus) by a technician.

artificial vagina   A device used to collect semen from a male while he mounts in a normal manner to copulate. The male ejaculates into this device, which simulates the vagina of the female in pressure, temperature, and sensation to the penis.

ascaris   Any of the genus (Ascaris) of parasitic roundworms.

as fed   Refers to feeding feeds that contain their normal amount of moisture.

assimilation   The process of transforming food into living tissue.

atherosclerosis   A form of arteriosclerosis involving fatty deposits in the inner walls of the arteries.

atrophy   Shrinking or wasting away of a tissue or organ.

autopsy   A postmortem examination in which the body is dissected to determine the cause of death.

avian   Refers to birds, including poultry.

balance sheet   A statement of assets owned and liabilities owed in dollar terms that shows the equity or net worth at a specific point in time, i.e., net worth statement.

Bang's disease   See *brucellosis*.

barren   Not capable of producing offspring.

barrow   A male swine that was castrated before reaching puberty.

basal metabolism   The chemical changes that occur in an animal's body when the animal is in a thermoneutral environment, resting, and in a postabsorptive state. It is usually determined by measuring oxygen consumption and carbon dioxide production.

beef   The meat from cattle (bovine species) other than calves (the meat from calves is called veal).

beri-beri   A disease caused by a deficiency of vitamin B_1.

biologicals   Pharmaceutical products used especially to diagnose and treat diseases.

blemish   Any defect or injury that mars the appearance of, but does not impair the usefulness of, an animal.

bloat   An abnormal condition in ruminants characterized by a distention of the rumen, usually seen on an animal's upper left side, owing to an accumulation of gases.

blood spots   Spots in the egg caused by a rupture of one or more blood vessels in the yolk follicle at the time of ovulation.

BLUP   Best Linear Unbiased Prediction method for estimating breeding values of breeding animals.

boar   A male swine of breeding age. This term sometimes denotes a male pig, which is called a boar pig.

bog spavin   A soft enlargement of the anterior, inner aspect of the hock.

bolus   (1) Regurgitated food. (2) A large pill for dosing animals.

bone spavin   A bony (hard) enlargement of the inner aspect of the hock.

bots   Any of a number of related flies whose larvae are parasitic in horses and sheep.

bovine   Refers to a general family grouping of cattle.

boxed beef   Cuts of beef put in boxes for shipping from packing plant to retailers. These primal and subprimal cuts are intermediate cuts between the carcass and retail cuts.

boxed lamb   See *boxed beef*. Similar process except lamb instead of beef.

break joint   Denotes the point on a lamb carcass where the foot and pastern are removed at the cartilaginous junction of the front leg.

bred   Pregnant.

breech   The buttocks. A breech presentation at birth is one in which the rear portion of the fetus is presented first.

breed   Animals of common origin and having characteristics that distinguish them from other groups within the same species.

breeding value   A genetic measure for one trait of an animal, calculated by combining into one number several performance values, that have been accumulated on the animal and the animal's relatives.

brisket disease   A noninfectious disease of cattle characterized by congestive right heart failure. It affects animals residing at high altitudes (usually above 7,000 ft.)

British Thermal Unit   The quantity of heat required to raise the temperature of 1 lb of water 1°F at or near 39.2°F.

brockle-faced   White-faced with other colors splotched on the face and head.

broiler   A young meat-type chicken of either sex (usually up to 8–10 weeks of age) weighing 3–6 lb. Also referred to as a fryer.

broken-mouth   Some teeth are missing or broken.

broodiness   The desire of a female bird to sit on eggs (incubate).

browse   Woody or brushy plants. Livestock feed on tender shoots or twigs.

brucellosis   A contagious bacterial disease that results in abortions; also called Bang's disease.

buck   A male sheep or goat. This term usually denotes animals of breeding age.

bulbourethral (Cowper's) gland   An accessory gland of the male that secretes a fluid which constitutes a portion of the semen.

bull   A bovine male. The term usually denotes animals of breeding age.

buller-steer syndrome   A behavior problem where a steer has a sexual attraction to other steers in the pen. The steer is ridden by the other steers, resulting in poor performance and injury.

bullock   A young bull, typically less than 20 months of age.

buttermilk   The fluid remaining after butter has been made from cream. By use of bacteria, cultured buttermilk is also produced from milk.

buttons   May refer to cartilage or dorsal processes of the thoracic vertebrae. Also see cotyledons.

by-product   A product of considerably less value than the major product. For example in U. S. meat animals, the hide, pelt, and offal are by-products whereas meat is the major product.

C-section   See *cesarian section*.

calf   A young male or female bovine animal under a year of age.

calorie   The amount of heat required to raise the temperature of 1 g of water from 15°C to 16°C.

calve   Giving birth to a calf. Same as parturition.

calving interval   The amount of time (days or months) between the birth of a calf and the birth of a subsequent calf, both from the same cow.

candling   The shining of a bright light through an egg to see if it contains a live embryo.

canter   A slow, easy gallop.

capon   Castrated male chicken. Castration usually occurs between 3 and 4 weeks of age.

capped hocks   Hocks that have hard growths that cover, or "cap", their points.

carbohydrates   Any foods, including starches, sugars, celluloses, and gums, that are broken down to simple sugars through digestion.

carcass merit   The value of a carcass for consumption.

carnivorous   Subsisting or feeding on animal tissues.

carotene   The orange pigment found in carrots, leafy plants, yellow corn, and other feeds, which can be broken down to form two molecules of vitamin A.

caruncle   (1) The red and blue fleshy, unfeathered area of skin on the upper region of the turkey's neck; (2) the "buttons" on the ruminant uterus where the cotyledons on the fetal membranes attach.

casein The major protein of milk.

cash flow statement A financial statement summarizing all cash receipts and disbursements over the period of time covered by the statement.

castrate (1) To remove the testicles. (2) An animal that has had its testicles removed.

cattalo A cross between domestic cattle and bison.

cecum (ceca) A blind pouch at the junction of the small and large intestine. Poultry have two ceca.

cervix The portion of the female reproductive tract between the vagina and the uterus. It is usually sealed by thick mucus except when the female is in estrus or delivering young.

cesarian section Delivery of fetus through an incision in abdominal and uterine walls.

chalaza A spiral band of thick albumen that helps hold the yolk of an egg in place.

chemotherapeutics Chemical agents used to prevent or treat diseases.

chevon Meat from goats.

chorion The outermost layer of fetal membranes.

chromosome A rodlike or stringlike body found in the nucleus of the cell that is darkly stained by chrome dyes. The chromosome contains the genes.

chyme The thick, liquid mixture of food that passes from the stomach to the small intestine.

chymotrypsin A milk-digesting enzyme secreted by the pancreas into the small intestine.

clitoris The ventral part of the vulva of the female reproductive tract that is homologous to the penis in the male. It is highly sensory.

cloaca Portion of the lower end of the avian digestive tract that provides a passageway for products of the urinary, digestive, and reproductive tracts.

closed face A condition in which sheep cannot see because wool covers their eyes.

clutch Eggs layed by a hen on consecutive days.

coccidia A protozoan organism that causes an intestinal disease called coccidiosis.

coccidiosis A morbid state caused by the presence of organisms called coccidia, which belong to a class of sporozoans.

cock A male chicken; also called a rooster.

cockerel Immature male chicken.

cod Scrotal area of steer remaining after castration.

coefficient of determination A percent of variation in one trait that is accounted for by variation in another trait.

colic A nonspecific pain of the digestive tract.

colon The large intestine from the end of the ileum and beginning with the cecum to the anus.

colostrum The first milk given by a female after delivery of her young. It is high in antibodies that give the young protection from invading microorganisms.

colt A young male of the horse or donkey species.

comb The fleshy outgrowth on the top of a chicken's head, usually red in color with varying sizes and shapes.

compensatory gain A faster than normal rate of gain after a period of restricted gain.

compensatory growth See *compensatory gain.*

composite breed A breed that has been formed by crossing two or more breeds.

concentrate A feed that is high in energy, low in fiber content, and highly digestible.

conception Fertilization of the ovum (egg).

conditioning The treatment of animals by vaccination and other means before putting them in the feedlot.

conformation The physical form of an animal; its shape and arrangement of parts.

contagious disease Infectious disease; a disease that is transmitted from one animal to another.

contemporaries A group of animals of the same sex and breed (or similar breeding) that have been raised under similar environmental conditions (same management group).

contracted heels A condition in which the heels of a horse are pulled in so that expansion of the heel cannot occur when the foot strikes the ground.

core samples Samples of wool or feed taken by a coring device to determine the composition of the sample.

corpus luteum A yellowish body in the mammalian ovary. The cells that were follicular cells develop into the corpus luteum, which secretes progesterone. It becomes yellow in color from the yellow lipids that are in the cells.

correlation coefficient A measure of the association of one trait with another.

cow A sexually mature female bovine animal—usually has produced a calf.

cow–calf operation A management unit that maintains a breeding herd and produces weaned calves.

cow hocked A condition in which the hocks are close together but the feet stand apart.

cotyledon An area of the placenta that contacts the uterine lining to allow nutrients and wastes to pass from the mother to the developing young. Sometimes refered to as button.

creep An enclosure in which young can enter to obtain feed but larger animals cannot enter. This process is called creep feeding.

crimp The waves, or kinks, in a wool fiber.

crossbred An animal produced by crossing two or more breeds.

crossbreeding Mating animals from genetically diverse groups (i.e., breeds) within a species.

crutching Clipping wool from the dock, udder, and vulva regions of the ewe prior to her lambing.

cryptorchidism The retention of one or both testicles in the abdominal cavity in animals that typically have the testicles hanging in a scrotal sac.

cud Bolus of feed a ruminant animal regurgitates for further chewing.

cull To eliminate one or more animals from the breeding herd or flock.

curb A hard swelling that occurs just below the point of the hock.

curd Coagulated milk.

cutability Fat, lean, and bone composition of meat animals. Used interchangeably with yield grade. (See *yield grade*.)

cutting chute A narrow chute that allows animals to go through in single file with gates such that animals can be directed into pens along the side of the chute.

cwt An abbreviation for hundredweight (100 lb).

cycling Infers that nonpregnant females have active estrous cycles.

dam Female parent.

dark cutter Color of the lean (muscle) in the carcass has a dark appearance usually caused by stress (excitement, etc.) to the animal before slaughter.

debeaking To remove the tip of the beak of chickens.

dehorn To remove the horns from an animal.

deoxyribonucleic acid (DNA) A complex molecule consisting of deoxyribose, phosphoric acid, and four bases (two purines, adenine and guanine and two pyrimidines, thymine and cytosine). It is the coding mechanism of inheritance.

depreciation An accounting procedure by which the purchase price of an asset with a useful life of more than 1 year is pro-rated over time.

dewclaws Hard horny structures above the hoof on the rear surface of the legs of cattle, swine, and sheep.

dewlap Loose skin under the chin and neck of cattle.

DHIA Dairy Herd Improvement Association, an association which dairy producers participate in keeping dairy records. Sanctioned by The National Cooperative Dairy Herd Improvement Program.

DHIR Dairy Herd Improvement Registry, a dairy record-keeping plan sponsored by the breed associations.

digestibility   The quality of being digestible. If a high percentage of a given food taken into the digestive tract is absorbed into the body, that food is said to have high digestibility.

digestion   The reduction in particle size of feed so that the feed becomes soluble and can pass across the gut wall into the vascular or lymph system.

diploid   Having the normal, paired chromosomes of somatic tissue as produced by the doubling of the primary chromosomes of the germ cells at fertilization.

disease   Any deviation from a normal state of health.

disinfect   To kill, or render ineffective, harmful microorganisms and parasites.

disinfectant   A chemical that destroys disease-producing microorganisms or parasites.

distal   Position that is distant from the point of attachment of an organ.

DM   See *dry matter.*

dock   (1) To cut off the tail. (2) The remaining portion of the tail of a sheep that has been docked. (3) To reduce or lower in value.

doe   A female goat or rabbit.

dominance   (1) A situation in which one gene of an allelic pair prevents the phenotypic expression of the other member of the allelic pair. (2) A type of social behavior in which an animal exerts influence over one or more other animals.

dominant gene   A gene that overpowers and prevents the expression of its recessive allele when the two alleles are present in a heterozygous individual.

down   Soft, fluffy type of feather located under the contour feathers. Serves as insulating material.

drake   Mature male duck.

drench   To give fluid by mouth; for example, medicated fluid is given to sheep for parasite control.

dressing percentage   The percentage of the live animal weight that becomes the carcass weight at slaughter. It is determined by dividing the carcass weight by the live weight then multiplying by 100.

drop   Body parts removed at slaughter—primarily hide (pelt), head, shanks, and offal.

drop credit   Value of the drop.

dry (cow, ewe, sow, mare)   Refers to a nonlactating female.

dry matter   Feed after water (moisture) has been removed (100% dry).

dubbing   The removal of part or all of the soft tissues (comb and wattles) of chickens.

dung   The feces (manure) of farm animals.

dwarfism   The state of being abnormally undersized. Two kinds of dwarfs are recognized; one is proportionate and the other is disproportionate.

dysentery   Severe diarrhea.

dystocia   Difficult birth.

ectoderm   The outermost layer of the three layers of the primitive embryo.

edema   Abnormal fluid accumulation in the intercellular tissue spaces of the body.

ejaculation   Discharge of semen from the male.

emaciation   Thinness; loss of flesh where bony structures (hips, ribs, and vertebrae) become prominant.

embryo   Very early stage of individual development within the uterus. The embryo grows and develops into a fetus. In poultry, the embryo develops within the eggshell.

embryo transfer   The transfer of fertilized eggs(s) from a donor female to one or more recipient females.

endocrine gland   A ductless gland that secretes a hormone into the bloodstream.

endoderm   The innermost layer of the three layers of the primitive embryo.

enterotoxemia   A disease of the intestinal tract caused by bacterial secretion of toxins. Its symptoms are characteristic of food poisoning.

entropion   Turned-in eyelids.

environment   The sum total of all external conditions that affect the well-being and performance of an animal.

enzyme   A complex protein produced by living cells that causes changes in other substances in the body without being changed itself and without becoming a part of the product.

EPD   See *Expected Progeny Difference.*

epididymis   The long, coiled tubule leading from the testis to the vas deferens.

epididymitis   An inflammation of the epididymis.

epiphysis   A piece of bone separated from a long bone in early life by cartilage, which later becomes part of the larger bone.

epistatis   A situation in which a gene or gene pair masks (or controls) the expression of another nonallelic pair of genes.

equine   Refers to horses.

equine encephalomyelitis   An inflammation of the brain of horses.

eruction (or eructation)   The elimination of gas by belching.

esophageal groove   A groove in the reticulum between the esophagus and omasum. Directs milk in the nursing young ruminant directly from the esophagus to the omasum.

essential nutrient   A nutrient that cannot by synthesized by the body and must be supplied in the diet.

estrogen   Any hormone (including estradiol, estriol, and estrone) that causes the female to come physiologically into heat and to be receptive to the male. Estrogens are produced by the follicle of the ovary and by the placenta.

estrous   An adjective meaning "heat," which modifies such words as "cycle." The estrous cycle is the heat cycle, or time from one heat to the next.

estrous synchronization   Controlling the estrous cycle so that a high percentage of the females in the herd express estrus at approximately the same time.

estrus   The period of mating activity in the female mammal. Same as heat.

ET   Abbreviation for embryo transfer.

ethology   Study of animal behavior.

eviserate   The removal of the internal organs during the slaughtering process.

ewe   A sexually mature female sheep. A ewe lamb is a female sheep before attaining sexual maturity.

exocrine gland   Gland which secretes fluid into a duct.

expected progeny difference (EPD)   One-half of the breeding value; the difference in performance to be expected from future progeny of a sire, compared with that expected from future progeny of an average bull in the same test.

family selection   Selection based on performance of a family.

farrow   To deliver, or give birth to, pigs.

fat   Adipose tissue.

feather picking   The picking of feathers from one bird by another.

feces   Bowel movements, excrement from the intestinal tract.

feed additive   Ingredient (such as an antibiotic or hormone-like substance) added to a diet to perform a specific role; for example, to improve gain or feed efficiency.

feed bunk   A trough or container used to feed farm animals.

feed efficiency   (1) The amount of feed required to produce a unit of weight gain or milk; for poultry, this term can also denote the amount of feed required to produce a given quantity of eggs. (2) The amount of gain made per unit of feed.

feeder   Animals (i.e., cattle, lambs, pigs) which need further feeding prior to slaughter.

felting   The process of pressing wool fibers together in conjunction with heat and moisture to produce a fabric.

femininity   Well-developed secondary female sex characteristics, udder development and refinement in head and neck.

fertility   The capacity to initiate, sustain, and support reproduction. With reference to poultry, the term typically refers to the percentage of eggs which, when incubated, show some degree of embryonic development.

fertilization   The process in which a sperm unites with an egg to produce a zygote.

fetus   Later stage of individual development within the uterus. Generally, the new individual is regarded as an embryo during the first half of pregnancy, and as a fetus during the last half.

fill   The contents of the digestive tract.

filly   A young female horse.

fineness   A term used to describe the diameter of wool fibers.

finish   The degree of fatness of an animal.

Finnsheep   A highly prolific breed of sheep introduced into the U. S. in 1968.

fistula   A running sore at the top of the withers of a horse, resulting from a bruise followed by invasion of microorganisms.

flank firmness   Firmness of the flank muscle in lamb carcass evaluation.

flank streaking   Streaks of fat in the flank muscle of lamb carcasses.

fleece   The wool from all parts of the sheep.

flehmen   A pattern of behavior expressed in some male animals (i.e., bull, ram, stallion) during sexual activity. The upper lip curls up and the animal inhales in the vicinity of the vulva or urine.

flock   A group of sheep or poultry.

flushing   Placing females (typically sheep) on a gaining level of nutrition before breeding to stimulate greater rates of ovulation.

fly strike   An infestation with large numbers of blowfly maggots.

foal   A young male or female horse (noun) or the act of giving birth (verb).

follicle   A blisterlike, fluid-filled structure in the ovary that contains the egg.

follicle-stimulating hormone (FSH)   A hormone produced and released by the anterior pituitary that stimulates the development of the follicle in the ovary.

footrot   A disease of the foot in sheep and cattle. In sheep it causes rotting of tissue between the horny part of the foot and the soft tissue underneath.

forb   Weedy or broad-leaf plants, as contrasted to grasses, that serve as pasture for animals

forging   The striking of the heel of the front foot with the toe of the hind foot by a horse in action.

founder   Nutritional ailment resulting from overeating. Lameness in front feet with excessive hoof growth usually occurs.

freemartin   Female born twin to a bull (approximately 9 of 10 will not conceive).

freshen   To give birth to young and initiate milk production. This term is usually used with reference to dairy cattle.

fryer   See *broiler*.

FSH   See *follicle-stimulating hormone*.

full-mouth   Animal has all permanent teeth fully exposed.

full sibs   Animals having the same sire and dam.

gallop   A three-beat gait in which each of the two front feet and both of the hind feet strike the ground at different times.

gametes   Male and female reproductive cells. The sperm and the egg.

gametogenesis   The process by which sperm and eggs are produced.

gander   Mature male goose.

gelding   A male horse that has been castrated.

gene   An active area in the chromosome that codes for a trait and determines how a trait will develop.

general combining ability The ability of individuals of one line or population to combine favorably or unfavorably with individuals of several other lines or populations.

generation interval Average age of the parents when offspring are born.

generation turnover Length of time from one generation of animals to the next generation.

genotype The genetic constitution, or makeup, of an individual. For any pair of alleles, three genotypes (e.g., *AA, Aa,* and *aa*) are possible.

gestation The time from breeding or conception of a female until she gives birth to her young.

gilt A young female swine prior to the time that she has produced her first litter.

goiter Enlargement of the thyroid gland, usually caused by iodine-deficient diets.

gonad The testis of the male; the ovary of the female.

gonadotrophin Hormone that stimulates the gonads.

gossypol A toxic product contained in cottonseed.

grading up The continued use of purebred sires of the same breed in a grade herd or flock.

grass tetany A disease of cattle and sheep marked by staggering, convulsions, coma, and frequently death, and caused by a mineral imbalance (magnesium) while grazing lush pasture.

grease wool Wool as it comes from the sheep and prior to cleaning. It contains the natural oils from the sheep.

gross energy The amount of heat, measured in calories, produced when a substance is completely oxidized. It does not reveal the amount of energy that an animal could derive from eating the substance.

growth The increase in protein over its loss in the animal body. Growth occurs by increases in cell numbers, cell size, or both.

habituation The gradual adaptation to a stimulus or to the environment.

half sib Animals having one common parent.

hand mating Same as hand breeding—bringing a female to a male for service (breeding), after which she is removed from the area where the male is located.

hank A measurement of the fineness of wool. A hank is 560 yards of yarn. More hanks of yarn are produced from fine wools than coarse wools.

haploid One-half of the diploid number of chromosomes for a given species, as found in the germ cells.

hatchability A term that indicates the percentage of a given number of eggs set from which viable young hatch, sometimes calculated specifically from the number of fertile eggs set.

heat See *estrus.*

heat increment The increase in heat production after consumption of feed when an animal is in a thermoneutral environment. It includes additional heat generated in fermentation, digestion, and nutrient metabolism.

heaves A respiratory defect in horses in which the animal has difficulty completing the exhalation of inhaled air.

heifer A young female bovine cow before the time that she has produced her first calf.

heiferette A heifer that has calved once then the heifer is fed for slaughter; the calf has usually died or been weaned at an early age.

hemoglobin The iron-containing pigment of the red blood cells. It carries oxygen from the lungs to the tissues.

hen An adult female domestic fowl, such as a chicken or turkey.

herbivorous Subsisting or feeding on plants.

herd A group of animals. Used with beef, dairy, or swine.

heritability The portion of the total variation or difference (phenotype) among animals that is due to heredity.

hernia The protrusion of some of the intestine through an opening in the body wall (also commonly called rupture). Two types, umbilical and scrotal, occur in farm animals.

heterosis Performance of offspring that is greater than the average of the parents. Usually referred to as the amount of superiority of the crossbred over the average of the parental breeds. Also referred to as hybrid vigor.

heterozygous A term designating an individual that possesses unlike genes for a particular trait.

hides Skins from animals such as cattle, horses and pigs.

hinny The offspring that results from crossing a stallion with a female donkey (jenny).

hobble To tie two of an animal's legs together. An animal is hobbled to prevent it from kicking or moving a long distance.

homeotherm A "warm-blooded" animal. An animal that maintains its characteristic body temperature even though environmental temperature varies.

homogenized Milk that has had the fat droplets broken into very small particles so that the milk fat stays in suspension in the milk fluids.

homologous Corresponding in type of structure and derived from a common primitive origin.

homologous chromosomes Chromosomes having the same size and shape that contain genes affecting the same characters. Homologous chromosomes occur in pairs in typical diploid cells.

homozygous A term designating an individual whose genes for a particular trait are alike.

hormone A chemical substance secreted by a ductless gland. Usually carried by the bloodstream to other places in the body where it has its specific effect on another organ.

hybrid vigor See *heterosis*.

hydrocephalus A condition characterized by an abnormal increase in the amount of cerebral fluid, accompanied by dilation of the cerebral ventricles.

hypertension High blood pressure.

hypothalamus A portion of the brain found in the floor of the third ventricle. It regulates reproduction, hunger, body temperature, and has other functions.

hypoxia A condition resulting from deficient oxygenation of the blood.

immunity The ability of an animal to resist or overcome an infection.

implant To graft or insert material to intact tissues.

implantation The attachment of the fertilized egg to the uterine wall.

imprinting Learning associated with maturational readiness.

inbreeding The mating of individuals who are more closely related than the average individuals in a population. Inbreeding increases homozygosity in the population but it does not change gene frequency.

incisor A front tooth.

incubation period The time that elapses from the time an egg is placed into an incubator until the young is hatched.

independent culling level Selection method in which minimum acceptable phenotypic levels are assigned to several traits.

index (1) An overall merit rating of an animal. (2) A method of predicting the milk-producing ability that a bull will transmit to his daughters.

infection Invasion of the body tissues by microbial agents or parasites other than insects.

infectious Capable of invading and growing in living tissues. Used to describe various pathogenic micro-organisms such as viruses, bacteria, protozoa, and fungi.

influenza A virus disease characterized by inflammation of the respiratory tract, high fever, and muscular pain.

ingest Anything taken into the stomach.

inheritance The transmission of genes from parents to offspring.

insemination Deposition of semen in the female reproductive tract.

instinct Inborn behavior.

integration The bringing together of all segments of a livestock production program under one centrally organized unit.

intelligence   The ability to learn to adjust successfully to situations.

interference   The striking of the supporting leg by the foot of the striding leg by a horse in action.

interstitial cells   The cells between the seminiferous tubules of the testicle that produce testosterone.

intravenous   Within the vein. An intravenous injection is an injection into a vein.

jack   A male donkey.

jackass   See *jack.*

jennet   A female donkey

jenny   A female donkey.

Karakul   A breed of fat-tailed sheep having coarse, wiry fur-like hair. Used to produce Persian lamb skins.

ked   An external parasite that affects sheep. Although commonly called "sheep tick," it is actually a wingless fly.

kemp   Coarse, opaque, hairlike fibers in wool.

ketosis   A condition (also called acetonemia) that is characterized by high concentration of ketone bodies in the body tissues and fluids.

kid   Young goat.

kilocalorie (kcal, Kcal)   An amount of heat equal to 1,000 calories [See *calorie* (Also called Calorie).]

kosher meat   Meat from ruminant animals with split hooves where the animals have been slaughtered according to Jewish law.

lactalbumin   A nutritive protein of milk.

lactation   The secretion and production of milk.

lactose   Milk sugar. When digested, it is broken down into one molecule of glucose and one of galactose.

lamb   (1) A young male or female sheep, usually an individual less than a year of age. (2) To deliver, or give birth to, a lamb.

lamb dysentery   See *dysentery.*

lambing   Act of giving birth. Same as parturition.

lambing jug   A small pen in which a ewe is put for lambing. It is also used for containing the ewe and her lamb until the lamb is strong enough to run with other ewes and lambs.

laminitis   Lameness associated with inflammation of the laminae that attaches the horse's hoof wall to the fleshy part of the foot. Typically related to founder.

layer   A hen that is kept for egg production.

legume   Any plant of the family *Leguminosae,* such as pea, bean, alfalfa, and clover.

leucocytes   White blood cells.

LH   See luteinizing hormone.

libido   Sex drive or the desire to mate on the part of the male.

lice   Small, flat, wingless insect with sucking mouth parts that is parasitic on the skin of animals.

line crossing   The crossing of inbred lines.

lipid   An organic substance that is soluble in alcohol or ether but insoluble in water; used interchangeably with the term fat.

litter   The young produced by multiparous females such as swine. The young in a litter are called litter mates.

liver flukes   A parasitic flatworm found in the liver.

locus   The place on a chromosome where a gene is located.

longevity   Life span of an animal, usually refers to a long life span.

luteinizing hormone (LH)   A protein hormone, produced and released by the anterior pituitary which stimulates the formation and retention of the corpus luteum. It also initiates ovulation.

lymph   Clear yellowish, slightly alkaline fluid contained in lymphatic vessels.

macroclimate   The large, general climate in which an animal exists.

macromineral   A mineral that is needed in the diet in relatively large amounts.

maintenance   A condition in which the body is maintained without an increase or decrease in body weight and with no production or work being done.

mammal   Warm-blooded animals that suckle their young.

mammary gland   Gland that secretes milk.

management   The act, art, or manner of managing, handling, controlling, or directing a resource or integrating several resources.

marbling   The distribution of fat in muscular tissue; intramuscular fat.

mare   A sexually developed female horse.

market class   Animals grouped according to the use to which they will be put, such as slaughter, feeder, or stocker.

market grade   Animals grouped within a market class according to their value.

masticate   To chew food.

mastitis   Inflammation of the mammary gland.

masturbation   Ejaculation by a male by some process other than sexual intercourse.

maternal breeding value (MBV)   A breeding value which measures primarily milk production in beef cattle.

mean   (1) Statistical term for average. (2) Term used to describe animals having bad behavior.

meat   The tissues of the animal body that are used for food.

meat spots   Spots in the egg that are blood spots which have changed color or tissue sloughed off from the reproductive organs of the hen.

meiosis   A special type of cell nuclear division that is undergone in the production of gametes (sperm in the male, ova in the female). As a result of meiosis, each gamete carries half the number of chromosomes of a typical body cell in that species.

mesoderm   The middle layer of the three layers of the primitive embryo.

messenger RNA   The ribonucleic acid that is the carrier of genetic information from nuclear DNA. It is important in protein synthesis.

metabolism   (1) The sum total of chemical changes in the body, including the "building up" and "breaking down" processes. (2) The transformation by which energy is made available for body uses.

metabolizable energy   Gross energy in the feed minus the sum of energy in feces, gaseous products of digestion, and energy in urine. Energy that is available for metabolism by the body.

metritis   Inflammation (infection) of the uterus.

microclimate   A small, special climate within a macroclimate created by the use of such devices as shelters, heat lamps, and bedding.

microcomputer   A small computer that has a smaller memory capacity than a larger or main-frame computer.

micromineral   A mineral that is needed in the diet in relatively small amounts. The quantity needed is so small that such a mineral is often called a trace mineral.

milk fat   The fat in milk; synonymous with butterfat.

milk fever   See *parturient paresis.*

milk letdown   The release of milk into the teat cisterns.

Milk Only Records   Dairy record system similar to DHI except no milk fat samples are taken.

minimum culling level   A selection method in which an animal must meet minimum standards for each trait desired in order to qualify for being retained for breeding purposes.

mites   Very small arachnids that are often parasitic upon animals.

mitosis   A process in which a cell divides to produce two daughter cells, each of which contains the same chromosome complement as the mother cell from which they came.

modifying genes   Genes that modify the expression of other genes.

mohair Fleece of the Angora goat.

monogastric Having only one stomach or only one compartment in the stomach. Examples are swine and poultry.

monoparous A term designating animals that usually produce only one offspring at each pregnancy. Horses and cattle are monoparous.

moon blindness Periodic blindness that occurs in horses.

morbidity Measurement of illness; morbidity rate is the number of individuals in a group that become ill during a specified time.

mortality rate Number of individuals that die from a disease during a specified time, usually 1 year.

mouthed The examination of an animal's teeth.

mule The hybrid that is produced by mating a male donkey with a female horse. They are usually sterile.

mulefoot Having one instead of the expected two toes, on one or more of the feet.

multiparous A term that designates animals that usually produce several young at each pregnancy. Swine are multiparous.

mutation A change in a gene.

mutton The meat from a sheep which is more than 1 year old.

muzzle The nose of horse, cattle, or sheep.

myofibrils The primary component part of muscle fibers.

navel The area where the umbilical cord was formerly attached to the body of the offspring.

necropsy Perform a postmortem examination.

net energy Metabolizable energy minus heat increment. The energy available to the animal for maintenance and production.

nicking The way in which certain lines, strains, or breeds perform when mated together. When outstanding offspring result, the parents are said to have nicked well.

nipple See *teat*.

nodular worm An internal parasitic worm that causes the formation of nodules in the intestines.

nonruminant Simple stomached or monagastric animal.

NPN (nonprotein nitrogen) Nitrogen in feeds from substances such as urea, amino acids, etc., but not from preformed proteins.

nucleotide Compound composed of phosphoric acid, sugar, and a base (purine or pyrimadine), all of which constitute a structural unit of nucleic acid.

nutrient (1) A substance that nourishes the metabolic processes of the body. (2) The end product of digestion.

nutrient density Amount of essential nutrients relative to the number of calories in a given amount of food.

obesity An excessive accumulation of fat in the body.

offal All organs and tissues removed from inside the animal during the slaughtering process.

omasum One of the stomach components of ruminant animals that has many folds.

omnivorous Feeding on both animal and vegetable substances.

on full feed A term that refers to animals that are receiving all the feed they will consume. *Ad libitum.*

oogenesis The process by which eggs, or ova, are produced.

open Refers to nonpregnant females.

open-faced Face of sheep that is free from wool, particularly around the eyes.

opportunity costs Returns given up if debt-free resources (for example, land, livestock, equipment) were used in their next best level of employment.

osteopetrosis Abnormal thickening, hardening, and fragility of bones, making them weaker.

osteoporosis An abnormal decrease in bone mass with an increased fragility of the bones.

outbreeding   The process of continuously mating females of the herd to unrelated males of the same breed.

outcrossing   The mating of an individual to another in the same breed that is not related to it. Outcrossing is a specific type of outbreeding system.

ova   Plural of ovum, meaning eggs.

ovary   The female reproductive gland in which the eggs are formed and progesterone and estrogenic hormones are produced.

overeating disease   A toxic condition caused by the presence of undigested carbohydrates in the intestine, which stimulates harmful bacteria to multiply. When the bacteria die, they release toxins. Called enterotoxemia in some animals.

overshot jaw   Upper jaw is longer than lower jaw. Also called *parrot mouth*.

oviduct   A duct leading from the ovary to the horn of the uterus.

ovine   Refers to sheep.

ovulation   The shedding, or release, of the egg from the follicle of the ovary.

ovum   The egg produced by a female.

Owner-Sampler Record   Dairy record system similar to DHI except milk weights and samples are taken by the dairy producer instead of a DHIA supervisor.

pace   A lateral two-beat gait in which the right rear and front feet hit the ground at one time and the left rear and front feet strike the ground at another time.

paddling   The outward swinging of the front feet of a horse that toes in.

pale, soft, exudative (PSE)   A genetically predisposed condition in swine in which the pork is very light colored, soft, and watery.

palpation   Feeling by hand.

parasite   An organism that lives a part of its life cycle in or on, and at the expense of, another organism. Parasites of farm animals live at the expense of the farm animals.

parity   Number of different times a female has had offspring.

parrot mouth   Upper jaw is longer than lower jaw.

parturient paresis   Partial paralysis that occurs at or near time of giving birth to young and beginning lactation. The mother mobilizes large amounts of calcium to produce milk to feed newborn, and blood calcium levels drop below the point necessary for impulse transmission along the nerve tracks. Commonly called "milk fever."

parturition   The process of giving birth.

pasteurization   The process of heating milk to 161°F and holding it at that temperature for 15 sec to destroy pathogenic microorganisms.

pasture rotation   The rotation of animals from one pasture to another so that some pasture areas have no livestock on them in certain periods.

pathogen   Biologic agent, i.e., bacteria, virus, protozoa, nematode, that may produce disease or illness.

paunch   Another name for rumen.

PD   See *predicted difference.*

pedigree   The record of the ancestry of an animal.

pelt   The natural, whole skin covering, including the wool, hair, or fur.

pendulous   Hanging loosely.

penis   The male organ of copulation. It serves both as a channel for passage of urine from the bladder as an extension of the urethra, and as a copulatory organ through which sperm are deposited into the female reproductive tract.

per capita   Per person.

performance test   The evaluation of an animal according to its performance.

pernicious anemia   A chronic type of mycrocitic anemia caused by a deficiency of vitamin B_{12} or a failure of intestinal absorption of vitamin B_{12}.

pharmaceutical   A medicinal drug.

phenotype   The characteristics of an animal that can be seen and/or measured. For example, the presence or absence of horns, the color, or the weight of an animal.

pheromones   Chemical substances which attract the opposite sex.

photoperiod   Time period when light is present.

physiology   The science that pertains to the functions of organs, organ systems, or the entire animal.

picking   The removal of feathers in dressing poultry.

pigeon toed   See *toeing in.*

pin bones   In cattle, the posterior ends of the pelvic bones that appear as two raised areas on either side of the tail head.

pink tooth   Congenital porphyria, teeth are pink gray and the animals tend to sunburn easily.

pinworms   A small nematode worm with unsegmented body found as a parasite in the rectum and large intestine of animals.

pituitary   Small endocrine gland located at the base of the brain.

pneumonia   Inflammation or infection of alveoli of the lungs caused by either bacteria or viruses.

poikilotherm   A "cold-blooded" animal. An animal whose body temperature varies with that of the environment.

polled   Naturally or genetically hornless.

poll evil   An abscess behind the ears of a horse.

Polypay   A synthetic breed of sheep developed in the U.S. by combining the Dorset, Targhee, Rambouillet, and Finnsheep breeds.

pork   The meat from swine.

postgastric fermentation   The fermentation of feed that occurs in the cecum, behind the area where digestion has occurred.

postnatal   See *postpartum.*

postpartum   After birth.

postpartum interval   The length of time from parturition until the dam is pregnant again.

poult   A young turkey of either sex, from hatching to approximately 10 weeks of age.

poultry   This term includes chickens, turkeys, geese, pigeons, peafowls, guineas, and game birds.

predicted difference (PD)   Estimate of genetic transmitting ability (i.e., one-half of the breeding value) of dairy bulls. Estimated amount by which daughters of a bull will differ from the breed average.

pregastric fermentation   Fermentation that occurs in the rumen of ruminant animals. It occurs before feed passes into the portion of the digestive tract in which digestion actually occurs.

pregnancy disease   A metabolic disease in late pregnancy affecting primarily ewes carrying twins or triplets. A form of ketosis. Also called pregnancy toxemia.

pregnancy testing   Evaluation of females for pregnancy through palpation or using a sonoray machine.

prenatal   Prior to being born. Before birth.

probe   A device used to measure backfat thickness in pigs and cattle.

production testing   An evaluation of an animal based on its production record.

progeny testing   An evaluation of an animal on the basis of performance of its offspring.

progesterone   A hormone produced by the corpus luteum that stimulates progestational proliferation in the uterus of the female.

prolapse   Abnormal protrusion of part of an organ, such as uterus or anus.

prostaglandins   Chemical mediators that control many physiological and biochemical functions in the body. One prostaglandin ($PGF_{2\alpha}$) can be used to synchronize estrus.

prostate   A gland of the male reproductive tract that is located just back of the bladder. It secretes a fluid that becomes part of semen at ejaculation.

protein   A substance made up of amino acids that contains approximately 16% nitrogen (based on molecular weight).

protein supplement   Any dietary component containing a high concentration (at least 25%) of protein.

PSE   See *pale, soft, and exudative.*

puberty   The age at which the reproductive organs become functionally operative.

pullet   Young female chicken from day of hatch through onset of egg production; sometimes the term is used through the first laying year.

purebred   An animal eligible for registry with a recognized breed association.

quality grades   Animals grouped according to value as prime, choice, etc., based on conformation and fatness of the animals.

rack   (1) A rapid four-beat gait of a horse. (2) A wholesale cut of lamb located between the shoulder and loin.

ram   A male sheep that is sexually mature.

ration   The quantity of feed fed to an animal over a given period of time.

reach   See *selection differential.*

realized heritability   The portion obtained of what is reached for in selection.

reasoning   The ability of an animal to respond correctly to a stimulus the first time that the animal encounters a new situation.

recessive gene   A gene that has its phenotype masked by its dominant allele when the two genes are present together in an individual.

reciprocal recurrent selection   The selection of breeding animals in two populations based on the performance of their offspring after animals from two populations are crossed.

rectal prolapse   Protrusion of part of large intestine through the anus.

recurrent selection   Selection for general combining ability by selecting males that sire outstanding offspring when mated to females from varying genetic backgrounds.

red meat   Meat from cattle, sheep, swine, and goats as contrasted to the so-called "white meat" of poultry.

registered   Recorded in the herdbook of a breed.

regurgitate   To cast up digested food to the mouth as is done by ruminants.

reinforcement   A reward for making the proper response to a stimulus or condition.

replicate   To duplicate, or make another exactly alike, the original.

reproduction   The production of live, normal offspring.

reticulum   One of the stomach components of ruminant animals that is lined with small compartments, giving a honeycomb appearance.

rhinitis   Inflammation of the mucous membranes lining the nasal passages.

ribonucleic acid (RNA)   An essential component of living cells, composed of long chains of phosphate, ribose sugar, and several bases.

rhinopneumionitis   Equine herpes virus-1. It produces acute catarrh upon primary infection.

rickets   A disease of disturbed ossification of the bones caused by a lack of vitamin D or unbalanced calcium/phosphorus ratio.

ridgling   Another term for cryptorchid.

ringbone   An ossification of the lateral cartilage of the foot of a horse all around the foot.

RNA   See *ribonucleic acid.*

roughage   A feed that is high in fiber, low in digestible nutrients, and low in energy. Such feeds as hay, straw, silage, and pasture are examples.

rumen   The large fermentation pouch of the ruminant animal in which bacteria and protozoa break down fibrous plant material that is swallowed by the animal, sometimes referred to as the paunch.

ruminant   A mammal whose stomach has four parts (rumen, reticulum, omasum, and abomasum). Cattle, sheep, goats, deer, and elk are ruminants.

rumination   The regurgitation of undigested food and chewing of it for a second time, after which it is again swallowed.

salmonella   Gram-positive, rod-shaped bacteria that causes various diseases such as food poisoning in animals.

scale   (1) Size. (2) Equipment on which an animal is weighed.

scoured wool   Wool that has been cleaned of grease and other foreign material.

scours   Diarrhea; a profuse watery discharge from the intestines.

screwworms   Larvae of several American flies that infest wounds of animals.

scrotum   A pouch that contains the testes. It is also a thermoregulatory organ that contracts when cold and relaxes when warm, thus tending to keep the testes at a lower temperature than that of the body.

scurs   Small growths of horn-like tissue attached to the skin of polled or dehorned animals.

scurvy   A deficiency disease in humans that causes spongy gums and loose teeth. It is caused by a lack of vitamin C (ascorbic acid).

seed stock   Breeding animals; sometimes used interchangeably with purebred.

selection   Differentially reproducing what one wants in a herd or flock.

selection differential   The difference between the average for a trait in selected animals and the average of the group from which they come; also called reach.

selection index   Selection method in which several traits are evaluated and expressed as one total score.

semen   The fluid containing the sperm that is ejaculated by the male. Secretions from the seminal vesicles, the prostate gland, the bulbourethral glands, and the urethral glands provide most of the fluid.

seminal vesicles   Accessory sex glands of the male that provide a portion of the fluid of semen.

seminiferous tubules   Minute tubules in the testicles in which sperm are produced. They comprise about 90% of the mass of the testes.

service   To breed or mate.

settle   To become pregnant.

sex-limited   Exists in only one sex, such as milk production in dairy cattle.

shearing   The process of removing the fleece (wool) from a sheep.

sheath rot   Inflammation of the prepuce in male sheep.

sheep bot   Any of a number of related flies whose larvae are parasitic in sheep. They usually are found in the sinuses.

shipping fever   A widespread respiratory disease of cattle and sheep.

shoat   A young pig of either sex.

shoe boil   Blemish of the horse caused by the horseshoe putting pressure on the elbow when the horse lies down.

shrink   Loss of weight—commonly used in the loss in liveweight when animals are marketed or loss in weight from grease wool to clean wool.

sib   A brother or sister.

sickle hocks   Hocks that have too much set, causing the hind feet to be too far forward and too far under the animal.

side bones   Ossification of the lateral cartilages of the foot of a horse.

sigmoid flexure   The S-curve in the penis of boars, rams, bucks, and bulls.

silage   Forage, corn fodder, or sorghum preserved by fermentation that produces acids similar to the acids that are used to make pickled foods for people.

sire   Male parent.

skins   See hides.

sleeping sickness   An infectious disease common in tropical Africa and transmitted by the bite of a tsetse fly.

slotted floor   Floor having any kind of openings through which excreta may fall.

SNF   See *solids-non-fat*.

snood   The relatively long, fleshy extension at the base of the turkey's beak.

software   Program instructions to make computer hardware function.

soilage   Green forage that is cut and brought to animals as food.

solids-non-fat   Total milk solids minus fat. It includes protein, lactose, and minerals.

somatotropin   The growth hormone from the anterior pituitary that stimulates nitrogen retention and growth.

sonoray   A machine that is used to measure fat thickness and ribeye area in swine and cattle. The machine sends sound waves into the back of the animal and records these waves as they bounce off the tissues. Different wave lengths are recorded for fat than for lean. Also used to diagnose pregnancy.

sore mouth   A virus-caused disease affecting primarily lambs.

sow   A female swine that has farrowed one litter or has reached 12 months of age.

spay   To remove the ovaries.

specific combining ability   The ability of a line or population to exhibit superiority or inferiority when combined with other lines or populations.

spermatid   The haploid germ cell prior to spermiogenesis.

spermatogenesis   The process by which spermatozoa are formed.

spermiogenesis   The process by which the spermatid loses most of its cytoplasm and develops a tail to become a mature sperm.

spinning count   The number of hanks of yarn that can be spun from a pound of clean wool. One method of evaluating fineness of wool.

splay footed   See *toeing out*.

spool joint   The joint where the foot and pastern are removed from the front leg. Used to identify a mutton carcass.

spur   A sharp projection on the back of a male bird's shank.

stags   Castrated male sheep, cattle, goats, or swine that have reached sexual maturity prior to castration.

stallion   A sexually mature male horse.

staple length   Length of wool fibers.

steer   A castrated bovine male that was castrated early in life before puberty.

sterility   Inability to produce offspring.

stifle   Joint of the hindleg between the femur and tibia.

stifled   Injury of the stifle joint.

stillborn   Offspring born dead without previously breathing.

stocker   Weaned cattle that are fed high-roughage diets (including grazing) before going into the feedlot.

stomach worms   *Haemonchus contortus,* or worms of the stomach of cattle, swine, sheep, and goats.

strangles   An infectious disease of horses, characterized by inflammation of the mucous membranes of the respiratory tract.

streptococcus   Sperical, gram-positive bacteria that divide in only one plane and occur in chains. Some species cause serious disease.

stress   Any force causing a change in an animal's function, structure, or behavior.

stringhalt   A sudden and extreme flexion of the back of a horse, producing a jerking motion of the hind leg in walking.

strongyles   Any of various roundworms living as parasites, especially in domestic animals.

stud Usually the same as stallion. Also a place where male animals are maintained, i.e., bull stud.

suckling gain The gain that a young animal makes from birth until it is weaned.

subcutaneous Situated beneath, or occurring beneath, the skin. A subcutaneous injection is an injection made under the skin.

sulfonamides A sulfa drug capable of killing bacteria.

superovulation The hormonally induced ovulation of a greater than normal number of eggs.

sweeny Atrophy of muscle (typically shoulder) in horses.

sweetbread An edible by-product also known as the pancreas.

switch The tuft of long hair at the end of tail (cattle and horses).

syndactyly Union of two or more digits—for example, in cattle, the two toes would be a solid hoof.

synthetic breeds See *composite breed.*

tags (1) Wool covered with manure. (2) Abbreviated form of ear tags, used for identification.

tallow The fat of cattle and sheep.

tandem selection Selection for one trait for a given period of time followed by selection for a second trait and continuing in this way until all important traits are selected.

TDN Total digestible nutrients; it includes the total amounts of digestible protein, nitrogen-free extract, fiber, and fat (multiplied by 2.25) all summed together.

teaser ram A ram made incapable of impregnating a ewe by vasectomy or by use of an apron to prevent copulation, which is used to find ewes in heat.

teasing The stallion in the presence of the mare to see if she will mate.

teat The protuberance of the udder through which milk is drawn.

tendon Tough, fibrous connective tissue at ends of muscle bundles that attach muscle to bones or cartilage structures.

testicle The male sex gland that produces sperm and testosterone.

testosterone The male sex hormone that stimulates the accessory sex glands, causes the male sex drive, and causes the development of masculine characteristics.

tetanus An acute infectious disease caused by toxin elaborated by the bacterium *Clostridium tetani,* in which tonic spasms of some of the voluntary muscles occur.

tetrad A group of four similar chromotids formed by the splitting longitudinally of a pair of homologous chromosomes during meiotic prophase.

thermoneutral zone (TNZ) Range in temperature where rate and efficiency of gain is maximized. Comfort zone.

thoroughpin A hard swelling that is located between the Achilles tendon and the bone of the hock joint.

toeing in Toes of front feet turn in. Also called piegon toed.

toeing out Toes of front feet turn out. Also called splay footed.

tom A male turkey.

TPI Total prediction index used in dairy cattle breeding. It includes the predicted differences for milk production, fat percentage, and type into one figure in a ratio of milk production X 3: fat percentage X 1: type X 1.

transcription The synthesis of RNA from DNA in the nucleus by matching the sequences of the bases.

transmissable gastroenteritis (TGE) A serious, contagious diarrhea disease in baby pigs.

tripe Edible product from walls of ruminant stomach.

trot A diagonal two-beat gait in which the right front and left rear feet strike the ground in unison, and the left front and right rear feet strike the ground in unison.

twist Vertical measurement from top of the rump to point where hind legs separate.

twitch To tightly squeeze the upper lip of a horse by means of a small rope that is twisted.

type (1) The physical conformation of an animal. (2) All those physical attributes that contribute to the value of an animal for a specific purpose.

udder The encased group of mammary glands of mammals.

umbilical cord A cord through which arteries and veins travel from the fetus to and from the placenta, respectively. This cord is broken when the young are born.

undershot jaw Lower jaw is longer than upper jaw.

unsoundness Any defect or injury that interferes with the usefulness of an animal.

urinary calculi Disease where mineral deposits crystallize in the urinary tract. The deposits may block the tract causing difficulty in urination.

uterus That portion of the female reproductive tract where the young develop during pregnancy.

vaccination The act of administering a vaccine or antigens.

vaccine Suspension of attenuated or killed microbes or toxins administered to induce active immunity.

vagina The copulatory portion of the female's reproductive tract. The vestibule portion of the vagina also serves for passage of urine during urination. The vagina also serves as a canal through which young pass when born.

variety meats Edible organ by-products, i.e., liver, heart, tongue, tripe, etc.

vas deferens Ducts that carry sperm from the epididymis to the urethra.

vasectomy The removal of a portion of the vas deferens. As a result, sperm are prevented from traveling from the testicles to become part of the semen.

veal The meat from very young cattle, under 3 months of age.

vein Vessel through which blood passes from various organs or parts back to the heart.

vermifuge A chemical substance given to the animals to kill internal parasitic worms.

VFA See *volatile fatty acids.*

villi Projections of the inner lining of the small intestine.

virus Ultra-microscopic bundle of genetic material capable of multiplying only in living cells. Viruses cause a wide range of disease in plants, animals, and humans, such as rabies and measles.

viscera Internal organs and glands contained in the thoracic and abdominal cavities.

vitamin An organic catalyst, or component thereof, which facilitates specific and necessary functions.

volatile fatty acids (VFA) A group of fatty acids produced from microbial action in the rumen; examples are acetic, propionic, and butyric acids.

vulva The external genitalia of a female mammal.

walk A four-beat gait of a horse in which each foot strikes the ground at a time different from each of the other three feet.

warble The larval stage of the heel fly that burrows out through the hide of cattle in springtime.

wattle Method of identification in cattle where strips of skin (3–6 inches) long are usually cut on the nose, jaw, throat, or brisket.

weaning Separating young animals from their dams so that the offspring can no longer suckle.

weaning breeding value (WBV) A breeding value that measures primarily preweaning growth in beef cattle.

wether A male sheep castrated before reaching puberty.

white muscle disease A muscular disease caused by a deficiency of selenium or vitamin E.

winking Indication of estrus in the mare where the vulva opens and closes.

withers Top of the shoulders.

wool The fibers that grow from the skin of sheep.

wool blindness Sheep cannot see owing to wool covering their eyes.

woolens Cloth made from short wool fibers that are intermingled in the making of the cloth by carding.

worsteds Cloth made from wool that is long enough to comb and spin into yarn. The finish of worsteds is harder than woolens, and worsted clothes hold a press better.

yearling Animals that are approximately 1 year old.

yearling breeding value (YBV) A breeding value that measures primarily postweaning growth in beef cattle.

yield Used interchangeably with dressing percentage.

yield grades The grouping of animals according to the estimated trimmed lean meat that their carcass would provide; cutability.

yolk (1) The yellow part of the egg. (2) The natural grease (lanolin) of wool.

yolk sac Layer of tissue encompassing the yolk of an egg.

zone of thermoneutrality The environmental temperature (about 65°F) at which heat production and heat elimination are approximately equal for most farm animals.

zygote (1) A cell formed by the union of two gametes. (2) An individual from the time of fertilization until death.